A PUBLICATION OF THE
NATIONAL RESEARCH COUNCIL OF CANADA MONOGRAPH PUBLISHING PROGRAM

Biology of *Populus* and its Implications for Management and Conservation

Edited by

R.F. Stettler
University of Washington, College of Forest Resources, Seattle, WA 98195, USA

H.D. Bradshaw, Jr.
*Center for Urban Horticulture, Box 354115, University of Washington,
Seattle, WA 98195-4115, USA*

P.E. Heilman
*Washington State University, WSU Puyallup Research & Extension Center,
Puyallup, WA 98371, USA*

and T.M. Hinckley
University of Washington, College of Forest Resources, Seattle, WA 98195, USA

NRC·CNRC

NRC RESEARCH PRESS
Ottawa 1996

NRC Monograph Publishing Program

Publication Proposals: Proposals for the NRC Monograph Publishing Program should be sent to Gerald J. Neville, Head, Monograph Publishing Program, National Research Council of Canada, NRC Research Press, 1200 Montreal Road, Building M-55, Ottawa, ON K1A 0R6, Canada. Telephone: (613) 993-1513; FAX: (613) 952-7656; e-mail: gerry.neville@nrc.ca

---

ISBN 0-660-16506-6
NRCC No. 40337

Canadian Cataloguing in Publication Data

Main entry under title:
Biology of *Populus* and its implications for management and conservation

Includes an abstract in French. Includes bibliographical references.
Issued by the National Research Council of Canada
ISBN 0-660-16506-6
1. Poplar. 2. Poplar — Genetics. 3. Poplar — Growth
I. Stettler, R.F. (Reinhard Friedrich). II. National Research Council of Canada

SD397.P85B56 1996   634.9'723   C96-980261-7

Correct citation for this publication:

Stettler, R.F., Bradshaw, H.D., Jr., Heilman, P.E., and Hinckley, T.M. 1996. Biology of *Populus* and its implications for management and conservation. NRC Research Press, Ottawa, Ontario, Canada. 539 p.

Table of Contents

Abstract/Résumé

Trees of the genus *Populus* (poplars, cottonwoods, aspens) are widely distributed over the northern hemisphere and planted in many other parts of the world. In their native ecosystems they play a major role in the recolonization of sites after disturbances and provide important habitat for fish and wildlife. Poplar is also increasingly recognized as an excellent model tree for the study of tree growth and its underlying physiology and genetics. Major advances have been made in the study of poplar biology, from the tree and stand level to the biology of host–pathogen and host–insect interactions. New insights have been gained that are highly relevant to plantation culture, to the development of improved cultivars, and to the conservation and restoration of poplar in its native environment. These advances are presented in 20 chapters, authored by an international group of 47 researchers. The book is structured in two sections, one dealing with systematics, genetics, genetic manipulation, and biotic interactions of *Populus*, the second with the physiology of growth and productivity and of stress response. Introductory overviews by the section editors help the reader to integrate the diverse topics. A concluding chapter of poplar culture offers perspectives on the application of traditional and novel concepts in different parts of the world. The book is aimed at a broad readership of scientists and professionals in forestry, agriculture, agro-forestry, biology, conservation ecology, and the environmental sciences.

Les arbres du genre *Populus* (peuplier, peuplier deltoïde, peuplier faux-tremble) sont largement répandus dans l'hémisphère nord et sont cultivés dans le monde entier. Au sein de leurs écosystèmes naturels, ils jouent un rôle crucial dans la reconstitution postperturbation des territoires et constituent un important habitat pour les poissons et la faune. Le peuplier est aussi de plus en plus reconnu comme un excellent modèle d'étude de la croissance des arbres ainsi que de ses aspects physiologiques et génétiques. On a noté d'importants progrès dans l'étude de la biologie du peuplier, depuis l'arbre et son peuplement jusqu'à la biologie des interactions hôte-pathogène et hôte-insecte. On a acquis des connaissances inestimables sur la culture de plantations, sur la création de cultivars améliorés et sur la conservation et la régénération du peuplier dans son environnement naturel. Ces progrès sont recensés dans un recueil de 20 chapitres, signé par un groupe de 47 chercheurs issus des quatre coins du monde. Il est divisé en deux volets : le premier porte sur la taxonomie, la génétique, la manipulation génétique et les interactions biotiques de *Populus*; le second, sur la physiologie de la croissance et de la productivité ainsi que sur la réaction de stress. Les sections sont précédées d'un survol rédigé par le directeur de la section en question, ceci afin que les lecteurs puissent intégrer les divers sujets traités. Le dernier chapitre sur la culture du peuplier brosse un tableau général du potentiel des applications de concepts traditionnels et de nouveaux concepts de par le monde. Ce recueil a été rédigé à l'intention des scientifiques et des spécialistes en foresterie, en agriculture, en agroforesterie, en biologie, en écologie de la conservation et en sciences de l'environnement.

Preface

R.F. Stettler, H.D. Bradshaw, Jr., P.E. Heilman, and T.M. Hinckley

Populus L. is a genus comprising about 30 species of poplars, cottonwoods, and aspens widely distributed over the northern hemisphere and planted in many parts of the world. Poplars, because of their fast growth, their multiple use as a source of fuel, fiber, lumber, and animal feed, and the ease with which they can be propagated vegetatively, have been closely associated with agriculture for a long time, in the Near and Middle East since antiquity. These trees are widely used in windbreaks, in protective stands to prevent soil erosion, and in row or gallery plantings to stabilize the banks of streams and canals. Indeed, there are few temperate-zone landscapes where the columnar silhouettes of *Populus nigra* cultivars do not grace the countryside. In recent years, poplars have received increasing attention as a renewable source of biomass for energy and short-fiber furnish for papermaking. Managed in short rotations and under intensive culture, they have shown impressive productivity and have become a highly promising crop option, especially for marginal agricultural land. With the systematic shrinkage of forest land worldwide and an expanding human population, poplars are likely to gain importance in providing needed wood products while at the same time contributing to a more favorable carbon balance.

No less significant is the role poplars play in natural populations of their native habitat. Riparian ecosystems especially have seen progressive degradation from expanding human habitation and land use. There is a growing awareness that the health of rivers is critically tied to the surrounding vegetation they sustain and that poplars fill a special niche in it. Adapted to the dynamic regimes of erosion and deposition, poplars are uniquely suited to benefit from these cyclic disturbances and to capture transitory sites while vacating others. However, to the extent that these processes are impeded by river regulation and dams, regeneration of natural populations is failing, leading to senescent stands and their eventual decline. Restoration of riparian habitats is becoming a widely recognized need of increasing urgency.

R.F. Stettler and T.M. Hinckley. University of Washington, College of Forest Resources, Seattle, WA 98195-2100, USA.
H.D. Bradshaw, Jr. Center for Urban Horticulture, Box 354115, University of Washington, Seattle, WA 98185-4115, USA.
P.E. Heilman. Washington State University, WSU Puyallup, WA 98371, USA.

Major advances have been made during recent years in the study of poplar biology. Driven by the need for alternative energy sources and predictable fiber supply, scientists have explored the natural reservoirs of genetic variation in a number of native poplar species. Ways have been studied to manipulate this variation through breeding, selection, and genetic engineering, and to enhance it through cultural techniques. Critical components have been identified that affect poplar productivity and tolerance to biotic and abiotic stresses. Most recently, the application of molecular tools to genetically informative pedigrees has begun to elucidate the genetic underpinnings of these critical components. In the process, poplar has revealed itself as an eminently useful model tree to help shed light on tree growth and development in general.

The purpose of this book is to showcase these advances in our understanding of the biology of *Populus*, to integrate them, and to interpret their relevance to practical aspects of poplar culture and conservation. As editors we have brought an undeniable bias to this task. The approach we have chosen reflects very much the orientation of our own research and involves many of the scientists who have been engaged in productive collaboration with us. Thus, genetics and physiology take center stage and serve as the book's organizing principle, as they have been the focal themes of the University of Washington/Washington State University Poplar Research Program since 1978. At the same time we have attempted to provide current views on poplar evolution, ecology, and systematics, and to point to promising new developments in our understanding of biotic interactions of poplar with both pathogens and insects. Clearly, a more extensive coverage would be needed to do full justice to these topics. A chapter on poplar culture concludes the book, offering perspectives on the diversity of cultural systems in selected parts of the world.

Most of the chapter authors participated in the International Poplar Symposium, *Biology of Populus and its Implications for Management and Conservation*, convened on the University of Washington campus in Seattle, August 20–25, 1995. The conference was co-sponsored by the International Union of Forest Research Organizations, the Poplar Council of the United States, and supported by the Northwest forest industry. Travel funds were provided by a grant to the University of Washington, Washington State University, and Iowa State University from the Collaborative Research in Plant Biology Program, jointly administered by the U.S. Department of Agriculture, Department of Energy (DOE), and National Science Foundation. We gratefully acknowledge research support from The Washington Technology Center, and the sustained funding of our research by the DOE Biofuels Feedstock Development Program which provided for continuity of effort so critical in work with long-life-cycle plants.

The content of this book has greatly benefited from the constructive criticism by the following reviewers: B. Barnes, K. Brown, B. Callan, M. Coleman, D. DeBell, M. de Block, D. Dickmann, D. Eissenstat, J. Farrar, P. Gale, J. Hardin, D. Harry,

J. Isebrands, W. Johnson, D. Karnosky, M. Krasny, P. Larson, Y. Linhart, T. Martin, S. Merkle, M. Ostry, S. Pallardy, S. Pezeshki, J. Pinon, P. Schulte, J. Stromberg, C. Tauer, M. Tjoelker, M. Villar, N. Wheeler, R. Wu, S. Wullschleger, and G. Wyckoff. We are indebted to them for their efforts. Special thanks go to Lynn Catlett who provided invaluable staff assistance from the beginning to the end of the project. Daniel and Andrea Stettler offered their graphics expertise for the design of the book cover. Finally, we express our gratitude to Gerry Neville and Nancy Daly at NRC Research Press for their patient guidance throughout all phases of the book's genesis.

R.S., T.B., P.H., and T.H.
Seattle, October 1996.

PART I. EVOLUTION, GENETICS, AND GENETIC MANIPULATION

Overview

R.F. Stettler and H.D. Bradshaw, Jr.

Geneticists inherently are fascinated with variability. With this in mind, it is not surprising that forest geneticists have been drawn to the genus *Populus*, which is replete with genetic variation at many different levels: among sections within the genus, as well as among species, provenances, populations, individuals, and genes. Part I of this book is devoted to an understanding of genetic variation at all these levels, both in nature and in human-controlled environments.

Understanding natural variation runs as a theme through many chapters. According to Eckenwalder (Chapter 1) the broad patterns of diversity in the genus, currently accommodated in six taxonomic sections, reflect the evolutionary history as reconstructed from the fossil record. The three most advanced sections, *Aigeiros, Tacamahaca,* and *Populus* (formerly *Leuce*), seem to have appeared last, having evolved via *Leucoides* from *Abaso* ancestors, the only modern representative of which is *P. mexicana.* Poplars from Sect. *Populus,* the most advanced group, display strong reproductive isolation from those of *Aigeiros* and *Tacamahaca* which, on the other hand, interbreed quite freely among each other (Chapters 4 and 5). Although aspens (Sect. *Populus*) share several floral features with *Salix,* molecular data show a clear separation between the two genera. Certainly one of the most consistent and evolutionarily significant distinctions between poplars and willows is their mode of pollination. Insect vectors, with their capacity to discriminate among floral phenotypes, tend to enhance assortative mating and have likely speeded up the process of speciation in the willows. By contrast, wind pollination in the poplars, combined with their dioecy, has had the opposite effect. Also, natural hybridization is common among poplar species (Chapters 1, 4, and 11) and, as indicated by fossil evidence (Chapter 1), may have for a long time provided for interspecies gene exchange. Taken together, it will not surprise us that the number of species in the two genera differ by an order of magnitude.

Airborne pollen and the outbreeding habit, in combination with effective seed dispersal, shape also the internal structure of genetic variation in *Populus* species, as discussed by Farmer (Chapter 2). Patterns of isozyme variation across several species in over 30 enzyme systems show a recurrent theme of little

1

differentiation among populations, with the bulk of variation (>90%) residing within populations. Common-garden studies have revealed that polymorphism within populations is common also in morphological and phenological traits and may well be adaptive in itself in an opportunistic pioneer tree. At the same time, there is also abundant evidence from these studies for natural selection having favored adaptation to climatic variables such as photoperiod, moisture, and temperature. Some of the physiological underpinnings of this variation are elucidated in chapters of Part II. Thus, whereas gene flow between and within species has been effective in maintaining high levels of variability, it has not prevented genetic response to strong and consistent selective pressures imposed by the environment.

The capacity for asexual reproduction, so highly developed in *Populus*, is another major determinant of natural variation. Clonal repeats, in aggregate or dispersed form, are common stand features in many poplars but especially prominent in those of Sect. *Populus*. The significance of clonal propagation for the natural perpetuation of genotypes, for the complementation of sexual reproduction, for the invasion of stressful sites, for the facilitation of gene exchange between co-existing species (by extending the lifespan of hybrids and compatible breeding partners), and for the potential exploitation of gender differences in niche partitioning, are discussed in Chapters 2, 3, and 4. The ease with which poplar can be vegetatively copied and grown is also one of the key characteristics that has made it useful to humans.

Genetic manipulation of natural variation is another focal theme for several chapters. We view genetic manipulation here as a means both to render poplar more suitable for human purposes, as well as to gain more insight into the nature and amount of genetic variation that is available in the first place. The two go hand in hand. Since phenotypic variation is only a partial reflection of the underlying genetics, experiments are required to resolve it into its causal components. The simplest experiments, i.e., growing clonal propagules of different genotypes side by side, have shown many morphological and phenological traits to be under sufficiently strong genetic control to allow reliance on phenotypic assessment (Chapters 2, 4, 5, 6, 7, and 10). This is what has given poplar culture one of its major economic incentives: to exploit the repeatability of specific phenotypes and to seek gains through clonal selection. Another widely practiced manipulation of natural variation has been interspecific hybridization. Hybrids are easily produced and often superior in growth to their parents (Chapters 4, 5, and 6). Cloning then permits variation to be "fixed" in the very best of them and expanded to commercial scale. The quick success of this combined approach may well have detracted attention from the need to develop a better understanding of the genetics of quantitative traits such as growth and form, and the gains to be made through intraspecific population improvement. Nor is there an adequate theoretical basis for predicting hybrid performance from that of the parents. As argued by Bisoffi and Gullberg (Chapter 6), Riemenschneider, Stelzer, and Foster (Chapter 7), and Stettler, Zsuffa, and Wu (Chapter 4), longer-term perspectives must be adopted in the development of poplar breeding and selection programs.

Poplar productivity in wood quantity and quality derives from trait combinations that are amenable to analysis. This "dissection" has been pursued at many scales, from the morphometric to the molecular, in genetically structured populations of genotypes. The finding that a modest number of quantitative trait loci (QTLs) can have a disproportionate influence on such complex traits as height, stem diameter, or leaf phenology (Bradshaw, Chapter 8), has given new insight into the genetic control of tree growth and development. As in the case of such crops as tomato, corn, and soybean, this provides greater precision and focus for genetic manipulation. The question now is, which of these QTLs are universally limiting in a species, which are limiting only in certain genetic backgrounds and/or environments, and where are the sources for favorable alleles? Marker assisted selection (MAS) of parents or progenies will, of course, greatly increase the efficiency of genetic improvement in a crop that is as space and time demanding as a tree. Application of this concept is most immediate in an area that has repeatedly presented poplar growers with some of their greatest challenges, namely to cope with diseases (Chapter 10). Direct genetic engineering, i.e., the targeted introduction of genes into poplar genomes, is a further avenue under intense investigation and with positive results in several laboratories (see Han, Gordon, and Strauss, Chapter 9). In distinction to breeding, it permits the incorporation of genetic material for which sexual transfer is impossible. Key to its potential role in poplar improvement is the rate at which successful transformants can be recovered. It has to be high to be of practical value, given the rapid cycles in poplar breeding. Insect and herbicide resistance, reproductive sterility, and lignin modification are target traits currently under study.

Possibly the most perplexing questions in poplar culture relate to the differences between natural populations and artificial plantations, especially those with exotic germplasm. How do we transfer lessons learned from one to the other? Which are most sensitive to scale? This is well illustrated by Newcombe in his discussion of fungal pathogens of *Populus* and their specificity (Chapter 10). Virtually every poplar-growing region in the world has had its array of diseases to contend with, notably in non-native poplar material. New, virulent strains of pathogens have evolved regularly where production clones had been selected for disease resistance. Yet, in the face of sustained pressure from introduced pathogens, to date no major epidemics have threatened native populations of indigenous poplar species. Is this a question of competitive interaction among pathogens for limited substrate, or of interaction between pathogens and other mediating organisms? Careful comparative study of natural and artificial populations with sharper tools (e.g., molecular markers) and more coordinated protocols will help to shed light on this puzzle. Here again, the goal will be an understanding of processes, beyond the (needed) descriptive phase.

Conservation and restoration of variation, a third theme, is relevant for all the varied roles poplars will play in future biological systems, whether natural or artificial. Since many of these species formed part of the natural vegetation along river courses and other water bodies, where humans tended to settle, they were early victims of exploitation and destruction. There is a disturbing paucity of contiguous stands left, let alone extensive natural populations, of native species of *Populus* in most of Asia and Europe. Similar impacts can be seen in the more populated areas of North America. As well-documented by Braatne, Heilman, and Rood in Chapter 3, river regulation, damming, livestock grazing, gravel mining, and other activities pose serious problems for the continued establishment and regeneration of natural populations of poplars even in the more northern latitudes of that continent.

Just how much the loss can be, in ecological, genetic, and evolutionary terms, but above all in *information*, is the message of Chapter 11, based on the incisive studies by Whitham, Floate, Martinsen, Driebel and Keim. What these workers have found is that natural zones of hybridization between *P. angustifolia* and *P. fremontii* in the U.S. Rocky Mountains are major centers of arthropod and avian diversity and play an important role in the evolution of host/pest interactions. Although these zones tend to be narrow, they are found in many river valleys, especially where higher elevation species overlap with those of lower elevations. But once subject to fragmentation and disconnection from the parental species, hybrid swarms would tend to disappear and their remnants would escape detection. Similar ecological functions have been described for hybrid zones of other species in other regions, such as willows, pinyon pines, and eucalypts (Chapter 11). Paradoxically, hybrids are often not recognized as deserving protection by the laws that pertain to endangered species. This chapter documents nicely what valuable insights can be gained from the study of natural populations in combination with carefully conducted experiments. Lessons learned from such cases may be extendable beyond their original geographic context and may serve as proxies for areas of lesser poplar abundance elsewhere.

It is evident from most chapters in this first part of the book that a better understanding of processes that shape patterns of variation in *Populus* species and their relationship to the physical and biotic environment is essential for the ecological and economical benefits we hope to expect from these trees in the future. This will be even more urgent as new conditions may challenge the adaptive capacity of poplars in unprecedented ways. The minimum prerequisites for such understanding are (1) the continued existence of near-natural populations in their native habitat, ideally uncontaminated by exotic germ plasm, and (2) a rich source of genetic diversity with which to experiment. Both demand vigorous efforts in conservation and in gene-resource management.

Poplar as a model tree: *Populus* is also rapidly becoming a model organism for the study of tree biology. The rationale for its adoption as a model, the materials used for the purpose, and early results obtained from such studies, are described in Chapters 4, 5, and 8 and in much of Part II of this book. Features that make poplar attractive include: the genetic diversity in the genus and the ease with which it can be combined and recombined in hybrids; the uniform chromosome number across all species ($2n = 38$) and the small size of the nuclear genome ($2C = 1.2$ pg); the convenience of sexual propagation in the greenhouse and the abundance of seed that can be obtained in 4–8 wk; the rapid growth and expression of a wide range of traits in morphology, anatomy, physiology, and pest susceptibility; the early sexual maturity of at least some species and hybrids (4–6 yr), permitting rapid progression into advanced generations in breeding; and most importantly, the ability to resprout and to be propagated vegetatively, which allows replication in time and space very much as with inbred lines of *Drosophila* or *Arabidopsis*. Furthermore, poplar is amenable to cell and tissue culture, as well as to genetic transformation.

One of the standard materials that is currently most widely used by researchers is a three-generation pedigree developed at the University of Washington which includes one parent each of *P. trichocarpa* and *P. deltoides*, two of their F_1 offspring, and an F_2 family ($N = 379$) derived from them. A subset of 54 F_2 trees has served for the initial QTL mapping of the *Populus* genome (Chapter 8) and now exists in many replicate copies in experimental plantations and laboratories across the U.S. and Canada. Quantitative trait loci (QTLs) controlling more than 100 traits, including growth, form, phenology, anatomy, physiological characters, and disease resistance, have been mapped in this F_2 family. As phenotypic and genetic data begin to accumulate, connections between scientific disciplines have been strengthened. For example, the finding that there are regions of the genome with simultaneous effects on stem volume growth and sylleptic branch leaf area provides a link between genetics and physiology, and this kind of complementary interdisciplinary study is reflected in Parts I and II of this book. It is apparent that researchers in many fields will take advantage of "standard" genetic stocks when these are available, and that the benefits of sharing genetic materials are substantial. Perhaps the greatest benefit comes from the "model system effect" that attracts talented people from many areas of science who might otherwise be reluctant to work on organisms as difficult as trees, but who feel comfortable contributing their expertise to augment the large knowledge base that already exists for some *Populus* pedigrees.

The coordinated collection and distribution of cloned *Populus* pedigrees, and information on them, were made possible in large measure by the Collaborative Research in Plant Biology Program, jointly administered by the U.S. Department of Agriculture, Department of Energy, and National Science Foundation. The mission of this program was not to sponsor research *per se*, but rather to provide scientific infrastructure, such as computer networks, databases, newsletters, and meetings, in support of multidisciplinary research. With the growing ease of communication and travel, mere physical proximity of investigators is no longer essential. It makes sense to foster collaborative research among investigators at geographically dispersed institutions to take advantage of their unique capabilities. The merit of this approach is well-illustrated by the breadth and depth of information presented in the book chapters that follow.

CHAPTER 1
Systematics and evolution of *Populus*

James E. Eckenwalder

Introduction

This chapter briefly summarizes our understanding of the classification of *Populus* and some of the evolutionary processes that have given rise to its present diversity. The distinguishing features of the genus are placed in an ecological context and are also used to discuss its relationships with other genera and families. Different approaches to classification at the sectional and species level are discussed and a conservatively defined group of 29 species are used to explore phylogenetic relationships within *Populus*. Results of the phylogenetic analysis are compared to the pattern of diversification observed in the fossil record. Finally, I consider the effects of natural hybridization among species belonging to the same or different sections.

Definition of the genus *Populus*

General features relating morphology to mode of life

Species of the genus *Populus* are all single-trunked, deciduous (or semievergreen) trees and most spread clonally by means of root-borne sucker shoots (soboliferous), a feature uncommon among trees and of some interest in the management of poplar plantations. They are among the fastest-growing temperate trees, a quality tied ecologically to their role as vegetational pioneers as well as functionally to their heterophyllous growth habit. In contrast to northern oaks, for example, whose entire growth consists solely of a spring flush of leaves that overwinter as well-formed primordia in the buds (preformed or early leaves), poplar shoots continue to grow after bud burst by initiating, expanding, and maturing leaves (neoformed or late leaves) throughout the growing season (Fig. 1; Critchfield 1960). Cessation of growth and bud formation is induced by photoperiod in at least some poplars (Pauley and Perry 1954), so there is some

James E. Eckenwalder. University of Toronto, Department of Botany, 25 Willcocks St., Toronto, ON M5S 3B2, Canada.
Correct citation: Eckenwalder, J.E. 1996. Systematics and evolution of *Populus*. *In* Biology of *Populus* and its implications for management and conservation. Part I, Chapter 1. *Edited by* R.F. Stettler, H.D. Bradshaw, Jr., P.E. Heilman, and T.M. Hinckley. NRC Research Press, National Research Council of Canada, Ottawa, ON, Canada. pp. 7–32.

Fig. 1. Diagrammatic representation of seasonal heterophylly. Each line represents a generalized morphogenetic trajectory for a single foliar organ (bud scale or leaf). Preformed leaves are initiated at the end of one growing season and expand at the beginning of the next. Neoformed leaves are initiated and expand in concert during a single growing season.

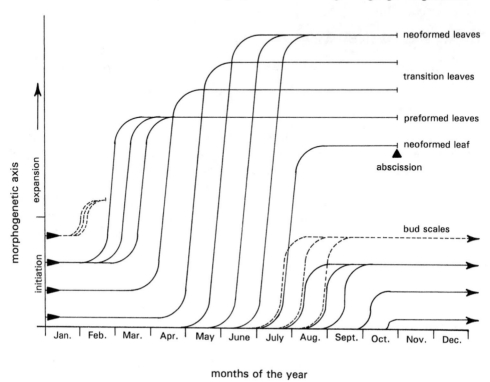

months of the year

risk that growing trees north of their provenance may inhibit hardening off in autumn and lead to winter damage to shoots. This is of little consequence where form is not important, although it may also lead to markedly reduced size.

Preformed and neoformed leaves often differ considerably in texture, shape, and toothing (Fig. 2). Preformed leaves are generally more diagnostic taxonomically than are neoformed leaves and tend to differ more among major groups (sections) of poplars than among species within sections. Neoformed leaves are relatively convergent among unrelated poplar species, with the exception of the unique (in poplar) lobed neoformed leaves of *P. alba* and its hybrids (Fig. 3). Preformed leaves, the only ones present in the frost-prone early spring, are typically tougher than the neoformed leaves. Since neoformed leaves experience predictably more favorable conditions than preformed leaves, they may possess structural and physiological traits promoting higher photosynthetic rates (Dickmann 1971; Donnelly 1974; Regehr et al. 1974).

There are separate male and female trees (they are dioecious) that (except in some subtropical species) flower before leaf emergence in spring from special-ized buds containing preformed inflorescences. This behavior is similar to that of other wind-pollinated temperate trees (which usually are monoecious, having separate male and female flowers on the same tree) for whom leaf emergence and canopy closure reduce wind speeds and interfere with the aerodynamics of pollination (Niklas 1992). In most localities, there is at least partial separation of flowering times in co-occurring species of poplars, but plants may be forced into flower in water culture for artificial pollination. The capsules and their airborne seeds (which have a readily detached coma of cottony hairs) mature with or after the overwintered preformed leaves. At that time, there is still usually the high soil moisture content (from spring floods in riparian species) required for immediate germination and establishment in these nondormant and unresistant seeds (Woolward 1907; Faust 1936). Males are often the preferred clones in horticulture and plantation culture both because of the perceived messi-ness of the falling capsules and seeds and their cottony tuft of hairs and because of the untested assumption that the greater resources required for seed production have a penalty in reduced growth of females compared to males. Because growth continues long after seed dispersal, this effect is not likely to be of much consequence. Details of flowers and inflorescences in both males and females

Fig. 2. Tooth sizes of preformed and neoformed leaves in successive leaf positions along shoots of several species of *Populus*. Depending on the shoot, leaf positions 5–9 represent transitions between preformed and neoformed leaves.

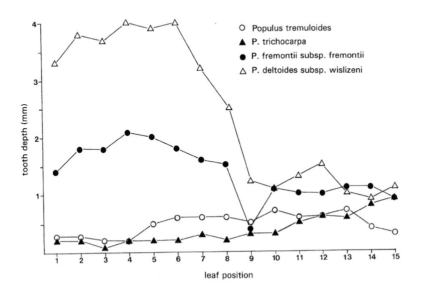

Fig. 3. Comparison of outlines of preformed and neoformed leaves of several species of *Populus*. A: *P. alba*. B: *P. deltoides* subsp. *monilifera*. C: *P. grandidentata*. D: *P. nigra* 'Italica.'

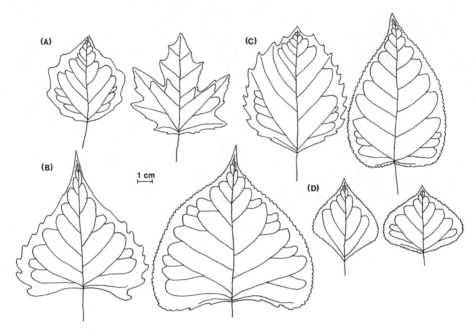

are the primary basis for defining sections of the genus (Table 1 and further discussion below).

Technical features

Poplars have been recognized as a group since very early times and have a unique combination of characteristics that distinguish them from all other genera of plants. The defining features are primarily in the reproductive structures. Like many other wind-pollinated trees, the flowers are borne in pendent racemes (catkins, aments) that vary in flower number and density among poplar species and sections but are generally similar for males and females of the same species. Many kinds of trees with male catkins, like oaks (*Quercus* spp.), have different arrangements of the female flowers. Among temperate trees with female catkins, only the poplars and willows (*Salix* spp.) have seeds with a coma of cottony hairs on parietal placentas in thin-walled capsules.

Individual flowers of both catkin types are subtended by thin bracteoles that fall as the catkins elongate during flowering. The caducous bracteoles distinguish the catkins of poplars from those of such trees as willows, alders (*Alnus* spp.), and birches (*Betula* spp.), whose bracts are either persistent or fall only with

10

Table 1. Sections of *Populus* and their morphological taxonomic characteristics.

Section	Abaso	Turanga	Leucoides	Aigeiros	Tacamahaca	Populus
			swamp poplars	cottonwoods, black poplar	balsam poplars	aspens, white poplars
Carpels	2	3	2–4	2–4	2–4	2
Seeds	5–15	20–40	4–12	4–25	2–30	2–15
Disks	deciduous laciniate irregular	deciduous laciniate irregular	persistent lobed regular	persistent entire regular	persistent entire regular	persistent entire or dentate oblique
Stamens	15–40 apiculate	25–30 apiculate	15–35 apiculate	12–60 emarginate	10–50 emarginate	5–25 truncate
Bracts	narrow not ciliate	narrow not ciliate	broad not ciliate	broad not ciliate	broad not ciliate	broad ciliate
Leaves	unifacial	unifacial	bifacial	unifacial or bifacial	bifacial	bifacial
	finely crenate	entire or coarsely dentate	finely crenate	finely or coarsely crenate	finely crenate	entire, finely crenate, sinuate or coarsely dentate
Buds	yellow dry	yellow dry	brown slightly resinous	tan or red resinous	red-brown resinous	brown slightly resinous
Species	1	3	3–5	2–8	6–15	7–13

release of the fruits or seeds. There are no ordinary petals or sepals, but the 5–60 stamens or the solitary pistil are borne on a more or less expanded floral disk. Willows have no such disk, but one or more tubular nectaries instead.

Overwintering vegetative and reproductive buds are covered by several bud scales, of which the lowest faces directly away from the stem, an unusual bud-scale arrangement (Trelease 1931). This lowest scale corresponds to the single scale covering willow buds. The buds are often covered with exudates that are rich in a variety of hydrophobic organic chemicals (Greenaway and Whatley 1990). These exudates presumably have some role in winter hardiness (and perhaps in protection from herbivores) since they are most prominent on buds of the more northerly poplars.

Corresponding to rapid, nearly continuous growth during the favorable season is a light, diffuse-porous wood structure whose annual increments are often very difficult to distinguish, making it difficult to determine tree age (Panshin and de Zeeuw 1970; Campbell 1981). Poplar wood differs from that of willows primarily in having homocellular rays, whereas willow wood is heterocellular. The wood differs from many similar light-colored woods primarily in lacking some more specialized features, like the marginal parenchyma of yellow poplar (*Liriodendron* spp.) or the spiral thickenings and multiseriate rays of basswood (*Tilia* spp.). The bark remains thin for a longer period than in most trees, especially in aspens, for which cortical photosynthesis makes a significant contribution to the annual energy budget (Strain and Johnson 1963).

The leaves are alternate, stipulate, petiolate (with the petiole often transversely flattened distally), and simple, with glandular teeth along the margin and often with glands at the junction of the blade and petiole as well. Variations in these features are of taxonomic significance in the genus. Although leaf shapes are predominantly variations on ovate in poplars and lanceolate in willows, there is considerable overlap, but the distinctive palmatopinnate venation of poplar leaves versus the strictly pinnate venation of willow leaves is diagnostic.

Relationships of *Populus*

Relationship to *Salix*

Populus and *Salix* comprise the family Salicaceae (along with some other genera for those who care to recognize segregates of either of these two genera). In phylogenetic taxonomy, the relationship between the two genera is important for their proper classification. If the common ancestor of all *Populus* and *Salix* species is not considered to belong to either genus, then the two genera may be considered monophyletic and hence truly natural groups. If, on the other hand, the common ancestor of either one is considered a member of the other genus, then the genus that gave rise to the other one is paraphyletic and not truly natural.

Formally, if the latest common ancestor of all poplars, for instance, were also the ancestor of some or all of the willows, then *Populus* would be paraphyletic and the definition of the genus should be modified, either to include the willows, or to be divided into a series of more restricted genera whose ancestors did not give rise to any willow species. We may be faced with this dilemma, because certain morphological features, notably prominent stipules (on some sucker shoots), a reduced number of bud scales, bracteoles ciliate, and sometimes entire, floral disk narrowly cup-shaped, reduced stamen numbers, narrowly ovoid capsules with just two carpels, simplified stigmas, and smaller, less numerous seeds, suggest a link between the aspens, section *Populus* (synonym, section *Leuce*), an advanced group of poplars (Table 1), and the willows. On the other hand, molecular evidence from both chloroplast DNA and nuclear ribosomal DNA suggests a deep gulf between the two genera that would not allow the willows to have emerged from within the genus *Populus* (Smith 1988). So, did the common features of willows and aspens evolve in parallel or did one group derive from the other? Species of *Salix* appear later in the fossil record than poplars (Collinson 1992), but even this is hardly decisive and the problem remains open. Analysis of additional genes in a careful selection of poplars and willows may help to settle the issue.

Relationship to other families

Salicaceae have traditionally been considered taxonomically isolated from other families, enough so as to merit recognition of the separate order Salicales. The position of this order has varied in different systems of angiosperm classification. Early workers were impressed by the catkins and included the Salicales in a group with other catkin-bearing trees, the Amentiferae. Most contemporary systems, however, recognize the fundamental difference in fruit structure between Salicaceae and other catkin-bearing trees and place the Salicales near the order Violales, containing the violet family and the tropical family Flacourtiaceae among others. The Violales, while not catkin-bearing, share capsules (or other fruit types) with parietal placentae (Cronquist 1981). A.D.J. Meeuse made a more radical suggestion, based on a detailed look at floral features, proposing inclusion of *Salix* and *Populus* within the Flacourtiaceae (Meeuse 1976). There is much to recommend this merger, including occurrence of the distinctive salicin glycosides in some members of the Flacourtiaceae (Hegnauer 1973, 1989). Unfortunately, the suggestion cannot yet be intelligently implemented because of difficulties in the classification of the Flacourtiaceae themselves and of the Violales generally. There are a number of subgroups (tribes) within the Flacourtiaceae that show similarities to *Populus* or *Salix*, and these tribes are not all considered closely related to one another (Sleumer 1980).

Much work needs to be done on the Flacourtiaceae and related families before the Salicaceae can take their proper place among them, but an important consequence of this connection is to challenge our usual view of the cool temperate origin of poplars. Although thought of as northern, because their greatest ecological importance is in the boreal forest region, the taxonomic diversity of poplars reaches its peak in warm temperate to subtropical latitudes across the northern hemisphere. Furthermore, those species (like *P. mexicana* and *P. ilicifolia*) that have the least reduced flowers and hence most closely resemble the Flacourtiaceae are tropical. And *P. mexicana* also most closely resembles the earliest known poplar fossils, which apparently lived under tropical conditions (see below).

Classification of *Populus*

Sectional classification

Compared to the willows, there are relatively few species of poplars but they clearly fall into a number of morphologically and ecologically quite distinct groups, traditionally recognized as sections. With few exceptions, there is a large measure of agreement in the literature on the characteristics and species composition of the sections, and the major barriers to hybridization in the genus lie between sections (Zsuffa 1975). With the description of section *Abaso* to accommodate *P. mexicana*, which only superficially resembles the *Aigeiros* cottonwoods with which it had previously been placed (Eckenwalder 1977a), the number of sections has been brought to six (Table 1). Additional sections have been described from time to time and then abandoned, such as *Tsavo* (Browicz 1966), erected for the east African *P. ilicifolia*, here included within section *Turanga*.

Another species of disputed position has been the Himalayan *P. ciliata*. Some breeding literature refers it to the swamp poplars of section *Leucoides* and then ascribes its crossability to balsam poplars of section *Tacamahaca* as an example of intersectional hybridization (Zsuffa 1975; Willing and Pryor 1976). Unfortunately, *P. ciliata* has none of the defining features of section *Leucoides*, like lobed floral disks, apiculate anthers, and a distinctive leaf venation. In these and other features, *P. ciliata* is a typical balsam poplar so its assignment to section *Leucoides* was simply a mistake. That is not to say that there are no difficulties in defining boundaries between sections *Leucoides* and *Tacamahaca*. Molecular work has shown a close similarity between the Chinese *P. lasiocarpa*, a usually unquestioned member of section *Leucoides*, and the sympatric *P. szechuanica*, equally unquestioned as a member of section *Tacamahaca* (Smith 1988). The situation clearly merits more extensive study.

The most interesting question of sectional affiliation concerns the relationships between sections *Aigeiros* (cottonwoods) and *Tacamahaca* (balsam poplars) and the status of *P. nigra*. These two sections are the only ones known to be freely intercrossable (Zsuffa 1975; Eckenwalder 1984b). Although vegetatively and ecologically readily distinguishable, there are no clear differences in flowers and inflorescences between cottonwoods and balsam poplars (Table 1), and they could be accommodated in a single section (as the aspens and white poplars are in section *Populus*), where they would have to be given separate subsections anyway. The balance of present evidence, including phylogenetic analyses (see below), seems to favor keeping them apart, but resolution of this issue affects the placement of *P. nigra*.

Populus nigra, the type species of section *Aigeiros*, is not clearly most similar to the North American cottonwoods placed with it in section *Aigeiros* (Eckenwalder 1977*a*). It is almost equally similar to some species of the balsam poplars, section *Tacamahaca*. It also has peculiar crossability relationships, successful only in one direction with both North American cottonwoods and balsam poplars (Zsuffa 1973; Eckenwalder 1982), unless rescued by embryo culture. Even more peculiarly, it has a chloroplast genome apparently captured from the sympatric white poplar, *P. alba* (or an ancestor of this oddest but undisputed aspen) and not at all like those of either the cottonwoods or balsam poplars (Smith and Sytsma 1990). Ultimately, the North American cottonwoods may need to be removed from section *Aigeiros* into their own section, but once again, a satisfactory resolution of this dilemma requires more research.

Classification at the species level

The biggest disagreements in poplar classification concern the number of species. Among modern workers, those in China and Russia (both of which have a rich diversity of poplars) have been "splitters" while North American and western European workers (with some notable exceptions) have held conservative (that is, broad) species concepts. Depending on individual predilections, world species counts could range from 22 to about 85 (29 species are provisionally accepted here; Table 2). The differences in number of species recognized are due to two main differences in interpretation. The more tractable of the two is the misinterpretation of some hybrids as species. Natural F_1 hybrids between particular combinations of parental species may have consistent morphologies and definite geographical ranges (Eckenwalder 1984*b*) but their coherence disappears with sexual reproduction (Rood et al. 1986; Keim et al. 1989) and they should therefore not be treated as species. In many cases, workers have simply not recognized the hybrid nature of the taxa they have described.

The second main cause of differences in species treatment by taxonomists is not so cut and dried because it is due to the differences in taxonomic philosophy between "splitters" who wish to emphasize variation and give formal recognition to any entities that can be recognized in the field and herbarium, and "lumpers" who accept a broader range of variation within species. The latter approach is preferable for both practical and philosophical reasons. A serious practical difficulty with a narrow species concept in *Populus* is that once one begins to name variants as species, there is essentially no end to it. For example, new Chinese

Table 2. Species of each section with their generalized distributions.

	Abaso	Turanga	Leucoides	Aigeiros	Tacamahaca	Populus
W. Eurasia + N. Africa	—	euphratica	—	nigra	laurifolia	alba tremula
E. Eurasia	—	euphratica pruinosa	lasiocarpa glauca, s.l.	nigra, s.l.	ciliata szechuanica yunnanensis suaveolens, s.l. simonii, s.l. laurifolia	alba adenopoda gamblei sieboldii tremula, s.l.
E. Africa	—	ilicifolia	—	—	—	—
N. America	mexicana	—	heterophylla	deltoides fremontii	balsamifera trichocarpa angustifolia	simaroa guzmananilensis monticola grandidentata tremuloides

Note: Species designated as s.l. (*sensu lato*) include others which are often recognized as distinct in the literature and which might be retained as subspecies. Some other species also contain additional subspecies or varieties.

species have already been added to the 60 recognized in the *Flora Reipublicae Popularis Sinicae* since its publication in 1984. Such narrowly defined species are often associated with a typological species concept and a reluctance to accept within-species variation. In contrast, a recent investigation of leaf morphology in Chinese aspens based on variability in population samples strongly supports abandoning most of the segregate species that have been described (Barnes and Han 1993). In addition to these practical considerations, philosophically, if species are real things in nature, not mere human constructs of convenience (but not all systematists will accept this premise), it is important to find their true boundaries and incorporate them into our classifications. Poplars, as long-lived, obligately outcrossed, and highly dispersible organisms, are typical of such organisms in storing a great deal of genetic variation both within and among populations (Weber and Stettler 1981; Cheliak and Dancik 1982; Farmer et al. 1988; Rajora et al. 1991). When species are broadly defined, we discover a pattern in which species of a section are generally widely distributed and largely allopatric so that species replace each other geographically. At any one spot, then, two or three local poplars typically represent different sections of the genus. The narrowly defined species that have been proposed in the genus often have scattered distributions that lack coherence and overlap broadly with their close relatives in the same section. As outlined below, this makes it difficult to make sense of poplar phylogeny and evolution in an ecological context.

The disparity in species concepts discussed above is slightly diminished if we look at classification below the species level. To a certain extent, lumpers may recognize additional geographical subdivisions of the species at the subspecies and variety level that are treated as species by splitters. This is the case, for example, with my classification of the North American *Aigeiros* cottonwoods (Eckenwalder 1977*a*) in which I recognize five taxa but treat them as subspecies of just two species, *P. deltoides* and *P. fremontii*. This is still fewer taxa than the seven species and additional varieties recognized by Sargent in his *Manual* (Sargent 1922), and the morphological and geographical boundaries among my subspecies are rather different from those of previous workers. Still, this kind of more complicated hierarchy partly bridges the gap betweeen the extremes of lumping and splitting. In Table 2, use of the abbreviation s.l. (*sensu lato*) after a name implies that it provisionally includes some additional, commonly recognized species. Some of these are likely to be retained as subspecies, or even reinstated as species, with further study. The classification adopted here is compared with one commonly used in the breeding literature (Zsuffa 1975) in Table 3.

Assuming a conservative species concept, there are strong disparities in the breadth of distribution and numbers of species in each of the sections (Table 2). The balsam poplars (section *Tacamahaca*) and aspens (section *Populus*) each have about 10 species, more equally distributed between the old and new worlds in the case of the aspens than of the balsam poplars. Both sections are far richer

Table 3. Comparison of the classification of *Populus* adopted here with one commonly used in the breeding literature (Zsuffa 1975).

Section	Zsuffa 1975	This chapter
Abaso Eckenwalder	—	*P. mexicana* Wesmael
Turanga Bunge	*P. euphratica* Olivier	the same
	—	*P. ilicifolia* (Engler) Rouleau
	P. pruinosa Schrenk	the same
Leucoides Spach	*P. lasiocarpa* Oliver	the same
	P. wilsonii Schneider	*P. glauca* Haines, s.l.
	P. heterophylla L.	the same
Aigeiros Duby	*P. nigra* L.	the same
	P. deltoides Marshall	the same
	P. fremontii S. Watson	the same
	P. sargentii Dode	*P. deltoides*, s.l.
	P. wislizenii Sargent	*P. deltoides*, s.l.
Tacamahaca Spach	*P. angustifolia* James	the same
	P. balsamifera L.	the same
	P. cathayana Rehder	*P. suaveolens*, s.l.
	P. ciliata Royle	the same
	P. koreana Rehder	*P. suaveolens*, s.l.
	P. laurifolia Ledebour	the same
	P. maximowiczii A. Henry	*P. suaveolens*, s.l.
	P. simonii Carrière	the same
	P. suaveolens Fischer	the same
	P. szechuanica Schneider	the same
	P. trichocarpa T. & G.	the same
	P. yunnanensis Dode	the same
Populus	*P. adenopoda* Maximowicz	the same
	P. alba L.	the same
	P. davidiana (Dode) Schneider	*P. tremula*, s.l.
	—	*P. gamblei* Haines
	P. grandidentata Michaux	the same
	—	*P. guzmanantlensis* Vazquez & Cuevas
	—	*P. monticola* Brandegee
	P. sieboldii Miquel	the same
	—	*P. simaroa* Rzedowski
	P. tomentosa Carrière	a hybrid
	P. tremula L.	the same
	P. tremuloides Michaux	the same

Note: Authorities are given for both species and sections.

in species in eastern Eurasia than in Europe and northern Africa. The other four sections have just one to three species each and two of them are confined to single continents. The two latter sections, *Turanga* and *Abaso*, with the most primitive floral characteristics in the genus, are very distinct in leaf form but much more similar reproductively. In that, they are somewhat comparable to section *Tacamahaca* versus section *Aigeiros*, or aspens versus white poplars in section *Populus*, but those two pairs tend to split latitudinally within each hemisphere rather than between old and new world continents. The swamp poplars of section *Leucoides* have a very restricted distribution confined to China (and the adjacent Himalaya) and eastern North America. Finally, section *Aigeiros* is widely distributed but has few species. If the cottonwoods are separated from *P. nigra* into their own section, each of the two resulting sections would span a single continent. At least some species of each of the three circumboreal sections have large populations, whereas those of the three more restricted sections tend to be small.

Evolution

Poplars are good subjects for a wide variety of evolutionary studies. Their modest number of species in diverse groups, general abundance, quick growth, ready vegetative propagation, easy cultivation, ability to flower and set seed on excised branches, obligate outcrossing, wide and varied crossability, cytological uniformity at the diploid level, extensive fossil record, and rich array of pests and diseases all contribute to their uncommon tractability among tree genera. Some of the many current studies relating to poplar evolution are discussed elsewhere in this volume (see, for example, Farmer, Chapter 2; Braatne et al., Chapter 3; Stettler et al., Chapter 4; and Whitham et al., Chapter 12). Here I concentrate on just three topics, each with a North American emphasis: the fossil record, phylogenetic studies, and the evolutionary consequences of natural hybridization in the genus.

Fossil record

Poplars generally have a high soil moisture requirement so most species grow along the floodplains of rivers and streams or along the shores of lakes and ponds. Even those species that are typically more upland in habitat often have riparian populations as well. Since the trees are deciduous, they are well placed to contribute annual large flushes of detached leaves to accumulating sediments (Spicer 1989). Although the trees may have brittle branches, and some species, at least, appear to be cladoptosic, with a regular abscission zone at the base of branches, shoots with attached leaves make up a minute fraction of the annual leaf fall and would rarely be expected to contribute to the sedimentary record.

Poplar organs other than leaves are, for one reason or another, much less likely to become fossilized. Pollen is produced in enormous quantities, but the exine is very thin, with little of the resistant sporopollenin that leads to pollen preservation in other taxa (Rowley and Erdtman 1967). Male catkins shrivel up and fall immediately after pollen release and quickly decay on the ground. Seeds, too, are very thin-walled and unresistant and either germinate immediately or decompose. Capsules vary in texture, but some are moderately thick-walled and accumulate with the fallen fruiting catkins beneath trees. The pedicel usually rots, but the detached capsules can then become fossilized, although never in the same volume as leaves. Finally, the wood lacks terpenoids and other heartwood extractives that resist decay, and large mature trees often have their centers well rotted even before they fall.

Given these factors, it is not surprising that isolated leaves make up the great bulk of the fossil record in poplars, with wood, twigs, pollen, and fruits much less common (Collinson 1992). Luckily, leaves (especially preformed leaves) have diagnostic value in *Populus*, at least at the sectional level, so this record is useful for tracing a variety of evolutionary patterns, including the history of the sections (Fig. 4).

Early literature recognized willows and poplars among Cretaceous leaf fossils but those fossils have all proven to belong to other taxa (e.g., *Cercidiphyllum*, Brown 1939). So Salicaceae are not among the oldest extant families, not surprising considering their highly derived floral features. Instead, the earliest recognizable Salicaceae are leaves of *Populus* sect. *Abaso* from western North America dating to the latest Paleocene, about 58 million years ago. These leaves are not identical to those of the extant *P. mexicana* but they are clearly related. They and their descendents remained in the fossil record for some 29 million years before disappearing after the early Oligocene, when tropical vegetation in North America was replaced by more temperate assemblages (Fig. 4).

Evidence shows that these ancestral poplars were widespread across North America (Fig. 5). The best-preserved and most interesting collection of material comes from the famous Eocene Green River Formation (MacGinitie 1969), where *P. meegsii* (listed as *P. cinnamomoides* and *P. wilmottae* by MacGinitie) was a local dominant. These fossils reveal a great deal about the biology of these 50-million-year-old trees. Some leaves have signs of herbivory, galls, and fungal diseases remarkably similar to those affecting *P. mexicana* today. One specimen includes a twig with leaves and an attached fruiting catkin (Manchester et al. 1986). Unlike the two-valved capsules of *P. mexicana*, capsules of the fossil are three-valved, like those of the related old world section *Turanga*, a condition that would be expected to be primitive based on its predominance in the Violales, including many Flacourtiaceae. The modern *P. mexicana* is strikingly heteroblastic, with willow-like juvenile leaves strongly differentiated from cottonwood-like adult leaves, and despite a lack of any fossils showing the two leaf types attached to the same twig (a very unlikely circumstance) the preponderance of evidence suggests that *P. meegsii* was equally strongly heteroblastic (Eckenwal-

Fig. 4. Generalized North American fossil record of sections of *Populus*. Many comparisons to extant poplars in the literature on poplar fossils are unreliable and the assignments here are based on published specimens or illustrations rather than on relationships proposed in the texts.

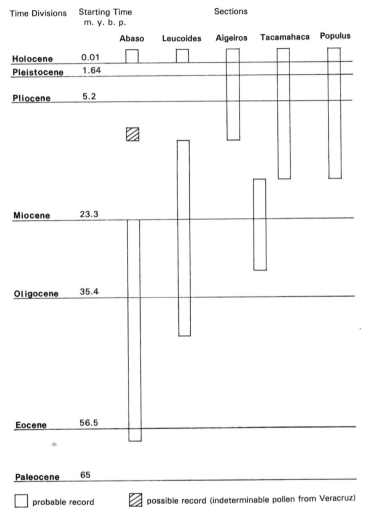

der 1980). There is no evidence for heteroblasty in the earlier Paleocene fossils and the phenomenon is lacking in related Flacourtiaceae, so it may have developed after the evolution of *Populus*. Perhaps the narrow juvenile leaves help small individuals retain their leaves when they are submerged in floods that do not reach the crowns of adults.

Fig. 5. Present and paleobotanical distribution of *Populus* section *Abaso* in North America.

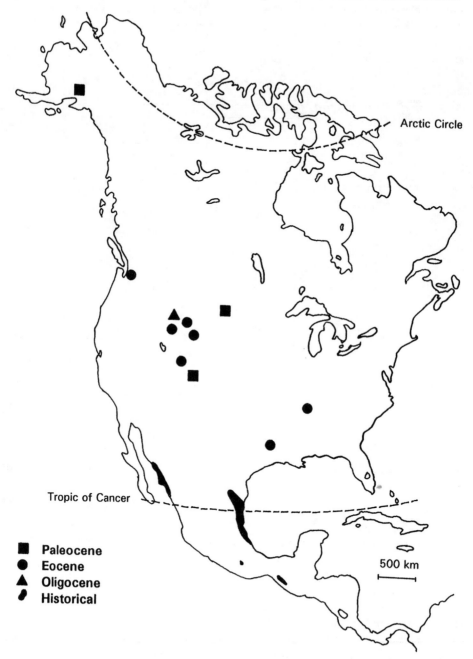

Arctic Circle

Tropic of Cancer

■ Paleocene
● Eocene
▲ Oligocene
◗ Historical

500 km

Members of section *Abaso* were the sole poplars in North America until the late Eocene, when precursors of other sections first make their appearance (Fig. 4) and the first Eurasian poplar fossils are recorded (Collinson 1992). The sequence of appearance is very roughly parallel to apparent primitiveness, both with respect to similarities to related Flacourtiaceae and to phylogenetic analysis (see below). Thus swamp poplars of section *Leucoides* are first to appear, in the late Eocene, and have a sparse fossil record. The extant North American species of the section, *P. heterophylla*, inhabits permanent swamps in the eastern United States, sites that are unlikely to provide good fossilization of leaves because a restricted input of mineral sediments leads to the development of highly organic accumulations (future coal deposits?).

The next developments occurred in the Oligocene, when fossils that have been assigned to section *Tacamahaca* occur in both the new and old worlds (Collinson 1992), but they seem more like precursors of both sections *Tacamahaca* and *Aigeiros*, especially specimens from the Creede flora of Colorado (Axelrod 1987; Wolfe and Schorn 1990). Unequivocal members of these two sections did not appear in North America until the Miocene, when definite aspens of section *Populus* are also first encountered (Fig. 4).

The detailed picture of the Neogene (Miocene and more recent) fossil record of *Populus* is fairly complicated (Eckenwalder 1977*b*). In those cases in which species within a section are too similar to distinguish easily by their leaves, like sections *Aigeiros* and *Tacamahaca*, it is not feasible to make overly refined comparisons of the fossils to extant taxa. On the other hand, the precursors of the modern aspens *P. grandidentata* and *P. tremuloides* are easily recognized in Miocene and Pliocene paleofloras. Before a more detailed interpretation of the later fossil record can be undertaken, much more detailed analyses of leaf architecture in extant poplars, and particularly its variability, needs to be completed.

The thousands of fossil leaf specimens available from hundreds of localities spanning millions of years offer a potentially rich source of data on a wide variety of evolutionary problems in *Populus*. Most of these have never been addressed and we are barely scratching the surface of others. For example, some inferences about the evolutionary importance of natural hybridization in poplars presented below have been enriched by fossil evidence.

Phylogenetic studies

Contemporary phylogenetics (or cladistics) stems from the work of the entomologist Willi Hennig, who sought a formal systematics founded in evolutionary theory (Hennig 1966). Anyone interested in biological classifications will know that tremendous debates about the proper bases for those classifications have rattled the complacency of systematists since the advent of cladistics. Whether or not one accepts the full program of phylogenetic systematics, most will agree that the technical methods of cladistics are powerful tools for inferring phylogenetic relationships using a wide variety of data. The Wagner parsimony method, coupled with outgroup analysis, is readily available in microcomputer-based implementations. The basic algorithm seeks a phylogenetic tree (cladogram) topology that minimizes the number of parallel originations or losses of features (Felsenstein 1993). Although the method is easy to apply, great care must be exercised in the choice and interpretation of characters and results, particularly when more than just a few species are involved since the results are often ambiguous in the sense that cladograms with different arrangements of the species may all be equally parsimonious (require the same number of character state changes). These discrepancies are usually addressed by constructing a consensus tree that emphasizes the phylogenetic information common to the different individual trees at the expense of loss of resolution in portraying relationships compared to any one of them.

The consensus cladogram presented here (Fig. 6) finds common groupings for the 29 provisional species listed in Table 2 among more than 800 different cladograms generated from Wagner analyses of different sets of taxa ranging from 20 to 64 "species," including the genus *Idesia* of the Flacourtiaceae as an outgroup. These analyses used 76 morphological characters of buds, leaves, inflorescences, male and female flowers, and fruits. Forty-two of the characters exhibited three or more conditions among the poplar species. Each was ordered by logical sequencing and outgroup analysis and converted to a minimal number of derived characters with only two states using the additive binary coding implemented in the "Factor" program of the Phylip package of programs for phylogenetic analysis (Felsenstein 1993). The criterion for including a group in the consensus is its occurrence in at least 80% of the original cladograms. The highly varied topologies of the individual cladograms lead to a much less resolved topology for their consensus, but there is still a considerable amount of structure held in common among them. There are three systematic issues to discuss that arise from an inspection of this cladogram: the monophyly of the sections, the relationships among them, and the relationships among species within sections.

The consensus cladogram provides support for the monophyly of all of the sections, except *Tacamahaca*, which consists here of two monophyletic groups (clades), one consisting of the typical balsam poplars, like *P. balsamifera* and *P. trichocarpa*, and the other of three somewhat anomalous, narrow-leaved and thin-twigged species resembling *P. angustifolia*. In many of the original cladograms, this narrow-leaved group joins the aspens of section *Populus* before they jointly group with the other balsam poplars, implying that section *Tacamahaca* is paraphyletic. Other trees, however, support a monophyletic section *Tacamahaca*. Monophyly could be correct; there is simply no compelling evidence for it in this dataset. The consensus does support the inclusion of *P. nigra* with the North American cottonwoods although some individual cladograms place it

Fig. 6. Consensus cladogram of the minimal species of *Populus* listed in Table 2. Numbers at the junctions represent the approximate percentage of trees containing that group among the 840 most parsimonious trees for 20–47 species.

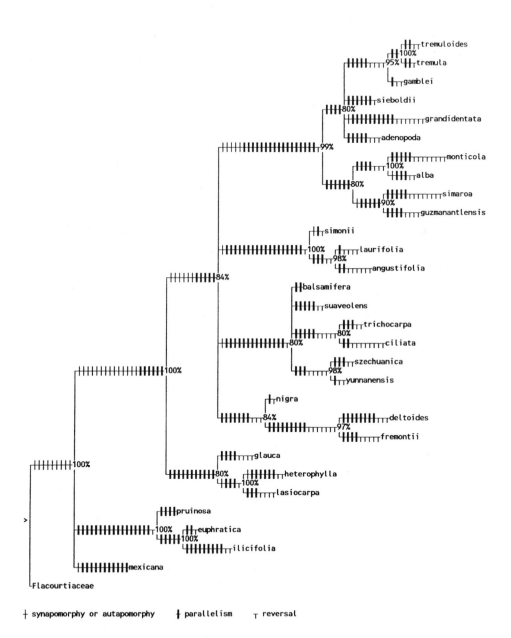

+ synapomorphy or autapomorphy   ‡ parallelism   ⊤ reversal

between the North American cottonwoods and the balsam poplars and aspens. The apparent monophyly of section *Aigeiros* here contrasts with the anomalous position of *P. nigra* in molecular analyses explained, in part, as the result of chloroplast capture through hybridization with a precursor of *P. alba* of section *Populus* (Smith and Sytsma 1990). Bud exudates, on the other hand, embed *P. nigra* within section *Tacamahaca* (Greenaway et al. 1990). Given also the strange crossing behavior mentioned above, the affinities of *P. nigra* certainly merit further study.

The relationships among the sections are less clear than their monophyly. The two most flacourt-like, partly tropical sections, *Abaso* and *Turanga*, are sister groups to the other four sections, although their relationships to each other vary among the trees. In most cases, they form a monophyletic group together, but in other cladograms one of the two is the sister group to the other sections and the second is the sister to that group. The sister group relationships of the swamp poplars of section *Leucoides* to the remaining three sections is also supported, but a few of the cladograms place them as sister group either to section *Aigeiros* or to section *Tacamahaca*, the latter more in accord with the limited available molecular data (Smith 1988). Finally, the four monophyletic clades of "advanced" poplars (sections *Aigeiros* and *Populus* and two clades of section *Tacamahaca*) are unresolved in their relationships to one another, although some of the possible resolutions are not found in any of the original cladograms. Most of these show one of three general patterns. If we call the clade of typical balsam poplars TAC1 and the narrow-leaved clade TAC2, these three arrangements can be represented as follows: AIG((TAC1 + TAC2)POP), AIG(TAC1(TAC2 + POP)), and (AIG(TAC1 + TAC2))POP. Overall, the phylogenetic relationships among the sections, both those that are resolved and those that are unresolved, closely parallel the sequence of appearance of the sections in the fossil record discussed above (Fig. 4).

Relationships among species within sections are also largely unresolved, particularly among the two larger sections, *Populus* and *Tacamahaca*. In each of the three sections (and the TAC2 clade) with just three species, one species appears consistently (although not necessarily exclusively) as the sister group to the other two. Among the typical balsam poplars (TAC1), the sister group relationships between *P. szechuanica* and *P. yunnanensis* and between *P. ciliata* and *P. trichocarpa* also break down into other arrangements in a few cladograms, but in less than 20% of them. The aspens (sect. *Populus*) show a little more structure. The Mexican endemics *P. simaroa* and *P. guzmanantlensis* are sister species to each other, as are *P. alba* and *P. monticola*. The resulting white poplar clade is sister group to the remaining typical aspens, among which *P. gamblei* joins with the tightly linked *P. tremula* and *P. tremuloides*, leaving other relationships unresolved. *Populus monticola* is a Mexican endemic that has frequently been confused with *P. alba* in earlier literature and in herbaria

(Eckenwalder 1977*b*), and *P. tremula* and *P. tremuloides* have even been considered conspecific as subspecies (Löve and Löve 1976).

These patterns of relationship within and between sections suggest a speculative evolutionary pattern involving phases of ecological radiation and geographical vicariance. Following the origin of the genus in the Paleocene and a spread of ancestral poplars from one hemisphere to the other (fossils favor a new world origin but the closest flacourt relatives are Asian), there was a vicariance that led to the separation of sections *Abaso* and *Turanga*, each confined to a single hemisphere. Invasion of temperate habitats was later initiated by an ancestral member of section *Leucoides*. This was followed in the Miocene by a rapid (effectively simultaneous) radiation into the distinct habitats that characterize the three advanced sections. The advanced sections then underwent relatively few cycles of rapid allopatric speciation (partly influenced by hybridization, as discussed below), leading to the mixture of resolved and unresolved relationships embodied in the consensus cladogram. The implication is that the two main phases of radiation, the emergence of the advanced sections and of species within them, were so rapid that the exact sequence of splittings (or speciations) is largely unrecoverable because of conflicting character distributions.

Natural hybridization

Intercrossability among species is, of course, one of the foundations of breeding work in the genus *Populus*. Attempts at comprehensive crossing programs have revealed both the wide extent of potential intercrossability and the very real limitations on it (Zsuffa 1975). Natural hybridization, and even spontaneous hybridization in cultivation, have also played roles in the generation of genetic material available for selection and utilization. One need only cite *P. ×canescens* (*P. alba × P. tremula*) and *P. ×canadensis* (*P. deltoides × P. nigra*), two of the earliest foundations of plantation culture in poplars. *Populus ×canadensis* (synonym *P. euramericana*) first arose spontaneously in Europe after the introduction of *P. deltoides* in the 18th century (or even the 17th), long before the deliberate breeding programs of our century. And the originations of *P. ×canescens*, although doubtless entirely natural in many places, was also encouraged by the anthropogenic spread of *P. alba* in very early times. Other poplars that have been grown for centuries, and were originally described as species, are also of natural or near natural hybrid origin, such as *P.* 'Balm-of-Gilead' (part of *P. ×jackii*: *P. balsamifera × P. deltoides*) and *P. ×tomentosa* (*P. alba × P. adenopoda*).

Natural hybridization in the absence of human interference is, as might be expected, also very common. Hybridization occurs freely within sections and between species in some combinations of sections (Fig. 7). *Populus mexicana* is allopatric to all other North American species and its crossing relationships

Fig. 7. Natural hybridization among North American species of *Populus*.

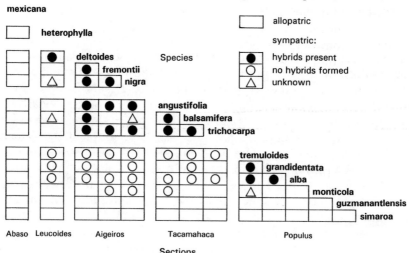

are unknown but among the four sections that show at least some sympatry in North America only section *Populus* is strongly reproductively isolated from the others, which accords well with the manipulations required to successfully artificially cross aspens with other poplars (Guries and Stettler 1976; Ronald 1982). With that exception, wherever two species of poplars (including introduced species) are sympatric in North America, there are hybrids between them to be found. That does not mean that hybrids overwhelm their parents; they are still restricted in distribution compared to their parents and the total population sizes of the parents are enormously greater than the numbers of hybrids.

The patterns of distribution of hybrids are different depending on whether they are of intrasectional or intersectional origin. Species within sections *Aigeiros* and *Tacamahaca* are basically geographical replacements for each other, with relatively narrow bands where they replace each other, so the relatively subtle hybridization between species within these sections has a limited geographical extent compared to the ranges of the parent species (Eckenwalder 1977a, 1977b). On the other hand, these sections are ecologically isolated from one another but broadly sympatric, so that hybridization occurs over large geographical areas but with a relatively narrow ecological range of overlap (Bennion et al. 1961; Eckenwalder 1984a, b). More complicated mixtures of these patterns arise in the more limited regions where three or more species are sympatric (Rood et al. 1986).

Although intersectional hybridization is often apparently restricted to the F_1 generation (Eckenwalder 1984a), examples of introgression are also known today (Keim et al. 1989). The ecological separation of sections *Aigeiros* and *Ta-*

camahaca has been maintained in the face of hybridization almost since their origin, as evidenced in the fossil record (Eckenwalder 1984*c*). There is, however, evidence for ancient introgression in reproductive characters shared by western versus eastern cottonwoods with sympatric balsam poplar species (Eckenwalder 1984*c*), so intersectional hybridization may have had a subtle role in the local adaptation of these two sections.

Intersectional hybridization between *P. heterophylla* (section *Leucoides*) and *P. deltoides* (section *Aigeiros*) is extremely rare and local in areas of the southeastern United States where the former reaches its greatest abundance. Other than noting the existence of these specimens, nothing is known about the circumstances of hybridization there.

The situation with respect to aspens is a little different from that of the other sections. Most North American species of section *Populus* do not even come into contact with one another today, but in northeastern North America two native aspens and the introduced *P. alba* are sympatric if partially ecologically separated and all three hybrid combinations occur (Barnes 1961; Spies and Barnes 1982). They have very different frequencies that appear to be linked to pollen competition rather than other possible explanations, like phenology or hybrid viability (Einspahr and Joranson 1960; Connolly and Eckenwalder 1985). As with the *Aigeiros–Tacamahaca* intersectional hybrids, there is evidence of aspen hybridization extending back into the Tertiary (Barnes 1967). One can speculate about possible consequences of this, like the origin of *P. monticola* through hybridization or the possibility that *P. tremuloides* might have acquired its enormous geographical distribution and ecological versatility as a compilospecies incorporating otherwise now vanished local Tertiary aspen species evident in the fossil record (Eckenwalder 1977*b*). Because of a lack of fixation of hybrid combinations through polyploidy, however, the long term consequences of hybridization in *Populus* generally appear to have involved exploitation of increased variability and novel gene combinations rather than the more dramatic hybrid speciation.

Natural poplar hybrids are commonly found away from one or both of their probable parents (Eckenwalder 1984*a*) and this has sometimes led to their description as new species (like *P. ×hinckleyana*, cf. Eckenwalder 1984*b*). On rare occasions, hybrids may be found hundreds of kilometers away from the nearest present locations of one parent, as, apparently, for *P. ×smithii* (*P. grandidentata* × *P. tremuloides*) in the Niobrara River drainage of western Nebraska. Such occurrences could represent ancient hybridizations under different climatic conditions because those hybrids capable of soboliferous clonal growth can potentially persist for thousands of years, if not the millions sometimes suggested (Barnes 1967). The alternative explanations involving long-distance dispersal of seeds or pollen are possible but less plausible, perhaps, particularly the latter.

A female aspen tree in my own yard is separated by a few kilometers from the nearest male clones but it appears to remain unpollinated and sets no viable seeds, a situation common in wind pollination (Niklas 1992). Hybridization seems to occur primarily at close range. Establishment of hybrid seed at a distance is more likely than effective long-distance transport of pollen, but nonhybrid seed is still a much larger part of the seed rain. All told, "relictual" hybrids pose many interesting questions worthy of investigation. Despite the many excellent studies of recent years, we have much yet to learn about natural hybridization in poplars in general, especially intrasectional hybridization.

Conclusions

Populus is a readily defined genus consisting of equally well-marked, ecologically coherent, species groups, called sections, that have differentiated sequentially since the origin of the genus in the Paleocene. *Populus* is tropical in origin and the greatest diversity of the genus still lies far south of the boreal region of its greatest ecological importance. Both the genus as a whole and its sections may be accepted provisionally as strictly monophyletic, although the evidence for this is conflicting. A broad species concept produces more readily interpretable distributional, ecological, and evolutionary patterns than the narrow one adopted by many taxonomists. Natural hybridization, while commonplace in the genus over at least the last several million years, has had little effect on speciation in comparison to processes of divergence.

References

Axelrod, D.I. 1987. The Late Oligocene Creede Flora, Colorado. Univ. Calif. Publ. Geol. Sci. **130**: 1–235.

Barnes, B.V. 1961. Hybrid aspens in the Lower Peninsula of Michigan. Rhodora, **63**: 311–324.

Barnes, B.V. 1967. Indications of possible mid-Cenozoic hybridization in the aspens of the Columbia Plateau. Rhodora, **69**: 70–81.

Barnes, B.V., and Han, F.-Q. 1993. Phenotypic variation of Chines aspens and their relationships to similar taxa in Europe and North America. Can. J. Bot. **71**: 799–815.

Bennion, G.C., Vickery, R.K., and Cottam, W.P. 1961. Hybridization of *Populus fremontii* and *Populus angustifolia* in Perry Canyon, Box Elder County, Utah. Proc. Utah Acad. Sci. Arts Lett. **38**: 31–35.

Browicz, K. 1966. *Populus ilicifolia* (Engler) Rouleau and its taxonomic position. Acta Soc. Bot. Poloniae, **35**: 325–335.

Brown, R.W. 1939. Fossil leaves, fruits, and seeds of *Cercidiphyllum*. J. Paleontol. **13**: 485–499, plates: 51–56.

Campbell, R.B., Jr. 1981. Field and laboratory methods for age determination of quaking aspen. USDA For. Serv. Res. Note INT.-314. pp. 1–5.

Cheliak, W.M., and Dancik, B.P. 1982. Genetic diversity of natural populations of a clone-forming tree, *Populus tremuloides*. Can. J. Genet. Cytol. **24**: 611–616.

Collinson, M.E. 1992. The early fossil history of the Salicaceae. Proc. Roy. Soc. Edinb. Sect. B, **98**: 155–167.

Connolly, V.A., and Eckenwalder, J.E. 1985. Hybridization between white poplar and native aspens in southern Ontario. Am. J. Bot. **72**: 947. (Abstr.)

Critchfield, W.B. 1960. Leaf dimorphism in Populus trichocarpa. Am. J. Bot. **47**: 699–711.

Cronquist, A. 1981. An integrated system of classification of flowering plants. Columbia University Press, New York, NY. 1262 p.

Dickmann, D.I. 1971. Photosynthesis and respiration by developing leaves of cottonwood (*Populus deltoides* Bartr.). Bot. Gaz. **132**: 253–259.

Donnelly, J.R. 1974. Seasonal change in photosynthate transport with elongating shoots of *Populus grandidentata*. Can. J. Bot. **52**: 2547–2559.

Eckenwalder, J.E. 1977*a*. North American cottonwoods (*Populus*, Salicaceae)of sections *Abaso* and *Aigeiros*. J. Arnold. Arbor. Harv. Univ. **58**: 193–208.

Eckenwalder, J.E. 1977*b*. Systematics of *Populus* L. (Salicaceae) in southwestern North America with special reference to sect. *Aigeiros* Duby. Ph.D. thesis, University of California, Berkeley, CA.

Eckenwalder, J.E. 1980. Foliar heteromorphism in *Populus* (Salicaceae), a source of confusion in the taxonomy of Tertiary leaf remains. Syst. Bot. **5**: 366–383.

Eckenwalder, J.E. 1982. *Populus* ×*inopina* hybr. nov. (Salicaceae), a natural hybrid between the native North American *P. fremontii* S. Watson and the introduced Eurasian *P. nigra* L. Madroño, **29**: 67–78.

Eckenwalder, J.E. 1984*a*. Natural intersectional hybridization between North American species of *Populus* (Salicaceae) in sections *Aigerios* and *Tacamahaca*. I. Population studies of *P.* ×*parryi*. Can. J. Bot. **62**: 317–324.

Eckenwalder, J.E. 1984*b*. Natural intersectional hybridization between North American species of *Populus* (Salicaceae) in sections *Aigerios* and *Tacamahaca*. II. Taxonomy. Can. J. Bot. **62**: 325–335.

Eckenwalder, J.E. 1984*c*. Natural intersectional hybridization between North American species of *Populus* (Salicaceae) in sections *Aigerios* and *Tacamahaca*. III. Paleobotany and evolution. Can. J. Bot. **62**: 336–342.

Einspahr, D.W., and Joranson, P.N. 1960. Late flowering in aspen and its relation to naturally occurring hybrids. For. Sci. **6**: 221–223.

Farmer, R.E., Jr., Cheliak, W.M., Perry, D.J., Knowles, P., Barrett, J., and Pitel, J.A. 1988. Isozyme variation in balsam poplar along a latitudinal transect in northwestern Ontario. Can. J. For. Res. **18**: 1078–1081.

Faust, M.E. 1936. Germination of *Populus grandidentata* and *P. tremuloides*, with particular reference to oxygen consumption. Bot. Gaz. **97**: 808–821.

Felsenstein, J. 1993. PHYLIP (Phylogeny Inference Package), Version 3.5. University of Washington, Seattle, WA. 190 p.

Greenaway, W., and Whatley, F.R. 1990. Resolution of complex mixtures of phenolics in poplar bud exudate by analysis of gas chromatography – mass spectrometry data. J. Chromatogr. **519**: 145–158.

Greenaway, W., English, S., and Whatley, F.R. 1990. Variation in bud exudate composition of *Populus nigra* assessed by gas chromatography –mass spectrometry. Z. Naturforsch. Sect. C. Biosci. **45**: 931–936.

Guries, R.P., and Stettler, R.F. 1976. Pre-fertilization barriers to hybridization in the poplars. Silvae Genet. **25**: 37–44.

Hegnauer, R. 1973. Chemotaxonomie der Pflanzen, Bd. 6. Birkhäuser Verlag, Basel, Switzerland. 882 p.

Hegnauer, R. 1989. Chemotaxonomie der Pflanzen, Bd. 8. Birkhäuser Verlag, Basel, Switzerland. 718 p.

Hennig, W. 1966. Phylogenetic systematics. University of Illinois Press, Urbana, IL. 263 p.

Keim, P., Paige, K.N., Whitham, T.G., and Lark, K.R. 1989. Genetic analysis of an interspecific hybrid swarm of *Populus*: occurrence of unidirectional introgression. Genetics, **123**: 557–565.

Löve, A., and Löve, D. 1976. Nomenclatural notes on arctic plants. Bot. Not. **128**: 497–523.

MacGinitie, H.D. 1969. Eocene Green River Flora of northwestern Colorado and northeastern Utah. Univ. Calif. Publ. Geol. Sci. **83**: 1–140, pl. 1–31.

Manchester, S.R., Dilcher, D.L., and Tidwell, W.D. 1986. Interconnected reproductive and vegetative remains of *Populus* from the Middle Eocene Green River Formation, northeastern Utah. Am. J. Bot. **73**: 156–160.

Meeuse, A.D.J. 1976. Taxonomic relationships of Salicaceae and Flacourtiaceae: their bearing on interpretative floral morphology and dileniid phylogeny. Acta Bot. Neerl. **24**: 437–457.

Niklas, K.J. 1992. Plant biomechanics; an engineering approach to plant form and function. University of Chicago Press, Chicago, IL. 607 p.

Panshin, A.J., and de Zeeuw, C. 1970. Textbook of wood technology, vol. 1. McGraw-Hill Inc., New York, NY. 705 p.

Pauley, S.S., and Perry, T.O. 1954. Ecotypic variation of the photoperiodic response in *Populus*. J. Arnold Arbor. Harv. Univ. **35**: 167–188.

Rajora, O.P., Zsuffa, L., and Dancik, B.P. 1991. Allozyme and leaf morphological variation of eastern cottonwood in Ontario. For. Sci. **37**: 688–702.

Regehr, D.L., Bazzaz, F.A., and Boggess, W.R. 1974. Photosynthesis, transpiration and leaf conductance of *Populus deltoides* in relation to flooding and drought. Photosynthetica, **9**: 52–61.

Ronald, W.G. 1982. Intersectional hybridization of *Populus* sections, *Leuce-Aigeiros* and *Leuce-Tacamahaca*. Silvae Genet. **31**: 94–99.

Rood, S.B., Campbell, J.S., and Despins, T. 1986. Natural poplar hybrids from southern Alberta. I. Continuous variation for foliar characteristics. Can. J. Bot. **64**: 1382–1388.

Rowley, J.R,. and Erdtman, G. 1967. Sporoderm in *Populus* and *Salix*. Grana Palynol. **7**: 517–567.

Sargent, C.S. 1922. Manual of the trees of North America, 2nd ed. Houghton Mifflin, New York, NY. 910 p.

Sleumer, H.O. 1980. Flacourtiaceae. Flora Neotrop. Monogr. **22**: 1–499.

Smith, R.L. 1988. Phylogenetics of *Populus* L. (Salicaceae) based on restriction site fragment analysis of cpDNA. M.Sc. thesis, University of Wisconsin, Madison, WI. 115 p.

Smith, R.L., and Sytsma, K.J. 1990. Evolution of *Populus nigra* (sect. *Aigeiros*: introgressive hybridization and the chloroplast contribution of *Populus alba* (sect. *Populus*). Am. J. Bot. **77**: 1176–1187.

Spicer, R.A. 1989. The formation and interpretation of plant fossil assemblages. Adv. Bot. Res. **16**: 96–191.

Spies, T.A., and Barnes, B.V. 1982. Natural hybridization between *Populus alba* L. and the native aspens in southeastern Michigan. Can. J. For. Res. **12**: 653–660.

Strain, B.R., and Johnson, P.L. 1963. Cortical photosynthesis and growth in *Populus tremuloides*. Ecology, **44**: 581–584.

Trelease, W. 1931. Winter Botany; an identification guide to native trees and shrubs. 3rd ed. Repr. Dover, New York, NY. 1961. 396 p.

Weber, J.C. and Stettler, R.F. 1981. Isoenzyme variation among ten populations of *Populus trichocarpa* Torr. et Gray in the Pacific Northwest. Silvae Genet. **30**: 82–87.

Willing, R.R., and Pryor, L.D. 1976. Interspecific hybridization in poplar. Theor. Appl. Genet. **47**: 141–151.

Wolfe, J.A., and Schorn, H.E. 1990. Taxonomic revision of the Spermatopsida of the Oligocene Creede Flora, southern Colorado. U.S. Geol. Surv. Bull. 1923: 1–40, plates: 1–13.

Woolward, F.H. 1907. The germination of poplars. J. Bot. **45**: 417–419, plate: 487.

Zsuffa, L. 1975. A summary review of interspecific breeding in the genus *Populus* L. *In* Proceedings of Can. Tree Improv. Assoc., 28–30 Aug. 1973, Fredericton, NB. **14**(2): 107–123.

CHAPTER 2
The genecology of *Populus*

Robert E. Farmer, Jr.

Introduction

Since becoming a tree about 40–50 million years ago *Populus* has occupied the range of soil-bearing terrestrial environments in the temperate and boreal parts of the northern hemisphere. During this period the genus has genetically differentiated into about 40 species the delineation of which is still occupying taxonomists. Some of this species differentiation appears to have an environmental basis; there are riparians and "dry-landers" and a few that do well in both places. But one of the outstanding features of the genus is that several important species have managed to hang together genetically over large land areas. This has required both major adaptations to local environments and continued genetic communication over long distances. I wish to examine the results of these processes. Earlier pertinent reviews include those of Pauley (1949), Muhle-Larsen (1970), Schreiner (1970), Jokela and Mohn (1976), and Mohrdiek (1983).

We believe that a balance of three processes is responsible for genetic differentiation across variable landscapes: migration, natural selection, and mutation. Briefly, migration introduces the species to establishment opportunities through seed movement, then continues to supply successful immigrants with genetic material via both pollen and seed. Natural selection acts upon immigrant populations to fit them with individuals that are adapted to local environmental conditions and in so doing changes, of course, their genetic structure. Selection is a continual process, adapting populations genetically to both local environmental uncertainty and long-term environmental change. If immigration is restricted and populations of colonists are small, then random changes in gene pool composition (genetic drift) or inbreeding effects may also cause differentiation that has no foundation in natural selection. Thus migration across populations opposes local differentiation (the consequences of adaptation and genetic drift) and tends to keep things glued together, genetically speaking. I will

R.E. Farmer, Jr. Faculty of Forestry, Lakehead University, Thunder Bay, ON P7B 5E1, Canada.
Correct citation: Farmer, R.E., Jr. 1996. The genecology of *Populus*. *In* Biology of *Populus* and its implications for management and conservation. Part I, Chapter 2. *Edited by* R.F. Stettler, H.D. Bradshaw, Jr., P.E. Heilman, and T.M. Hinckley. NRC Research Press, National Research Council of Canada, Ottawa, ON, Canada. pp. 33–55.

examine *Populus* in light of the above model, with special attention to natural selection.

The equipment and devices by which material moves around in a gene pool and through which genetic pitfalls are avoided are collectively called a "genetic system." This system includes the nature of reproductive structures, pollen dispersal features, flowering phenology and fertilization devices which are successful in generating genetically variable progeny. These features, together with silvical and physiological characteristics, constitute a species "strategy," to use the jargon of population ecologists. Its components are, of course, genetically controlled and the product of natural selection. *Populus* has a very "fit" combination of features. First, it is dioecious, a characteristic which promotes outcrossing. Pollen is airborne in a most effective fashion and used efficiently; the seed/pollen ratio is high relative to many other wind-pollinated trees. Practically all poplar species begin reproduction early in ontogeny, and there are subsequent abundant seed crops almost annually. At least parts of these crops can be moved for relatively long distances, i.e., several kilometres. While establishment requirements are strict and in some cases the required conditions are rare, the strategy of producing seed every year makes up for this limitation. Once trees are established under the right sort of conditions their growth is aggressive and opportunistic. Moreover, species typically occupying areas deficient in good seed regeneration sites (e.g., aspens, balsam poplar) have evolved vegetative propagation modes that render successful genotypes potentially immortal.

We have reached our present state of knowledge in *Populus* via the combined application of techniques from classical genecology, taxonomy, quantitative genetics, physiological ecology, and, most recently, molecular genetics. By "classical genecology" techniques I refer mostly to use of "common garden" or provenance tests in which propagules from diverse habitats are grown together in such a way that genetic differences (i.e., the degree of genetic differentiation) can be evaluated. Special attention is usually given to characters presumed to have adaptive value. Over the last 40 years, use of a quantitative genetics approach in forest genetics has led to increasingly refined common garden test designs. These allow good estimates of genetic variance among and within populations to give a more complete picture of variation patterns. This approach has also led to better knowledge of the degree of genetic control over quantitative characters. Many morphological (e.g., bud and leaf features) and phenological characters have been shown to be under such strong genetic control that simple surveys of phenotypic variation in them have been useful. A few attempts at determining the physiological basis of geographically variable quantitative characters have been made. Finally, surveys of allozyme variation have focused on genetic differentiation in enzyme systems that are probably not subject to much selection pressure. Observations from this mix are beginning to fuse into coherence.

Adaptation to the growing period

Aspens

Species occupying a large range in the northern hemisphere must adapt to widely variable growing periods associated with geography. We first learned about this adaptation in trees from studies of the wide-ranging European aspen, *P. tremula*. In a common garden test Sylven (1940) grew seedlings originating between 55°54′N and 65°50′N on sites at 56°N, 62°30′N, and 65°50′N. Movement of material south resulted in abnormally early growth cessation. (The reports by Johnsson (1956) and Ekberg et al. (1976) are useful English accounts of this work for those who do not have access to the older volumes of Swedish journals.) Vaartaja (1960) subsequently demonstrated, via controlled environment studies, that photoperiod was the operant.

Briefly, plants adapted to the early autumns of high latitudes (50°N +) set buds under the relatively long photoperiods there; more southerly plants require a shorter day to do the same. Because this response is under strong genetic control, when moved around latitudinally plants react to the local photoperiod. Major (>4°) movement south produces a conservative, stunted, noncompetitive phenotype; movement north guarantees profligate shoot elongation into the certainty of fall freezing. This reaction to movement has now been observed in numerous *Populus* common garden experiments which I review below. Most of the geographically related genetic differentiation in *Populus* probably has its basis in this relationship. Surprisingly *P. tremula,* the "organism of discovery," has received little investigative attention in this area since Sylven's time. The initial confirming evidence was provided by Pauley and Perry (1954) and Pauley et al. (1963*a, b*) who essentially repeated Sylven's study with small, latitudinally diverse populations of *P. deltoides, P. trichocarpa, P. balsamifera, P. tremuloides,* and *P. tremula* grown in Massachusetts (42°N). Brissette and Barnes (1984) have reported the only replicated study of *P. tremuloides* in which small populations from Alaska, Alberta, Utah, and Michigan were compared at a Michigan location; this material exhibited the classical response with northern and high-elevation western sources stopping shoot growth early in the growing season, the Alaska source by June 15.

Eastern cottonwood

After the establishment of *P. deltoides* breeding programs in the early 1960s and the 1964 range-wide seed collection by J.J. Jokela, large common garden experiments were established at several locations. The designs usually allowed evaluation of both intra- and inter-population genetic variance. While detailed shoot elongation measurements were not common, assessment of height and survival in these studies provide indirect evidence of adaptive shoot elongation patterns in the main portion of the *P. deltoides* range.

The first southern test was established at 34°N (Stoneville, MS) and included 4 clones from each of 24 open-pollinated families collected at 6 locations on a latitudinal transect along the Mississippi River from Louisiana (31°N) to Minnesota (45°N). The first-year observations reported by Rockwood (1968) are based on the only analyzed data from the test. The most northern provenances (41°N +) all stopped growth by mid-August at a height of about 3.5 m. Provenances from 31°N–38°N added an additional 0.3–0.5 m in late August and early September. While the reported analysis did not include variance components, I estimate (from data in Rockwood's thesis) that provenances accounted for about 35% of the genetic variance in first-year height, families/provenances 2% and clones/families 63%. With exception of the Minnesota–Wisconsin material which began growth about the same time as Mississippi clones (April 12), northern provenances flushed later in the spring of the second year (April 15–16). Given only one season's weed control and their relatively poor subsequent growth, the northern provenances were lost to local competing vegetation as this study was abandoned without further reports.

A second southern test, replicated at Stoneville and Starkville, Mississippi, contained open-pollinated seedling progeny from parent trees in 15 locations throughout the southern United States. Collections were by D.T. Cooper and T. Hicks of the U.S. Forest Service. During its second year, bud break dates were recorded and shoot elongation was followed via biweekly height measurements (Friend 1981). Southern provenances (30°N) began growth 6 d earlier than northern ones (35°N) on average, though there was wide family variation within provenances. Northern families exhibited slower shoot elongation than southern provenances during August and September, but all of them continued growth into October. Thus, results were in contrast to those from several other studies in which shoot growth cessation time clearly accounted for major differences in height. The southern provenances, especially those from the southwest (e.g., south Texas), were significantly taller at the end of both first and second years.

A 7-yr test of the Jokela collection (498 clones from 116 families) was conducted by Ying and Bagley (1976) in eastern Nebraska (40°N). Here mean provenance height increased from northern to southern provenances to the southern Illinois (37°N) material; provenances below 37°N were severely damaged by freezing. Thus material from about 2° south of the planting site was moved there with impunity. In this test about 33% of the genetic variance in height was associated with provenance, 27% with families/provenance, and 40% with clones/families. The sequence of spring leaf flush was from northwestern to southeastern provenances, and roughly equal amounts of genetic variance were partitioned among the three sources of variation. Other work (Chandler and Thielges 1973) with these populations suggests that there is no clinal trend in chilling requirement for bud break, but that there is clonal variation in chilling requirement within provenances. Northern trees dropped leaves earlier in the fall than southern ones. In the smaller study of Jokela et al. (1982) planted at 37°26′N, trees from

Louisiana, Mississippi, and Texas all exhibited shoot dieback and basal stem lesions resulting from acclimation failure. Eldridge et al. (1972) tested clones from major sections of the range at Canberra (35°S) where photoperiodic response differences resulted in superior growth of Louisiana and Texas sources.

A replicated Minnesota planting at 45°N, which includes 108 families from central and northern provenances, has been followed for 17 yr (Mohn and Pauley 1969; Dihr and Mohn 1974; Mohn and Radsliff 1983). Early results (to 6 yr) suggested that sources 3°–4° south of the planting site might be climatically adapted to the growing season there. However, survival at 17 yr was less than 60% for most of the mid-latitude provenances (38°–42°N); the cause of mortality during the period from 6 to 17 yr was ambiguous. Since it is the oldest common-garden test for which we have a report, these 17-yr data must be given substantial weight, at least with respect to limits of adaption at the northern edge of the species range.

A second set of *P. deltoides* studies has been conducted with samples from the southwestern portion of the species' range from southern Texas to southeastern Colorado. These short-term nursery tests planted centrally in Oklahoma (Posey 1969; Nelson and Tauer 1987) demonstrated a southeast to northwest cline of decreasing growth, with northwestern material adapted to the shorter growing seasons at higher elevation and latitude there. Elevation differences in phenology and resulting growth pattern have also been noted in a common garden study with *P. angustifolia* in Colorado (Ernst and Fechner 1981), and they are analogous to latitudinal differences.

Balsam poplar

Except for the early work of Pauley and Perry (1954) studies of adaptive variation in *P. balsamifera* have been limited to a latitudinal cross section of its range at around 90°W. Using material from 45° to 49°N, Riemenschneider et al. (1992) noted only modest population differentiation in bud-set date and shoot growth at a common garden site near 46°N, though expected trends were evident. In a latitudinally broader sample planted at 48°N, Farmer (1993) observed that clones from 53°N stopped growth in mid-July while those from 45° to 46°N continued until September. This phenological difference partly resulted in provenances accounting for 85–90% of genetic variance in height growth. Clones from 53°N were about 55% as tall on average as those from 45° to 46°N, which have grown at 48°N for 7 yr without freezing damage. In controlled environment tests using this material Charrette (1990) further observed that growth cessation under short days (e.g., 6 h) was more rapid for northern provenances than southern ones. In contrast to bud-set, spring bud-break varied little among sources, though clones from 45° to 46°N began growth 2–3 d earlier on average than the rest of the provenances. Clones within provenances accounted for most of the genetic variance in bud-break. Frost hardiness in the spring was almost entirely related to the degree of shoot development, with susceptibility developing only after bud-break (Watson 1990). Thus provenance variation in frost susceptibility was a function of differential spring phenology. Immediately prior to spring bud-break plants are in a state of imposed dormancy, all provenances above 45°N having completed chilling requirements by early January (Farmer and Reinholt 1985). Therefore the wide within-population variance in time of spring leaf flush is probably a function of variable response to spring temperature.

Black cottonwood

Populations of *P. trichocarpa* along the western coast of North America have also differentiated photoperiodically as first suggested by the observations of Pauley and Perry (1954). The most geographically complete sample (35°–60°N) has been studied in a common garden in Scotland (55°N) by Cannell and Willett (1976). In addition to observing the usual latitudinal trend in growth cessation they noted its consequences for first year shoot/root ratios which were lower for northern provenances. In the central part of the species coastal range (44°–49°N) there is a modest latitudinal trend in photoperiodic response and associated branch and leaf characteristics (Heilman and Stettler 1985; Weber et al. 1985; Rogers et al. 1989). However, perhaps the most interesting aspect of these latter studies is the high degree of genetic variance in phenology and related shoot growth within populations. Most of this variation is accounted for by clonal differences within open-pollinated families. Weber et al. (1985) note that this within-population variation may be a selective response to the temporally variable environment in which the species occurs in the northwestern United States.

European and Asian species

Except for the early *P. tremula* work, patterns of adaptive variation for European and Asian species have generally not been reported. This is particularly unfortunate with respect to *P. nigra* since there are now probably only scattered remnants of the species in wild stands. Opportunities still exist in many of the Asian species, at least two of which are subjects of major current investigations: *P. ciliata* (Khurana 1995), and *P. tomentosa* (Zhu 1988). The work with *P. ciliata* is particularly important genecologically since substantial natural Himalayan populations are still widespread over a variety of ecological situations from about 1300–3000 m elevation. A third studied species is *P. maximowiczii,* the adaptive genecology of which Chiba (1984) has studied on the island of Hokkaido, 42°–45°N, 140°–144°W. Latitude is confounded with elevation on the island, with natural populations occuring up to 1000 m in elevation in the north. The experimental population of 129 clones included equal representation from each of four quarters (regions) of the island; common garden tests were conducted in each region. There was strong genetic control over time of growth cessation, with northern provenances stopping first. Subsequent freezing tests demonstrated that northern material developed hardiness earlier than other provenances. Spring bud-break of southern provenances was slightly earlier than flushing of northern ones, and during the early stages of bud-break all clones survived to –15°C. Hardiness of all material was lost by the leaf expansion stage of flushing.

Adaptation to climate

The occurrence of *Populus* species across major climatic and edaphic gradients has tempted physiological ecologists to examine possible intraspecific physiological adaptations, mainly to different temperature and moisture regimes. The southeast to northwest gradient in moisture conditions in the southwestern portion of *P. deltoides'* range has stimulated investigation of geographic variation in drought resistance there. Kelliher and Tauer (1980) examined the response to moisture stress of four clones from the region, two from an eastern moist site and two from a western dry site. Under adequate soil moisture, dry-site clones had lower stomatal resistance (2–5 s/cm) than wet-site plants (3–12 s/cm). Dry-site plants stopped height growth after 23 d of drought; growth of wet-site clones was reduced under moderate stress and ceased at 21 d when stomatal resistance was 55 s/cm. The authors concluded that "wet site plants possess little or no ability to deal with drought." However, working with a randomly selected population of 30 *P. deltoides* clones from lower Mississippi Valley mesic sites Farmer (1970) noted substantial genetic variance in response to drought. Further north, the drought resistance of one Ohio clone was compared with two western clones of *P. deltoides* by Gebre and Kuhns (1991) who reported a significant superiority in acclimation of the western clones to stress. In the central part of its range, local differentiation in flood and drought tolerance of Illinois *P. deltoides* has been examined by McGee et al. (1981) who sampled "several trees" from sand dunes, a strip-mined area and a flood plain "plantation" and evaluated them in soils from the collection areas. Dune plants had a heavier root mass and were less susceptible to flooding in terms of photosynthesis and transpiration reduction. Photosynthesis of plants from the strip-mined area was less affected by drought than assimilation rate of plants from other sites. The authors note that "apparently habitat selection is strong enough to overcome gene flow by seed dispersal, especially in populations that occur in close proximity to each other."

Regional and local differentiation in drought and flood tolerance and related assimilation characteristics have also been the subject of several *P. trichocarpa* studies all incorporating similar (and adequate) sampling methods along river systems in mesic and xeric parts of Oregon and Washington. It has been established that there is little population differentiation in flood tolerance and associated physiological parameters, but substantial within-population variance in response to flooding (Smit 1988). On the other hand, observations of growth (Dunlap et al. 1994), shoot and leaf morphology (Dunlap et al. 1995), and photosynthesis (Dunlap et al. 1993) all point to genetic differentiation among populations on climatically contrasting river systems and along climatic gradients within systems. Evidence to date suggests that these differences can be reasonably attributed to selection for drought resistance. Detailed examination of the assimilation physiology of single clones from these xeric and mesic regions (Bassman and Zwier 1991) support the idea of climatic adaption.

Adaptation to temperature regimes in *P. trichocarpa* has been the subject of one study using single clones from coastal Alaska and Montana (Drew and Chapman 1992); the higher photosynthesis of the Alaskan clone over a range of temperatures may have been related to its higher stomatal conductance, a characteristic of coastal provenances. In a photosynthesis study with *P. deltoides,* Drew and Bazzaz (1979) observed that plants from Louisiana responded to low (4°C) night temperatures more negatively than did northern (Wisconsin, northern Illinois) provenances. In this work based on two plants per provenance, night temperature regulated the rate at which stomata opened. The authors concluded that "night temperature in particular exerts strong control over population differentiation in *P. deltoides.*" However, this provenance difference (Louisiana vs. Wisconsin) in photosynthesis did not translate into differences in Relative Growth Rate; and net photosynthesis rates (based on 2–3 plants per provenance) under a range of light intensities was essentially equal for northern Illinois and Wisconsin provenances (Drew and Bazzaz 1978). Lack of latitudinal difference in assimilation rate has also been observed for juvenile *P. balsamifera* using populations from 45° to 53°N about 90°W (Schnekenburger and Farmer 1989). In this study, functional growth analysis (Hunt 1978) was used to examine Unit Leaf Rate and other growth characteristics under long and short photosynthetic periods imposed under a long photoperiod to keep plants growing. The most outstanding feature of the study was the relatively high degree of clonal variance within provenances. These results and other observations of the same material (Farmer and Reinholt 1985; Farmer et al. 1989) led the authors to suggest that "photoperiodic response may be the only major adaptive mechanism responsible for genetic differentiation in growth along most of this latitudinal gradient."

A case for adaptation to climatic stress in *Populus* arises from work on ozone tolerance of *P. tremuloides.* Berrang et al. (1986, 1989) first noted that populations of clones (11–14) from ozone polluted national parks in the northeastern U.S. were less susceptible to ozone fumigation than those from unpolluted parks. This relationship has been subsequently observed in a nationwide sample (Berrang et al. 1991). The most sensitive groups of clones (from unpolluted areas) had the greatest degree of clonal variation in susceptibility to ozone, suggesting they had not been "ozone rogued." Karnosky et al. (1992) have shown that relatively ozone-intolerant clones exhibit reduced growth relative to tolerant ones given seasonal ozone exposures in open-top outdoor chambers. These workers suggest that while ozone may not kill susceptible clones, it does reduce their competitive ability thereby enhancing the position of tolerant clones.

In brief, while much of the evidence on genecologically important variation in stress resistance and assimilation characteristics is limited because of inadequate sample size, genetic differentiation in response to chronic moisture stress appears to be a strong possibility. Moreover, the nature of gene flow in *Populus* considered together with the observed patterns of variation suggests that selection may be rapid. Though evidence for selection in response to edaphic factors and temperature regimes is even less than sketchy, combined with other information (e.g., the aspen–ozone relationship) it suggests that some well designed direct tests of natural selection might be quickly productive. Given the generally high level of within-population variance in *Populus,* substantially more than a few clones per provenance will usually be needed for studies of physiological differences.

Morphology

Since Barnes (1966) first reported putative natural clones of *P. tremuloides* and *P. grandidentata,* wide within-population phenotypic variance in morphological and reproductive characteristics has been noted in aspens (e.g., Lester 1963; Farmer and Barnes 1978; Steneker 1974). Data on geographical variation are rarer. Hence Barnes' (1975) study of leaf morphology in 1257 natural *P. tremuloides* clones in western North America is especially noteworthy. His sample extended from mostly unglaciated areas in Utah and Colorado northwest to Vancouver Island. Within this region there is a clinal south–north gradient in leaf shape, size, and tooth number. Multivariate analyses differentiated at least four groups of populations along this gradient, with the smaller leaves of the southeastern clones resembling fossil Pliocene and Miocene aspens. Barnes suggests that "perhaps a few are the very clones that were established in early time and propagated vegetatively to the present." The more northerly populations in the west resemble eastern *P. tremuloides* in their morphology. Barnes and Han (1993) have extended the examination of aspen leaf morphology to Europe and China where their studies suggest that in Eurasia there may be a single, highly polymorphic species, viz. *P. tremula,* which includes *P. davidiana* and *P. rotundifolia.* This conclusion invites further work using common-garden comparisons and techniques of molecular genetics.

Aspen bark characteristics are so variable that in the older literature "varieties," "races," and "forms" were distinguished by it. Barnes (1966) showed that populations are just locally polymorphic for this character. The only geographically related phenotypic differences have been observed on an elevational transect in New Mexico where there is a negative correlation between elevation and bark chlorophyll content (Covington 1975).

Data from several types of studies show that leaf size in *P. deltoides* decreases from south and southeast to north and west. The most extensive study of phenotypic variation is that of Sokal et al. (1986) who used spatial autocorrelation analysis and leaves from 522 trees in 302 localities throughout the range. While they noted the above geographical trend, their regression analyses of leaf morphology and climatic variables indicated "the patterns we observed do not seem to be accounted for by the apparent underlying climatic variation." Assuming the observed differences have a genetic basis (very likely), the pattern of their data also tended to rule out genetic drift as a differentiating mechanism. Ying and Bagley's (1976) common-garden test of 498 clones in Nebraska included material from the northern half of the range and established a genetic basis to the southeast–northwest decrease in leaf size. Branchiness increased from southeast to northwest in this study. Friend's (1981) nursery study of the Cooper–Hicks southern (30°–35°) collection revealed an east (82°W) to west (96°W) decrease in leaf size and length:width ratio, wide variation within populations in these characters, and a high degree of genetic control over juvenile leaf morphology. North–south trends were not significant in this study because of wide provenance variation within latitudes. In southwestern North America, branchiness of *P. deltoides* increases and leaf size decreases with longitude along a gradient of decreasing water availability (Posey 1969; Koster 1976). Multivariate analysis of leaf characters in nine southern-Ontario populations did not suggest clinal variation patterns related to either latitude or longitude at the northeastern edge of *P. deltoides'* range (Rajora et al. 1991).

In *P. balsamifera,* leaf size clearly decreases, specific leaf weight increases, and there is no trend in stomatal density or length with latitude across the center of its range (Penfold 1991). There is wide genetically controlled variation in all of these characters within provenances. Geographical variation is, however, not readily associated with an environmental gradient that might be related to selection of leaf morphology.

An increase in leaf size and decrease in branchiness with latitude has been reported for *P. trichocarpa* west of the Cascade mountains in Oregon and Washington (Weber et al. 1985; Rogers et al. 1989). This trend parallels one of decreasing water availability, from north to south in this instance. Studies of populations from contrasting river systems in this region also demonstrate that leaves are smaller and branchiness greater under xeric conditions (Dunlap et al. 1995). Together the studies of *P. deltoides* and *P. trichocarpa* suggest the possibility of selection for leaf size and shoot morphology in response to moisture stress.

The data on geographic variation in wood properties (Einspahr and Benson 1967; Posey et al. 1969; Wilcox and Farmer 1968) are to date all from studies of phenotypic variation and are too limited for even tentative conclusions. They are best noted as points of departure for future research in an area which has been essentially neglected.

Reproductive characteristics

Perhaps because we frequently abandon test trees before they reach sexual maturity, there are few data on reproductive aspects of genecology. However, Ying and Bagley (1976) noted that 7-yr-old flowering trees of *P. deltoides* had a 1:1 sex ratio across the northern portion of its range and that catkin size decreased from southeast to northwest. The 1:1 sex ratio has also been observed in natural stands in the lower Mississippi Valley (Farmer 1964) where there is substantial within-population variation in flowering and seed dispersal time (Farmer 1966). The wide range in local dispersal time (June–August) is especially interesting ecologically since it ensures seed deposition on alluvial "new ground" gradually exposed by summer's falling water levels on the Mississippi River. Whether this variance is an adaptation must remain a speculative matter.

Reported sex ratios for trembling aspen range from male predominance to equity (see Grant and Mitton 1979 for review). The only geographically related difference in ratio has been noted in Colorado where Grant and Mitton (1979) observed an increase in percent of males from low to high elevations. The radial growth rate of female clones was generally greater than that of males, but this difference decreased with elevation. This relationship, which suggests a sex-related difference in mortality at high elevations, begs investigation at other locations.

The literature on vegetative propagation of poplars by cuttings is enormous, and there are a substantial number of reported studies of within-population variance and heritability. However, patterns of geographical variation have been the subject of only three papers, all involving North American species: Ying and Bagley (1977) used material from 192 *P. deltoides* clones to evaluate rooting of seven provenances from southern Ohio to Minnesota. A key element of this type of test is that cutting material be grown in a common garden prior to evaluating rooting in order to reduce environmental conditioning, i.e., "C" effects. Ying and Bagley met this requirement and noted that cuttings from northwestern clones produced more roots than southern Ohio clones under both greenhouse and nursery conditions. However, there was no cline in rooting capacity since other provenances had roughly equally rooting percents. In a *P. balsamifera* test of similar design Farmer et al. (1989) reported no latitudinal effect in the degree of rooting which was generally high. Ernst and Fechner (1981) evaluated phenotypic variation in rootability of cuttings from 92 *P. angustifolia* trees in a variety of Colorado locations and observed no geographical differences. Substantially more studies of rooting will be required before drawing conclusions on geographical variation patterns. These studies are of interest because one might reasonably presume that preformed root initials (characteristic of Aigeiros and Tacamahaca species) are adaptations to siltation frequently associated with riparian habitats and therefore might vary with habitat.

Diseases and insects

With perhaps two exceptions the large body of information on poplar diseases and insects contains little data on patterns of natural variation in host–pathogen and host–insect relationships. The first exception involves *Melampsora medusae,* a leaf rust the urediospores of which move annually north to south in North America from northern conifer aecial hosts (Widin and Schipper 1980). Urediospores, which infect poplar leaves, may also overwinter at southern locations. Susceptibility of *P. deltoides* to *Melampsora* rust is highly variable and under strong genetic control (Jokela 1966; Thielges and Adams 1975). Cooper and Filer (1977), Friend (1981), Hamelin et al. (1994), and Thielges and Adams (1975) have reported on common-garden tests in which there is a clinal decrease in susceptibility from northern to southern provenances. There is also wide genetic variation within provenances. The physiological basis of genetic differences in resistance is not clear, though Cooper and Filer (1977) present some interesting hypotheses.

The second exception is a thorough, rangewide study of phenotypic co-variation in *P. deltoides* vegetative features and characteristics of gall-forming aphids in the genus *Pemphigus* (Sokal and Unnasch 1988). Spatial autocorrelation analyses revealed geographic variation in both organisms, but correlations between *Populus* and *Pemphigus* variables were low. The authors conclude that "the undoubtedly close *Pemphigus–Populus* coevolution over geological time is not reflected in the microevolutionary variation over geographical space." Though no relationship was found between host–parasite variables, the study highlights the need for co-evaluation of variation in both organisms and sets a pattern for such studies.

In *P. tremuloides,* phenotypic variation in susceptibility to *Hypoxylon mammatum* has been observed within populations in northern Michigan (Copony and Barnes 1974), and there is some evidence of a north–south increase in susceptibility (French and Hart 1978) which may have some genetic basis. Variation in susceptibility among nine New York clones to five *Hypoxylon* isolates of differing virulence has been demonstrated in a replicated study (Griffin et al. 1984).

Molecular genetics

During the past decade the techniques of molecular genetics have been successfully used in *Populus* species delineation (e.g., Liu and Furnier 1993; Barrett et al. 1993; Rajora and Zsuffa 1990; Rajora and Dancik 1992*b*), clonal identification (e.g., Rajora 1988; Rajora 1989 *a, b,* and *c*) and studies of isozyme and DNA inheritance (e.g., Hyun et al. 1987*a*; Rajora 1990; Rajora et al. 1992; Rajora and Dancik 1992*a*), all topics considered in detail elsewhere in this volume. Additionally, isozyme analysis and more recently restriction fragment length polymorphism (RFLP) and random amplified polymorphic DNAs (RAPD) have been used to directly assess patterns of genetic variation in enzyme systems and nucleic acids. This information may be helpful in establishing the nature of population genetics for a species even when enzyme systems are under little or no selection pressure. However, they are more potentially interesting in my estimation because in some organisms functional differences with adaptive value are related to, for example, allozyme variation (Powers et al. 1991). It is therefore of particular interest that Bradshaw and Stettler (1995) have recently reported relationships between quantitative trait loci (QTL's) and genetic variance of adaptive and growth characteristics in an F_2 generation of *Populus*.

In the first of the population studies in *Populus*, Mitton and Grant (1980) assessed isozymes at three loci using 107 natural clones of *P. tremuloides* occurring from 1700 to 3200 m elevation in Colorado. They reported a positive relationship between heterozygosity and clone growth rate, but little variation related to elevational source. This was followed by an examination of 26 loci in 222 clones from seven locations in Alberta (Cheliak and Dancik 1982). Though these investigators (Cheliak and Dancik) did not report on the degree of population differentiation, they did observe a high level of genic diversity which they argue might be related to aspen's asexual reproduction. Hyun et al. (1987*b*) then sampled 15 loci in 200 clones from eight widely separated regions in Ontario (42°–51°N), and reported a much lower general level of heterozygosity (H = 0.125) than Cheliak and Dancik and a very modest level of differentiation among populations (about 6.8% of genetic diversity). They concluded that, with the possible exception of a northern population on James Bay, there was unrestricted gene flow throughout the range in Ontario. Results from a more geographically restricted sample of 347 trees in nine Minnesota populations (Lund et al. 1992) are generally similar to those of Hyun et al. Variation at 10 loci produced an observed heterozygosity of 0.217. F statistics and Nei's genetic distances all indicated no population differentiation, high levels of gene flow among populations and little inbreeding. Higher levels of genetic variation in both *P. tremuloides* and *P. grandidentata* in this region were revealed by RAPD markers, but the pattern of variation was similar to that noted with isozymes (Liu and Furnier 1993). In contrast, the work of Jelinski and Cheliak (1992) with *P. tremuloides*, in a geographically restricted but ecologically diverse area in southern Alberta, suggests a moderate amount of heterogeneity among populations. Six populations totaling 156 clones were examined, using 14 enzyme systems. However, F statistics indicated that 97% of genetic variation is within populations and average Nei's genetic distance was low (D = 0.013).

Two studies of *P. deltoides* have been reported, both utilizing multivariate analysis. One was an examination of 33 loci in nine populations (84 trees) from southern Ontario, at the northern edge of the species' range (Rajora et al. 1991). Forty-two percent of the loci were polymorphic, heterozygosity was low (H =

0.063) and the mean Nei's genetic distance was 0.007, suggesting no major barriers to gene flow. The authors used principal component analysis (PCA) to separate the sample into five groups with a slight east–west trend, but there was no strong evidence of population differentiation. Population groupings were nonconcordant with PCA grouping using leaf characters from the same sample.

In contrast, the 22 loci studied by Marty (1984) were from 21 *P. deltoides* populations located throughout the north–south range (30°–47°N) on several river systems. Thirty-four percent of loci were polymorphic, mean heterozygosity was 0.085 and the average Nei's genetic distance, 0.004, values not strongly different from those of Rjora et al. (1991). However, though about 94% of variation was within populations, genetic distance was positively correlated with geographical distance, frequencies of alleles for 4 loci were correlated with latitude (and its related climatic parameters) and PCA revealed differentiation of populations in the lower Mississippi Valley from those in the north.

Weber and Stettler (1981) used a large sample (500 trees) from 10 populations of *P. trichocarpa* in an area of western Washington and Oregon similar in size to the one sampled by Rajora et al. (1991). From 33 to 39% of loci were polymorphic depending upon population, and 94% of total genetic diversity was found within populations. A dendrogram based on Roger's genetic similarity coefficients also suggested a slight correspondence between genetic and geographic distance (a north–south gradient), but the relationship was not statistically significant. Thus the study provided little evidence for enzyme differentiation, and "the high overall genetic similarity probably reflects the occurrence of periodic gene flow throughout" the study area. In contrast to most of the other *Populus* isozyme studies however, about 13% of the loci deviated from the expected Hardy-Weinberg proportions, mainly due to an excess of homozygotes in the more southern populations.

The only study of *P. balsamifera* isozymes (Farmer et al. 1988) included sample populations along a latitudinal transect used in a variety of experiments noted above. In the examination of eight polymorphic loci, Nei's genetic distances ($\overline{MD}$ = 0.0007) and F statistics ($\overline{MF}_{SI}$ = 0.014) indicated "that there is little genetic differentiation among populations in terms of the observed enzyme systems." These populations were widely differentiated in their photoperiodic response (Charrette 1990) and resultant growth patterns (Farmer 1993).

One can conclude on the basis of these studies that there is probably substantial gene flow among geographically diverse populations and little population differentiation in the studied enzyme systems. Further, most of the populations appear to be in Hardy–Weinberg equilibrium with respect to these systems. This lack of population differentiation in genetically diverse species is congruent with results of the Hamrick et al. (1992) literature survey which indicated that tree species with large geographic ranges, outcrossing breeding systems, and wind dispersal of pollen and seed have less genetic variation among than within populations. It is significant that 35 enzyme systems were included in the investigations, each study (excluding Mitton and Grant 1980) examined an average of 10 systems, none of the systems was used in all studies, only three were used in at least six of the eight tests, and resulting data were subjected to various sorts of analysis. This diversity in experimental method adds strength to the above general conclusions, but suggests that some findings unique to individual studies require more investigative attention.

Natural variation and wood production

Given the well established photoperiodic influence upon shoot growth periodicity in *Populus,* one would expect that movement of plants 2°–3° N of origin would translate into a growth advantage over local material. Namkoong (1969) has referred to this movement potential as one aspect of the "nonoptimality of local races." One can more generally hypothesize that movement into a gradient of increasing stress might be possible with consequent increase in productivity relative to local stock. Here I review experimental evidence pertinent to this hypothesis.

Several major studies in *P. deltoides* focus upon this relationship. In the largest of these Foster (1986) used 1440 clones developed by D.T. Cooper from seedlings in 36 natural stands located along the Mississippi River from Louisiana (30°30′N) to Tennessee (34°55′N). A 6 × 6 lattice experiment planted at 32°45′N was evaluated at 1, 2, 4, and 7 years, when height averaged 18.4 m and diameter 16.9 cm. In the early years of the test, stands accounted for over 40% of genetic variance in growth, but by year 7 this had decreased to 12%. At this time there was no relationship between latitude of stand (provenance) and its performance. In fact the most northerly stand ranked first and the most southerly second. Wide clonal variation within stands was thus the dominant feature of the test, which provided good data on inheritance and potential selection gains (see also Riemenschneider et al., Chapter 7). A second test (Tang 1988) of 15 provenances covering the entire southern portion of *P. deltoides'* range (the Cooper–Hicks collection), was planted at the same location (i.e., 32°45′N). Its nested design included 8 clones/3 parent trees/2 stands/provenance (720 clones). Fifth year heights ranged significantly from 4 m for an eastern North Carolina provenance (35°30′N) to 10.5 m for the southern Texas source (30°30′N); diameter and volume followed this northeast to southwest trend. Eighty percent of the genetic variance in height was attributable to provenance, 5% to families within provenances and 15% to clones within families. Mean height of the southern Texas source was 50% greater than that of the local Mississippi source; I estimate (from data in the report) the volume advantage to be about 170%. At a common garden further up the river at Cairo, Illinois, Mississippi clones outperformed local material without being damaged by cold (Randall 1973). On the other hand, in the Nebraska test of wide-ranging provenances (Ying and Bagley 1976), only a Missouri provenance was clearly more productive at 7 yr than local material, and the estimated volume advantage was about 27%. Further north in Minnesota none of the more southern provenances kept up with local material in terms of height and diameter, and most met mysterious deaths by age 17 (Mohn and Radsliff 1983).

In the southwest, the environmental gradient from Arkansas to southeastern Colorado investigated by Posey (1969) and Nelson and Tauer (1987) contains the confounded effects of temperature and moisture stress. Here results are based on very young plants, but suggest that productivity gains could be obtained by moving plants northwest. The same pattern of increasing productivity from xeric to mesic provenances has also been observed for 2-yr-old *P. trichocarpa* growing on a mesic common-garden site (Dunlap et al. 1994, 1995). Carl Mohn (personal communication) has observed a clear reduction in growth potential from eastern to western Minnesota provenances growing in central Minnesota.

Other geographical trends in productivity have been noted for *P. trichocarpa* in the Oregon, Washington, southern British Columbia region and are a function of test location. First, material from southern and west central Oregon has performed better than more northern clones in tests at around 50°N in Europe (Schulzke and Weisgerber 1984) and in southern Washington (Heilman and Stettler 1985). Second, at a site in northwestern Washington (49°N) local clones have outperformed southern ones (44°–45°N) (Rogers et al. 1989). Finally, *P. balsamifera* from 45° to 53°N in the central part of its range has performed in classical fashion, with provenances 2°–3° south of the test site (48°N) growing larger than local or more northern sources (Farmer 1993). The degree of the height difference (i.e., 45° vs. 48°) varied from 22 to 82% depending upon soil conditions at the planting site and age of trees.

Viewed in the context of increasing wood production through application of genecological information, results to date vary from encouraging to confusing to bleak. However, except for movement into the northern edge of species range, moving material a few degrees north does appear to be advantageous more frequently than not if other factors are not limiting. Movement from xeric to mesic sites and vice versa requires more observation before firm predictions of effect can be made.

Conclusions

The above reported information has led me to the following general conclusions about the degree and pattern of variation in relation to processes effecting genetic change. They are, perhaps unfortunately, mostly applicable to North American species since there are large gaps in our published knowledge about natural populations of *P. nigra*, *P. tremula*, and several major Asian species.

First, data from molecular genetics investigations confirm our expectation that periodic gene flow (i.e., migration) has been sufficient to prevent genetic drift, inbreeding, and other processes that might cause geographical differentiation unrelated to selection. Additionally, these studies almost uniformly report a substantial degree of isozyme variation within local populations, as noted by Hamrick et al. (1992) for many long-lived species. One must bear in mind, however, that these conclusions stem from simple observations of present genetic structure, not direct observation of pertinent genetic processes. Further, in no case has geographical variation in isozymes been congruent with morphometric characters; we have not commonly included morphometric characters and related enzyme systems in the same study.

Second, the evidence for adaptive geographical differentiation via selection, while still mostly circumstantial (we have again observed products, not process), is substantial. Adaptation to local growing seasons via genetic variation in photoperiodic response is well documented in several species. Selection for this adaptation is probably rapid, as mortality data from many provenance tests suggest. Most adaptation to available growing periods is apparently grounded in timely fall hardiness induction; available data on genetic variance in spring bud break, considered together, are ambiguous, perhaps due to provenance × test location interactions which we have not yet formally examined. If one assumes that genetic differences in leaf and branch morphology are reliable indicators of selection response to moisture stress, then there is a fair case for geographic differentiation along moisture gradients. The case for differentiation in assimilation characteristics is presently shaky because of some gross sampling deficiencies. One of the more interesting aspects of these putatively adaptive characteristics is that within-population genetic variance is usually high. Even variation in photoperiodically controlled shoot elongation patterns is very high among individuals in some populations. This suggests that variance per se may be of adaptive value under conditions of temporal uncertainty in stress, an idea advanced for seed germination polymorphism by Westoby (1981) and noted for poplar by Weber et al. (1985).

Third, variation in some characters is clinal but unrelated to likely differentiating factors. Resistance to *Melampsora* rust has this pattern, the cause of which probably lies in the current physiological relationship between host and pathogen, a topic not much investigated yet. On the other hand some clinal variation in leaf morphology may be the result of geological events and/or genetic processes which have left little evidence of when and how they occurred.

Finally, one must note that neither common-garden tests, comparison of physiological characters, or molecular genetic surveys can tell us how and why genetic differentiation has or has not taken place across environmental gradients, though they provide important clues. To obtain this information we must move to the next stage in the evolution of *Populus* genecological research: selection experiments. The work on aspen and ozone may represent the first step. Some existing provenance tests may present opportunities to observe the genetic consequences of mother nature's roguing. Ultimately, however, we must place substantial samples of sexually active natural populations under environmental conditions selected for their ecological significance and observe the genetic changes that take place over long stretches of time.

Topics deserving further study

- Patterns of genetic variation in relatively uninvestigated Asian species.

- Genetic variation in adaptively important physiological processes.

- The molecular genetic basis of natural variation in function.

- Natural selection processes which have resulted in population differentiation.

- Natural variation in wood properties.

- Natural variation in reproductive characteristics.

Acknowledgements

I thank Burton Barnes, Carl Mohn, and Reinhard Stettler for helpful reviews.

References

Bassman, J.H., and Zwier, J.C. 1991. Gas exchange characteristics of *Populus trichocarpa, Populus deltoides* and *P. trichocarpa* × *P. deltoides* clones. Tree Physiol. **8**: 145–159.

Barnes, B.V. 1966. The clonal growth habit of American aspens. Ecology, **47**: 439–449.

Barnes, B.V. 1975. Phenotypic variation of trembling aspen in western North America. For. Sci. **21**: 319–328.

Barnes, B.V., and Han, F. 1993. Phenotypic variation of Chinese aspens and their relationships to similar taxa in Europe and North America. Can. J. Bot. **71**: 799–815.

Barrett, J.W., Rajora, O.P., Yeh, F.C.H., and Dancik, B.P. 1993. Mitochondrial DNA variation and genetic relationships of *Populus* species. Genome, **36**: 87–93.

Berrang, P., Karnosky, D.F., Mickler, R.A., and Bennett, J.P. 1986. Natural selection for ozone tolerance in *Populus tremuloides*. Can. J. For. Res. **16**: 1214–1216.

Berrang, P., Karnosky, D.F., and Bennett, J.P. 1989. Natural selection for ozone tolerance in *Populus tremuloides*: field verification. Can. J. For. Res. **19**: 519–522.

Berrang, P., Karnosky, D.F., and Bennett, J.P. 1991. Natural selection for ozone tolerance in *Populus tremuloides:* an evaluation of nationwide trends. Can. J. For. Res. **21**: 1091–1097.

Bradshaw, H.D., and Stettler, R.F. 1995. Molecular genetics of growth and development in *Populus*. IV: mapping QTLs with large effects on growth, form, and phenology traits in a forest tree. Genetics, **139**: 963–973.

Brissette, J.C., and Barnes, B.V. 1984. Comparisons of phenology and growth of Michigan and western North American sources of *Populus tremuloides*. Can. J. For. Res. **14**: 789–793.

Cannell, M.G.R., and Willett, S.C. 1976. Shoot growth phenology, dry matter distribution and root:shoot ratios of provenances of *Populus trichocarpa, Picea sitchensis* and *Pinus contorta* growing in Scotland. Silvae Genet. **25**: 49–59.

Chandler, J.W., and Thielges, B.A. 1973. Chilling and photoperiod affect dormancy of cottonwood cuttings. Proc. South. For. Tree Improve. Conf. **12**: 200–205.

Charrette, P. 1990. The effect of photoperiod on apical growth cessation in tamarack (*Larix laricina*) and balsam poplar (*Populus balsamifera*) provenances from northern Ontario. M. Sc. thesis, Lakehead University, Thunder Bay, ON. 80 p.

Cheliak, W.M., and Dancik, B.P. 1982. Genic diversity of natural populations of a clone-forming tree, *Populus tremuloides*. Can. J. Genet. Cytol. **24**: 611–616.

Chiba, S. 1984. Provenance selection and cross breeding of *Populus maximowiczii* in northern Japan. Paper presented at 17th Session of Int. Poplar Comm., Ottawa, ON. 80 p.

Cooper, D.T., and Filer, T.H., Jr. 1977. Geographic variation in *Melampsora* rust resistance in eastern cottonwood in the Lower Mississippi Valley. *In* Proceedings of the 10th Central States Forest Tree Improvement Conference, Purdue University, West Lafayette, IN. pp. 146–151.

Copony, J.A., and Barnes, B.V. 1974. Clonal variation in the incidence of *Hypoxylon* canker on trembling aspen. Can. J. Bot. **52**: 1475–1481.

Covington, W.W. 1975. Altitudinal variation of chlorophyll concentration and reflectance of the bark of *Populus tremuloides*. Ecology, **56**: 715–720.

Dhir, N.K., and Mohn, C.A. 1974. Growth and flowering of NC–99 cottonwood seed sources in Minnesota. Minn. For. Res. Notes, 253. 4 p.

Drew, A.P., and Bazzaz, F.A. 1978. Variation in distribution of assimilate among plant parts in three populations of *Populus deltoides*. Silvae Genet. **27**: 189–193.

Drew, A.P., and Bazzaz, F.A. 1979. Response of stomatal resistance and photosynthesis to night temperature in *Populus deltoides*. Oecologia, **41**: 89–98.

Drew, A.P., and Chapman, J.A. 1992. Inheritance of temperature adaptation in intra- and inter-specific *Populus* crosses. Can. J. For. Res. **22**: 62–67.

Dunlap, J.M., Braatne, J.H., Hinkley, T.M., and Stettler, R.F. 1993. Intraspecific variation in photosynthetic traits of *Populus trichocarpa*. Can. J. Bot. **71**: 1304–1311.

Dunlap, J.M., Heilman, P.E., and Stettler, R.F. 1994. Genetic variation and productivity of *Populus trichocarpa* and its hybrids. VII. Two-year survival and growth of native black cottonwood clones from four river valleys in Washington. Can. J. For. Res. **24**: 1539–1549.

Dunlap, J.M., Heilman, P.E., and Stettler, R.F. 1995. Genetic variation and productivity of *Populus trichocarpa* and its hybrids. VIII. Leaf and crown morphology of native black cottonwood clones from four river valleys in Washington. Can. J. For. Res. **25**: 1710–1724.

Einspahr, D.W., and Benson, M.K. 1967. Geographic variation of quaking aspen in Wisconsin and upper Michigan. Silvae Genet. **16**: 106–112.

Ekberg, I., Dormling, I., Eriksson, G., and von Wettstein, D. 1976. Inheritance of the photoperiodic response in forest trees. *In* Tree physiology and yield improvement. *Edited by* M.G.R. Cannell and F.T. Last. Academic Press, New York, NY. pp. 207–221.

Eldridge, K.G., Rout, A.R., and Turnbull, J.W. 1972. Provenance variation in the growth pattern of *Populus deltoides*. Aust. For. Res. **5**: 45–50.

Ernst, S.G., and Fechner, G.H. 1981. Variation in rooting and juvenile growth phenology of narrowleaf cottonwood in Colorado. *In* Proc. of the 2nd North Central Tree Improvement Conference, University of Nebraska, Lincoln, NB. pp. 111–118.

Farmer, M.M., and Barnes, B.V. 1978. Morphological variation of families of trembling aspen in southeastern Michigan. Mich. Bot. **17**: 141–153.

Farmer, R.E., Jr. 1964. Sex ratio and sex-related characteristics in eastern cottonwood. Silvae Genet. **13**: 116–118.

Farmer, R.E., Jr. 1966. Variation in time of flowering and seed dispersal of eastern cottonwood in the lower Mississippi Valley. For. Sci. **12**: 343–347.

Farmer, R.E., Jr. 1970. Variation and inheritance of eastern cottonwood growth and wood properties under two soil moisture regimes. Silvae Genet. **19**: 5–8.

Farmer, R.E., Jr. 1993. Latitudinal variation in height and phenology of balsam poplar. Silvae Genet. **42**: 148–153.

Farmer, R.E., Jr., and Reinholt, R.W. 1985. Genetic variation in dormancy relations of balsam poplar along a latitudinal transect in northwestern Ontario. Silvae Genet. **35**: 38–42.

Farmer, R.E., Jr., Cheliak, W.M., Perry, D.J., Knowles, P., Barrett, J., and Pitel, J. 1988. Isozyme variation in balsam poplar along a latitudinal transect in northwestern Ontario. Can. J. For. Res. **18**: 1078–1081.

Farmer, R.E., Jr., Freitag, M., and Garlick, K. 1989. Genetic variance and "C" effects in balsam poplar rooting. Silvae Genet. **38**: 62–65.

Foster, G.S. 1986. Provenance variation of eastern cottonwood in the lower Mississippi Valley. Silvae Genet. **35**: 32–38.

French, J.R., and Hart, J.H. 1978. Variation in resistance of trembling aspen to *Hypoxylon mammatum* identified by inoculating naturally occurring clones. Phytopathology, **68**: 485–489.

Friend, M.M. 1981. Genetic variation in juvenile traits of eastern cottonwood from the southern United States. M. Sc. thesis, Mississippi State University, Mississippi State, MS. 117 p.

Gebre, G.M., and Kuhns, M.R. 1991. Seasonal and clonal variations in drought tolerance of *Populus deltoides*. Can. J. For. Res. **21**: 910–916.

Grant, M.C., and Mitton, J.B. 1979. Elevational gradients in adult sex ratios and sexual differentiation in vegetative growth rates of *Populus tremuloides* Michx. Evolution, **33**: 914–918.

Griffin, D.H., Marion, P.D., Valentine, F.A., and Gustavson, L. 1984. Canker elongation, branch death, and callus formation as resistance or susceptibility responses in *Populus tremuloides* and virulence or avirulence characteristics of *Hypoxylon mammatum*. Phytopathology, **74**: 683–687.

Hamelin, R.C., Ferriss, R.S., Shain, L., and Thielges, B.A. 1994. Prediction of poplar leaf rust epidemics from a leaf-disk assay. Can. J. For. Res. **24**: 2085–2088.

Hamrick, J.L., Godt, M.J.W., and Sherman-Broyles, S.L. 1992. Factors influencing levels of genetic diversity in woody plant species. New For. **6**: 95–124.

Heilman, P.E., and Stettler, R.F. 1985. Genetic variation and productivity of *Populus trichocarpa* and its hybrids. II. Biomass production in a 4-year plantation. Can. J. For. Res. **15**: 384–388.

Hunt, R. 1978. Plant growth analysis. Edward Arnold Ltd., London, UK. 67 p.

Hyun, J.O., Rajora, O.P., and Zsuffa, L. 1987a. Inheritance and linkage of isozymes in *Populus tremuloides*. Genome, **29**: 384–388.

Hyun, J.O., Rajora, O.P., and Zsuffa, L. 1987b. Genetic variation in trembling aspen in Ontario based on isozyme studies. Can. J. For. Res. **17**: 1134–1138.

Jelinski, D.W., and Cheliak, W.M. 1992. Genetic diversity and spatial subdivision of *Populus tremuloides* (Salicaceae) in a heterogeneous landscape. Amer. J. Bot. **79**: 728–736.

Johnsson, H. 1956. Genetics and breeding of poplars. *In* Poplars in forestry and land use. FAO Forestry and Forest Products Studies No. 12. pp. 360–394.

Jokela, J.J. 1966. Incidence and heritability of Melampsora rust in *Populus deltoides* Bartr. *In* Breeding rust resistant trees. *Edited by* H.D. Gerhold, R.E. McDermott, E.J. Schreiner, and J.A. Winieski. Pergamon Press, New York, NY. pp. 111–117.

Jokela, J.J., and Mohn, C.A. 1976. Geographic variation in eastern cottonwood. *In* Proceedings of the Symposium on Eastern Cottonwood and Related Species, 28 Sept. – 2 Oct. 1976, Greenville, Miss. *Edited by* B.A. Thielges and S.B. Land. Louisiana State University, Baton Rouge, LA. pp. 109–125.

Jokela, J.J., Melick, R.A., and Cooper, D.T. 1982. Five-year results from the cottonwood evaluation plantation in southern Illinois. *In* Proceedings of the 19th Annual Meeting of the North American Poplar Council, Rhinelander, WI. Kansas State University of Agriculture and Applied Science, Manhattan, KS. pp. 69–73.

Karnosky, D.F., Gagnon, Z.E., Reed, D.D., and Witter, J.A. 1992. Growth and biomass allocation of symptomatic and asymptomatic *Populus tremuloides* clones in response to seasonal ozone exposures. Can. J. For. Res. **22**: 1785–1788

Kelliher, F.M., and Tauer, C.G. 1980. Stomatal resistance and growth of drought-stressed eastern cottonwood from a wet and dry site. Silvae Genet. **29**: 166–171.

Khurana, D.K. 1995. *Populus ciliata*—its culture and genecology. [Available from Y.S. Parmar, University of Horticulture and Forestry, Solan, India.]

Koster, R. 1976. Observations on *Populus deltoides* provenances grown in Holland. *In* Proceedings of the Symposium on Eastern Cottonwood and Related Species, 28 Sept. – 2 Oct. 1976, Greenville, MS. *Edited by* B.A. Thielges and S.B. Land, Jr. Louisiana State University, Baton Rouge, LA. pp. 126–133.

Lester, D.T. 1963. Floral initiation and development in quaking aspen. For. Sci. **9**: 323–329.

Liu, Z., and Furnier, G.R. 1993. Comparison of allozyme, RFLP, and RAPD markers for revealing genetic variation within and between trembling aspen and bigtooth aspen. Theor. Appl. Genet. **87**: 97–105.

Lund, S.T., Furnier, G.R., and Mohn, C.A. 1992. Isozyme variation in quaking aspen in Minnesota. Can. J. For. Res. **22**: 521–524.

Marty, T.L. 1984. Population variability and genetic diversity of eastern cottonwood (*Populus deltoides* Bartr.). M. Sc. thesis, University of Wisconsin, Madison, WI. 64 p.

McGee, A.B., Schmierback, M.R., and Bazzaz, F.A. 1981. Photosynthesis and growth in populations of *Populus deltoides* from contrasting habitats. Am. Midl. Nat. **105**: 305–311.

Mitton, J.B., and Grant, M.C. 1980. Observations on the ecology and evolution of quaking aspen, *Populus tremuloides,* in the Colorado Front Range. Am. J. Bot. **67**: 202–209.

Mohn, C.A., and Pauley, S.S. 1969. Early performance of cottonwood seed sources in Minnesota. Minn. For. Notes, 207. 4 p.

Mohn, C.A., and Radsliff, W. 1983. Geographic variation in the Rosemount, Minnesota NC-99 eastern cottonwood provenance test: Final report. Proceedings of the 3rd North Central Forest Tree Imp. Assoc. Conf., Wooster, Ohio. pp. 62–70.

Mohrdiek, O. 1983. Discussion: Future possibilities for poplar breeding. Can. J. For. Res. **13**: 465–471.

Muhle-Larsen, C. 1970. Recent advances in poplar breeding. Int. Rev. For. Res. **3**: 1–67.

Namkoong, G. 1969. Nonoptimality of local races. Proceedings of the South. Forest Tree Improve. Conf. **10**: 149–153.

Nelson, C.D., and Tauer, C.G. 1987. Genetic variation in juvenile characters of *Populus deltoides* Bartr. from the southern Great Plains. Silvae Genet. **35**: 216–221.

Pauley, S.S. 1949. Forest-tree genetics research: *Populus* L. Econ. Bot. **3**: 299–330.

Pauley, S.S., and Perry, T.O. 1954. Ecotypic variation of the photo-periodic response in *Populus.* J. Arnold Arbor. Har. Univ. **25**: 167–188.

Pauley, S.S., Johnson, A.G., and Santamour, F.S., Jr. 1963*a*. Results of aspen screening tests: I. Seed sources of quaking aspen (*P. tremuloides Michx*). Minn. For. Res. Notes, 136. 2 p.

Pauley, S.S., Johnson, A.G., and Santamour, F.S., Jr. 1963*b*. Results of aspen screening tests: II. Seed sources of European aspen (*Populus tremula* Linnaeus). Minn. For. Res. Notes, 137. 2 p.

Penfold, C.S. 1991. Genetic variation in traits affecting the water relations of balsam poplar along a latitudinal transect in northwestern Ontario. M. Sc. thesis, Lakehead University, Thunder Bay, ON. 62 p.

Posey, C.E. 1969. Phenotypic and genotypic variation in eastern cottonwood in the southern Great Plains. Proceedings of South. For. Tree Improve Conf. **10**: 130–135.

Posey, C.E., Bridgewater, F.E., and Buxton, J.A. 1969. Natural variation in specific gravity, fiber length, and growth rate of eastern cottonwood in the southern Great Plains. TAPPI, **52**: 1508–1511.

Powers, D.A., Lauerman, T., Crawford, D., and DiMichele, L. 1991. Genetic mechanisms for adapting to a changing environment. Annu. Rev. Genet. **25**: 629–659.

Rajora, O.P. 1988. Allozymes as aids for identification and differentiation of some *Populus maximowiczii* Henry clonal varieties. Biochem. Syst. Ecol. **16**: 635–640.

Rajora, O.P. 1989*a*. Characterization of 43 *Populus nigra* L. clones representing selections, cultivars and botanical varieties based on their multilocus allozyme genotypes. Euphytica, **43**: 197–206.

Rajora, O.P. 1989*b*. Identification of some *Populus deltoides* Marsh. × *P. nigra* L. clones developed in North America with the aid of allozymes. Euphytica, **43**: 207–213.

Rajora, O.P. 1989*c*. Genetic structure and identification of *Populus deltoides* clones based on allozymes. Genome, **32**: 440–448.

Rajora, O.P. 1990. Genetics of allozymes in *Populus deltoides* Marsh., *P. nigra* and *P. maximowiczii* Henry. J. Hered. **81**: 301–308.

Rajora, O.P., and Dancik, B.P. 1992*a*. Chloroplast inheritance in *Populus*. Theor. Appl. Genet. **84**: 280–285.

Rajora, O.P., and Dancik, B.P. 1992*b*. Genetic characterization and relationships of *Populus alba, P. tremula* and *P. canescens,* and their clones. Theor. Appl. Genet. **84**: 291–298.

Rajora, O.P., and Zsuffa, L. 1990. Allozyme divergence and evolutionary relationships among *Populus deltoides, P. nigra,* and *P. maximowiczii*. Genome, **33**: 44–49.

Rajora, O.P., Zsuffa, L., and Dancik, B.P. 1991. Allozyme and leaf morphological variation of eastern cottonwood at the northern limits of its range in Ontario. For. Sci. **37**: 688–702.

Rajora, O.P., Barrett, J.W., Dancik, B.P, and Strobeck, C. 1992. Maternal transmission of mitochondrial DNA in interspecific hybrids of *Populus*. Curr. Genet. **22**: 141–145.

Randall, W.K. 1973. Mississippi cottonwoods outperform local clones near Cairo, Illinois. U.S. For. Serv. South. For. Range Exp. Stn. Res. Note SO-164. 2 p.

Riemenschneider, D.E., McMahon, B.G., and Ostry, M.E. 1992. Use of selection indices to increase tree height and to control damaging agents in 2-year-old balsam poplar. Can. J. For. Res. **22**: 561–567.

Rockwood, D.L. 1968. Variation within eastern cottonwood along the course of the Mississippi River. M. Sc. thesis, University of Illinois, Urbana, IL. 50 p.

Rogers, D.L., Stettler, R.F., and Heilman, P.E. 1989. Genetic variation and productivity of *Populus trichocarpa* and its hybrids. III. Structure and pattern of variation in a 3-year field test. Can. J. For. Res. **19**: 372–377.

Schnekenburger, F., and Farmer, R.E., Jr. 1989. Genetic variance in growth of balsam poplar under 16- and 8-hour photosynthetic periods. For. Sci. **35**: 903–919.

Schreiner, E.J. 1970. Genetics of eastern cottonwood. U.S. For. Serv. Wash. Off. Res. Pap. WO-11. 24 p.

Schulzke, R., and Weisgerber, H. 1984. The international *Populus trichocarpa* provenance trial 1973/1975 after 10 years of observation. Paper presented at 17th Session of the Int. Poplar Comm., Ottawa, ON. 10 p.

Smit, B.A. 1988. Selection of flood-resistant and susceptible seedlings of *Populus trichocarpa* Torr. et Gray. Can. J. For. Res. **18**: 271–275.

Sokal, R.R., and Unnasch, R.S. 1988. Geographic covariation of hosts and parasites: evidence from *Populus* and *Pemphigus*. Z. Zool. Syst. Evolutionsforsch. **26**: 73–88.

Sokal, R.R., Crovello, T.J., and Unnasch, R.S. 1986. Geographic variation of vegetative characters of *Populus deltoides*. Syst. Bot. **11**: 419–432.

Steneker, G.A. 1974. Factors affecting the suckering of trembling aspen. For. Chron. **50**: 32–34.

Sylven, N. 1940. Lang-och Kordagstyper av de svenska skogstraden. Sven. Papperstidn. 17-19. [Cited in Ekberg et al. 1976.]

Tang, Z. 1988. Geographic variation and genetic parameters for growth and stem form of five-year-old eastern cottonwood from the southern United States. M. Sc. thesis, Mississippi State University, Mississippi State, MS. 77 p.

Thielges, B.A., and Adams, J.C. 1975. Genetic variation and heritability of *Melampsora* leaf rust in eastern cottonwood. For. Sci. **21**: 278–282.

Vaartaja, O. 1960. Ecotypic variation of photoperiodic response in trees especially in two *Populus* species. For. Sci. **6**: 200–206.

Watson, S.R. 1990. Frost hardiness of balsam poplar (*Populus balsamifera* L.) during the spring dehardening period. M. Sc. thesis, Lakehead University, Thunder Bay, ON. 125 p.

Weber, J.C., and Stettler, R.F. 1981. Isoenzyme variation among ten populations of *Populus trichocarpa* Torr. et Gray in the Pacific Northwest. Silvae Genet. **30**: 82–87.

Weber, J.C., Stettler, R.F., and Heilman, P.E. 1985. Genetic variation and productivity of *Populus trichocarpa* and its hybrids. I. Morphology and phenology for 50 native clones. Can. J. For. Res. **15**: 376–383.

Westoby, M. 1981. How diversified seed germination behavior is selected. Am. Nat. **118**: 882–885.

Widin, K.D., and Schipper, A.L., Jr. 1980. Epidemiology of *Melampsora medusae* leaf rust of poplars in the north central United States. Can. J. For. Res. **10**: 257–263.

Wilcox, J.R., and Farmer, R.E., Jr. 1968. Variation of fiber length of eastern cottonwood in the lower Mississippi Valley. TAPPI, **51**: 574–576.

Ying, C.C., and Bagley, W.T. 1976. Genetic variation of eastern cottonwood in an eastern Nebraska provenance test. Silvae Genet. **25**: 67–73.

Ying, C.C., and Bagley, W.T. 1977. Variation in rooting capability of *Populus deltoides*. Silvae Genet. **26**: 204–207.

Zhu, Z. 1988. Collection, conservation and breeding studies of gene resources of *Populus tomentosa* in China. Proceedings of the 18th Session of the International Poplar Commission, Beijing, China. 31 p.

CHAPTER 3
Life history, ecology, and conservation of riparian cottonwoods in North America

Jeffrey H. Braatne, Stewart B. Rood, and Paul E. Heilman

Introduction

The life history and ecology of plants are closely related to the natural dynamics of their environment. In the case of poplars, the life history and ecology of riparian cottonwoods are interrelated with the patterns and processes of riverine systems. In this chapter, we describe some of the key features of riverine environments and the life history of riparian cottonwoods from seed dispersal and germination through maturity and senescence. Our intent is to reveal the fundamental ecological relationships between riparian cottonwoods and the alluvial floodplains they inhabit. On the basis of these relationships, we propose some approaches to the conservation and restoration of riparian cottonwoods.

The riparian cottonwoods of North America include: *P. angustifolia, P. balsamifera,* and *P. trichocarpa* from the *Tacamahaca* section; and *P. deltoides,* and *P. fremontii* from the *Aigeiros* section. These dioecious species are widely distributed throughout North America (see Eckenwalder in Chapter 1). In general, members of the *Tacamahaca* section are found at higher elevations and latitudes (i.e., high gradient riverine systems of montane and young piedmont valley floodplains), whereas members of the *Aigeiros* section are primarily limited to lower elevations and latitudes (i.e., lower gradient riverine systems of mature piedmont valley floodplains). Contact zones between species, particularly between members of different sections, occur at critical ecotones (see Whitham in

J.H. Braatne. College of Forest Resources, University of Washington, Seattle, WA 98195, USA.
Stewart B. Rood. Department of Biological Sciences, University of Lethbridge, Lethbridge, AB T1K 3M4, Canada.
Paul E. Heilman. Washington State University, Agricultural Research and Extension Center, Puyallup, WA 98371, USA.
Correct citation: Braatne, J.H., Rood, S.B., and Heilman, P.E. 1996. Life history, ecology, and conservation of riparian cottonwoods in North America. *In* Biology of *Populus* and its implications for management and conservation. Part I, Chapter 3. *Edited by* R.F. Stettler, H.D. Bradshaw, Jr., P.E. Heilman, and T.M. Hinckley. NRC Research Press, National Research Council of Canada, Ottawa, ON, Canada. pp. 57–85.

Chapter 11) where contact gives rise to a broad range of natural hybrid complexes (see Eckenwalder in Chapter 1).

Environmental characteristics of riverine systems

The river systems and alluvial floodplains inhabited by riparian cottonwoods are the product of a complex array of interrelated fluvial geomorphic processes (Leopold 1994; Leopold et al. 1964). Given the complexity of fluvial processes, this chapter can present only a brief description of the hydrogeomorphic features of riverine systems. A more in-depth treatment and summary of fluvial geomorphology can be found in Leopold et al. (1964), Dunne and Leopold (1978), Leopold (1994), and Rosgen (1994).

The physical appearance and character of a river and its floodplain are a product of the continual modification of the river channel by streamflow and sediment regime (Leopold 1994; Rosgen 1994). Local and regional differences in fluvial geomorphology result in a broad range of river types (i.e., meandering to braided channels) with variable width, depth, and rates of lateral migration (Rosgen 1994). As shown in a generalized view (Fig. 1), river channels meander within the alluvial floodplain. These meanders reflect a balance between erosional and depositional processes within the river channel. Erosion of the concave bank is balanced by deposition on the adjacent convex bank (Fig. 1). As the concave bank recedes due to erosion, the point bar builds outward from the convex bank into the channel (Fig. 1). As a result, the form of the channel remains, but its position changes. Floodplains are typically built and continually modified by this process of point-bar extension (Leopold 1994).

Although portions of the floodplain may be far removed from active channel processes, they remain hydrologically linked to the main channel by the alluvial water table. The floodplain is directly linked with main channel processes (erosion and accretion of sediments) primarily during flooding events (flows exceeding bankfull stage, Fig. 1). Historical patterns of channel movement and processes of floodplain formation are readily apparent in aerial photos of extant floodplains and riparian forests (Figs. 2 and 8).

Most erosional and depositional events that affect channel morphology occur in high flow periods, typically during spring snowmelt and periodic stormflows. Such flows are characterized as either bankfull stage (Fig. 1), which occurs approximately every other year (1.5–2 yr intervals), or the less frequent overbank flooding events (5–10+ yr intervals; Leopold 1994; Rosgen 1994). After high flow events, water levels decline, exposing bare mineral soils within the alluvial floodplain, commonly along point and gravel bars (see Fig. 1). These barren, yet moist, alluvial soils are critical microsites for colonization by cottonwoods via wind-and water-dispersed seed (see Figs. 1 and 6). These

Fig. 1. Generalized view of river channel and floodplain dynamics. Diagrammatic plan view and cross section indicating the relative balance between erosion of the concave bank and accretion of material in a building point bar. Bankfull condition shows that the level of the floodplain is the same as the top of the point bar. (Modified from Leopold 1994.)

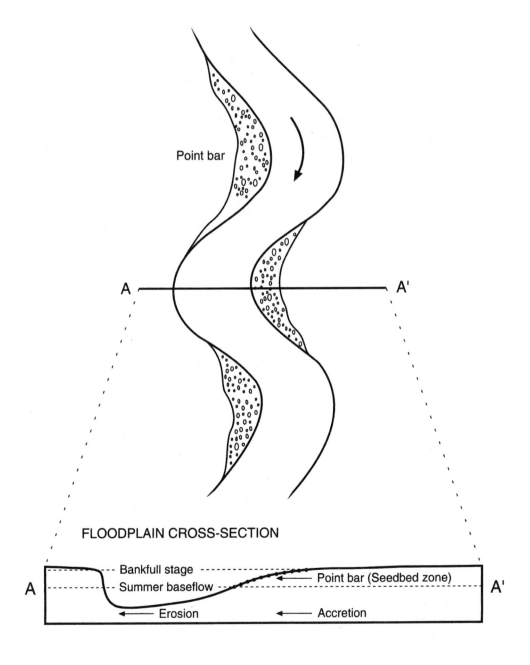

Fig. 2. Aerial photograph of Fraser River Island, British Columbia showing sequential patterns of establishment for Black Cottonwood (*Populus trichocarpa*) in relation to historical routes of river channel movement. (Photo by Scott Paper Ltd.)

microsites are also vulnerable to subsequent scouring and depositional processes during flooding.

Regional variation in climatic patterns modifies fluvial geomorphic processes. For example, many arid and semi-arid regions of the western U.S. are dominated by higher levels of precipitation in winter relative to summer months, while humid regions of the Midwest have a more uniform distribution of precipitation throughout the year. Major mountain systems, such as the Cascades and the Rocky Mountains, further modify these regional precipitation patterns. This climatic variation affects many of the fluvial processes that control river channel morphology (Leopold et al. 1964; Dunne and Leopold 1978; Leopold 1994). For example, sporadic heavy rains are more prevalent in arid regions. Such sporadic, yet intense storms accelerate terrace erosion (i.e., channel widening) and/or valley evacuation (i.e., channel downcutting) (Huckleberry 1994). In humid regions, small, light rainstorms are more prevalent and promote valley deposition (i.e., alluviation). In the arid Southwest, geologic evidence shows that the relative dominance of these geomorphic processes shifted as periods dominated by arid conditions (1880–1920) evolved towards more humid climatic conditions (1950–1980) (Leopold 1994).

Climate and watershed position strongly influence alluvial water table characteristics. As shown in Fig. 3, alluvial water tables occur at shallower depths

Fig. 3. Generalized relationship between water table depth and distance from streambank for different climatic regimes and fluvial geomorphic settings. (Modified from Reichenbacher 1984.)

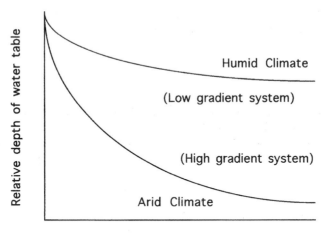

under humid than arid conditions. Water table characteristics also differ within a watershed, as valley and channel slope vary from steep, narrow floodplains at high elevations (i.e., high riverine gradients of montane floodplains) to relatively flat and wide alluvial floodplains at low elevations (i.e., lower riverine gradients of mature piedmont valley floodplains). In steep gradient reaches, water tables may fluctuate rapidly, whereas water levels remain more constant in broad, flat floodplains. Large, alluvial floodplains are also closely-linked with regional aquifers that tend to moderate water table fluctuations (Leopold 1994; Leopold et al. 1964). As a result of these linkages, larger floodplains are relatively complex hydrogeomorphic systems in which interactions between fluvial processes and alluvial groundwater systems (i.e., hyporheic zones) may have profound effects upon riparian cottonwood forests (Stanford and Ward 1993); they warrant further investigation.

The environmental features of riverine systems provide a variable water supply and periodic, yet repeated, disturbances to which riparian cottonwoods have adapted. But there is variability in fluvial processes from year to year, and it has significant effects on the long-term health and vigor of riparian cottonwood forests.

Life history and ecological properties of riparian cottonwoods

Comprehensive life history and demographic studies of riparian cottonwoods are limited, yet general patterns emerge that link the nature and timing of fluvial processes with the expression of life history traits in these species (Rood and Mahoney 1990; Johnson 1992, 1994; Stromberg 1993; Stromberg et al. 1991, 1993, 1996*a*; Scott et al. 1996*a*, *b*). Key life history and ecological properties of riparian cottonwoods are summarized in Tables 1 and 2, while relationships between the expression of these traits and stream discharge are shown in Fig. 4.

Table 1. Life history traits and ecological properties of *Populus angustifolia, P. balsamifera,* and *P. trichocarpa* (*Tacamahaca* section).

Life history traits/ecological properties	Species characteristics
Reproduction:	
Flowering time	Apr.–May (*P. balsamifera*)[6]
	Mar.–May (*P. trichocarpa*)[5]
Seed dispersal time	May–July (*P. balsamifera*)[6]
	May–June (*P. trichocarpa*)[5]
Seed weight	0.3 mg (*P. balsamifera*)[6]
Dispersal agents/distance	Air and water/several km (All spp.)
Asexual traits	Cladoptosis (*P. trichocarpa*)[5]
	Root suckering and crown breakage
Germination/establishment:	
Seed viability (natural conditions)	1–2 wk (*P. balsamifera*)[6]
	1–2 wk (*P. trichocarpa*)[5]
Seed germination	24 h/moist, bare soil (All spp.)
Seedling root growth rates	6–8 mm/d (*P. balsamifera*)[1]
	4–12+ mm/d (*P. trichocarpa*)[4,8]
Soil pH	6–8 (*P. balsamifera*)[6]
	5–7 (*P. trichocarpa*)[5]
Growth/maturation:	
Age at reproductive maturity	8–10 yr (*P. balsamifera*)[6]
	7–10 yr (*P. trichocarpa*)[5]
Lifespan	100–200 yr (*P. angustifolia*)[7]
	100–200 yr (*P. balsamifera*)[1,6]
	100–200+ yr (*P. trichocarpa*)[5]
Plant height at reprod. maturity	8–13.5 m (*P. balsamifera*)[1]
	10–16.8 m (*P. trichocarpa*)[5]
Plant dbh at reprod. maturity	8–11.7 cm (*P. balsamifera*)[1]
	12–20 cm (*P. trichocarpa*)[5]
Mature stand density (trees/ha)	38.3–91.5/ha (*P. angustifolia*)[2,3]
	88.9–120/ha (*P. balsamifera*)[1,3]
	110–294/ha (*P. trichocarpa*)[5]
Rooting depths of mature stands	3–5+ m (All spp.)

Sources: [1]Peterson and Peterson (1992); [2]Szaro (1990); [3]Shaw (1991); [4]Reed (1995); [5]DeBell (1990); Dewit and Reid (1992); [6]Zasada and Phipps (1990); [7]Baker (1990); [8]Mahoney and Rood (1991, 1992).

Sexual reproduction and establishment

Being dioecious, cottonwood trees are either male or female. In both sexes, the flowers are clustered in catkins, which tend to be borne in the upper tree crown. Male and female catkins are readily distinguished from one another, as male catkins are typically smaller and reddish-purple, whereas female flowers and catkins are significantly larger and greenish in appearance. Males commonly initiate flowering before females and both sexes flower approximately 1–2 wk

Table 2. Life history traits and ecological properties of *Populus deltoides* and *Populus fremontii* (*Aigeiros* section).

Life history traits/ecological properties	Species characteristics
Reproduction:	
Flowering time	Mar.–Apr. (*P. deltoides*)[8–10]
	Feb.–Mar. (*P. fremontii*)[6,12,13]
Seed dispersal time	May–Aug. (*P. deltoides*)[8–10]
	Mar.–Apr. (*P. fremontii*)[6]
Seed weight	0.3–0.6 mg (*P. deltoides*)[4,9,10]
Seeds/tree/yr	25+ million (*P. deltoides*)[5,9,10]
Dispersal agents/distance	Air and water/several km (All spp.)
Asexual traits	Limited to crown breakage and flood-related disturbance
Germination/establishment:	
Seed viability (natural conditions)	1–2 wk (*P. deltoides*)[9,10]
	1–3 wk (*P. fremontii*)[7]
Seed germination	24 h/bare
Seedling root growth rates	4–6 mm/d (*P. deltoides*)[16]
	4–12 mm/d (*P. fremontii*)[6,7,15]
Soil pH	5.5–8 (*P. deltoides* var. *delt.*)[9]
Soil salinity	0–1500 mg/L (*P. fremontii*)[17]
Growth/maturation:	
Age at reproductive maturity	5–10 yr (*P. d.* var. *delt.*)[5,9]
	10 yr (*P. d.* var. *occ.*)[5,10]
	5–10 yr (*P. fremontii*)[6]
Lifespan	130+ yr (*P. fremontii*)[3,19]
	100–150+ yr (*P. deltoides*)[9,10]
Plant height at reprod. maturity	10–15 m (*P. deltoides*)[9,10]
Plant dbh at reprod. maturity	12–20 cm (*P. deltoides*)[9,10]
Mature stand density (trees/ha)	192/ha (*P. deltoides*)[11]
	50–400+/ha (*P. fremontii*)[1,6,19]
Rooting depths of mature stands	3–5+ m (All spp.)[14,18]

Sources: [1]Strahan (1983); [2]Szaro (1990); [3]Shanfield (1983); Howe and Knopf (1991); [4]Bessey (1904); [5]Schreiner (1974); [6]Reichenbacher (1984); [7]Horton et al. (1960) and Fenner et al. (1984); [8]Farmer (1966); [9]Cooper (1990); [10]Van Haverbeke (1990); [11]Johnson et al. (1976); [12]Asplund and Gooch (1988); [13]Stromberg et al. (1991); [14]Jackson et al. (1987); [15]McBride et al. (1988); [16]Segelquist et al. (1993) and Stromberg et al. (1993, 1996a); [17]Jackson et al. (1990) and Shafroth et al. (1995b); [18]Stromberg et al. (1996b); [19]Hunter et al. (1987) and Szaro (1989).

Fig. 4. Generalized timing and duration of reproductive events for riparian cottonwoods in relation to the annual pattern of stream discharge.

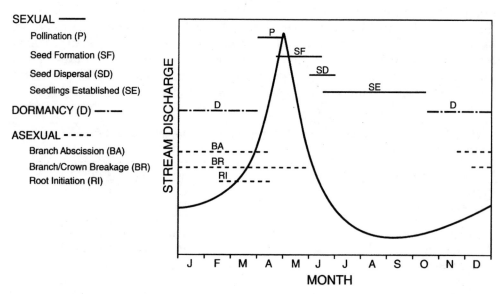

prior to leaf initiation in the early spring (March–April). Flowering and pollination thus coincide with springtime peaks in riverine flow (Fig. 4), though significant variation in the timing and duration of flowering exists within and between species (Table 2). In part, phenological variation within and between species reflects environmental differences between plants growing at different elevations and latitudes (Dunlap 1991; Farmer 1993). Under the cool, shortgrowing season of high latitudes and elevations, flowering may not begin until late May, while at lower elevations and latitudes flowering will have ceased in most populations by mid-April. In the arid Southwest, flowering by *P. fremontii* is over by late February to early March (Reichenbacher 1984; Asplund and Gooch 1988; Stromberg et al. 1991). Significant phenological variation in flowering period has also been reported within populations (Dunlap 1991; Farmer 1993), occasionally spanning a range of more than 2 months. Such intrapopulation variance in flowering appears greater in *P. deltoides* than in other species (Farmer 1966; Brian Stanton, personal communication). Additional studies are needed to clarify genetic-environmental interactions controlling the timing of flowering and pollination (see Stanton and Villar in Chapter 5).

Wind-dispersed pollen fertilizes the ovule within 24 h of landing upon the receptive stigma (see Stettler et al. in Chapter 4 and Stanton and Villar in Chapter 5). The process of ovule ripening and seed maturation is temperature-dependent, with seed formation and dispersal occurring within 3–6 wk following fertilization (Fig. 4, Tables 1 and 2; see also Stettler et al. in Chapter 4). Upon maturation, cottonwood seeds are extremely small, weighing approximately 0.3–0.6 mg per seed (Schreiner 1974; Hardin 1984; Zasada and Phipps 1990), and contain little or no endosperm. Females produce a large and dependable crop of seed, more or less annually. Estimates of annual seed production by large, mature individuals of *P. deltoides* have been reported to exceed 25 million seeds per tree (Bessey 1904; Schreiner 1974). Yet age- and size-specific studies of seed production have been limited (Hardin 1984) and merit further investigation.

Cottonwood seeds, borne by numerous fluffy, cotton-like hairs, are dispersed long distances by wind and water. Although no studies have specifically documented dispersal distance, general observations suggest that most of the seed is deposited within a few hundred metres of the mother plant. The potential for long-range disperal (several km or more) via convective wind currents clearly exists. However, methodological constraints associated with the large, air-borne seed crop of these species have prevented quantitative studies on the nature and timing of long-distance dispersal. The best evidence for effective dispersal (and associated gene flow) is the common lack of genetic differentiation among populations, with the bulk of genetic variation (e.g., in isozymes >90%) being found within populations (see Farmer in Chapter 2).

Seed dispersal typically coincides with declining river flows following springtime snowmelt and stormflows (Fig. 4), thereby increasing the probability of seeds landing in favorable microsites along the river channel. In some instances, seed dispersal may persist well into the summer months. For example, seed dispersal has been observed in mid-July among populations of *P. deltoides* along the upper Missouri River (Johnson et al. 1976) and Central Platte River (Johnson 1994) and late-August in the lower Mississippi Valley (Farmer 1966; Brian Stanton, personal communication). This late shedding of seed by *P. deltoides* may reflect an adaptation to summer rainfall and periodic summer flooding on rivers within its natural range.

Seed viability is very short, generally lasting only 1–2 wk under natural conditions (Tables 1 and 2; Horton et al. 1960; Fenner et al. 1984; Cooper 1990; Debell 1990; VanHaverbeke 1990; Zasada and Phipps 1990). Once a seed becomes wet, viability will be lost in 2–3 d if a favorable microsite is not encountered. Low seed viability has also been reported in relation to high levels of air humidity. The short-term viability of seeds is clearly a limiting factor in the life cycle of cottonwoods, as germination must occur within a relatively short time period. In some cases, seeds may not be fully viable when dispersed from the mother plant (*P. fremontii,* Fenner et al. 1984). These seeds typically become viable within a few days following dispersal; however, the pattern and mechanism of post-dispersal seed viability requires additional study.

On appropriate microsites, germination is rapid. The root radicle emerges from the seed, enters the soil, and cotyledons begin to expand within 24 h (Reed 1995). Young roots are noted for their development of "collet hairs" at the base

Fig. 5. All cottonwoods are prolific seed producers and initial viability is almost complete. Consequently, if seeds land on moist, mineral soils, extensive mats of seedlings result. However, almost all of these seedlings die due to drought stress and complete mortality often follows dewatering when stream flows are diverted for irrigation or other uses. (Photo by S.B. Rood.)

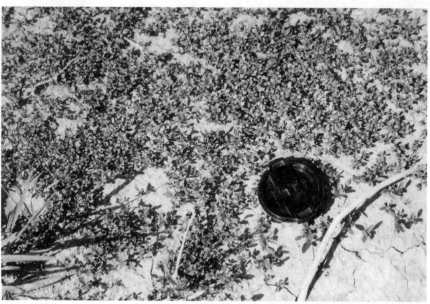

of the hypocotyl (Moss 1938; Noble 1979). These hairs are anatomically distinct from root hairs and attach quickly to sand and silt particles to provide anchorage and absorption (Noble 1979; Johnson 1994).

In late spring and early summer, germinating seeds and seedlings are commonly found in large numbers along point bars as well as other moist, exposed substrates within alluvial floodplains (Figs. 5 and 6). Seedling densities have been reported to range from as few as 20 to more than 4000 per square metre (Strahan 1983; McBride and Strahan 1984; Lee et al. 1991; Virginillo et al. 1991; Johnson 1994; Reed 1995; Stromberg et al. 1991, 1993). Temporal and spatial variation in favorable microsites appears to be the primary determinant of seedling recruitment (see mortality zones in Fig. 6). The growth and development of seedlings is closely correlated with the relative abundance of light and soil moisture (Rood and Mahoney 1990; Mahoney and Rood 1991, 1992). Given the lack of endosperm, full sunlight is critical as seedlings are highly dependent upon photosynthate derived from cotyledons and juvenile leaves for sustained growth and development. As a result, cottonwood seedlings are poor competitors in vegetated sites (Johnson et al. 1976; Fenner et al. 1984; Johnson 1994). The soil must also be moist throughout the early stages of seedling establishment (1–2 wk), and seasonal declines in water tables regulate patterns of seedling recruitment

Fig. 6. Patterns of seed dispersal, germination, and establishment in relation to microtopographic position and river stage of a meandering river. (Modified from Bradley and Smith 1986.)

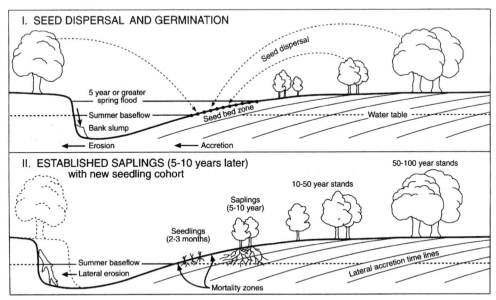

throughout the first growing season (Mahoney and Rood 1991, 1992; Segelquist et al. 1993; Johnson 1994). If the rate of water table decline exceeds the rate of root growth (4–6 mm/d for most species, but up to 12 mm/d in *P. trichocarpa*, Tables 1 and 2), water deficits lead to seedling mortality. Early germinating seedlings can tap groundwater at depths of 75 cm by the end of the growing season, and in some instances reach depths greater than 150 cm (Johnson 1994; Reed 1995). Rates of root growth and seedling establishment are also higher in fine, silty sands than coarse, gravelly soils (Kocsis et al. 1991). The impact of high surface temperatures is partially moderated by narrow juvenile leaves, yet periods of high evaporative demand accentuate seedling water deficits and drought-induced mortality (Johnson 1994). Vulnerability to drought persists until sapling roots reach alluvial water tables at depths of two metres or more (Fig. 6).

While cottonwood seedlings and saplings are intolerant of drought, they are tolerant of inundation and siltation (Fig. 7, Smit 1988; Rood and Mahoney 1990; Mahoney and Rood 1992). This tolerance helps cottonwoods survive extended periods of inundation (3–4 wk or more) during establishment and in subsequent years. Inundation and siltation eliminate many competitors, thus aiding seedling/sapling growth by keeping recruitment zones relatively open. While seedlings are tolerant of inundation, springtime flooding also eliminates many seedlings adjacent to the main channel by physical scouring (Figs. 6 and 7, Bradley and Smith 1986; Rood and Mahoney 1990). The lack of cottonwoods

Fig. 7. Cottonwood saplings in the flooded Oldman River at Lethbridge, Alberta. The diversity of sapling form reflects the occurrence and hybridization of three species, the Plains Cottonwood, *Populus deltoides* var. *occidentalis*, the narrowleaf cottonwood, *P. angustifolia*, and the black cottonwood, *P. trichocarpa*, or balsam poplar, *P. balsamifera*, that are difficult to distinquish without female flowers. All riparian cottonwoods are very flood-tolerant, capable of surviving weeks or even months of inundation. (Photo by S.B. Rood)

along some steep-gradient reaches and small watersheds may be due to post-dispersal scouring. In some instances, scouring by winter ice also leads to extensive seedling and sapling mortality (Johnson 1994).

The complexity of interactions between fluvial processes and seedling recruitment reveals a critical "bottleneck" in the life history of riparian cottonwoods. Low flows during seed dispersal are necessary to expose open, moist microsites for germination and recruitment. In contrast, higher "peak" flows during the dispersal phase may prevent exposure of microsites for recruitment until after the seeds have been dispersed or lost their viability. Higher flows following the dispersal phase may also bury or scour newly germinated seedlings. As a result, the location of germinating seedlings relative to the main channel influences subsequent patterns of seedling recruitment and mortality (see mortality zones in Figs. 6 and 8, Bradley and Smith 1986; Segelquist et al. 1993; Johnson 1994; Stromberg et al. 1991, 1993, 1996a). Under conditions of low riverine flows and seasonal drought, proximity to the main channel would be advantageous as seedlings not established in moist soils would likely succumb to drought. Yet, these seedlings would be vulnerable to physical scouring in subsequent years. A reversal in recruitment success may occur in years dominated by higher riverine flows and cool, moist growing conditions that enable recruitment above the scour zone of the main channel. However, these seedlings would remain vulnerable to drought-induced mortality. Given the nature of these interactions, it is apparent that the conditions essential for seedling recruitment are not met on an annual basis (Baker 1990; Rood and Mahoney 1990; Johnson 1994). In fact, suitable conditions occur irregularly (Barnes 1985; Johnson 1994), on intervals of 5–10 yr or longer (Figs. 6 and 8, Bradley and Smith 1986; Baker 1990; Stromberg et al. 1991, 1993; Hughes 1994; Johnson 1994; Scott et al. 1996*a, b*).

Major differences in river channel morphology may also influence spatial and temporal patterns of seedling recruitment, as the distribution of suitable microsites change in relation to the dominant fluvial processes of a given river system (Johnson 1994; Scott et al. 1996*a, b*). For example, along low-gradient meandering streams in the arid Southwest, small floods allow for frequent (≤5 yr) episodes of seedling establishment on point bars in what has been described as an incremental-replenishment model (Hughes 1994). On less sinuous arid-region rivers, such as the Hassayampa River, cottonwoods establish in large numbers at infrequent intervals, after large, erosional floods that scour sediment from terraces (Stromberg et al. 1993). This pattern has been referred to as a

Fig. 8. Successful seedlings and some clonal saplings originate in arcuate bands that track specific elevations along meandering rivers and especially, at point bars at the end of meander lobes. The curving bands of even-aged saplings or trees originated from specific flood events and provide a hydrological history of the river. Here, sapling bands of narrowleaf cottonwoods and plains cottonwoods occur at a meander lobe along the Oldman River near Lethbridge, Alberta. (Photo by S.B. Rood.)

general-replenishment model, characterized by infrequent (ca. 30–50 yr recurrence intervals), large floods that set up recruitment conditions over large areas of the floodplain (Stromberg et al. 1993; Hughes 1994). In effect, temporal and spatial variation in fluvial processes result in a highly variable environmental regime for riparian cottonwoods, yet life history traits associated with reproduction and establishment converge upon the sequence of fluvial events following springtime snowmelt and stormflows. Future studies should explore the dynamic nature of these events and seek to quantify the physiological and genetic components of life-history variation relative to the timing and duration of fluvial geomorphic processes.

Asexual reproduction

Asexual reproduction is widespread among riparian cottonwoods (Tables 1 and 2). The most common mode of asexual reproduction is crown breakage and tree fall during wind storms and flooding events (Fig. 4). Broken branches can become buried in sediment, where they subsequently sprout and develop vigorous shoots. Crown damage and disturbance of shallow roots may also promote root suckering in some species, though this form of suckering is less common among *Aigeiros* species (Rood et al. 1994). Cladoptosis, the shedding of branchlets via formation of an abscission layer during winter months or following pollen release in the spring (Fig. 4), is a unique form of asexual reproduction common to *Tacamahaca* species but absent in *Aigeiros* species (Galloway and Worrall 1979; Dewit and Reid 1992).

Within native stands, the proportion of trees established from sexual vs. asexual propagules will vary with species, climatic condition, and drainage basin. In some instances, asexual propagules may outcompete seedlings, though few studies have sought to document the relative role of sexual vs. asexual propagules in native stands (Rood et al. 1994; Stromberg et al. 1996a). In either case, asexual propagation offers an alternative pathway to establishment in a highly variable riverine environment and, as a result, significantly influences the genetic structure of riparian cottonwood populations (Rood et al.1994; Reed 1995; Stromberg et al. 1996a).

Establishment of riparian cottonwoods in nonalluvial habitats

Many of the ecological and life history properties associated with the establishment of cottonwoods in riparian habitats can result in their colonization of nonalluvial environments. During early spring and summer, the bare, moist mineral soils required for germination and establishment are readily found in adjacent agricultural fields and forest clearings. Given sufficient precipitation during the growing season, cottonwood seedlings can establish in great numbers in these disturbed environments. As a result, the establishment of *P. trichocarpa* and *P. deltoides* in agricultural fields and forest clearings is common in moister regions of North America, such as the Pacific northwest and midwest, especially lands cleared along upper floodplain terraces (J.H. Braatne and P.E. Heilman, personal observations). In the Great Lakes region, seedlings and young stands of *P. deltoides* and *P. balsamifera* are also commonly observed along the margins of lakes and wetlands. In general, these isolated individuals and/or small stands of riparian cottonwood represent an opportunistic event as they are invaded and eventually dominated by secondary successional forest species.

Growth and maturation

Similar as in other plants, the nature and timing of environmental stress determine relative rates of growth and development in riparian cottonwoods. Early stages of sapling growth and stand development are influenced by seasonal flooding, drought, grazing, fire, and other site-specific conditions. Vulnerability to drought persists until sapling roots reach the moist soil associated with late-season alluvial water tables at depths of two metres or more (Fig. 6). Major flooding events (10- to 50-yr floods) eliminate young saplings as well as mature trees, though losses associated with these floods are often compensated by additional seedling recruitment on newly, exposed microsites. In general, *Aigeiros* species are more drought-tolerant than *Tacamahaca* species (see Blake et al. in Chapter 17). In fact, morphological and physiological adaptations to a warmer, drier climatic regime may account for the relative dominance of *Aigeiros* species at lower elevations and latitudes relative to the *Tacamahaca* species (Braatne et al. 1992; Hinckely et al. 1992, see Eckenwalder in Chapter 1). For example, stomata of *Aigeiros* species close rapidly in response to increasing air and soil-water deficits, while leaf orientation (ca. perpendicular to the sun) minimizes heat loads (Hinckley et al. 1992; see also Ceulemans and Isebrands in Chapter 16 and Blake et al. in Chapter 17). Differences in frost tolerance and susceptibility to xylem embolism may also account for major differences in species distributional patterns (Tyree et al. 1994; also see Chapters 16–18). Additional ecophysiological studies are needed to assess relationships between physiological adaptation to environmental stress and species distributional patterns.

Height growth during the early stages of sapling development may be limited, as energy is preferentially allocated to rapidly-growing roots. Two- to three-year-old cohorts of *P. fremontii* ranged from 5 to 50 cm tall (Stromberg et al. 1991, 1996a), while similar age classes of other species typically range from 25 to more than 60 cm tall (Cordes 1991; Stobbs et al.1991; Peterson and Peterson 1992; Reed 1995). Once their root systems have become established, height growth is rapid and may reach 10–15 m upon the attainment of reproductive maturity (Tables 1 and 2).

Age of reproductive maturity and the lifespan of riparian cottonwoods vary among species, though quantitative studies of these demographic parameters are generally lacking for natural populations. The age at which reproductive maturity is attained ranges from 5 to 10 yr for most cottonwoods (Tables 1 and 2), though *Aigeiros* species typically reach reproductive maturity earlier than *Tacamahaca* species (Reichenbacher 1984; DeBell 1990; Cooper 1990). On average, *Aigeiros* poplars also appear to be shorter-lived than *Tacamahaca* poplars. Ages of older trees from 100 to 150 yr have been observed for *Aigeiros* species, while *Tacamahaca* species may live more than 200 yr (Cooper 1990; DeBell 1990; VanHaverbeke 1990; Zasada and Phipps 1990; Stromberg 1993). Although these data provide a general background on patterns of reproductive maturity and longevity, these life history and demographic parameters require more extensive study in natural populations. *after this their natural life span is 100-200 years.*

Studies of sex ratios in natural populations of riparian cottonwoods have been limited. Balanced sex ratios (1:1) have been reported for populations of *P. deltoides* growing along the lower Mississippi River (Farmer 1964) and the Hocking River in Ohio (Hardin 1983). In contrast, Comtois et al. (1986) documented skewed sex ratios among *P. balsamifera* populations in northern Quebec. In this study, males were more common in extreme environments, whereas females typically dominated more protected and nutrient-rich environments. Other researchers have also observed skewed sex ratios in riparian cottonwoods in western North America. In some populations of *P. trichocarpa*, males dominate warmer and drier habitats, whereas other populations are either completely male or female and seemingly independent of environmental conditions (J.H. Braatne, unpublished data and personal observations). Given the widespread habitat partitioning between genders in *Salix* spp. (Dawson and Bliss 1989, 1993) and *Acer negundo* (Dawson and Ehleringer 1993), the possibility for habitat partitioning between male and female cottonwoods deserves more attention. Further research is needed to determine the cause and consequences of skewed sex ratios in riparian cottonwood populations.

Mature stand densities are highly variable within and among species (Tables 1 and 2); reported values range from 40 to 192+/ha for *Aigeiros* species (Johnson et al. 1976; Strahan 1983; Reichenbacher 1984; Szaro 1989; Stromberg et al. 1991, 1993) to 82–294+/ha for *Tacamahaca* species (DeBell 1990; Zasada and Phipps 1990; Peterson and Peterson 1992). Given seedling and sapling requirements for light, no regeneration by seed occurs within cottonwood stands. Any small shoots observed within mature stands are due to asexual propagation (root suckering and/or rooting of broken branches). As a result, riparian cottonwood forests often appear as linear strips of even-age/size stands; each stand representing a discrete period of propagule establishment and growth (Figs. 6 and 8). In places where the river channel has moved systematically in a uniform direction, an age/size gradient develops with young stands of small trees nearest the river and older stands of larger trees found farther from the channel (Figs. 6 and

Fig. 9. In very dry areas of the American Southwest, riparian cottonwoods provide a sharp contrast to the adjacent xeric landscapes. Here, a few Fremont cottonwoods, *Populus fremontii*, persist along the lower Truckee River near Reno, Nevada, after a century of river damming, water diversion, and tree harvesting. Although only a half dozen cottonwoods exist in this view, they still provide many woodland attributes. (Photo by S.B. Rood.)

8). Mature stands of cottonwood may remain within the active floodplain or occur at slightly higher elevations on secondary terraces. As stands mature and become increasingly isolated from active fluvial processes, sex ratios may become skewed (J.H. Braatne, unpublished data), and stands are invaded by secondary successional forest species (Johnson et al. 1976). Of increasing concern is the relative dominance of older, relictual cottonwoods and lack of younger stands of riparian cottonwood throughout western North America (Fig. 9). The causes and consequences of these recent declines in riparian cottonwoods are discussed in the following section.

Conservation and restoration of riparian cottonwood forests

As noted in previous sections, cottonwoods are often the dominant forest species in many of the riparian habitats in western North America. These riparian forests have special importance for humans and are extremely rich wildlife habitats (Fig. 10, Finch and Ruggiero 1993). For example, although riparian vegetation occurs on less than 1% of the western North American landscape, it provides habitat for more bird species than all other vegetation types combined (Knopf et al. 1988).

Fig. 10. A riparian woodland with narrowleaf, *Populus angustifolia*, and black cottonwoods, *P. trichocarpa*, along the Oldman River in southwestern Alberta. In semi-arid regions, the riparian cottonwoods provide welcome aesthetic, recreational, and environmental relief from the otherwise treeless regions; these riparian woodlands harbor the region's richest wildlife habitats, providing environmental value much beyond that of the wood resource. (Photo by S.B. Rood.)

Due to their rapid growth and sometimes ragged appearance (due to cavitation-induced branch and crown dieback, Tyree et al. 1994), riparian cottonwoods have sometimes been considered as undesirable weeds ('cottonweeds'). However, these trees often serve as the foundation for the riparian forest ecosystem and are especially valued in the otherwise treeless semi-arid regions of western North America. Unlike wetter areas to the east (Wilson 1970) and west (Szaro 1990), a loss of riparian cottonwoods in many semiarid riparian areas is not compensated by enrichment from other tree species. If these cottonwoods die, so does the riparian forest ecosystem.

Causes of decline in riparian cottonwood populations

Only small remnants of once abundant riparian cottonwood forests survive in most regions of the southwestern United States (Fig. 9). Estimates of the magnitude of riparian vegetation loss and degradation range from 70 to 95% for the Southwest (Johnson and Haight 1984; National Research Council 1992). Even more severe declines have been experienced in the heavily developed areas of California, such as the Sacramento Valley, which has lost about 98.5% of the riparian forests that existed in 1850 (Sands and Howe 1977). Losses in the more

Table 3. Negative impacts on riparian cottonwood forests across western North America.

Factor	Comment
1. Livestock grazing	Cattle graze and trample seedlings. Overgrazed regions are characterized by a deficiency of seedlings and saplings, and forests decline as older trees die out.
2. Water diversion	Following river damming or the construction of diversion weirs, water is diverted offstream for irrigation. Subsequent instream flows are often insufficient, creating drought stress and accelerating mortality.
3. Domestic settlement	Clearing for homes, roads, bridges, and various other uses. Pressure is generally proportional to human population density.
4. Exotic plants	Characterized by natural (and artificial) disturbance, riparian areas are especially vulnerable to encroachment of exotic plants. Introduced trees include the salt cedar (*Tamarix pentandra*) and Russian olive (*Elaeagnus angustifolia*), and aggressive noxious weeds such as leafy spurge (*Euphorbia esula*) also occur.
5. Onstream reservoirs	Many riparian areas have been flooded by reservoirs. The rate of dam construction in the United States has declined over the past two decades but some damming is likely to continue in Canada.
6. Channelization	In many areas, extensive programs have attempted to straighten rivers and armor banks. Such actions inhibit the dynamic meandering of rivers that is essential for cottonwood replenishment.
7. Agricultural clearing	Clearing for pasture or crop production occurs where the proximity of floodplains to river water provides inexpensive irrigation. Agricultural clearing was more extensive in the early 1900s, and little net change has occurred in many areas since 1950.
8. Gravel mining	River valleys are prime areas for sand and gravel extraction. In addition to the areas excavated, roads, buildings, and screening plants often involve forest clearing. Although aesthetically offensive, abandoned gravel pits are sometimes areas of cottonwood recruitment, particularly through root suckering.
9. Direct harvesting	During early white settlement of western North America, poplars were harvested to provide building materials for forts and homes as well as fuel wood for heating and riverboat engines. Such use is presently minor in most regions.
10. Beavers	Beavers are a natural component of many riparian ecosystems and contribute to various processes, including cottonwood rejuvenation after beaver harvesting. However, an imbalance between beavers and trees may result from the loss of natural predators of beavers and the loss of some trees. The present consumer preference away from natural furs has reduced trapping, an artificial measure that controlled beaver populations through the past century.

Note: Impacts are listed in likely descending order of importance. The ranking would vary across river systems (revised from Rood and Mahoney 1990).

northerly areas of Colorado, Idaho, Wyoming, Montana, and Alberta have lagged behind the decline in California, although similar patterns are emerging. The causes of declines in riparian cottonwoods are numerous; similar types of impacts occur across different areas, but their relative significance differs among regions (see Table 3).

In many areas in western North America the heaviest pressure on riparian cottonwoods is related to livestock grazing (Table 3). Cattle browse and trample seedlings and saplings, thereby preventing replenishment of the forest. Management efforts to control livestock grazing include rotational grazing and exclusion fencing. This limits cattle use of riparian areas for short periods of time (ca. 5 yr) to allow younger trees to outgrow their most vulnerable stage.

Another major cause of the decline of riparian cottonwoods is river damming and water diversion (Tables 3, 4, and 5). Declines of cottonwoods downstream from dams in semi-arid regions of North America are well documented (Tables 4 and 5). Fortunately, these impacts are site specific since it is largely the pattern of downstream flow regulation, rather than simply the presence or absence of dams that determines the effect on riparian ecosystems. Although cottonwood decline has been common, occasional increases of cottonwoods have occurred following damming and stream flow modification, thus confirming that river type and flow patterns are critical factors in influencing riparian cottonwood forests (Rood and Mahoney 1990; Johnson 1994; Scott et al. 1996a).

In many semi-arid areas, onstream reservoirs are managed to conserve spring snowmelt which will later be diverted offstream for irrigation during summer. Dam operation can result in abrupt reductions in flow and sedimentation in late spring and early summer as well as in insufficient flows through the hot, dry period of midsummer. Both the abrupt flow reduction (Mahoney and Rood 1991) and the low summer flows probably contribute to drought stress which results in cottonwood die-back and mortality. Cottonwood seedlings are particularly vulnerable since they have limited, shallow root systems. The retention of sediments by dams also decreases the potential availability of microsites for seedling establishment. Due to the vulnerability of seedlings, recruitment of replacement trees is significantly impacted by river damming and water diversion. Without periodic recruitment, the cottonwood forest will suffer gradual decline as previously established trees age and die. Older trees appear more vulnerable to drought stress as they are physiologically decrepit and generally located on higher terraces where their root systems become isolated from alluvial waters due to excessive surface water diversion and groundwater extraction.

Cottonwoods are especially vulnerable to drought-induced xylem cavitation (Tyree et al. 1994). In some instances, vulnerability to cavitation may contribute favorably to natural systems since it results in shoot pruning during drought periods to reduce transpirational water loss. However, in flow-reduced systems,

Table 4. Reports of negative impacts of river damming on downstream riparian cottonwood forests in western North America.

Author (date)	River	Region	*Populus*	Comments
Johnson et al. (1976)	Missouri	N. Dakota	*P. deltoides*	Reduced tree growth and fewer seedlings
Brown et al. (1977)	Various	Arizona	*P. fremontii, P. angustifolia*	Reduced abundance
Ohmart et al. (1977)	Colorado	California	*P. fremontii*	Reduced abundance, absence of seedlings
Crouch (1979)	South Platte	Colorado	*P. deltoides*	Reduced abundance
Behan (1981)	Missouri	Montana	*P. deltoides*	Reduced abundance, absence of seedlings
Reily and Johnson (1982)	Missouri	N. Dakota	*P. deltoides*	Reduced tree growth
Brothers (1984)	Owens	California	*P. fremontii*	Reduced abundance
Stine et al. (1984)	Rush Creek	California	*P. balsamifera*	Reduced abundance
Strahan (1984)	Sacramento	California	*P. fremontii*	Fewer seedlings
Fenner et al. (1985)	Salt	Arizona	*P. fremontii*	Conditions unsuitable for seeding establishment
Bradley and Smith (1986)	Milk	Alberta/ Montana	*P. deltoides*	Reduced abundance, fewer saplings
Akashi (1988)	Bighorn	Wyoming	*P. deltoides*	Reduced abundance
Rood and Heinze-Milne (1989)	St. Mary, Waterton, and Belly	Alberta	*P. deltoides, P. trichocarpa[a], P. angustifolia*	Reduced abundance
Howe and Knopf (1991)	Rio Grande	New Mexico	*P. fremontii*	Absence of seedlings
Smith et al. (1991)	Bishop Creek	California	*P. fremontii, P. balsamifera*	Smaller leaves, lower transpiration, and H_2O potential
Snyder and Miller (1991)	Arkansas	Colorado	*P. deltoides*	Reduced abundance
Stromberg and Patten (1991)	Bishop Creek	California	*P. fremontii, P. balsamifera*	Reduced tree diameter growth, crown cover, and survival
Stromberg and Patten (1992)	Bishop and Pine Creeks	California	*P. trichocarpa*	Increased mortality, reduced growth
Johnson (1992)	Missouri	North Dakota	*P. deltoides*	Fewer saplings
Rood et al. (1995)	St. Mary	Alberta	*P. deltoides, P. trichocarpa, P. angustifolia*	Reduced abundance, absence of seedlings
Rood and Mahoney (1996)	Marias	Montana	*P. deltoides, P. trichocarpa, P. angustifolia*	Absence of seedlings

Note: This is a chronological listing based on a table in Rood and Mahoney (1990), revised and expanded here.

[a]Discrimination of *P. balsamifera* and *P. trichocarpa* is difficult, particularly in areas where both species co-occur and hybridize.

Table 5. Contributing factors to the decline of western riparian cottonwood forests following river damming or water pumping from wells (revised from Rood and Mahoney 1990).

Proposed cause	Comments	References
I. Hydrological changes:		
A. Reduced water availability	Diversion of water offstream or well pumping creates a water deficit, resulting in drought stress, slow growth, and increased mortality	Brown et al. (1977); Brothers (1984); Stine et al. (1984); Hardy BBT Ltd. (1988); Rood et al. (1989); Reily and Johnson (1982); Smith et al. (1991); Snyder and Miller (1991); Stromberg and Patten (1991); Stromberg and Patten (1992); Rood et al. (1995); Scott et al. (1996)
B. Reduced flooding	Spring flooding is essential to create moist seedbeds for seedling establishment	Brown et al. (1977); Ohmart et al. (1977); Johnson et al. (1976); Reily and Johnson (1982); Johnson (1992); Rood et al. (1995)
C. Stabilized flows	Dynamic flows are essential for seedling establishment	Strahan (1984); Fenner et al. (1985); Howe and Knopf (1991); Johnson (1992); Rood and Mahoney (1996)
II. Geomorphological changes resulting from hydrological alterations:		
A. Reduced meandering and channelization	With reduced flooding, channel migration and the creation of seedbeds are reduced	Ohmart et al. (1977); Johnson et al. (1976); Bradley and Smith (1986); Howe and Knopf (1991); Snyder and Miller (1991); Johnson (1992)

cavitation can result in significant shoot mortality and crown die-back, particularly if drought stress is prolonged (Albertson and Weaver 1945). These consequences are typical of problems along flow-reduced dammed rivers, although different patterns of flow regulation can also create other problems for riparian cottonwood forests (Table 5). Temporal and spatial patterns in cottonwood decline can be diagnostic in revealing the specific negative impact(s) of river damming and flow diversion.

Conservation and restoration strategies

An important prerequisite for the conservation of riparian cottonwoods is a greater recognition of both their value and their vulnerability. (See Whitham in Chapter 11, for a special reference on the need to protect natural cottonwood hybrids and riparian hybrid zones.) Back in the 1950s and 1960s, there were active programs to clear riparian woodlands in an effort to reduce transpirational water loss in semi-arid regions (National Research Council 1992; U.S. Department of Interior 1994). The ineffectiveness of such 'phreatophyte control' programs was soon revealed since the loss of the stabilizing streamside vegetation resulted in increased erosion, reduced retention of rainfall, and subsequently falling rather than rising alluvial water tables. Although phreatophyte control programs are seldom practiced anymore, other state and federal resource management programs often fail to promote conservation of riparian woodlands (National Research Council 1992; U.S. Department of Interior 1994; Shafroth et al. 1995*a*).

With greater appreciation of riparian cottonwoods, more attention should be directed to protect unregulated streams and rivers. Rather than allocating riparian lands to various purposes and later protecting selective reaches of a river system, conservation planning should begin by identifying valuable riparian zones and ensuring their protection prior to further regional development.

Along dammed streams and rivers, firmly legislated commitments are required to ensure the delivery of sufficient flows and sediment to allow the survival, growth, and reproduction of riparian cottonwoods and other riparian vegetation. The U.S. Department of Interior has recently sought to counteract damming effects along the Colorado River by mimicking natural flood and sedimentation regimes. Such efforts are essential for the development of new approaches to the restoration of vegetation along regulated streams and rivers. However, it will also be critical to guarantee sufficient instream flows during drought years. While it may be difficult to justify instream flows solely for riparian vegetation, maintenance of these flow regimes also improves water quality and fisheries as well as other aesthetic and recreational river resources, benefitting human as well as ecosystem health.

Riparian restoration programs have been promoted along various rivers in western North America (National Research Council 1992; Friedman et al. 1995; Scott et al. 1996*a*), yet the role of riparian cottonwoods in many of these revegetation programs remains unclear. Riparian restoration requires the integrated management of both the land and water, creating complexities in both biological and physical resources and complicating administration which invariably involves multiple private, regional, state or provincial, and federal participants (National Research Council 1992; U.S. Department of Interior 1994).

Restoration programs will involve a combination of approaches related to flow regulation, land-use policies, and intervention to promote stream channel restoration and revegetation (National Research Council 1992). Revegetation measures may include deliberate seeding of suitable riparian zones where a shallow water table would promote seedling survival. In some instances, plowing and sod removal in combination with limited irrigation may also be required to promote natural seedling establishment (Friedman et al. 1995). In areas lacking native seed sources, rooted seedlings, and saplings as well as unrooted cuttings, whips, and poles may be propagated and transplanted into suitable riparian habitats (Hoag 1993; Briggs 1994). In revegetation programs, native species should be used, and a range of genotypes should be propagated in a manner that encourages biodiversity. Site scarification, the mechanical disturbance of the substrate, may also be used to propagate cottonwoods that already exist on site. Scarification will promote root suckering and thus may be most effective for *Tacamahaca* species including *P. trichocarpa*, *P. balsamifera* and *P. angustifolia*. Although root suckering of the *P. deltoides* and *P. fremontii* (Sect. *Aigeiros*) is uncommon, even these species appear to respond to some mechanical disturbance.

In spite of active cottonwood restoration programs along various rivers and streams, only a limited number of published reports on these activities exist, particularly in refereed journals (see Friedman et al. 1995; Shafroth et al. 1995*a*). There is a serious information deficiency on these topics. Researchers and riparian resource managers who have experience in cottonwood restoration are encouraged to publish their findings as they will be of considerable interest for riparian cottonwood restoration not only in western North America, but also in other parts of the world.

Acknowledgements

The authors are grateful for the critical review and suggestions provided by Juliet C. Stromberg, W. Carter Johnson, Lawrence C. Bliss, and Reinhard F. Stettler.

References

Albertson, F., and Weaver, J. 1945. Injury and death or recovery of trees in prairie climate. Ecol. Monogr. **15**: 395–433.

Akashi, Y. 1988. Riparian vegetation dynamics along the Bighorn River, Wyoming. M.Sc. thesis, University of Wyoming, Laramie, WY. 245 p.

Asplund, K.K., and Gooch, M.T. 1988. Geomorphology and the distributional ecology of Fremont Cottonwood (*Populus fremontii*) in a desert riparian canyon. Desert Plants, **9**: 17–27.

Baker, W.L. 1990. Climatic and hydrologic effects on the regeneration of *Populus angustifolia* along the Animas River, CO. J. Biogeo. **17**: 59–73.

Barnes, W. 1985. Population dynamics of woody plants on a river island. Can. J. Bot. **63**: 647–655.

Behan, M. 1981. The Missouri's stately cottonwoods: How can we save them? Montana Magazine, September, pp. 76–77.

Bessey, C.E. 1904. The number and weight of cottonwood seeds. Science (Washington, DC), **20**: 118.

Braatne, J.H., Hinckley, T.M., and Stettler, R.F. 1992. Influence of soil water supply on the physiological and morphological components of plant water balance in *Populus trichocarpa, Populus deltoides* and their F$_1$ hybrids. Tree Physiol. **11**: 325–340.

Bradley, C.E., and Smith, D.G. 1986. Plains cottonwood recruitment and survival on a prairie meandering river floodplain, Milk River, southern Alberta and northern Montana. Can. J. Bot. **64**: 1433–1442.

Briggs, M.K. 1994. Repairing degraded riparian ecosystems: a guidebook for resource managers in dry climates. Rincon Institute. Tucson, AZ.

Brothers, T.S. 1984. Historical vegetation change in the owens river riparian woodland. *In* California riparian systems: ecology, conservation and productive management. *Edited by* R. Warner and C. Hendricks. University of California Press. Berkeley, CA. pp. 75–84.

Brown, D.E., Lowe, C.H., and Hausler, J.F. 1977. Southwestern riparian communities: their biotic importance and management in Arizona. *In* Importance, preservation and management of riparian habitat: a symposium. *Edited by* R.R. Johnson and D.A. Jones. Tucson, AZ. July 9, 1977. pp. 201–211.

Comtois, P., Simon, J.P,. and Payett, S. 1986. Clonal constitution and sex ratio in northern populations of balsam poplar, *Populus balsamifera*. Holarct. Ecol. **9**: 251–260.

Cooper, D.T. 1990. *Populus deltoides* Bartr. ex Marsh.var. deltoides: Eastern cottonwood. *In* Silvics of North America: hardwoods. Vol. 2. *Edited by* R.M. Burns and B.H. Honkala. U.S. Dep. Agric. Agric. Handb. 654. pp. 530–535.

Cordes, L. 1991. The distribution and age structure of cotttonwood stands along the lower Bow River. *In* The biology and management of southern Alberta cottonwoods. *Edited by* S.B. Rood and J.M. Mahoney. University of Lethbridge, Lethbridge, AB. pp. 13–23.

Crouch, G. 1979. Changes in the vegetation complex of a cottonwood ecosystem on the South Platte River, Great Plains. Agric. Council Pub. **91**: 19–22.

Dawson, T.E., and Bliss, L.C. 1989. Patterns of water use and tissue water relations in the dioecious shrub, *Salix arctica*: the physiological basis of habitat partitioning between the sexes. Oecologia, **79**: 332–343.

Dawson, T.E., and Bliss, L.C. 1993. Plants as mosaics: leaf-, ramet- and gender-level variation in the physiology of the dwarf willow, *Salix arctica*. Funct. Ecol. **7**: 293–304.

Dawson, T.E., and J.R. Ehleringer. 1993. Gender-specific physiology, carbon isotope discrimination and habitat distribution in boxelder, *Acer negundo*. Ecology, **74**: 798–815.

DeBell, D.S. 1990. *Populus trichocarpa* Torr. & Gray: black cottonwood. *In* Silvics of North America: hardwoods. Vol. 2. *Edited by* R.M. Burns and B.H. Honkala. U.S. Dep. Agric. Agric. Handb. 654. pp. 570–576.

Dewit, L., and Reid, D.M. 1992. Branch abscission in balsam poplar (*Populus balsamifera*): characterization of the phenomenon and the influence of wind. Int. J. Plant Sci. **153**: 556–564.

Dunlap, J.M. 1991. Genetic variation in natural populations of *Populus trichocarpa* T. & G. from four river valleys in Washington. Ph.D. dissertation, University of Washington, Seattle, WA. 447 p.

Dunne, T., and Leopold, L.B. 1978. Water in environmental planning. WH. Freeman, New York, NY. 818 p.

Farmer, R.E., Jr. 1964. Sex ratio and sex-related characteristics in eastern cottonwood. Silvae Genet. **13**: 116–118.

Farmer, R.E., Jr. 1966. Variation in times of flowering and seed dispersal of eastern cottonwood in the lower Mississippi Valley. For. Sci. **12**: 343–7.

Farmer, R.E., Jr. 1993. Latitudinal variation in height and phenology of balsam poplar. Silvae Genet. **42**: 148–153.

Fenner, P., Brady, W.W., and Patton, D.R. 1984. Observations on seeds and seedlings of Fremont cottonwood. Desert Plants, **6**: 55–58.

Fenner, P., Brady, W., and Patton, D. 1985. Effects of regulated water flows on regeneration of Fremont cottonwood. J. Range Manage. **38**: 135–138.

Friedman, J.M., Scott, M.L., and Lewis, W.M. 1995. Restoration of riparian forest using irrigation, artificial disturbance, and natural seedfall. Environ. Manage. **19**: 547–557

Galloway, G., and Worrall, J. 1979. Cladoptosis: a reproductive strategy in black cottonwood? Can. J. For. Res. **9**: 122–125.

Hardin, E.D. 1983. Patterns in floodplain herbaceous vegetation and some aspects of the population biology of *Populus deltoides* on the Hocking River, Ohio. Ph.D. dissertation, Ohio State University, Columbus, OH. 279 p.

Hardin, E.D. 1984. Variation in seed weight, number per capsule and germination in *Populus deltoides* trees in southeastern Ohio. Am. Midl. Nat. **112**: 29–34.

Hardy BBT Limited. 1988. Cottonwood mortality assessment — Police Point Park. Prepared for the City of Medicine Hat, AB. 21 p.

Hinckley, T.M., Braatne, J.H., Ceulemans, R., Clum, P., Dunlap, J., Newman, D., Smit, B., Scarascia-Mugnozza, G., and Van Volkenburgh, E. 1992. Growth dynamics and canopy structure of fast-growing trees. *In* Ecophysiology of short rotation forest crops. *Edited by* P.K. Mitchell, L. Sennerby-Forsee, and T. Hinckley. Elsevier Applied Science, London and New York. pp. 1–34.

Hoag, J.C. 1993. How to plant willows and cottonwoods for riparian restoration. U.S. Dep. Agric., Technical Notes No. 23. Natural Resources Conservation Service Plant Materials Center, Aberdeen, ID.

Horton, J.S., Mounts, F.C., and Kraft, J.M. 1960. Seed germination and seedling establishment of phreatophyte species. U.S. For. Serv. Rocky Mt. For. Range Exp. Stn. Pap. 48.

Howe, W.H., and Knopf, F.L. 1991. On the imminent decline of Rio Grande cottonwoods in central New Mexico. SW Naturalist, **36**: 218–224.

Huckleberry, G. 1994. Contrasting channel response to floods on the middle Gila River, Arizona. Geology, **22**: 1083–1086.

Hughes, F.M.R. 1994. Environmental change, disturbance, and regeneration in semi-arid floodplain forests. *In* Environmental change in drylands: biogeographical and geomorphological perspectives. *Edited by* A.C. Millington and K. Pye. John Wiley and Sons, Ltd., New York, NY. pp. 321–345.

Hunter, W.C., Anderson, B.S., and Ohmart, R.D. 1987. Avian community structure changes in a mature floodplain forest after extensive flooding. J. Wild. Manage. **51**: 495–502.

Jackson, W., Martinez, T., Cuplin, P., Mickley, W.L., Shelby, B., Summers, P., McGlothlin, C., and Van Haveren, B. 1987. Assessment of water conditions and management opportunities in support of riparian values: BLM San Pedro River Properties. Denver, CO: U.S. Bureau of Land Management (BLM-YA-PT-88).

Jackson, J., Ball, J.T., and Rose, M.R. 1990. Assessment of the salinity tolerance of eight Sonoran Desert trees and shrubs. Desert Research Institute, Reno, NV.

Johnson, R.R., and Haight, L.T. 1984. Riparian problems and initiative in the American Southwest: a regional perspective. *In* California riparian systems: ecology, conservation and productive management. *Edited by* R. Warner and C. Hendricks. University of California Press, Berkeley, CA. pp. 404–412.

Johnson, W.C. 1992. Dams and riparian forests: case study from the upper Missouri River. Rivers, **3**: 229–242.

Johnson, W.C. 1994. Woodland expansion in the Platte River, Nebraska: patterns and causes. Ecol. Monogr. **16**: 45–84.

Johnson, W.C., Burgess, R.L., and Keammerer, W.R. 1976. Forest overstory vegetation and environment on the Missouri River floodplain in North Dakota. Ecol. Monogr. **46**: 59–84.

Kocsis, M., Mahoney, J.M., and Rood, S.B. 1991. Effects of substrate texture and rate of water table decline on transpiration and survival of poplar species. *In* The biology and management of southern Alberta cottonwoods. *Edited by* S.B. Rood and J.M. Mahoney. University of Lethbridge, Lethbridge, AB. pp. 63–67.

Knopf, F.L., Johnson, R.R., Rich, T., Samson, F.B., and Szaro, R.C. 1988. Conservation of riparian ecosystems in the United States. Wilson. Bull. **100**: 272–284.

Lee, C., Mahoney, J.M., and Rood, S.B. 1991. Poplar seeds and seedlings along the St. Mary, Belly and Waterton Rivers, Alberta. *In* The biology and management of southern Alberta cottonwoods. *Edited by* S.B. Rood and J.M. Mahoney. University of Lethbridge, Lethbridge, AB. pp. 85–90.

Leopold, Luna B. 1994. A view of the river. Havard University Press, Cambridge, MA. 298 p.

Leopold, Luna B., Wolman, M.G., and Miller, J.P. 1964. Fluvial processes in geomorphology. W.H. Freeman, San Francisco, CA. 511 p.

McBride, J.R., and Strahan, J. 1984. Establishment and survival of woody riparian species on gravel bars of an intermittent stream. Am. Midl. Nat. **112**: 235–245.

McBride, J.R., Sugihara, N., and Nordberg, E. 1988. Growth and survival of three riparian woodland species in relation to simulated water table dynamics. Unpublished report to Pacific Gas and Electric Co., San Ramon, CA.

Mahoney, J.M., and Rood, S.B. 1991. A device for studying the influence of declining water table on poplar growth and survival. Tree Physiol. **8**: 305–314.

Mahoney, J.M., and Rood, S.B. 1992. Response of hybrid poplar to water table declines in different substrates. For. Ecol. Manage. **54**: 141–156.

Moss, E.H. 1938. Longevity of seed and establishment of seedlings in species of *Populus*. Bot. Gaz. **99**: 529–542.

National Research Council. 1992. Restoration of aquatic ecosystems. Committee on the Restoration of Aquatic Ecosystems: Science, Technology and Public Policy. National Academy Press, Washington, DC.

Noble, M.G. 1979. The origin of *Populus deltoides* and *Salix interior* zones on point bars along the Minnesota River. Am. Midl. Nat. **102**: 59–67.

Ohmart, R.D., Deason, W.O., and Burke, C. 1977. A riparian case history: the Colorado River. *In* Importance, preservation and management of riparian habitat: a symposium. *Edited by* R.R. Johnson and D.A. Jones. July 9, 1977, Tucson, AZ. pp. 35–47.

Peterson, E.B., and Peterson, N.M. 1992. Ecology, management and use of aspen and balsam poplar in the Prairie provinces. Special Report No. 1, Forestry Canada, Northern Forestry Centre, Victoria, BC. 252 p.

Reed, J.P. 1995. Factors affecting the genetic structure of black cottonwood populations. M.S. thesis, University of Washington, Seattle, WA. 115 p.

Reichenbacher, F.W. 1984. Ecology and evoluation of southwestern riparian plant communities. Desert Plants, **6**: 15–23.

Reily, P.W., and Johnson, W.C. 1982. The effects of altered hydrologic regime on tree growth along the Missour River in North Dakota. Can. J. Bot. **60**: 2410–2423.

Rood, S.B, and Heinze-Milne, S. 1989. Abrupt riparian forest decline following river damming in southern Alberta. Can. J. Bot. **67**: 1744–1749.

Rood, S.B., and Mahoney, J.M. 1990. Collapse of riparian poplar forests downstream from dams in western prairies: probable causes and prospects for mitigation. Environ. Manage. **14**: 451–464.

Rood, S.B., and Mahoney, J.M. 1996. River damming and riparian cottonwoods along the Marias River, Montana. Rivers, **5**(3): 195–207.

Rood, S.B., Hillman, C., Sanche, T., and Mahoney, J.M. 1994. Clonal reproduction of riparian cottonwoods in southern Alberta. Can. J. Bot. **72**: 1766–1774.

Rood, S.B., Mahoney, J.M., Reid, D.E., and Zilm, L. 1995. Instream flows and the decline of riparian cottonwoods along the St. Mary River, Alberta. Can. J. Bot. **73**: 1250–1260.

Rosgen. D.L. 1994. A classification of natural rivers. Catena, **22**: 169–199.

Sands, A., and Howe, G. 1977. An overview of riparian forests in California: their ecology and conservation. *In* Importance, preservation and management of riparian habitat: a symposium. *Edited by* R.R. Johnson and D.A. Jones. July, 1977, Tucson, AZ. pp. 35–47.

Schreiner, E.J. 1974. *Populus* L. Poplar. *In* Seeds of woody plants in the United States. *Edited by* C.S. Schopmeyer. U.S. Dep. Agric. Agric. Handb. 450. pp. 645–655.

Scott, M.L., Friedman, J.M., and Auble, G.T. 1996*a*. Fluvial process and the establishment of bottomland trees. Geomorphology, **14**: 327–333.

Scott, M.L., Friedman, J.M., and Auble, G.T. 1996*b*. Flood dependency of cottonwood establishment along the Missouri River, Montana, USA. Ecol. Appl. In press.

Segelquist, C.A., Scott, M.L., and Auble, G.T. 1993. Establishment of *Populus deltoides* under simulated alluvial groundwater declines. Am. Midl. Nat. **130**: 274–285.

Shafroth, P.B., Auble, G.T., and Scott, M.L. 1995*a*. Germination and establishment of native plains cottonwood (*Populus deltoides* subsp. *monilifera*) and the exotic Russian olive (*Elaeagnus angustifolia*). Conserv. Biol. **9**: 1169–1175.

Shafroth, P.B., Friedman, J.M., and Ischinger, L.S. 1995*b*. Effects of salinity on establishment of *Populus fremontii* (cottonwood) and *Tamarix ramosissima* (saltcedar) in southwestern United States. Great Basin Nat. **55**: 58–65.

Shanfield, A.N. 1983. Alder, cottonwood and sycamore distribution and regeneration along the Nacimiento River, Californina. *In* California riparian systems. *Edited by* R.E. Warner and K.M. Hendrix. University of California, Davis, CA. pp. 196–202.

Shaw, K. 1991. Ecology of the riverbottom forest on St. Mary River, Lee Creek and Belly River in southwestern Alberta. *In* The biology and management of southern Alberta cottonwoods. *Edited by* S.B. Rood and J.M. Mahoney. University of Lethbridge, Lethbridge, AB. pp. 79–84.

Smit, B. 1988. Selection of flood resistant and susceptible seedlings of *Populus trichocarpa*. Can. J. For. Res. **18**: 271–275.

Smith, S.D., Wellington, A.B., Nachlinger, J.L,. and Fox, C.A. 1991. Functional responses of riparian vegetation to streamflow diversion in the eastern Sierra Nevada. Ecol. Appl. **1**: 89–97.

Snyder, W.D., and Miller, G.C. 1991. Changes in plains cottonwoods along the Arkansas and South Platte Rivers - Eastern Colorado. Prairie Nat. **23**: 165–176.

Stanford, J., and Ward, J.A. 1993. An ecosystem perspective of alluvial rivers: connectivity and the hyporheic corridor. J. North Am. Benthol. Soc. **12**: 48–60.

Stine, S., Gaines, D., and Vorster, P. 1984. Destruction of riparian systems due to water development in the Mono Lake watershed. *In* California riparian systems: ecology, conservation and productive management. *Edited by* R. Warner and C. Hendricks. University of California Press, Berkeley, CA. pp. 528–533.

Stobbs, K., Corbiere, A., Mahoney, J.M., and Rood, S.B. 1991. Influence of rate of water table decline on establishment and survival of hybrid poplar seedlings. *In* The biology and management of southern Alberta cottonwoods. *Edited by* S.B. Rood and J.M. Mahoney. University of Lethbridge, Lethbridge, AB. pp. 47–53.

Strahan, J. 1983. Regeneration of riparian forests of the Central Valley. *In* California riparian systems. *Edited by* R.E. Warner and K.M. Hendrix. University of California, Davis, CA. pp. 58–67.

Stromberg, J.C. 1993. Fremont cottonwood-Goodding willow riparian forests: a review of their ecology, threats, and recovery potential. J. Ariz.-Nev. Acad. Sci. **27**: 97–110.

Stromberg J.C., and Patten, D.T. 1991. Instream flow requirements for cottonwoods at Bishop Creek, Inyo County, California. Rivers, **2**: 1–11.

Stromberg, J.C., Patten, D.T., and Richter, B.D. 1991. Flood flows and dynamics of Sonoran riparian forests. Rivers, **2**: 221–235.

Stromberg, J.C., Richter, B.D., Patten, D.T., and Wolden, L.G. 1993. Response of a Sonoran riparian forest to a 10-yr return flood. Great Basin Nat. **53**: 118–130.

Stromberg, J.C., J. Fry, and D.T. Patten. 1996*a*. Vegetation and geomorphic change after large floods in an alluvial, semi-arid region river. Wetlands. In press.

Stromberg, J.C., Tiller, R. and Richter, B. 1996*b*. Effects of groundwater decline on riparian vegetation of semiarid regions: San Pedro, Arizona. Ecol. Appl. **6**: 113–131.

Szaro, R. 1989. Riparian scrubland and community types of Arizona and New Mexico. Desert Plants, **9**: 1–138.

Szaro, R. 1990. Southwestern riparian plant communities: site characteristics, tree species distributions, and size-class structures. For. Ecol. Manage. **33**: 325–334.

Tyree, M.T., Kolb, K.J., Rood, S.B., and Patino, S. 1994. Vulnerability to drought-induced cavitation of riparian cottonwoods in Alberta: a possible factor in the decline of the ecosystem? Tree Physiol. **14**: 455–466.

U.S. Department of Interior. 1994. The impact of federal programs on wetlands. Vol. II. A report to Congress by the Secretary of Interior, Washington, DC.

VanHaverbeke, D.F. 1990. *Populus deltoides* var. *occidentalis* Rydb.: Plains cottonwood. *In* Silvics of North America: hardwoods. Vol. 2. *Edited by* R.M. Burns and B.H. Honkala. U.S. Dep. Agric. Agric. Handb. 654. pp. 536–543.

Virginillo, M., Mahoney, J.M., and Rood, S.B. 1991. Establishment and survival of poplar seedlings along the Oldman River, Southern Alberta. *In* The biology and management of southern Alberta cottonwoods. *Edited by* S.B. Rood and J.M. Mahoney. University of Lethbridge, Lethbridge, AB. pp. 55–61.

Wilson, R. 1970. Succession in stands of *Populus deltoides* along the Missouri River in SE South Dakota. Am. Midl. Nat. **83**: 330–342.

Zasada, J.C., and Phipps, H.M. 1990. *Populus balsamifera* L.: balsam poplar. *In* Silvics of North America: hardwoods. Vol. 2. *Edited by* R.M. Burns and B.H. Honkala. U.S. Dep. Agric. Agric. Handb. 654. pp. 518–529.

CHAPTER 4
The role of hybridization in the genetic manipulation of *Populus*

R.F. Stettler, L. Zsuffa, and R. Wu

Introduction

Domestication of poplar has been largely a process of interspecific hybridization and clonal selection. It is hybrids that have given poplars their prominence, and it is safe to say that they will be the mainstay of poplar culture in the near future. In fact, in no other planted forest tree have hybrids played such a pivotal role as in *Populus*.

Several reasons help to explain this. Crossability among many of the 30-odd species is high (Zsuffa 1975; Willing and Pryor 1976). The ease with which most poplars can be propagated vegetatively allows the bulk of varied hybrid progenies to be ignored for the benefit of a select few that can be multiplied and perpetuated at will. Rapid juvenile growth has made poplars a tree of choice to farmers who for centuries have planted them for a wide variety of uses (FAO 1980). Emphasis on the early part of their life cycle and on favorable cultural growing conditions — high nutrient levels, control of weeds and herbivores, irrigation — have tended to selectively enhance rapidly growing hybrid clones and to compensate for their shortcomings in sexual reproduction and allocation to defense. Thus, whereas hybrids might face considerable fitness challenges from their parental species in the wild, they can display their commercially attractive features under artificial conditions to full advantage.

We foresee this trend toward poplar domestication, if anything, to be stepped up in the future. A steadily growing human population worldwide, combined with a shrinking landbase and an increased sequestering of forest land for protection and recreation purposes, call for wood-, fiber-, and energy-production

R.F. Stettler and R. Wu. College of Forest Resources, University of Washington, Seattle, WA 98105, USA.
L. Zsuffa. Faculty of Forestry, University of Toronto, Toronto, ON M5S 3B3, Canada.
Correct citation: Stettler, R.F., Zsuffa, L., and Wu, R. 1996. The role of hybridization in the genetic manipulation of *Populus*. *In* Biology of *Populus* and its implications for management and conservation. Part I, Chapter 4. *Edited by* R.F. Stettler, H.D. Bradshaw, Jr., P.E. Heilman, and T.M. Hinckley. NRC Research Press, National Research Council of Canada, Ottawa, ON, Canada. pp. 87–112.

systems with greater output and efficiency. Poplars can meet many of these demands more effectively than any other tree in the temperate zone (Ranney et al. 1987; Abelson 1991). Thus, they are likely to undergo further changes commensurate with the domestication process, such as enhancement of productivity, improvement of the harvest index, and increased dependence on cultural regimes specifically tailored to their needs. Hybridization will be an integral part of this process and will draw on the collective gene pool of several species, just as in the domestication of wheat and potato (Simpson and Conner-Ogorzaly 1986).

There is an additional reason why the mixing and recombining of multiple genomes, or parts thereof, can be argued for more forcefully today. Over the past 10 yr, we have learned much about the molecular genetics of animal and plant development. What has emerged is a much more coherent concept of critical processes around a few unifying themes, namely (*i*) the recurrence of developmental patterns, such as polarities, symmetries, etc. in many unrelated organisms; (*ii*) the disproportionate influence of a subset of genes that regulate patterned development; and (*iii*) the remarkable conservatism of shared DNA-sequences in many such genes across phyletic space (see reviews in Alberts et al. 1989; Meyerowitz 1994). Apparently, there is more of a common denominator that unites species than previously assumed, thus making species from an evolutionary perspective more ad hoc units with a linked past, rather than taxa of unique adaptive merit. The targeted reconstitution of genomes through hybridization, recombination and selection is, therefore, an attractive and efficient way to combine desirable traits in poplar, with the hope of seeing these new genetic constructs functioning well in typical production environments.

In this chapter we show that hybridization is a natural process that can be enhanced and more widely applied, not only to give more desirable production material, but also to grant more insight into the genetic basis of species differences. We will treat the topic in a conceptual fashion rather than exhaustively, illustrating it with selected examples, and refer the reader to earlier reviews on poplar by Muhle Larsen (1970), Zsuffa (1975), Dickmann and Stuart (1983), and Mohrdiek (1983), and on hybridization in other forest trees by Nikles (1993). We will refer to other chapters in this book wherever appropriate throughout the text.

1. Natural hybridization

As in many other wind-pollinated, allogamous forest trees, natural hybridization is common in *Populus* and well documented (see Eckenwalder, Chapter 1, and Whitham et al., Chapter 11). Hybrids are regularly found wherever species of Sections *Aigeiros* and *Tacamahaca* are sympatric, for example in the contact zones of *P. angustifolia*, *P. trichocarpa*, and *P. balsamifera* (Brayshaw 1965).

Similarly, species of section *Populus* hybridize naturally, such as *P. alba* with *P. tremula* (Krembs 1956), or *P. grandidentata* with *P. tremuloides* (Barnes 1961), but not with species from other sections.

The readiness with which poplars and aspens reproduce asexually further facilitates gene flow among species, by perpetuating hybrid plants and prolonging their role as agents of introgression. This may well have been going on for a long time, as there is evidence for ancient introgression between North American cottonwoods and sympatric balsam poplar species (Eckenwalder 1984*a, b, c*), as well among North American aspens (Barnes 1967). Thus, it is not unreasonable to assume that there have been abundant opportunities for gene exchange among sympatric species, even intersectionally. Yet, as further elaborated by Eckenwalder in Chapter 1, the ecological differentiation of sections *Aigeiros* and *Tacamahaca* has been maintained in the face of hybridization.

In spite of the commonness of hybridization among sympatric poplar species, hybrid zones are generally narrow. Foliar morphology and chemotaxonomic analysis of trees in such hybrid zones have typically shown continuous or near-continuous variation patterns (Brayshaw 1965; Rood et al. 1986; Campbell et al. 1993). This conforms to Stebbins' (1950) prediction that backcrosses should be more prevalent than F_1 hybrids in hybrid swarms, given the relative abundance of parental-species pollen. However, backcrossing is sometimes unidirectional, rather than bidirectional, as reported by Keim et al. (1989) for a hybrid swarm in the contact zone of *P. angustifolia* and *P. fremontii* in Utah. Making use of restriction fragment length polymorphisms (RFLPs), these researchers were able to show that the hybrid population consisted of either F_1 hybrids or backcrosses to *P. angustifolia*. No trees were found that could be attributed to crosses among F_1 or between F_1s and *P. fremontii*. Controlled backcrosses to *P. fremontii* led to early death of seedlings (Keim et al. 1989), possibly due to developmental incongruity (Hogenboom 1984).

Hybrid populations of *P. angustifolia* and *P. fremontii* have also been the object of a long series of studies by Whitham and co-workers, focusing on poplar-herbivore interactions (summarized by Whitham et al. in Chapter 11). These studies, too, have documented the restricted extent of hybrid zones and the skewed representation of hybrids therein. But the group has also been able to show a remarkable abundance and diversity of insect and avian fauna in these zones and has pointed to their ecological and evolutionary significance (e.g., Whitham 1989; Floate and Whitham 1993).

How can we distinguish between introgression (the transfer of genes across an incompletely-formed interspecific barrier; Stebbins 1950) and convergent evolution? A specific case may help to illustrate the dilemma. On the east slope of the Cascade Range in Washington State, the Yakima River descends from the cool, moist, mountain climate to the hot, arid environment of the Columbia River basin. The vegetation surrounding the river reflects the climate, changing from montane conifer forest to xeric shrub steppe. This change happens at about 65 km from the river's source and over a relatively short distance. *P. trichocarpa* populations bordering the river show concomitant changes in morphology, anatomy, and physiology. As revealed by common-garden studies and laboratory experiments, these differences have a genetic basis and seem to indicate adaptation in response to divergent selection pressures (Dunlap et al. 1993, 1994, 1995). What remains to be elucidated is, whether the adaptive changes were mediated by introgression, and if so, to what degree. *P. deltoides* along the Columbia River would have been the likely source, and hybrids between the two species have been identified up to 50 km above the two rivers' confluence (R. F. Stettler, unpublished data). However, a zone of 80 km separates the hybrids from the nearest extant *P. trichocarpa* stands upstream, and trees in those stands have the diagnostic characteristics of *P. trichocarpa*. Yet their leaves are smaller, have more stomata and a slightly greener abaxial surface; and their photosynthetic rates are higher at high light intensity and temperature than those of *P. trichocarpa* clones from the cooler sources upriver (Dunlap et al. 1993, 1995). A modest number of mutations would suffice to cause such changes (Gottlieb 1984). On the other hand, these trait modifications may be the result of past gene flow from *P. deltoides*, since they all deviate in the direction of that species. A molecular comparison of potential donor and recipient populations (or at least, of their descendants) remains to be done to help clarify the case. The outcome bears on the inherent capacity of a poplar species to evolve adaptations on a local scale; it also has implications for conservation and restoration.

Spontaneous hybridization is also a common feature wherever poplar cultivars are planted in the vicinity of native populations. This was early detected after the introduction of *P. deltoides* in Europe in the 18th century (Houtzagers 1937) and played no minor role in the spread of euramerican hybrid cultivars, combining *P. deltoides* and *P. nigra* genomes, in poplar culture. By the same token, the widespread use of these and later euramerican hybrids has subsequently compromised the genetic integrity of native *P. nigra* stands. The rapid shrinkage of those stands has further exacerbated the problem and has called for vigorous gene conservation measures in that species (Frison et al. 1995). Here again, molecular techniques have added a welcome tool for the diagnosis and quantification of genetic contamination (e.g., Faivre-Rampant et al. 1995). Other examples of genetic input into native species by introduced cultivars include the diverse hybrids generated by *P. alba* with *P. tremula* (= *P. ×canescens*) in Europe, with *P. adenopoda* (= *P. ×tomentosa*) in Asia, and with *P. tremuloides* and *P. grandidentata* in North America. Similarly, many semifastigate hybrid trees in the vicinity of *P. nigra* var. *italica* (Lombardy poplar) plantings attest to the Darwinian fitness of that male clone in various species' backgrounds.

2. Crossability

Controlled crosses among species of poplars have greatly augmented the hybrid combinations beyond those from spontaneous hybridization. If the objective is to generate a few hybrid clones, almost any species combination will work, even if special measures are occasionally necessary to obtain them. This is one of the remarkable features of the genus *Populus* and has both tempted and rewarded breeders for a long time. However, if the objective of hybridization is to capture a sufficiently inclusive and unbiased genetic representation and to generate large hybrid families — as deemed necessary in a long-term breeding strategy (see Bisoffi and Gullberg, Chapter 6) , then the many impediments to full reproductive compatibility set major constraints. Thus, from a breeding point of view, we have to consider crossability within the context for which crosses are intended.

Each mating, whether inter- or intraspecific, involves the combination of all chromosomes of two sex cells, i.e., two complete haploid genomes. The ultimate objective of this mating, however, may differ: in one case it may be the production of a well-performing F_1 hybrid: in another it may be only the first of several steps to transfer a single resistance gene into a recipient genetic background. A variety of barriers can intercede in this process, especially in intersectional matings. From a breeder's point of view we may distinguish three types:

a) Prefertilization barriers

Prefertilization barriers refer to dysfunctional pollen-pistil interactions that prevent the pollen tube from delivering its gametes to the ovule. They are commonly associated with early capsule or catkin abscission (e.g., Guries and Stettler 1976; Villar et al. 1993).

b) Postzygotic barriers

Postzygotic barriers manifest themselves in the early abortion of the embryo (Melchior and Seitz 1968; Steenackers 1970) resulting in empty seed, or in the disruption of embryo maturation through premature capsule dehiscence (Villar and Gaget-Faurobert 1996).

c) Hybrid inviability

Hybrid inviability becomes evident during the first months or years of seedling growth and manifests itself in the progressive decline or developmental disharmony of the young plant.

Although the genetic mechanisms underlying these barriers are still poorly understood, it is reasonable to assume that they become more complex from (a) to (c). Single mutations are sufficient to disrupt the "dialogue" between pollen and pistil, as shown in numerous studies of sporophytic and gametophytic incompatibility systems (reviewed by de Nettancourt 1984). By contrast, postzygotic barriers may involve several separate functions. Given the recent interest in parent–offspring conflicts in plants and the allocation strategies they engender (reviewed by Shaanker et al. 1988), it may be appropriate to view seed maturation as a competition for limited resources. Accordingly, the strategy of the developing embryo should be to maximize resource acquisition at the expense of the maternal phenotype. The maternal strategy, on the other hand, should be to limit allocation to seed in any given year so as to maximize seed output over a lifetime. Catkin, capsule, and placenta contain strictly the maternal genome and may, thus, set constraints for their response to signals from the developing embryo, half of whose genome represents the male parent. The schedule of capsule development may, thus, be partly disconnected from that of embryo development, although the two will coincide in a normal, compatible mating. In interspecific crosses, especially those that combine species with different seed maturation schedules, the asynchrony between the two processes would present a major barrier and, as a consequence, an effective isolating mechanism.

In both pre- and postzygotic barriers one would expect to find natural variation among individuals, especially in female parents — which is indeed the case, as attested to by most breeders. Hybrid inviability likely reflects genomic disharmony, or "incongruity" at many loci (Hogenboom 1984), as an expression of evolutionary divergence, and involving many processes in growth and development.

Crossing relationships between members of any pair of poplar species cover the range from complete compatibility to the combined action of multiple barriers (a)–(c). Intrasectional crosses tend to be fully fertile (Zsuffa 1975; Willing and Pryor 1976). However, the direction of the cross may matter: *P. deltoides* and *P. nigra* genomes combine well if *P. nigra* is used as the male parent. The reciprocal cross seems hampered by postzygotic barriers (Melchior and Seitz 1968; Steenackers 1970; Villar and Gaget-Faurobert 1996), possibly due to nuclear/cytoplasmic incompatibility. *P. nigra*, as the only member of *Aigeiros* poplars, has a chloroplast genome similar to *P. alba*, or an ancestor thereof (Smith and Sytsma 1990), which may render the cytoplasm a dysfunctional host for the hybrid nucleus. There exist, however, other hybrids such as *P. nigra* × *P. trichocarpa and P. nigra* × *P. laurifolia* (Dickmann and Stuart 1983) which show that *P. nigra* chloroplasts are capable of functioning with nuclear genomes from certain other species.

Intersectional crosses between *Tacamahaca* and *Aigeiros* are generally rated as compatible in most combinations (Zsuffa 1975; Willing and Pryor 1976) and have given rise to many vigorous production clones (see listings in Dickmann and Stuart 1983, and Padro 1992). However, fertility in these crosses tends to be well below intraspecific matings. Postzygotic barriers, especially premature capsule dehiscence, may further hamper seed production, as, e.g., in *P. tricho-carpa* × *P. deltoides* crosses (but typically not in their reciprocals; B. Stanton, personal communication). Embryo rescue has proved successful, if tedious, to overcome this problem (Stanton and Villar, Chapter 5). Given the generally favorable results from combining *Tacamahaca* and *Aigeiros* genomes, or parts thereof, it seems worth the effort to improve crossability between these sections to a fertility level that approaches intraspecific crosses. Crossing relationships between *Leucoides* and both *Aigeiros* and *Tacamahaca* are similar to those between the latter two (Willing and Pryor 1976).

Intersectional crosses that involve section *Populus* are generally rated as incompatible (Zsuffa 1975; Willing and Pryor 1976) largely due to type (a) and (c) barriers. The former can be overcome by pollen or stigmatic treatment (reviewed by Willing and Pryor 1976; Stettler and Ager 1984; Villar and Gaget-Faurobert 1996), whereas the latter constitute a more formidable impediment. Viable hybrids from manipulated crosses have been reported (e.g., von Wettstein 1933; Sekawin 1963*a*; Stettler 1968; Zsuffa 1968; Knox et al. 1972; Willing and Pryor 1976) but their numbers have been sporadic. This is not surprising and reflects the evolutionary divergence of the section *Populus* poplars from those in even the two closest sections, *Aigeiros* and *Tacamahaca* (see Eckenwalder, Chapter 1). Some crosses between these three sections have also succeeded when hybrids were used as parents (Zsuffa 1975; Lemoine 1988). But all in all, it seems prudent to view *Populus* members as potential donors, or recipients, of only individual genes or small gene blocks, rather than entire genomes, to/from poplars of other sections, and to consider other methods than conventional crossing, e.g., genetic transformation, as promising avenues for gene transfer (see Han et al., Chapter 9). Little is known about crossing relationships of *P. mexicana*, the sole species in section *Abaso* (Eckenwalder, Chapter 1).

We conclude this part with the cautionary remark that crossability results tend to be underreported and underdocumented; most breeders prefer to report successful efforts and often attribute failed attempts to inadequate techniques, unfavorable material or unsuitable environmental conditions.

3. Hybridization for increased productivity

The three main motives for hybridization in this context are to (*i*) combine desirable traits from different species (*ii*) capture heterosis, or hybrid vigor, and (*iii*) obtain increased developmental homeostasis, i.e., greater phenotypic stability in varied environments. The three cannot always be separated, as will become obvious in the sections below.

a) Combination of desirable traits

A number of trait combinations, arrived at by hybridization, may serve as examples.

Rootability: Rootability of stem cuttings, an important trait in plantation culture, varies widely in *P. deltoides* and can be significantly improved in crosses with the well-rooting *Tacamahaca* poplars, *P. balsamifera, P. maximowiczii, P. simonii,* and *P. trichocarpa,* as well as with *P. nigra* (Zsuffa 1975; Dickmann and Stuart 1983; Wu et al. 1992). In the poplars of section *Populus,* the only species to propagate with stem cuttings, *P. alba,* has served to introduce rooting genes into advanced generation hybrids with *P. grandidentata, P. tremuloides, P. tremula, P. davidiana,* and *P. sieboldii* (Heimburger 1968; Zsuffa 1979).

Stem growth: The proportionate allocation to height relative to diameter in juvenile stem growth is greater in *P. trichocarpa* than in *P. deltoides.* Four-year-old F_1 hybrids, tested in the native environment of *P. trichocarpa,* matched the height of the latter but significantly exceeded its radial growth (Stettler et al. 1988). The differential contribution by the two parental species to the components of stem growth has been elucidated at the molecular genetic level (Bradshaw and Stettler 1995).

Branching: Sylleptic branches, derived from nondormant buds on current year's shoots, are common in *P. trichocarpa* but rare or absent in *P. deltoides.* Their numbers vary in F_1, F_2, and B_1 hybrids between the two species and are positively correlated with radial growth (Ceulemans et al. 1989; Wu and Stettler 1994; Bradshaw and Stettler 1995).

Leaf traits: Leaves of *P. trichocarpa* are ovate, bifacial with abaxial stomata, have relatively large epidermal cells, and are held more or less horizontally by short, terete petioles. Those of *P. deltoides* are more deltoid, unifacial, amphistomatous, have a larger number of relatively small epidermal cells, and are held at tilted angles by long, flattened petioles. F_1 hybrid leaves are approximately twice the size of either species, by combining larger cell number with large cell size (Ridge et al. 1986), are amphistomatous but bifacial (Hinckley et al. 1989) and are held by intermediate petioles in a slightly hanging but untilted orientation (Isebrands et al. 1988). The combined features have shown to be correlated with higher productivity in the F_1 hybrids at the individual tree and stand level in short rotation plantation regimes (Ceulemans et al. 1992; Hinckley et al. 1989).

Phenology: Combining *P. trichocarpa* from the Pacific Northwest (44–49°N) with *P. deltoides* from the southern United States (30–33°N) has significantly extended leaf area duration in F_1 hybrids grown at 46–47°N (Ceulemans et al. 1992) and thereby prolonged the effective production period (Isebrands et al. 1988).

Disease resistance: Poplars vary in their disease resistance at the level of the species, the geographic origin and the individual (see Newcombe, Chapter 11). Resistance genes to *Marssonina*, *Septoria*, and *Dothichiza populea* are more likely to be found in *P. deltoides*, whereas those against *Xanthomonas populi*, in *P. nigra* (Steenackers 1972). However, resistance may not necessarily be expressed in F_1 hybrids between susceptible and resistant parents because it may be under the control of recessive gene(s), as shown in many crop plants (e.g., Kolmer et al. 1994). The continual evolution of new pathotypes in intensively cultivated *Populus* (Pinon 1995) will demand the recruitment of resistance or tolerance genes from many sources.

b) Heterosis

Hybrid vigor has been early observed in poplar and has been one of the driving forces in poplar breeding (Muhle Larsen 1970). However, the degree to which hybrids are actually superior to their nonhybrid counterparts, and the underlying cause for such superiority have been less clear. Heterosis, as originally coined by Shull in 1914 (see Sinha and Khanna 1975), refers to the genetic mechanisms causing the superiority of offspring over the midparent value (average between the two parents) or over the better parent, as often observed in F_1 hybrids. In common usage, however, heterosis has become synonymous with the phenotypic expression, i.e., the product of this mechanism (Sinha and Khanna 1975). Heterosis generally refers to increased size or productiveness but can also be used to describe superiority in other traits (Allard 1960). In crosses between parents from inbred lines, heterosis is relatively easy to quantify since (*i*) the parental genotypes are likely to be a representative sample of the parental generation, (*ii*) within-line parent-offspring correlations are high, often obviating the need for control-crosses, and (*iii*) the F_1 generation is genetically uniform (Allard 1960).

In the case of poplar, as in many other forest trees, high average heterozygosity (Jelinski and Cheliak 1992; Weber and Stettler 1981) violates every one of these conditions and makes the estimation of heterosis more complicated. Valid estimates require a sufficient number of parentals to be recruited and mated both within and between species, and their progenies to be tested in sufficient numbers and locations, replicated in space and time. Geographic variation within species adds a further layer of complexity in experimental design on the choice of parents, cross combinations, and test locations. Field-test designs demand extra precaution to minimize competitive interactions among heterotic hybrids, parentals and intraspecific progenies. Water and nutrient supply should allow genotypes to fully express their genetic potential. Few, if any, tests satisfy all these requirements. The most common tests are operationally oriented and tend to compare the top clones of hybrid families with one or both of the parents and/or with selected clones of the native parent species. Finally, as further tive genetic models to the prediction of hybrid performance in interspecific crosses, may be flawed by the violation of such basic assumptions as linkage equilibrium and random mating. As a consequence, all statistical estimates of heterosis in poplar hybrids must be viewed as approximations with greater or lesser error margins, depending on the experimental design from which they were derived.

Rather than passing judgment on the accuracy of various estimates reported, we conclude that there is sufficient evidence for heterotic hybrids in specific species combinations (see listings in Zsuffa 1975 and Padro 1992) to explain why hybridization has played such an important role in poplar breeding. At the same time, we encourage breeders to establish adequate tests to validate earlier studies and to further refine the quantification of the phenomenon. More importantly, we urge researchers to study *why* hybrids are superior to parental species when they are, i.e., to elucidate the physiological mechanisms of heterosis and their underlying genetics (see below).

Given the abundance of superior hybrid clones in breeding programs, their lack of prominence in natural hybrid swarms has puzzled students of natural populations. Campbell et al. (1993) addressed this question by studying a wide range of naturally occurring hybrids in the overlap zone of *P. angustifolia*, *P. balsamifera*, and *P. deltoides* in southwestern Alberta. Cuttings taken from the material and grown for three years in a common nursery under favorable growing conditions, failed to show heterosis. The same regime favored the expression of growth vigor in comparable but artificial F_1 hybrids derived from different sources. The authors attributed this discrepancy to the difference in inherent genetic distance in the two materials. According to their argument, the continual introgression over generations in the Alberta material had eroded the original genetic differences between species, thereby reducing the level of heterozygosity in the natural hybrids (Campbell et al. 1993). The most recent molecular evidence that is pertinent to this case questions this interpretation, without rejecting it outright. Bradshaw and Stettler (1995) in the study of a *P. trichocarpa* × *P. deltoides* F_2 generation — which is further described below — found no correlation between overall marker heterozygosity and 2-year growth. It was heterozygosity at a finite set of specific markers that seemed to account for superior growth. Another hypothesis for the absence of heterotic hybrids in natural stands recognizes the stressful conditions prevailing in these peripheral zones. Natural selection may put a premium on survival rather than growth. Additional pressure by herbivores, as shown by Whitham et al. (Chapter 11), may divert allocation of photosynthates to defense. This is a rich area for research and deserves more attention. The rapid decline of natural populations and hybrid swarms due to human activity gives it greater urgency.

c) The physiology and genetics of heterosis

The past 10 years have given us important insights into the physiological basis of heterosis, its relationship to developmental homeostasis, and its underlying genetics. Since much of this is covered in greater detail by Bradshaw in Chapter 8, as well as in Part II of this book, we will confine ourselves here to a summary treatment.

Most of the information on the components of hybrid vigor has been generated through a comprehensive research program at the University of Washington and Washington State University, which has focused on the study of *P. trichocarpa* × *P. deltoides* hybrids and the two parental species. In a series of parallel studies, conducted in replicated field trials, and under greenhouse and growth-chamber conditions, hybrids and parentals were systematically analyzed at the level of leaf, branch, and the whole tree. In a first summary of those studies, Hinckley et al. (1989) described nine important physiological, morphological, and ana-tomical factors that helped explain the superior productivity of hybrids. They included (*i*) the more efficient ability of stomata to respond to decreasing soil moisture and/or increasing evaporative demand; (*ii*) the coupling between leaf morphological and anatomical features and the response of stomata to drought; (*iii*) the larger leaf size, combining large cells with large cell numbers; (*iv*) osmotic maintenance of foliar cell turgor and cell wall extensibility; (*v*) the favorable combination of leaf diurnal growth patterns; (*vi*) the interaction be-tween crown architecture, leaf phenology, and distribution, photosynthesis, and carbon allocation; (*vii*) the nature and rate of root development as a function of carbon allocation; (*viii*) root structure and function in the response to either anoxia or drought stress; and (*ix*) the role of gibberellic acid in heterosis. In an important follow up on the impact of soil–water loss on whole plants, Braatne et al. (1992) further found that *P. trichocarpa* × *P. deltoides* hybrids were better able to maintain the balance between growth and water loss than either parent species. Increased productivity of the hybrids was, thus, in part a function of their greater developmental stability.

In 1991, these studies were further refined by adopting a three-generation pedi-gree, including *P. trichocarpa* and *P. deltoides* parentals, F_1 hybrids and their F_2 offsprings, as study material. All genotypes were established in a replicated field plantation in Puyallup. In 1993, this material was augmented and planted, with industrial support, in two additional field trials in two contrasting environ-ments, east and west of the Cascade Range. The main purpose of the three trials was to provide data for the mapping of quantitative trait loci (QTLs) of important commercial traits, and to study their expression in different environments. Re-sults from these studies have shown that (*i*) contrary to the polygenic model, individual QTLs account for as much as 21–55% of the genetic variation in such quantitative traits as height, basal area, or spring phenology, (*ii*) favorable alleles are contributed by both parental species, but differently in the various traits, (*iii*) dominance, rather than overdominance,[1] explains heterosis in these hybrids, and

---

[1]Under *overdominance*, the phenotype of the heterozygote exceeds that of either homozygote, rather than equaling the better of the two (*dominance*).

(*iv*) clustering of QTLs on the linkage map indicates functional correlations, possibly pleiotropy,[2] between such traits as total leaf area on sylleptic branches and basal area growth (Bradshaw and Stettler 1995).

In summary, heterosis in this pedigree can be attributed to complementation between the two parental species in many traits, the interaction of these traits, and the compounding effects these interactions have on the total plant phenotype and its developmental stability over time. Hopefully, similar analyses will be performed of other hybrid combinations to test the generality of this model. It will be especially interesting to verify the absence of any locus-specific over-dominance, since this would question the efficacy of recurrent F_1 breeding in maximizing genetic gains. A recent study by Xiao et al. (1995) showed dominance to be the major genetic basis of heterosis in rice, thus differing from maize where overdominance seems to be the prominent factor explaining it (Stuber et al. 1992). Since rice and maize are closely enough related to share many genes of common ancestry (Ahn and Tanksley 1993), the discrepancy in mechanisms has been called to attention. One explanation offered is that the overdominant gene action in maize may actually be pseudo-overdominance, i.e., due to the occurrence of favorable dominant and unfavorable recessive alleles in coupling at closely linked loci (Xiao et al. 1995). Fine mapping will help to resolve this important question.

d) Recurrent F₁-breeding

Few breeding programs stop at the first F_1 generation — although in regions of abundant natural poplar material there is a greater opportunity to restart by making crosses with newly recruited parents from the wild. However, with the prospect of incremental gains to be made through successive breeding/selection cycles, the two major options are (*i*) recurrent F_1 breeding, in which the breeding populations of two species are systematically refined through within-species breeding/selection in parallel with hybrid breeding/selection (Fig. 1); and (*ii*) advanced-generation hybridization, either through successive F_2–F_n breeding, through backcrossing to one of the parental species, or through multiple-species hybridization (Fig. 2).

Recurrent F_1 breeding is aimed at capturing both additive and nonadditive genetic variance. It maximizes heterozygosity and, thus, captures heterosis due to dominance and overdominance. On the other hand it cannot take advantage of favorable recessives, present at low frequency in one of the parental populations which may be important, e.g., in disease resistance. It sets progressively moving standards against which to measure genetic gain through hybridization. Two pathways can be distinguished: (*i*) recurrent selection for general combining

---

[2]*Pleiotropy* refers to a single gene having a number of distinct, seemingly unrelated phenotypic effects.

Fig. 1. The main steps in two different recurrent F_1 breeding schemes.

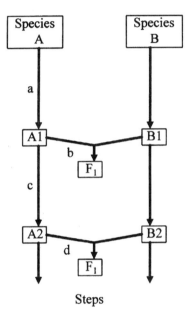

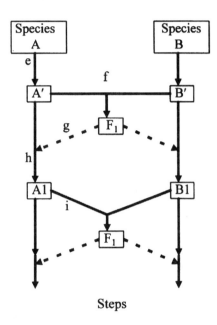

Steps

a. Mating and progeny testing within species

b. Selection of parents with high GCA and inter-specific crossing

c. Intraspecific mating among parents with high GCA and testing of progenies

d. Selection of progenies to Serve as parents for next round of interspecific crossing

e. Phenotypic selection of parents

f. Interspecific crossing of selected parents

g. Testing of F_1 families and identification of parents with high GHA

h. Intraspecific mating among parents with high GHA

i. Selection of progenies to serve as parents for next round of interspecific crossing

Fig. 2. Possible pathways of advanced generation hybridization.

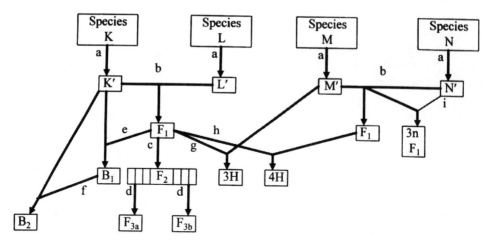

Pathways:

a. Phenotypic selection of parents

b. Interspecific crossing among selected parents

c. Selection and mating among F_1 individuals

d. Selection and mating among F_2 individuals, possible sub-lining

e. Selection and backcrossing to parental species

f. Selection and continued back-crossing to parental species

g. Selection and crossing with third species, resulting in 3-way hybrids

h. Crossing among selected F_1 to generate 4-way hybrids

i. Unreduced gametes from parent N' result in triploid F_1 hybrids (MNN)

ability (RS-GCA[3]; Fig. 1a–d), and (*ii*) reciprocal recurrent selection for general hybridizing ability (RRS-GHA; Fig. 1e–i; after Nikles 1993).

RS-GCA is initiated with within-species breeding in both species to identify parents with high GCA; these are then crossed to give rise to F_1 families, within which cloned testing will identify selections for mass propagation. At the same

---

[3]GCA vs. GHA: The relative ability of an individual to transmit genetic superiority to its offspring when crossed with a number of individuals of the same species (GCA), vs. with those of another species (GHA).

time, intraspecific matings among high GCA parents will produce the next improved generation from which to recruit parents for the second F_1 hybrids. By contrast, RRS-GHA is initiated by crossing nonprogeny-tested (but possibly clonally tested) parents from the two species (Fig. 1e). Family testing of F_1 hybrids will identify parents with high GHA (Nikles 1993). Intraspecific mating among these parents will give rise to the next improved generation in each species from which parents will be chosen for the second F_1 crosses. Gains through RRS-GHA will be greater than through RS-GCA, the greater the ratio of nonadditive to additive genetic variance, and the lower the GCA/GHA correlation. But foundation populations for RRS-GHA need to be large if crossabilities and GCA/GHA correlations are low (Nikles 1993). A detailed example of the RRS-GHA pathway is given by Bisoffi and Gullberg in Fig. 2 of Chapter 6, describing the Italian breeding program for *P. deltoides* and *P. nigra*. Both pathways involve work with two species as well as hybrids and are thus demanding of resources. In return they offer three different types of increasingly refined materials which may be useful not only in production silviculture, but also in conservation.

e) Advanced-generation hybridization

i) F_2–F_n breeding and backcrossing

In contrast to F_1 breeding, where species-specific linkages are left intact, this group of pathways (Fig. 2a–f) emphasizes recombination and maximizes genetic variance. It captures transgressive variation which is especially attractive in a clonable tree. It also allows for the combination of heterozygosity at some loci with homozygosity at others and, thus, to the maximization of hybrid superiority by tapping dominance and overdominance without eliminating or masking favorable recessives. However, for this to be done efficiently, molecular analysis of an adequate number of progenies is necessary (see Bradshaw, Chapter 8). Backcrossing permits introgression of selected genomic portions of a donor species into the genetic background of a recipient species. This procedure, too, has become remarkably streamlined through molecular tools, reducing the process of eliminating the undesirable bulk of the donor genome from 8 to 3 generations (Tanksley et al. 1989).

Backcrossing is especially attractive if a productive native species is available, which can be enhanced in targeted suboptimal traits while keeping the fraction of interspecific recombinants within bounds. A good example would be *P. deltoides* in the southern United States which offers commercially desirable qualities in most traits but frequently falls short in rooting (see Dickmann and Stuart 1983). This deficiency could be easily remedied by a transfer of rooting genes from *P. trichocarpa*, which could be quickly separated from portions of the genome that contribute undesirable alleles for phenology, water use, and disease resistance.

Both pathways can be pursued via inbreeding (full-sib, half-sib matings, or backcrossing to original parents) or via outcrossing. The former will tend to result in some inbreeding depression (ID[4]) until deleterious recessives have been purged. ID has been recently estimated in 30 families of advanced generation hybrids between *P. trichocarpa* and *P. deltoides* (R.F. Stettler and R. Wu, unpublished results). F_2 families derived from full-sib crosses (FS) showed average ID in 2-yr height and basal area of 12–19% as compared to 2% in half-sib crosses (HS, expressed as a percentage of unrelated crosses). More severe reductions were incurred by backcrossing to one of the original parents (average ID in height: 26%; in basal area: 49%), as compared to unrelated trees of the recurrent species. Additional direct evidence for deleterious recessives in this material has been presented by Bradshaw and Stettler (1994). Inbreeding within, and subsequent mating among, sublines will tend to fix favorable alleles and will help to arrive at new composites that combine useful traits, although the time required will be substantial. Further delays may be incurred because of later onset of reproductive age or a higher incidence of abnormal flowering in advanced-generation hybrids.

ii) Multi-species hybrids

Given the wide crossability among species and the reasonable, if reduced, fertility many hybrids display, multispecies hybrids can be produced in various combinations and have a legitimate role to play in the development of high productivity cultivars. They are one of several options in the assembly of desirable traits from different sources (Fig. 2g–h).

If a *P. trichocarpa* × *P. maximowiczii* (or TM) hybrid is crossed with a third species, say *P. deltoides* (or D), the resulting trihybrids will carry 50% of the D genome and, on average, 25% each of the T and M genomes. The same three genomes can be assembled in two additional ways, giving each of the T and M genomes a 50% representation. The three different dosages of the three genomes may be used to more closely tailor hybrids to given environments. For example, in the Pacific Northwest, cool, maritime locations close to the coast may favor hybrids with 50% M, whereas locations on the eastside of Vancouver Island may be more suitable for hybrids with 50% T, or again warmer locations in the Willamette Valley of Oregon may call for higher D representation. But even within a particular cross combination, e.g., TM × D, the relative proportion of T and M genomes will vary from individual to individual because of variable degrees of recombination. This offers a second level of variation to the selector of production hybrids. That variation may also be exploited in further backcrossing. For example, if in that hypothetical cross the T genome were to serve primarily as source of rooting genes, whereas the M genome as a source of

---

[4]ID = reduction in the performance of offspring from related, as compared to unrelated, matings.

resistance to *Venturia*, selecting parents with high T percentage would be more efficient, since the former trait would likely involve more genes than the latter. Molecular markers may assist in identifying the appropriate individuals to be mated.

In the development of multiple hybrids, breeders will be guided by the same trends that apply to dihybrids, described under Section 2, Crossabilty. That is, they can expect productive combinations among species within sections and between sections *Aigeiros* and *Tacamahaca*. But to the extent that hybrids often are more successful parents for difficult crosses than pure species (Zsuffa 1975), intersectional transfer of genes from *Populus*, *Leucoides*, and *Turanga* sections to the other three may be feasible, too, and deserves further exploration.

iii) Polyploid hybrids

Ever since Nilsson-Ehle (1936) discovered a triploid[5] *P. tremula*, there has been a continual interest in exploiting polyploidy in poplar improvement (reviewed in Muhle Larsen 1970). Aspens have played a prominent role in the study of polyploidy and triploids have shown the greatest promise, e.g., in *P. tremula* (Johnsson 1956) and in *P. tremuloides* (van Buijtenen 1958; Einspahr 1965). Attractive features of triploids have been their greater volume growth and their longer and wider fibers (Johnsson 1956; van Buijtenen 1958). One of the currently widely known triploid aspens is the cultivar "Astria," a *P. tremula* × *P. tremuloides* hybrid developed by Weisgerber (1983) and commercially mass propagated through micropropagation.

Hybridity adds one more dimension to polyploidy and, at the same time, makes use of it in regulating genomic dosages. Triploidy arises from the involvement of diploid male or female gametes in a mating (Fig. 2i), either as a result of nondisjunction during meiosis (e.g., Muhle Larsen 1963; Bradshaw and Stettler 1993) or from the use of a tetraploid parent (Sekawin 1963*b*). The hybrid phenotype in triploids will be biased toward the parent from which it carries two genomes. For example, crosses using a mixture of diploid and haploid pollen from the euramerican hybrid "Heidemij" on *P. deltoides* resulted in many triploid clones with high rooting capacity, obviously derived from the male parent (Muhle Larsen 1970). Conversely, triploid *P. trichocarpa* × *P. deltoides* hybrids, as shown by RFLPs, have consistently revealed the genomic configuration of TTD and displayed leaf-cell and fall-phenology characteristics skewed toward values typical of the female parent (Bradshaw and Stettler 1993). Thus, triploid hybrids approximate the genomic dosage of a backcross (67% vs. 75%), but

---

[5]*Triploids* have three, *tetraploids* four, haploid sets of chromosomes, *allotetraploids* two each from two different species. *Aneuploids* have other than exact multiples of chromosome sets, e.g., $3n - 1$.

requiring one less generation to arrive at it. Their genomic constitution can again play a role in the choice of site for which they are selected.

Triploid hybrids may be more common than anticipated. Of 15 female *P. tricho-carpa* parents used in crosses with *P. deltoides*, 10 produced tri/aneuploid hybrids, several of them in high proportions (up to 100%) (Bradshaw and Stettler 1993). It has been hypothesized that intra-capsule competition may favor TTD embryos over their TD counterparts, since the former benefit more from the more rapid maturation rate conferred by the *P. trichocarpa* parent (R.F. Stettler, unpublished results). It may be worth screening other hybrid combinations for the occurrence of triploids (conveniently done through flow cytometry or molecular markers), especially in cases where hybrid phenotypes are skewed toward one parent. It may not only help explain the sterility of such hybrids but, in fact, may take advantage of such material for deployment, where genetic contamination of native populations should be minimized.

In contrast to other genera where allotetraplody is a pathway to reconstitute fertile species from hybridization (Stebbins 1950), tetraploidy has shown little promise in *Populus*. This may reflect the fact that the universal chromosome number of *Populus*, $2n = 38$, represents already a derived, higher-than-diploid number (Raven 1975), which is also borne out by the duplication of a significant proportion (11%) of loci in the genetic map (Bradshaw et al. 1994).

4. Hybridization as analytical tool

The Mendelian approach to genetic analysis, namely to cross distinctive parents that are true breeding and to analyze their transmission of traits in their F_2 offspring, can be extended to the species level. This is particularly attractive in *Populus*, where many species are crossable, yet distinctive, and where F_1 fertility is sufficient to offer adequate-size F_2 families. An immediate benefit from this approach may be that new recombinant material becomes available that will directly benefit an ongoing breeding program (see Section 3e, above). A second benefit — especially if supported through QTL analysis — will be the information generated on the nature (e.g., mono-, oligo-, polygenic) and direction (dominance relationships) of trait transmission from the two parents and its consequences on breeding strategy. A third advantage of this approach is that through heightening the differences between the two parent species, new genes come to interact that "jolt" traditional trait states by eliciting new developmental expressions. We may view it as a genetic perturbation analysis, in which previously canalized growth patterns — through adaptation, but also through chance — become altered through the introduction of foreign "probes." From this analysis we may, at the same time, recognize certain relationships among growth components to be stable across genetic backgrounds of different species.

Fig. 3. Distribution of phenotypes for 1st (1991) and 2nd year (1992) height increment (HTI), basal area increment (BAI), and diameter/height ratio (DHR) *for Populus trichocarpa × P. deltoides* (T × D) hybrids F_2 Family 331 (black bars) and backcross (TD × D) Family 342 (cross-hatched bars). Means for the two original parentals (T and D), the F_1 parents, and the two families are indicated (from Wu and Stettler 1994).

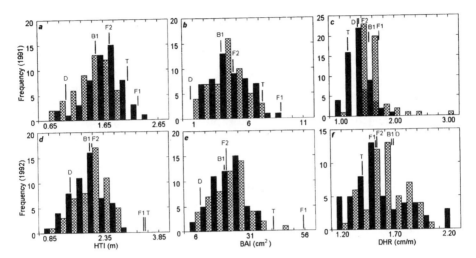

An example of this approach has been the abovementioned study of a 3-generation pedigree plantation, combining *P. deltoides* and *P. trichocarpa* parents and their F_1 and F_2 offspring, as well as their backcross generation to the *P. deltoides* parent. The pair of species chosen is well suited for such a study as it shows contrasts at many levels, from the individual leaf to the canopy architecture of the whole tree. Quantitative-genetic analysis of leaf, branch, and stem traits over the first three years of growth led to the following major findings (Wu and Stettler 1994, 1996): (*i*) in growth-related traits, the mean of the F_1 was equal to or greater than the better parent (T), in other traits intermediate between parents; means of F_2 and B_1 were intermediate, the latter more skewed towards the recurrent parent (D); (*ii*) in the F_2, all traits but leaf pigmentation showed continuous, and in some traits trangressive, variation (see Fig. 3); (*iii*) broad-sense heritabilities (H^2) ranged from 0.50 to 0.80 but varied among branch types (sylleptics vs. proleptics) and crown positions (upper vs. lower) and time; in year 2, H^2 values were higher for sylleptics than proleptics, and for upper than lower crown positions; but by year 3, positions within crown became even more important; (*iv*) genetic correlations between branch/leaf traits and growth indicated that sylleptics are important in early radial growth, whereas proleptics are more associated with height growth, but by year 3 also with radial growth.

Equally important to the general relationships between canopy components and growth revealed by the study, was the opportunity to identify recombinant phenotypes of unusual, or contrasting, trait combinations. For example, since the T parent contributes more enhancing alleles to height than radial growth, whereas the reverse is true for the D parent (Bradshaw and Stettler 1995), extremely slender individuals were found in the F_2 next to stocky ones, both breaking conventional height/diameter allometrics (Niklas 1992). Another set of contrasting recombinants showed how the same stem dry weight could be produced by remarkably different leaf areas and branch dry weights (Wu and Stettler 1996). Such materials help "dissect" growth into its components and may give direction to breeding and selection. For example, the existence of recombinants with lower branch/stem dry weight ratio (i.e., greater harvest index) begs the question, how these genotypes function, and in which traits they are more efficient and at what cost. Clonal propagation of such contrasting types in replicated experiments will offer physiologists a highly informative study material to help elucidate the question. This, in turn, may provide some concrete models for the formulation of ideotypes, i.e., idealized phenotypes, as targets for future poplar breeding (Dickmann 1985; see also Dickmann and Keathley in Chap. 19). The somewhat exaggerated contrasts of interspecific recombinants may, thus, alert researchers to critical growth components that may vary more subtly within species. Subsequent, more directed analysis with more refined tools, can then deal with that variation at the intraspecific level. However, to the extent that intergenomic interactions at certain loci may create new variation, the transferability from inter- to intraspecific information will be limited.

5. Conclusions and implications

From the foregoing treatment we have distilled a number of statements that may serve as guidelines for the practice of intensive poplar culture and for efforts aimed at conservation and restoration. We have also identified a number of research questions that, in our view, deserve high priority.

- Hybridization among species will continue to be a major avenue to the genetic improvement of poplars. It can tap sources of variation that are unavailable within species and can offer them in recombined form in new cultivars.

- Our understanding of the role of hybridization will greatly benefit from more balanced mating designs, larger numbers of families, adequately controlled testing, and closer comparison with intraspecific breeding.

- Recurrent F_1 breeding should be viewed as only one of several options of hybridization. Increasing emphasis will be placed on advanced-generation hybridization and on the introgression of selected genes into the genetic background of individual species.

- Molecular tools will become a routine adjunct to breeding and will greatly increase its precision and efficiency.

- Hybridization, more than any other genetic improvement, depends on international collaboration in breeding and testing, long-term maintenance of collections, and data sharing and, as a consequence, on joint protocols and program continuity.

- Natural populations of indigenous species are the ultimate source of genetic variation. They need to be conserved, whenever possible restored, and protected from genetic contamination through nonnative germ plasm. They should be complemented by a network of *ex-situ* collections in designated arboreta.

- Special attention should be given to the preservation of natural hybrid zones of sympatric species. They may also serve as sources of material for the restoration of similar populations that have been degraded or eliminated through human activity.

- Wherever production plantations are proximate to natural populations, preference should be given to the use of sterile hybrids (triploids, or genetically engineered sterile clones) so as to minimize genetic contamination.

- Deliberate but carefully monitored experiments should be conducted to test sterile hybrid clones as part of restoration projects, in mixture with native material. Their greater adaptability and more rapid growth may claim the site more readily and may provide protection for other vegetation.

- With each additional generation of domestication we foresee a diminishing contamination pressure on native populations since "hybrid" seed will be increasingly dependent on cultural regimes and inviable in the wild.

Questions deserving study

- What are the correlations between clonal performance and components of parental performance such as general combining ability (GCA), general hybridizing ability (GHA), and specific hybridizing ability[6] (SHA)? Can these be used to accelerate progress in breeding via testcrosses or other shortcuts? Are there geographic trends in these correlations for particular species combinations in particular environments?

- What is the physiological and genetic basis of heterosis and why do we not observe it in natural hybrids?

---

[6]SHA = the relative ability of an individual to transmit genetic superiority to its offspring when crossed with a specific individual from another species.

- Are there a number of gene loci that account for a disproportionate share of phenotypic variation in commercially important traits? Which among them are found in more than one hybrid combination and in a range of environments?

- What is the genetic basis for nondisjunction in male/female meiosis and how can it be enhanced, or better exploited, for the production of triploid hybrids?

Acknowledgements

The senior author expresses his gratitude to Toby Bradshaw for stimulating discussions throughout the preparation of this chapter, to Marc Villar for review of an earlier draft, and to the U.S. Department of Energy Biofuels Feedstock Development Program for sustained support of his research since 1978.

References

Abelson, P. 1991. Improved yields of biomass. Science (Washington, DC), **252**: 1469.

Ahn, S., and Tanksley, S.D. 1993. Comparative linkage maps of the rice and maize genomes. Proc. Natl. Acad. Sci. U.S.A. **90**: 7980–7984.

Alberts, B., Bray, D., Lewis, J., Raff, M., Roberts, K., and Watson, J.D. 1989. Molecular biology of the cell. 2nd ed. Garland Publishing Inc., New York, NY.

Allard, R.W. 1960. Principles of plant breeding. John Wiley & Sons, Inc., New York, London.

Barnes, B.V. 1961. Hybrid aspens in the Lower Peninsula of Michigan. Rhodora, **63**: 311–324.

Barnes, B.V. 1967. Indications of possible mid-Cenozoic hybridization in the aspens of the Columbia Plateau. Rhodora, **69**: 70–81.

Braatne, J.H., Hinckley, T.M., and Stettler, R.F. 1992 Influence of soil water on the physiological and morphological components of plant water balance in *Populus trichocarpa*, *P. deltoides* and their F$_1$ hybrids. Tree Physiol. **11**: 325–339.

Bradshaw H.D., Jr., Stettler, R.F. 1993. Molecular genetics of growth and development in *Populus*. I. Triploidy in hybrid poplars. Theor. Appl. Genet. **86**: 301–307.

Bradshaw, H.D., Jr., and Stettler, R.F. 1994. Molecular genetics of growth and development in *Populus*. II. Segregation distortion due to genetic load. Theor. Appl. Genet. **89**: 551–558.

Bradshaw, H.D., Jr., and Stettler, R.F. 1995. Molecular genetics of growth and development in *Populus*. IV. Mapping QTLs with large effects on growth, form, and phenology traits in a forest tree. Genetics, **139**: 963–973.

Bradshaw, H.D., Jr., Villar, M., Watson, B.D., Otto, K.G., Stewart, S., and Stettler, R.F. 1994. Molecular genetics of growth and development in *Populus*. III. A genetic linkage map of a hybrid poplar composed of RFLP, STS, and RAPD markers. Theor. Appl. Genet. **89**: 551–558.

Brayshaw, T.C. 1965. Native poplars of southern Alberta and their hybrids. Can. For. Serv. Publ. 1109, Ottawa, ON.

Campbell, J.S., Mahoney, J.M., and Rood, S.B. 1993. A lack of heterosis in natural poplar hybrids from southern Alberta. Can. J. Bot. **71**: 37–42.

Ceulemans, R.J., Hinckley, T.M., Heilman, P.E., Isebrands, J.G., and Stettler, R.F. 1989. Crown architecture in relation to productivity of *Populus* clones in the Pacific Northwest, USA. Ann. Sci. For. **46**: 199s–201s.

Ceulemans, R.J., Scarascia-Mugnozza, G., Wiard, B.M., Braatne, J.H., Hinckley, T.M., Stettler, R.F., Isebrands, G., and Heilman, P.E. 1992. Production physiology and morphology of *Populus* species and their hybrids grown under short rotation. I. Clonal comparisons of four-year growth and phenology. Can. J. For. Res. **22**: 1937–1948.

de Nettancourt, D. 1984. Incompatibility. *In* Cellular interactions. *Edited by* H.F. Linskens and J. Heslop-Harrison. Encycl. of Plant Physiol. New Ser. **17**: 624–639.

Dickmann, D.I. 1985. The ideotype concept as applied to forest trees. *In* Attributes of trees as crop plants. *Edited by* M.G.R. Cannell and J.E. Jackson, Titus Wilson and Son, Ltd., Kendal, Cumbria, UK. pp. 89–101.

Dickmann, D.I., and Stuart, K.W. 1983. The culture of poplars in eastern North America. Michigan State University, East Lansing, MI.

Dunlap, J.M., Braatne, J.H., Hinckley, T.M., and Stettler, R.F. 1993. Intraspecific variation in photosynthetic traits of *Populus trichocarpa*. Can. J. Bot. **71**: 1304–1311.

Dunlap, J.M., Heilman, P.E., and Stettler, R.F. 1994. Genetic variation and productivity of *Populus trichocarpa* and its hybrids. VII. Two-year survival and growth of native black cottonwood clones from four river valleys in Washington. Can. J. For. Res. **24**: 1539–1549.

Dunlap, J.M., Heilman, P.E., and Stettler, R.F. 1994. Genetic variation and productivity of *Populus trichocarpa* and its hybrids. VIII. Leaf and crown morphology of native *P.trichocarpa* clones from four river valleys in Washington. Can. J. For. Res. **25**: 1710–1724.

Eckenwalder, J.E. 1984*a*. Natural intersectional hybridization between North American species of *Populus* (Salicaceae) in sections *Aigeiros* and *Tacamahaca*. I. Population studies of *P. parryi*. Can. J. Bot. **62**: 317–324.

Eckenwalder, J.E. 1984*b*. Natural intersectional hybridization between North American species of *Populus* (Salicaceae) in sections *Aigeiros* and *Tacamahaca*. II. Taxonomy. Can. J. Bot. **62**: 325–335.

Eckenwalder, J.E. 1984*c*. Natural intersectional hybridization between North American species of *Populus* (Salicaceae) in sections *Aigeiros* and *Tacamahaca*. III. Paleobotany. Can. J. Bot. **62**: 336–342.

Einspahr, D.W. 1965. Colchicine treatment of newly formed embryos of quaking aspen. For. Sci. **11**: 456–459.

Faivre-Rampant, P., Castiglione, S., Le Guerroue, B., Bisoffi, S., Lefèvre, F., and Villar, M. 1995. Molecular approaches to the study of poplar systematics. *In* Proceedings of the International Poplar Symposium, Seattle, WA, 20–25 Aug. 1995. 45 p.

Floate, K.D., and Whitham, T.G. 1993. The "hybrid bridge" hypothesis: Host shifting via plant hybrid swarms. Am. Nat. **141**: 651–662.

Food and Agriculture Organization of the United Nations. 1980. Poplars and willows in wood production and land use. FAO For. Ser. 10.

Frison, E., Lefèvre, F., De Vries, S., and Turok, J. 1995. *Populus nigra* Network. Report of the 1st Meeting, 3–5 Oct. 1994, Izmit, Turkey. International Plant Genetic Resources Institute, Rome, Italy.

Gottlieb, L.D. 1984. Genetics and morphological evolution in plants. Am. Nat. **123**: 681–709.

Guries, R.P., and Stettler, R.F. 1976. Pre-fertilization barriers to hybridization in the poplars. Silvae Genet. **25**: 37–44.

Heimburger, C. 1968. Poplar breeding in Canada. *In* Growth and utilization of poplars in Canada. *Edited by* J.S. Maini and J.H. Cayford. Can. Dep. For. Rural Dev. For. Branch Dep. Publ. 1205. pp. 88–100.

Hinckley, T.M., Ceulemans, R., Dunlap, J.M., Figliola, A., Heilman, P.E., Isebrands, J.G., Scarascia-Mugnozza, G., Schulte, P.J., Smit, B., Stettler, R.F., van Volkenburgh, E., and Wiard, M. 1989. Physiological, morphological and anatomical components of hybrid vigor in *Populus*. *In* Structural and functional responses to environmental stresses. *Edited by* K.H. Kreeb, H. Richter, and T.M. Hinckley. SPB Academic Publishing, The Hague. pp. 199–217.

Hogenboom, N.G. 1984. Incongruity: Non-functioning of intercellular and intracellular partner relationships through non-matching information. *In* Cellular interactions. *Edited by* H.F. Linskens and J. Heslop-Harrison. Springer-Verlag, Berlin. pp. 640–654.

Houtzagers, G. 1937. Het geslacht *Populus* in verband met zijn betekenis voor de houttelt. Veenman & Sons, Wageningen, The Netherlands.

Isebrands, J.G., Ceulemans, R., and Wiard, B. 1988. Genetic variation in photosynthetic traits among *Populus* clones in relation to yield. Plant Physiol. Biochem. **26**: 427–437.

Jelinski, D.E., and Cheliak, W.M. 1992. Genetic diversity and spatial subdivision of *Populus tremuloides* (Salicaceae) in a heterogeneous landscape. Am. J. Bot. **79**: 728–736.

Johnsson, H. 1956. Génétique et amélioration des peupliers. *In* Les peupliers dans la production du bois et l'utilisation des terres. FAO Publ. 12, Rome, Italy. pp. 372–410.

Keim, P., Paige, K.N., Whitham, T.G., and Lark, K.R. 1989. Genetic analysis of an interspecific hybrid swarm of *Populus*: Occurrence of unidirectional introgression. Genetics, **123**: 557–565.

Knox, R.B., Willing, R.R., and Pryor, L.D. 1972. Interspecific hybridization in poplars using recognition pollen. Silvae Genet. **21**: 65–69.

Kolmer, J.A., and Dyck, P.L. 1994. Gene expression in the *Triticum aestivum–Puccinia recondita f. sp. tritici* gene-for-gene system. Phytopathology, **84**: 437–440.

Krembs, O. 1956. Die Graupappel in den Donau-Auen. Allg. Forstzeitschr. **11**: 345–347.

Lemoine, M. 1988. Hybrides intersectionaux chez le Peuplier. Proceedings of the 18th Session of the International Poplar Commission. 5–8 Sept. 1988, Beijing, China. Food and Agriculture Organization of the United Nations, Rome, Italy. 5 p.

Melchior, G.H., and Seitz, F.W. 1968. Interspezifische Kreuzungssterilität innerhalb der Pappelsektion Aigeiros. Silvae Genet. **17**: 88–93.

Meyerowitz, E.M. 1994. Pattern formation in plant development: four vignettes. Curr. Opin. Genet. Dev. **4**: 602–608.

Mohrdiek, O. 1983. Discussion: future possibilities for poplar breeding. Can. J. For. Res. **13**: 465–471.

Muhle Larsen, C. 1963. Considérations sur l'amélioration du genre *Populus* et specialement sur la section Aegeiros. FAO/Forgen 26/9.

Muhle Larsen, C. 1970. Recent advances in poplar breeding. Int. Rev. For. Res. **3**: 1–67.

Niklas, K.J. 1992. Plant biomechanics. The University of Chicago Press, Chicago and London.

Nikles, D.G. 1993. Breeding methods for production of interspecific hybrids in clonal selection and mass propagation programmes in the tropics and subtropics. *In* Recent advances in mass clonal multiplication of forest trees for plantation programmes. *Edited by* J. Davidson. Food and Agriculture Organization of the United National, Los Baños, The Philippines. pp. 218–252.

Nilsson-Ehle, H. 1936. Über eine in der Natur gefundene Gigasform von *Populus tremula*. Hereditas, **21**: 379–382.

Padro, A. 1992. Poplar and willow growing in combination with agriculture. *In* Proceedings of the 19th Session of the International Poplar Commission. Zaragoza, Spain.

Pinon, J. 1995. Poplar diseases — European perspective. *In* Proceedings of the International Poplar Symposium, 20–25 Aug. 1995, University of Washington, Seattle, WA. p. 26.

Ranney, J.W., Wright, L.L., and Layton, P.A. 1987. Hardwood energy crops: the technology of intensive culture. J. For. **85**: 17–28.

Raven, P.H. 1975. The bases of angiosperm phylogeny: Cytology. Ann. Mo. Bot. Gard. **62**: 724–764.

Ridge, C., Hinckley, T.M., Stettler, R.F., and van Volkenburgh, E. 1986. Leaf growth characteristics of fast growing poplar hybrids *Populus trichocarpa* × *P. deltoides*. Tree Physiol. **1**: 209–216.

Rood, S.B., Campbell, J.S., and Despins, T. 1986. Natural poplar hybrids from southern Alberta. I. Continuous variation for foliar characteristics. Can. J. Bot. **64**: 1382–1388.

Sekawin, M. 1963a. Génétique des peupliers. FAO/Forgen 4/4: 1–6.

Sekawin, M. 1963b. Etude d'un peuplier tétraploide obtenu artificiellement et de sa descendance. FAO/Forgen 1/4: 1–11.

Shaanker, R.V., Ganeshaiah, K.N., and Bawa, K.S. 1988. Parent-offspring conflict, sibling rivalry, and brood size patterns in plants. Ann. Rev. Ecol. Syst. **19**: 177–205.

Simpson, B.B., and Conner-Ogorzaly, M. 1986. Economic biology: Plants in our world. McGraw-Hill Inc., New York, NY.

Sinha, S.K., and Khanna, S. 1975. Physiological, biochemical, and genetic basis of heterosis. Adv. Agron. **27**: 123–174.

Smith, R.L., and Sytsma, K.J. 1990. Evolution of *Populus nigra* (sect. *Aigeiros*): introgressive hybridization and the chloroplast contribution of *Populus alba* (sect. *Populus*). Am. J. Bot. **77**: 1176–1187.

Stebbins, G.L., Jr. 1950. Variation and evolution in plants. Columbia University Press, New York, NY.

Steenackers, V. 1970. La populiculture actuelle. Bull. Soc. R. For. Belg. (Oct 1970). 38 p.

Steenackers, V. 1972. Breeding poplars resistant to various diseases. *In* Biology of rust resistance in forest trees. Misc Publ. U.S. Dep. Agric. No. 1221. pp. 599–607.

Stettler, R.F. 1968. Irradiated mentor pollen: its use in remote hybridization of black cottonwood. Nature (London), **219**: 746–747.

Stettler, R.F., and Ager, A.A. 1984. Mentor effects in pollen interactions. Encycl. Plant Physiol. New Ser. **17**: 609–621.

Stettler, R.F., Fenn, R.C., Heilman, P.E., and Stanton, B.J. 1988. *Populus trichocarpa* × *Populus deltoides* hybrids for short rotation culture: variation patterns and four-year field performance. Can. J. For. Res. **18**: 745–753.

Stuber, C.W., Lincoln, S.E., Wolff, D.W., Helentjaris, T., and Lander, E.S. 1992. Identification of genetic factors contributing to heterosis in a hybrid from two elite maize inbred lines using molecular markers. Genetics, **132**: 823–839.

van Buijtenen, J.P., Joranson, P.N., and Einspahr, D.W. 1958. Diploid versus triploid aspen as pulpwood sources with reference to growth, chemical, physical and pulping differences. Tappi, **41**: 170–175.

Villar, M., Gaget, M., Rougier, M., and Dumas, C. 1993. Pollen-pistil interactions in Populus: β-galactosidase activity associated with pollen tube growth during the crosses *Populus nigra* × *P. nigra* and *P. nigra* × *P. alba*. Sex. Plant Reprod. **6**: 249–256.

Villar, M., and Gaget-Faurobert, M. 1996. Mentor effects in pistil-mediated pollen-pollen interactions. *In* Pollen biotechnology for crop production and improvement. *Edited by* K.R. Shivanna and V.K. Sawhney. Cambridge University Press, New York, NY. In press.

von Wettstein, W. 1933. Die Züchtung von *Populus* II. Züchter, **5**: 280–281.

Weber, J., and Stettler, R.F. 1981. Isoenzyme variation among ten populations of *Populus trichocarpa* (Torr. & Gray) in the Pacific Northwest. Silvae Genet. **30**: 82–87.

Weisgerber, H. 1983. Forstpflanzenzüchtung, Aufgaben, Ergebnisse und Ziel von Züchtungsarbeiten mit Waldbäumen in Hessen. Mitt. Hess. Landesforstverw.

Whitham, T.G. 1989. Plant hybrid zones as sinks for pests. Science (Washington, DC), **244**: 1490–1493.

Willing, R.R., and Pryor, L.D. 1976. Interspecific hybridization in poplar. Theor. Appl. Genet. **47**: 141–151.

Wu, R., and Stettler, R.F. 1994. Quantitative genetics of growth and development in *Populus*. I. A three-generation comparison of tree architecture during the first two years of growth. Theor. Appl. Genet. **89**: 1046–1054.

Wu, R., and Stettler, R.F. 1996. The genetic resolution of juvenile canopy structure and function in a three-generation pedigree of *Populus*. Trees. In press.

Wu, R.L., Wang, M.X., and Huang, M.R. 1992. Quantitative genetics of yield breeding for *Populus* short rotation culture. I. Dynamics of genetic control and selection model of yield traits. Can. J. For. Res. **22**: 175–182.

Xiao, J., Li, J., Yuan, L., and Tanksley, S.D. 1995. Dominance is the major genetic basis of heterosis in rice as revealed by QTL analysis using molecular markers. Genetics, **140**: 745–754.

Zsuffa, L. 1968. The present work on poplar breeding in Ontario. Proceedings of the 13th Session of the International Poplar Commission, Montreal, QC. FO/CIP/13/39. 18 p.

Zsuffa, L. 1975. A summary review of interspecific breeding in the genus *Populus*. *In* Proceedings of the 14th Meeting of the Canadian Tree Improvement Association, Part 2. Canadian Forest Service, Ottawa, ON. pp. 107–123.

Zsuffa, L. 1979. Vegetative propagation of forest trees: problems and programs in Ontario. *In* Proceedings of the COJFRC Tree Improvement Symposium, O-P-7, 12–21 Sept. 1978. Toronto, ON. pp. 95–104.

CHAPTER 5
Controlled reproduction of *Populus*

Brian J. Stanton and Marc Villar

Introduction

Populus domestication owes much to the ease with which the genus adapts to an agronomic style of intensive plantation culture but, perhaps even more so, to the significant gains in productivity that have been achieved through controlled breeding and clonal selection. Towards this end, it is the breeder's good fortune that many desirable features of the group's reproductive biology allow its members to be artificially bred quite readily.

Some of these are obvious and include unisexual inflorescences, a relatively brief juvenile phase, flowering regularity, *ex situ* propagation and reproduction of floral scions, and the production of substantial seed quantities that mature quickly and germinate without stratification. Each is quite valuable to the production of pedigreed seedling populations. What are often overlooked, however, are the specific peculiarities of each aspect of the controlled breeding process that frequently require varying and exacting procedures and conditions dependent upon the species and type of cross in question. This chapter brings together current information on each of the key steps in the controlled reproduction of the major *Populus* species.

Experimental *Populus* breeding actually has a relatively long history within forest genetics. Henry (1914) first reported the technique used in the hybridization of *P. deltoides, P. nigra,* and *P. trichocarpa* at the Kew Botanical Gardens in England between 1912 and 1914. A first assessment of the range of potential species combinations among and within the *Populus* sections was made in 1926 (Stout and Schreiner 1933). Accounts of successful artificial crossing procedures for *P. tremula* and *P. alba* adapted from work with *Salix* soon followed (Wettstein-Westersheim 1933; Al'benskii and Delitsina 1934). Within the next

B.J. Stanton. James River Corporation, Northwest Fiber Supply Division, Camas, WA 98607-2090, USA.
M. Villar. INRA, Station d'Amélioration des Arbres Forestiers, 45160 Ardon, France.
Correct citation: Stanton, B.J., and Villar, M. 1996. Controlled reproduction of *Populus*. *In* Biology of *Populus* and its implications for management and conservation. Part I, Chapter 5. *Edited by* R.F. Stettler, H.D. Bradshaw, Jr., P.E. Heilman, and T.M. Hinckley. NRC Research Press, National Research Council of Canada, Ottawa, ON, Canada. pp. 113–138.

three decades, detailed instructions for the controlled reproduction of *P. deltoides, P. nigra, P. tremula,* and *P. tremuloides* followed (Johnson 1945; Bergman and Lantz 1958; Mulhe Larsen 1960; Farmer and Nance 1967).

Today the cultivation of *Populus* is progressing at a rapid pace as plantings become evermore popular as important sources of fiber, biomass, and veneer. As such, there is a renewal of genetic improvement efforts that require efficient and reliable controlled breeding techniques critical to the study of the inheritance of commercial traits and the implementation of full-sib family selection programs. The importance of controlled crossing extends well beyond, however, to the creation of interspecific hybrid populations on which many operations rely and to the vegetative propagation of intensively selected clones of known parentage. Full pedigree information is needed if such clones are to be deployed and rotated in a way that uses their genetic diversity to manage the risk of planting failures.

Collection of reproductive material

Maturity

Members of the *Populus* genus ordinarily reach sexual maturity at a relatively early age with individuals of the section *Populus* typically passing through the juvenile phase in less time than that required by those of the *Tacamahaca* and *Aigeiros* sections (Schreiner 1974). Although flowering 3- and 4-year-old *P. tremuloides* and *P. deltoides* seedlings (Pauley 1950; Cooper 1990), and 6-year-old *P. deltoides* ramets (cited in Hardin 1984), have been reported, it is safe to say that reproductive maturity is probably reached no sooner than the 10th or 15th year in most cases (Schreiner 1974). Yet, when planning controlled breeding projects using wild selections from native stands, the age at which the juvenile phase is completed is often misleading. As examples, *P. trichocarpa* from the Pacific Northwest generally does not flower at an adequate level for artificial breeding until age 20–25. Furthermore, *P. deltoides* from the lower Mississippi River valley usually does not produce measurable flower and seed crops until well after age 10 (Maisenhelder 1960). Similarly, substantial flowering of *P. tremuloides* and *P. grandidentata* does not begin until the 10th through 20th years (Laidly 1990; Perala 1990).

On the other hand, in well-managed test plantations or breeding orchards, ample flowering is regularly reached at an earlier age. For instance, Barnes (1958) observed the flowering of both sexes of *P. tremula* ×*tremuloides* hybrids in test plantings during their 7th year. Plantation-grown *P. tremuloides* from the Lake States (Einspahr and Winton 1976) and *P. deltoides* from the lower Mississippi River delta have both been bred as early as their 8th year. Moreover, *P. tremula, P. trichocarpa,* and *P. deltoides* plantings in France have been bred as early as their 4th, 8th, and 10th years, respectively. In the Pacific Northwest, intensively managed clone tests of first generation hybrid (*P. trichocarpa* ×*deltoides*) populations normally produce a sufficient quantity of flowers during their 5th growing season to sustain advanced generation breeding efforts. In Japan, however, plantings of *P. maximowiczii* do not become reproductively mature until their 15th through 20th years (H. Kohda, personal communication).

Phenology

Once chilling requirements have been fulfilled, the seasonal timing of flower emergence is regulated by warming spring temperatures (Pauley 1950; Pauley and Perry 1954). For instance, temperature increases above 12°C over a period of 6 d are the controlling factor in the emergence of *P. tremuloides* flowers (Perala 1990). Photoperiod likely does not function to any great extent in the regulation of spring floral phenology (Pauley 1950; Pauley and Perry 1954; Vaartaja 1960; Lester 1961; Farmer and Reinholt 1986; Howe et al. 1995).

A cline has been reported in the timing of anthesis of *P. deltoides* in which flowering begins in populations from the more northerly and westerly portions of the range and progresses through the southern and eastern populations (Ying and Bagley 1976). Even though geographic variation in anthesis has not been studied in other *Populus* species, similar patterns of adaptation of the flowering phenomenon might exist based on evidence of latitudinal adaptation of other phenological events (mostly timing of shoot growth cessation and leaf abscission) in *P. trichocarpa* (Weber et al. 1985), *P. deltoides* (Nelson and Tauer 1987), and *P. balsamifera* (Farmer 1993; Riemenschneider and McMahon 1993). The early flowering response of more northerly populations could represent acclimation to lower springtime temperatures as well as lower chilling requirements (Farmer 1964a; Farmer and Reinholt 1986).

Of equal importance to such broad trends in anthesis are those substantial differences that have been locally observed. Considerable phenotypic variation in flowering times have been recorded among and within local populations of *P. deltoides* (Farmer 1966a), *P. trichocarpa* (Boes and Strauss 1994), *P. tremuloides* (Perala 1990), and *P. grandidentata* (Barnes 1961). These differences are likely under strong genetic control (Valentine 1975) and, as such, must be taken into account when scheduling the controlled breeding of a population of diverse as well as related selections. To illustrate, *P. grandidentata* habitually flowers 10–12 d later than *P. tremuloides* in the same locales (Einspahr and Joranson 1960; Pregitzer and Barnes 1980) while half-sibs of *P. deltoides* can vary in their timing of anthesis by as much as 9 d (Ying and Bagley 1976). Except for the section *Populus* that flowers in intervals of 2–4 year (Laidly 1990; Perala 1990), flowering periodicity is no doubt of little concern in the design and execution of most controlled *Populus* breeding projects (Schreiner 1974).

Sexing

Identification of the sexes can be accomplished easily during anthesis. Staminate inflorescences are distinguished from their pistillate counterparts by their generally greater abundance and the promptness with which they are cast off following the release of pollen. The sexes of the *Tacamahaca* can also be identified by the reddish hue of the staminate inflorescence and the pistillate inflorescence's green color. To a lesser extent flower color is also helpful in sex identification in *P. deltoides*; males can have red, green, or yellow inflorescences that are not always distinguishable from the green pistillate ones. The staminate inflorescences of *P. tremuloides* and *P. grandidentata* vary from red to reddish brown; the variations are, in many cases, sufficiently distinct to be diagnostic of individual clones (G. Wyckoff, personal communication).

Prior to anthesis, however, sex identification on the basis of dormant floral buds is more problematic, but may be completed with reasonable accuracy once some experience has been gained. A case in point is Farmer (1964b), who was able to correctly sex 87% of a sample of dormant *P. deltoides* from the ground using binoculars. Dormant floral buds are situated in axillary positions on 1-year-old shoots. They are often clustered at the base of long shoots but on short shoots they appear to subtend the terminal vegetative bud (Pregitzer and Barnes 1980). Staminate buds are noticeably larger and plumper than both pistillate and vegetative buds, but the magnitude of the difference varies appreciably among the various *Populus* species, at times complicating the ease with which the sexes can be distinguished. In *P. deltoides*, for instance, pistillate and axillary vegetative buds are comparable in volume and are, on average, about one-quarter to one-third the size of the staminate buds. On the other hand, the floral buds of the two sexes are not as distinct in *P. trichocarpa*, oftentimes intergrading in size and shape to the extent that sexing with binoculars is time consuming and frequently imprecise. Hence, when attempted during the dormant phase, a tree's sex should be confirmed by examination of the floral primordia using a hand lens after the bud scales and floral bracts have been removed. Here, a multiplicity of anthers rising from a shallow basal disk, the remnant of the perianth, characterizes the staminate flowers, whereas the pistillate flower is identified by a cup-shaped disk that encloses most of the ovary and allows for a protruding lobed stigma (Nagaraj 1952; Lester 1963a; Fechner 1972).

Phenotypic differences between the sexes in other anatomical features (height, stem diameter, stem form, crown architecture, wood specific gravity, etc.) have generally not been substantiated in either *P. tremuloides* (Einspahr 1960), *P. ciliata* (Khosla et al. 1979; Khurana and Khosla 1982), or *P. deltoides* (Farmer 1964b; Walters and Bruckmann 1965). Neither was genetic variation in growth and morphology of *P. deltoides* found to be associated with the sexes (Ying and Bagley 1976). Therefore such contrasts are seemingly of little assistance in sex identification. Correspondingly, sexual identification of juvenile phenotypes using paper partition chromatography of vegetative bud extracts has shown inconsistent results and is evidently not too promising either (Blake et al. 1960; Lester 1963*b*; Blake et al. 1967).

Sex ratio

It has long been observed that the sexes within various *Populus* species do not always conform to a hypothesized 1:1 ratio with males regularly outnumbering females in native stands (Williamson 1913; Pauley and Mennel 1957). Most designed sampling trials have not, however, found sufficient evidence to dispute a balanced sex ratio (Einspahr 1960; Lester 1963*b*; Farmer 1964*b*; Ying and Bagley 1976). Yet actual interpretations of sex ratios in *Populus* are apparently complicated by the effect of site, age, and parentage. As illustrations, it has been concluded that, on sites of more favorable fertility (Comtois et al. 1986) and moisture (Dolgosheev 1968; Il'in 1974), and at lower elevations (Grant and Mitton 1979), the balance between the sexes may be shifted in favor of the female component of *P. balsamifera*, *P. tremula*, and *P. tremuloides* populations, respectively. The age at which a population is inspected may also skew the sex ratio as, for example, in *P. tremuloides* and *P. deltoides* populations where males seem to reach sexual maturity earlier than females (Kaul and Kaul 1984; Valentine 1975). In like fashion, the ecological stage of stand development has also been associated with an imbalance in the division between staminate and pistillate trees of *P. tremula* (Falinski 1980). The sex ratio is also likely to vary appreciably among a population's full-sib families (Valentine 1975).

Material handling

Reproductive materials (dormant floral branches) are usually collected as close to the onset of anthesis as practical. Adequate chilling either met before or, artificially imposed once the collections have been made, is important for successful flowering of both sexes. Thirty to 45 d of storage at 4°C will more than likely satisfy the chilling requirements of winter collections from lower latitudes. On the other hand, 30 d storage at 5°C made possible the flowering of autumn branch collections of the northerly species, *P. tremuloides* (Lester 1961). Late summer staminate branch collections of *P. tremula* were forced after chilling at 4 to –5°C, while autumn collections required exposure to –20°C before anthesis occurred (Seitz 1958).

Generally speaking, branches should be large enough to supply cuttings approximately 1–2 m in length if rooting is the intended female propagation technique or scions 20–30 cm long if grafting is to be pursued (Fig. 1). The best propagation results are often achieved when collections are taken from the most vigorous upper portion of the crown. In the case of *P. deltoides* female selections whose adventitious rooting ability is often troublesome, the added provision that the bases of the cuttings be composed of 3- to 4-year-old wood will further guarantee propagation success (Allen and McComb 1956).

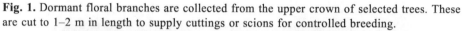

Fig. 1. Dormant floral branches are collected from the upper crown of selected trees. These are cut to 1–2 m in length to supply cuttings or scions for controlled breeding.

While in the field, the severed branch bases are wrapped in moistened sphagnum moss to prevent desiccation during transport. The branch bases should be recut, placed in pails of water and then stored under refrigeration at 0–4°C (storage at −2°C may be necessary to prevent floral bud development during protracted storage). Placing the branches in sealed plastic bags to retard moisture loss for extended periods often promotes the growth of mold and is used only as an interim measure until the collections can be properly stored.

Pollen management

Forcing

Pollen is routinely forced from 1 to 2 m cuttings maintained in water culture in the greenhouse or laboratory before forcing of the female selections (Johnson 1945; Einspahr and Benson 1964; Farmer and Nance 1967). Under conditions of proper temperature, humidity, and ample movement of water into the cuttings, the staminate inflorescences will emerge and lengthen, followed by maturation and dehiscence of their anthers. Pollen is then collected, dried, and refrigerated for short-term storage or frozen for long-term preservation. When done properly, *Populus* pollen reliably maintains satisfactory short-term viability for 1–2, perhaps 3 mo, and at least 1 year after long-term storage (Yang and Villar 1989; Cagelli 1995).

At the outset of the process, male floral cuttings are placed in water pails of 7–20 L capacity after the basal 3–5 cm have been trimmed. A moderate thinning of floral buds has been reported to improve pollen quality in a variety of *Populus* species (Tkachenko 1984). Basal one centimeter disks are cut from the cuttings 2–4 times weekly and the water changed that often (Wyckoff 1975). Worsley (1959) recommends that cutting bases be cut under water. Iced water has also been recommended as a precaution against vascular plugging that may impede water uptake (Bergman and Lantz 1958). The use of distilled water is also commonly advised. Intermittent overhead mist application is sometimes used to further prevent desiccation of the cuttings and premature abscission of floral buds up to, but not beyond, anthesis (Stanton 1990).

Temperatures used to induce anthesis of *P. deltoides* male selections from southern latitudes, have ranged from 24 to 29°C (Farmer and Nance 1967) to 21°C (Miller 1972). In contrast, species and provenances of higher latitudes generally force better at cooler temperatures. For example, a constant temperature setting of 18°C was used in accelerating the flowering of *P. tremuloides* (Einspahr and Benson 1964) while *P. trichocarpa* from the Pacific Northwest has been forced under temperature settings that fluctuate between 16 h at 18°C and 8 h at 8°C (Stanton 1990). Successful emergence of the staminate inflorescences is further ensured by relative humidity in the range of 80–100% (Worsley 1959). Avoidance of excessive heat and aridity (especially during the last 12 h preceding dehiscence) is critical to normal floral maturation as well as ensuring pollen quantity and quality (Larson 1958; Yang and Villar 1989). Anthesis and dehiscence of *P. deltoides,* *P. trichocarpa*, and *P. tremuloides* are typically complete between 1 and 2 wk under greenhouse conditions (Farmer and Nance 1967; Einspahr and Winton 1976; Stanton 1990). Upon maturation the anthers appear bright yellow as the ripening and shedding of pollen commences. Finally, the anthers blacken as dehiscence terminates.

Collection

Two techniques are favored in pollen collection. Wood-framed plastic and muslin tents or glass compartments roughly 0.5–1 m³ in volume are routinely used to house an entire cutting collection of individual selections (Barner and Christiansen 1958). Agitation of the ripened stamens dislodges pollen that falls into cardboard boxes or glass vessels placed beneath individual inflorescences. Thorough misting within the cages prevents pollen contamination of the following lot when the extraction of a second male is started. Alternatively, glassine or parchment bags (approximately 152 × 89 × 330 mm) have been used but to a lesser extent in *Populus* pollen collections (Stanton 1990). The bags are secured with a wire tie around a twig supporting one or more inflorescences and are removed and replaced with each collection until dehiscence is complete. Gathering of pollen by either of these methods should be scheduled once or twice daily to maintain high viability.

A second technique involves the removal of nearly ripened whole inflorescences at the point of incipient pollen shed. These are then placed in a warm (20–22°C), well-ventilated area of reduced humidity (40–60%) for 24 h to allow the anthers to complete their ripening and release of pollen (Einspahr and Winton 1976; Gras et al. 1987; Cagelli 1995). Cross-lot contamination can be controlled by drying the flowers in a reasonably large cardboard box (30 dm³) whose walls and ceiling contain holes (18.5 cm in diameter) covered with 20–25 µm paper filters. This technique works moderately well with *P. trichocarpa* and *P. deltoides* when individual inflorescences are detached for drying after one-third to one-half of their flowers have fully ripened. However, *in vitro* tests have indicated lower viability for lots collected in this fashion relative to those collected from inflorescences maintaining their branch attachment, though *in vivo* comparisons have not substantiated such a distinct effect (Cagelli 1995).

Processing

Pollen lots are sieved to remove bud scales and floral bracts using an 80 or 100 mesh screen (sieve openings of 177 and 149 µm, respectively) immediately upon collection. Cross-lot contamination can be minimized during this process by working beneath a fume hood the opening of which is fitted with a slotted plastic wind screen and by misting between extraction of different lots. The cleaned lots are then air-dried to an approximate moisture content of 6–10% (Yang and Villar 1989). Drying can usually be completed in 24 h at 21–24°C and 20–30% relative humidity. In actuality, however, the duration of the drying phase will be determined by the fresh moisture content of the lots that can vary appreciably among species, genotypes, collection days, and collection sites (i.e., greenhouse vs. laboratory) (Cagelli 1995). Ordinarily, such variations can be reliably accommodated by air-drying for as long as 48 h, although lengthening the process appears to reduce pollen viability. Dehydration of pollen over a desiccant at 4°C for 12 h is therefore more commonly used as it is believed to be a more exacting technique capable of handling pollen collections of varying moisture. For lots of large quantity, drying times can be tailored to specific fresh moisture contents of individual genotypes using an electronic thermobalance to accurately gauge drying rate and end-point moisture content (Yang and Villar 1989).

Dried lots are placed into cold storage in rubber- or cotton-stoppered glass vials (the latter placed inside desiccators) or cryogenic tubes in amounts metered for a single day's pollen application. Refrigeration at 1–4°C appears more than sufficient for 1 or 2 month's storage but not for extended periods. *P. ciliata* pollen, for instance, stored at 4°C, maintained viability no longer than 4–6 mo (Dhir et al. 1982). *P. deltoides* and *P. nigra* pollen lots preserved in this way were found inviable after one year's storage (Cagelli 1995). Storage at –18°C is therefore an absolute recommendation for pollen that is to be kept up to a year or longer. *P. tremula* ×*tremuloides* pollen has been stored at –18°C for 5 years with just a slight reduction in seed production (Herrmann 1976).

Preparation for use

Pollen lots removed from short-term storage are typically rehydrated just before use by exposure to a humid environment (60–70% relative humidity) for 1 or 2 h at 20°C (Worsley 1959). Those lots removed from long-term storage are initially rehydrated for 1 h at 4°C followed by a 2nd h at 20°C (Yand and Villar 1989).

Germination rates of 10–55% are sufficient for adequate seed set in most breeding work (Rajora and Zsuffa 1986). This potential can be assessed using *in vitro* pollen viability tests though the results have not always correlated well with the number of sound seed set (Cagelli 1995). Nevertheless, a recommended test involves germination counts of 200-grain samples on a suitable artificial medium (Brewbaker and Kwack 1963) after 24 h incubation at 24°C. The grains are lightly dusted onto the medium's surface through a 100-mesh sieve to achieve a uniform distribution. Viability is expressed in terms of the percentage of grains that initiate growth of the pollen tube. Light conditions do not seem to be critical but the optimum sucrose concentration may differ by as much as twofold for different species (Klaehn and Neu 1960; Rajora and Zsuffa 1986; Cagelli 1995). Pollen viability may also be monitored using the Fluorochromatic Reaction Test (Heslop-Harrison and Heslop-Harrison 1970) or by staining with tetrazolium chloride (Rajora and Zsuffa 1986).

Seed production

The key steps involved in the production of viable seed are the propagation of female floral cuttings or grafted scions, the application of pollen to receptive pistillate flowers, and the maturation of the seed capsules (Fig. 2). The first of

Fig. 2. Reproductive structures: (a) staminate inflorescence, *P. deltoides*, (b) pistillate inflorescence, *P. trichocarpa*, (c) fruiting capsules, *P. trichocarpa*, (d) dehiscing capsules, *P. deltoides*.

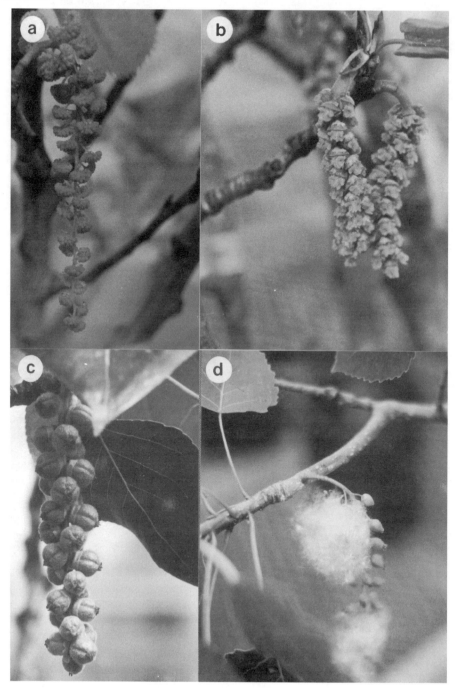

these is the most critical step and here a breeder's skill more often than not will determine the overall outcome of the breeding project.

Breeding stock propagation

Female reproductive materials are propagated by 1 of 3 methods, the choice of which is determined primarily by the length of a species' seed maturation period and, secondarily, by its ability to form adventitious roots. As might be expected, each method is more or less tailored to one of the three main *Populus* sections.

The simplest is the "cut-branch" technique commonly used with members of the *Populus* section and described by Johnson (1945), Bergman and Lantz (1958), Einspahr and Benson (1964), and Gladysz (1983). Essentially a duplication of the technique used in forcing pollen, cuttings 1–2 m in length are cultured in pails containing 7–20 L of distilled water that is changed daily or several times weekly. Trimming the base of the cuttings whenever the water is replenished will further reduce bacterial plugging of the vascular elements as will the addition of ice to the pails. Icing female *P. tremula* cuttings 3 times a day lowered water temperatures to 2.5°C and provided larger catkins that yielded seed of increased fresh weight (Bergman and Lantz 1958). Nutrient solutions such as Hoagland's and Knopp's have been tested but do not seem necessary (Al'benskii and Delitsina 1934; Johnson 1945; Gladysz 1983). A comparatively small seed catkin and a brief 2–3 wk span between pollination and seed maturation makes a functional root system unnecessary and allows the "cut-branch" technique to be recommended with good results for *P. tremuloides, P. canescens,* and *P. tremula,* with moderate success with *P. alba,* and with a measure of difficulty with *P. grandidentata* (Benson 1972).

A much larger catkin and a moderately long maturation period (6–8 wk) are characteristic of the *Tacamahaca* and, presumably, dictate a higher moisture and nutrient demand and the need for a working root system. Hence, the "cut-branch" technique oftentimes yields inconsistent results although it has been used in breeding *P. maximowiczii* in Japan (H. Kohda, personal communication). Alternatively, *Tacamahaca* selections can still be easily bred on cuttings in soil culture owing to the presence of preformed root primordia (Schier and Campbell 1976; Cunningham and Farmer 1984) (Fig. 3). The "large-slip-in-pot" method takes advantage of this (Joennoz and Vallee 1972). Briefly, floral cuttings 90–150 cm long are stuck in pots, the lower one quarter of which is filled with moistened perlite. A peat and sand mix is added to the remainder. Uptake of moisture through the base of the cuttings maintains the stock until ample rooting occurs in the soil. In *P. trichocarpa,* this takes place within 2 wk. Although the source of cutting material and the season in which it is collected have been documented as affecting adventitious rooting in *P. balsamifera* (Farmer et al. 1989; Houle and Babeux 1993), and, to a lesser extent, *P. trichocarpa*

Fig. 3. Indoor *P. trichocarpa* breeding orchard. Cuttings are potted in buckets providing a lower reservoir to sustain the crop until adventitious rooting occurs in the upper soil compartment.

(Bloomberg 1959), experience suggests that neither topophysis nor cyclophysis are too important a determinant in the success of using this method.

Some members of the *Aigeiros* section such as *P. nigra* have been bred at times using either the "cut-branch" (Muhle Larsen 1960) or the "large-slip-in-pot" technique. Then again, other members such as *P. deltoides* have proven quite recalcitrant with either of these approaches. Here a combination of an uncertain and laggard adventitious rooting capacity (Maisenhelder 1960), and the longest seed maturation period (12–20 wk for southern provenances) dictate the use of the "bottle-graft" technique (Muhle Larsen 1960; Farmer and Nance 1967; Koster 1968; Gras et al. 1987; Strobl 1992). Bottle grafts are necessary to maintain the moisture status of floral scions until graft unions are formed with the rootstock. Hybrid (*P. deltoides* ×*trichocarpa* and *P. deltoides* ×*nigra*) rootstock is preferred when grafting *P. deltoides* scions as both the survival and developmental rates of the fruit crop are increased (Stettler et al. 1980; Strobl 1992). Typically, 1-year-old nursery stock is lifted in the autumn, potted in containers, and top-pruned to a height of 75 cm (Strobl 1992). Held in cold frames over the winter, the rootstock is brought into the greenhouse and allowed to flush just before the grafting of 20–30 cm-long scions. Farmer and Nance (1967) have recommended grafting during the autumn to ensure a functional cambial union at the time the breeding stock is brought into the greenhouse.

Bottle grafting is a comparatively expensive procedure requiring considerable skill and experience without which an appreciable number of graft failures may occur (Farmer and Nance 1967). Accordingly, attempts to breed *P. deltoides* using procedures that rely on the production of adventitious roots by floral cuttings have been tried but have often resulted in excessive catkin abscission (Larsson 1976; Miller 1972). Undoubtedly these results have been compounded by pronounced clone, topophytic, and age effects (Cunningham 1953; Farmer 1966*b*; Wilcox and Farmer 1968). More recently, it has been demonstrated that first generation *P. deltoides* clones, developed in part for their adventitious rooting ability, can be bred up to age 15 using a variation of the "large-slip-in-pot" technique that incorporates the use of a holding solution.

Holding solutions are combinations of sucrose and bactericides (e.g., 8-hydroxyquinoline sulfate) that prevent vascular plugging of floral cuttings while providing nourishment. Their utility was first demonstrated with the controlled reproduction of *P. canescens* ×*tremuloides* hybrids (Einspahr and Benson 1971). They have since come to be used in large scale breeding programs of the *Populus* section (Wyckoff 1975), and in the modification of the "large-slip-in-pot" method. In the latter case, planting containers are fitted with a false Styrofoam bottom to which a standard potting mix is added. The volume below the false bottom is partially filled with a holding solution. The bases of the floral cuttings have access to this reservoir through holes cut in the disk and are thereby ensured of an adequate supply of moisture and nutrition for the prolonged period during which rooting occurs.

Nowadays, commercial floral preservatives that contain sucrose, a bactericide, and an acidifying agent to control fungi, are readily available. One and one-half percent solutions replenished twice weekly using a stationary fill-tube assists the controlled breeding of selected *P. deltoides* clones (Stanton 1990). Also, prompt adventitious rooting of *P. deltoides* floral cuttings may be promoted by a 24-h basal soak in a 50 ppm indolebutyric acid solution prior to potting (Farmer 1966*b*), and by the use of cuttings whose basal wood is no older than 4 years (Allen and McComb 1956). Maintaining potted breeding stock in the dark at moderately cool temperatures (4–10°C) for 4–6 wk prior to the induction of forcing may further enhance rooting success by advancing the initiation of adventitious root callus (Larsson 1976; Hall et al. 1991). Ultimately, advanced generation *P. deltoides* breeding programs that rely on adventitious rooting probably will enjoy much improved success as many first generation clonal improvement programs currently emphasize the selection of this highly heritable trait (Wilcox and Farmer 1968; Ying and Bagley 1977). Field crossing in breeding orchards is another technique that advanced generation *P. deltoides* programs could pursue (Dhir and Mohn 1976).

Forcing pistillate flowers

Female selections of species and provenances from higher latitudes such as *P. tremuloides* and *P. balsamifera* force well within a temperature range of 18–21°C (Benson 1972; Larsson 1976). Likewise, *P. trichocarpa* is forced using a daytime temperature setting of 18°C and a 7°C nighttime setting. In contrast, southern provenances of *P. deltoides* force well over temperature ranges of 18–24°C during the day and 15–18°C throughout the night. Maintenance of a daytime relative humidity level of 60–80% seems conducive to vigorous floral development. During this forcing period, female breeding stock is sometimes thinned of excess floral buds as a way of insuring that the remaining inflorescences ripen to full term. Gladysz (1983) recommends thinning *P. tremula* inflorescences to a 30:1 cutting-to-inflorescence fresh weight ratio. Thinning has been recommended for the *Populus* section (Ronald 1982) and *Aigeiros* sections (Strobl 1992) but does not appear necessary when breeding the rapid-rooting *Tacamahaca*.

Pollination

Individual flowers may remain receptive to pollen for just a few days, as in *P. tremuloides* (Einspahr and Joranson 1960) or, as in *P. trichocarpa*, from several days to an entire week dependent upon the individual clone. *P. trichocarpa* and *P. deltoides* can both be best judged receptive when the floral bracts have fallen from the inflorescences and the bright green stigmata have extended well beyond the perianth cup. A bright red stigmatic surface signals a receptive pistil in *P. tremuloides* (Einspahr and Winton 1976), while receptive stigmata of *P. alba* can vary in color from yellow-green to rose (cited in Barnes 1961).

Because of the progression in which whole inflorescences and individual flowers of the same inflorescence develop, pollinations are scheduled to coincide with the peak period of receptivity. In *P. tremuloides* and *P. grandidentata*, the most pronounced within-clone variation in flowering occurs among inflorescences on the same shoot (Pregitzer and Barnes 1980). The range in flowering time also varies among individual selections but will usually span no more than a 1–2 wk period when chilling requirements have been fully met (Johnson 1945; Strobl 1992). One exception is *P. deltoides* from the lower Mississippi River valley that flowers over a 6-wk period in its native habitat (Farmer 1966a). This latter variation can be effectively reduced by extending the chilling period for several weeks.

Pistillate inflorescences are isolated well before separation of the bud scales to prevent unintentional open pollinations. Isolation is most commonly achieved by placing the cuttings or grafts of a single female in compartments to which the pollen of a single male selection is applied; variation in the timing of receptivity may preclude the placement of more than one female's reproductive material in a compartment at any given time. The compartments are misted or washed thoroughly with water before being reused with the pollen of another male. Transpiration stress may occur within these compartments and careful monitoring with temperature gauges and heat mitigation with filtered ventilation, shade cloth, and misting have all been recommended to ward off premature abscission of the developing flower crop (Koster 1968; Strobl 1992). An alternative to cages is the use of translucent or clear glassine and parchment bags to isolate individual or whole clusters of inflorescences (Pohl 1962). Gras et al. (1987) describe a polyacetate cellulose bag (10 × 30 cm) used to isolate entire potted *Aigeiros* grafts for pollination. While laborious, bagging allows greater flexibility in the execution of mating designs and more precise matching of pollination with the timing of floral receptivity.

Pollen applications are normally scheduled for the drier, midportion of the day to minimize clumping of the grains. However pollinations should be delayed until the late afternoon hours if midday greenhouse temperatures should rise to excessive levels. In this way, the timing of pollen germination will coincide with nighttime cooling that promotes rapid growth of the pollen tube (Villar et. al. 1993). Pollen is applied with either a bulb syringe or camel's hair brush and repeated once or twice at 24-h intervals to cover the variation in receptivity among the flowers of any one inflorescence (Fig. 4). Repeated pollinations have been reported to cause more rapid fruit ripening and higher seed yields in interspecific crosses with *P. nigra* (Li 1960). Both brushes and syringes are sterilized with alcohol before use with another pollen source. Wyckoff (1980) describes a unique Plexiglas cylinder that uses air pressure to circulate a pollen cloud around entire cutting or graft collections for mass production of hybrid *P. tremuloides* ×*tremula* seed. Isolation of the flowers maintained for 48–72 h following the final application will ensure the paternity of the cross, as pollen germination and subsequent fertilization are likely to occur well within this time for *Aigeiros, Populus,* and *Tacamahaca* (Winton 1968; Fechner 1972; Knox et al. 1972; Stettler and Guries 1976).

Fruit development

The development of the seed capsules normally occurs quite rapidly following effective pollination and fertilization. Inflorescences will, nonetheless, continue to elongate without fertilization (Fechner 1976), but abscission will usually occur within 7–10 d in the absence of embryogenesis. Stigmata of the section *Populus* will wither within 48 h if pollination is ineffective. When fertilization is effective, fruit maturation occurs within 2–3 weeks in the *Populus* section (Einspahr and Benson 1964; Fechner 1972; Brown 1989), 4–6 weeks in *P. nigra*, 6–8 weeks in *P. balsamifera* (Larsson 1976), *P. trichocarpa*, *P. maximowiczii* (H. Kohda, personal communication), 3–5 months in southern provenances of *P. deltoides* (Farmer 1966a; Koster 1968), and within 4–6 weeks in *P. deltoides* from northern sources (Hall et al. 1991). Diurnal temperatures are often increased to 25–30°C to foster the growth and development of the seed

Fig. 4. Controlled pollination with a bulb syringe.

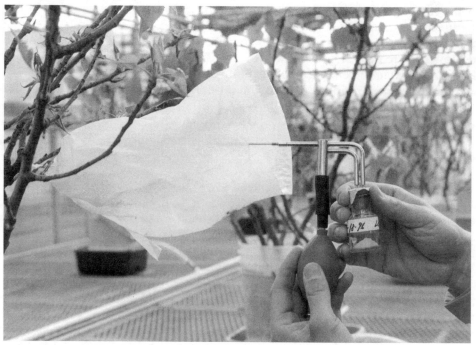

capsules of *Tacamahaca* and *Aigeiros* selections so that enough time remains for seedling propagation and proper hardening. Fertilization of cuttings and grafts of these two sections is also initiated on a 7 or 14 d schedule. Formulations approximating 100 ppm nitrogen, 75 ppm phosphorus, and 100 ppm potassium may be close to the amount required for optimum crop development (Bonner and Broadfoot 1967).

Progeny propagation

Seed collection, processing, and storage

During dehiscence as the capsules begin to dry, their carpels will separate along the sutures to release the fully-formed seed. The process moves more quickly during hot and dry periods and usually completes within a week's time. As it is released, seed should be collected on a daily or twice weekly schedule and refrigerated for three weeks until cleaned. Another common practice is to collect the seed catkins as they initiate dehiscence followed by air drying at room temperature for a 3–5 d period to force the shedding of seed (Faust 1936; Johnson 1946; denHeyer and Seymour 1978; Singh and Gupta 1981; Singh and Arya 1987; Fung and Hamel 1993). The cotton fibers are separated from the seed by tumbling the mixture through a series of standard sieves under air pressure (Einspahr and Schlafke 1957; Roe and McCain 1962; denHeyer and Seymour 1978; Fung and Hamel 1993). For example, 30, 8, 12, 16, and 60 mesh screens (0.59, 2.38, 1.68, 1.19, and 0.25 mm openings, respectively) arranged top to bottom with the cotton and seed placed between the top two screens, work well with the extraction of *P. trichocarpa* and *P. deltoides* seed. The arrangement of 20, 20, 40, and 60 mesh screens (0.84, 0.84, 0.42, and 0.25 mm openings, respectively) is appropriate for the cleaning of the smaller seed of the section *Populus* (Einspahr and Schlafke 1957).

Populus seed is microbiotic and will rapidly lose viability when kept at room temperature and humidity (Engstrom 1948; McComb and Lovestead 1954; Hamilton 1966; Fechner et al. 1981; Singh and Gupta 1981). Hellum (1973) reported a 12% reduction in the germination percentage of *P. balsamifera* seed stored at 7°C for 4 mo. Yet viability can be maintained for several years when seed is stored at below-freezing temperatures in a dry atmosphere. *P. balsamifera* seed has been kept for 3 years with only a slight decrease in germination percentage when stored between –5°C and –10°C and 10% relative humidity (Moss 1938; Zasada and Densmore 1980). For shorter storage, conditions for other members of *Tacamahaca* range from 3°C and 30% relative humidity for 10 mo storage of *P. maximowiczii* seed (Asakawa 1982) to –1°C for 6 mo storage of *P. trichocarpa* seed (Hamilton 1966). In *Aigeiros*, *P. deltoides* seed has been reported to store well for 1 year at –12°C and 25% relative humidity (McComb and Lovestead 1954). Tauer (1979) also found that freezing temperature (–20°C) was conducive to maintenance of high viability of *P. deltoides* seed for an extended period, especially when seed moisture content was reduced to 6–10% prior to storage.

The seed of the *Populus* section is noticeably smaller in size. Nevertheless, it will maintain viability quite well for 1 year when stored over desiccant at 4°C (Wyckoff 1975). Johnson (1946) reported that *P. grandidentata* and *P. tremuloides* seed, stored at 20% relative humidity and 20°C, maintained appreciable viability for up to 5 mo. Over the long term however, viability of seed of the *Populus* section is best maintained at low temperatures. Fechner et al. (1981) recommended storage at –18°C for *P. tremuloides* seed kept for 2 years. Benson and Harder (1972) reported good results for *P. tremuloides* ×*canescens* and *P. tremuloides* ×*grandidentata* hybrid seed undergoing 4 year's storage at –24°C. *P. tremula* seed has been stored at –20°C for 44 mo with good results (Simak 1980).

Germination and establishment

Germination and early seedling development of most *Populus* species is rapid and nearly complete over a temperature range of 20–30°C with adequate moisture (Faust 1936; Farmer and Bonner 1967; McDonough 1979; Fechner et al. 1981). Although apparently unaffected by light (Faust 1936), germination is most successful when seed is not covered with soil (McDonough 1979; Singh et al. 1985). However, a light sandy-loam top dressing over the seed may at times be warranted to control algae without inhibiting germination (Benson and Einspahr 1962). Seed size too is an important determinant of germination success. Twofold increases in germination percentages accompanying lots of increased seed volume have been reported for *P. deltoides* (Farmer and Bonner 1967), and *P. grandidentata* and *P. tremuloides* (Faust 1936). Increasing seed weight has also been associated with larger seedlings in *P. tremula* (Bergman and Lantz 1958; Gladysz 1983) and *P. deltoides* (Hardin 1984). A like relationship between seed weight and seedling height could not be detected in *P. tremula, P. tremuloides*, and their hybrids beyond the third month after germination (Gallo 1985). Removal of the cotton fibers is similarly important, as their attachment to the seed can prevent close soil contact, thereby hampering both imbibition and early seedling development (McDonough 1979; Myers and Fechner 1980; Singh and Arya 1987). During the initial growth stage, *Populus* seedlings may be susceptible to damping-off fungi. Effective control can be provided through the use of boiling water soil drenches and the use of finely ground sphagnum moss or silica sand germination media (Shea and Kuntz 1956). Sterilization by heating moistened germination media to 85°C for 30 min is also highly effective (Hartman and Kester 1975) as are a variety of commercial fungicidal soil drenches.

Although the practice of nursery production of bare-root *Populus* seedling stock was developed some time ago (Engstrom 1948; Einspahr 1959; Wycoff 1960; Gammage and Maisenhelder 1962), control-bred seed lots usually have sufficiently high value to justify greenhouse containerized propagation using a variety of soilless mixes. In many interspecific crosses, limited seed quantities are the rule. Such seed is often germinated in flats and then later transplanted to containers when seedlings have grown to a height of 1 cm to conserve a maximum number of progeny (Burr 1986). When abundant seed yield is the norm, seed is sown directly into containers and thinned 3–4 wk later as competition becomes keen.

Populus seedling growth is optimized in greenhouse environments approximating a temperature setting of 24°C during 18 h of daylight (supplemental lighting provided by a bank of alternating incandescent and fluorescent lights set 1.2 m above bench level) and 18°C during the night hours (Faltonson 1982). Liquid fertilizer applications in concentrations approximating 200 ppm nitrogen, 88 ppm phosphorus, and 166 ppm potassium applied once or twice weekly seem suitable for the *Aigeiros* and *Tacamahaca* sections (Faltonson 1982). Seedlings of the *Populus* section have lower nutrient requirements; optimum concentrations are 75–100 ppm nitrogen, 22 ppm phosphorus, 62–77 ppm potassium, 46 ppm calcium, and 21 ppm magnesium (Einspahr 1971). Either overhead mist or subsurface irrigation is required for the first 4–6 wk period following sowing as the crop may prove too succulent to withstand heavy overhead watering.

Embryo culture

Although seed propagation of *Populus* is straightforward and generally quite reliable, there are conditions under which its outcome is not well assured. For example, the premature splitting and abscission of capsules in response to embryo abortion is commonplace in hybridization of *P. trichocarpa* and *P. deltoides* in which the former serves as the female parent. Moreover, in the reciprocal cross, hybrid seed lots oftentimes exhibit poor germination due to a moderately high frequency of immature embryos. Obstacles such as these have fostered the development of *in vitro Populus* embryo culture techniques about 12 years ago to improve the recovery rates of such heretofore sporadic hybrid genotypes. More recently, the technique has been applied to those crosses whose lengthy periods of capsule maturation pose considerable difficulty for seed propagation. One good example involves intraspecific crosses of southern provenances of *P. deltoides* in which the 12–20 wk period between pollination and dehiscence often does not allow for adequate time for growing and hardening the stock during a single season. Here *in vitro* culture has been used to step up the timing of germination by months so as to expedite the ensuing seedling rearing and hardening-off stages.

Originally *in vitro* procedures involved the removal of individual embryos or whole ovules several weeks after pollination (Kouider et al. 1984; Li and Li 1985; Noh et al. 1986). These could be cultured on Murashige-Skoog (MS) salt media under a 16-h photoperiod of low light intensity at 20–22°C to induce shoot development. Excised shoots were then subcultured on modified MS medium to induce rooting, resulting in plantlets that could be gradually acclimated to soil culture while maintaining a high-humidity environment. Since that time, a one-step method of liquid suspension culture has been recommended by Savka et al. (1987). Further improvements have included the culture of half capsules or individual carpels followed by the subculture of germinated embryos (Raquin et al. 1993). This technique is rapid, applicable to a wide array of intra- and interspecific crosses, and eliminates the injury that can occur when individual embryos are excised. More recently, a technique of culturing a capsule's internal contents *en masse* has been successfully worked out and used in large-scale *Populus* improvement programs (T. Chen, personal communication).

Concluding comments

Floral branch collection, pollen extraction, capsule fruition of female cuttings or grafts, and progeny propagation from seed or embryos are the four major steps in the controlled crossing of *Populus*. Success hinges on attention to six critical functions: the handling and chilling of floral branches, the regulation of humidity and temperature during staminate anthesis, dehiscence, and pollen extraction, the maintenance of the moisture status of female breeding stock, the prevention of open pollination, the control of temperature and humidity during seed processing and storage, and the moderation of temperature and soil moisture and nutrition during seed germination and establishment.

Improvements in the overall process are especially warranted where reciprocal interspecific crosses exhibit large differences in success like the *P. nigra* ×*deltoides* (Melchior and Seitz 1968) and *P. trichocarpa* ×*deltoides* combinations. Such reproductive failures may prevent implementation of parental population improvement efforts (reciprocal recurrent selection for example) on which long-term, first generation hybrid breeding strategies may come to rely (Bisoffi 1989). Here the refinement of *in vitro* techniques that target, perhaps, the earliest embryogenic stages would be valuable to the extent that they lead to more broadly based estimates of breeding value and, thereby, parental population advancements.

Equally important is the identification of techniques to stimulate early flowering, of which no reliable methods currently exist for *Populus* (Meilan and Strauss 1996). This would ensure that the generation cycle is coincident with the testing cylce, the latter possibly shortened by applications of molecular techniques (Bradshaw and Stettler 1995). Related is the extended flowering of potted grafts and rooted cuttings that has been observed at times (Pauley 1949). Investigations here are meaningful too as they may yet provide an alternative to the high cost of establishing and maintaining traditional breeding orchards.

References

Al'benskii, A.V., and Delitsina, A.V. 1934. (Experiments on hybridizing poplars in the laboratory) Experiments and investigations. All-Union Inst. For. Cult. and For. Melioration, Moscow, 2nd issue. pp. 107–119 (Plant Breeding Abst. 5: 62. 1934).

Allen, R.M., and McComb, A.L. 1956. Rooting of cottonwood cuttings. U.S. For. Serv. South. For. Exp. Stn. Occas. Pap. 151. 10 p.

Asakawa, S. 1982. Storage of *Populus maximowiczii* seeds. *In* Proceedings of International Symposium on Forest Tree Seed Storage. *Edited by* B.S.P. Wang and J.A. Pitel. Canadian Forestry Service, Chalk River, ON, Canada. pp. 136–141.

Barner, H., and Christiansen, H. 1958. On the extraction of forest-tree pollen from inflorescences forced in a specially designed house. Silvae Genet. **7**: 19–24.

Barnes, B.V. 1958. Erste aufnahme eines sechsjahrigen bestandes von aspenhybriden. [First survey of a six-year-old stand of hybrid aspen, translated from German.] Silvae Genet. **7**: 98–102.

Barnes, B.V. 1961. Hybrid aspens in the lower peninsula of Michigan. Rhodora, **63**: 311–324.

Benson, M.K. 1972. Breeding and establishment—and promising hybrids. *In* Proceedings of Aspen Symposium, College of Forestry, University of Minnesota, Minneapolis, MN. U.S. For. Serv. Gen. Tech. Rep. NC-1. pp. 88–96.

Benson, M.K., and Einspahr, D.W. 1962. Improved method for nursery production of quaking aspen seedlings. U.S. For. Serv. Tree Plant. Notes, **53**: 11–13.

Benson, M.K., and Harder, M.L. 1972. Storage of aspen seed. Institiue of Paper Chemistry. Genet. Physiol. Notes, **11**. pp. 1–4.

Bergman, F., and Lantz, A. 1958. Ein versuch zum treiben von kreuzungsreisern von aspen (*Populus tremula* L.) bei niedriger temperatur. [Attempt at forcing branches of aspen (*Populus tremula* L.) at low temperatures for crossing purposes, translated from German.] Silvae Genet. **7**: 155–159.

Bisoffi, S. 1989. Recent developments of poplar breeding in Italy. *In* Proceedings of a meeting of the International Union of Forestry Research Organizations, Working Party S2.02.10, Hann. Recent developments in poplar selection and propagation techniques. Münden, Germany. pp. 18–45.

Blake, G.M., Hossfeld, R.L., and Pauley, S.S. 1960. A technique for determining sex from vegetative buds of quaking aspen. For. Sci. **6**: 363–364.

Blake, G.M., Hossfeld, R.L., and Pauley, S.S. 1967. More on the determination of sex from the vegetative buds of aspen. For. Sci. **13**: 89.

Bloomberg, W.J. 1959. Root formation of black cottonwood cuttings in relation to region of parent shoot. For. Chron. **35**: 13–17.

Boes, T.K., and Strauss, S.H. 1994. Floral phenology and morphology of black cottonwood, *Populus trichocarpa* (*Salicaceae*). Am. J. Bot. **81**: 562–567.

Bonner, F.T., and Broadfoot, W.M. 1967. Growth response of eastern cottonwood to nutrients in sand culture. U.S. For. Serv. South. For. Exp. Stn., Res. Note SO-65. 4 p.

Bradshaw, H.D., Jr., and Stettler, R.F. 1995. Molecular genetics of growth and development in *Populus*. IV. Mapping QTLs with large effects on growth, form, and phenology traits in a forest tree. Genetics, **139**: 963–973.

Brewbaker, J.L., and Kwack, B.H. 1963. The essential role of calcium ion in pollen germination and pollen tube growth. Am. J. Bot. **50**: 859–864.

Brown, K.R. 1989. Catkin growth, seed production, and development of seed germinability in quaking aspen in central Alberta. U.S. For. Serv. Tree Plant. Notes, **40**: 25–29.

Burr, K.E. 1986. Greenhouse production of quaking aspen seedlings. U.S. For. Serv. Rocky Mt. For. Range Exp. Stn. Gen. Tech. Rep. RM-125. pp. 31–37.

Cagelli, L. 1995. Task 3 — Improvement of breeding tools, long-term preservation of germplasm. Project: interdisciplinary research for poplar improvement. Istituto di Sperimentazione per la Pioppicoltura (SAF-ENCC). Casale Monferrato. 3 p.

Comtois, P., Simon, J.P., and Payette, S. 1986. Clonal constitution and sex ratio in northern populations of balsam poplar *Populus balsamifera*. Holarct. Ecol. **9**: 251–260.

Cooper, D.T. 1990. *Populus deltoides* Bartr. ex Marsh. var. *deltoides* eastern cottonwood. *Salicaceae* Willow family. *In* Silvics of North America. Vol. 2, Hardwoods. *Technical coordinators*: R.M. Burns and B.H. Honkala. U.S. Dep. Agric. Agric. Handb. 654. pp. 530–537.

Cunningham, F.E. 1953. Rooting ability of native cottonwood depends on the clone used. U.S. For. Serv. Northeast. For. Exp. Stn. Res. Note 26. 2 p.

Cunningham, T.W., and Farmer, R.E., Jr. 1984. Seasonal variation in propagability of dormant balsam poplar cuttings. The Plant Propagator, **30**: 13–15.

denHeyer, J., and Seymour, N. 1978. Aspen and balsam poplar seed collection and storage. U.S. For. Serv. Tree Plant. Notes, **29**: 35.

Dhir, N.K., and Mohn, C.A. 1976. A comparative study of crosses between and within two geographically diverse sources of eastern cottonwood. Can. J. For. Res. **6**: 400–405.

Dhir, K.K., Chark, K.S., Khurana, D.K., and Dua I.S. 1982. Changes in the protein bands in pollen grains of *Populus ciliata* during storage and its effect on their viability and germination. Silvae Genet. **31**: 6–8.

Dolgosheev, V.M. 1968. Correlation between masculine and feminine aspen individuals in forests of Kirov district [from author's English summary]. Lesovedenie, **4**: 97–99.

Einspahr, D.W. 1959. Nursery production of aspen seedlings. U.S. For. Serv. Tree Plant. Notes, **35**: 22–24.

Einspahr, D.W. 1960. Sex ratio in quaking aspen and possible sex-related characteristics. *In* Proceedings of the Fifth World Forestry Congress, **2**: 747–750.

Einspahr, D.W. 1971. Growth and nutrient uptake of aspen hybrids using sand culture techniques. Silvae Genet. **20**: 132–137.

Einspahr, D.W., and Benson,M.K. 1964. Production and evaluation of aspen hybrids. J. For. **62**: 806–809.

Einspahr, D.W., and Benson, M.K. 1971. Development of a holding solution for use in crossing cottonwood. *In* Progress Report No. 23 to the Lake States Aspen Genetics and Tree Improvement Group. Genetic improvement of aspen — Basic and applied studies during 1971. Institute of Paper Chemistry, Appleton, WI, Project 1800. pp. 33–36.

Einspahr, D.W., and Joranson, P.N. 1960. Late flowering in aspen and its relation to naturally occurring hybrids. For. Sci. **6**: 221–224.

Einspahr, D., and Schlafke, D. 1957. A method for aspen and cottonwood seed extraction. U.S. For. Serv. Tree Plant. Notes, **28**: 10.

Einspahr, D.W., and Winton, L.L. 1976. Genetics of quaking aspen. U.S. For. Serv. Res. Pap. WO-25. 23 p.

Engstrom, A. 1948. Growing cottonwood from seed. J. For. **46**: 130–132.

Falinski, J.B. 1980. Vegetation dynamics and sex structure of the populations of pioneer dioecious woody plants. Vegetatio, **43**: 23–38.

Faltonson, R. 1982. Controlled-environment culture of *Populus* clones. *In* Methods of rapid, early selection of poplar clones for maximum yield potential: A manual of procedures. U.S. For. Serv. North Central For. Exp. Stn. Gen. Tech. Rep. NC-81. pp. 12–16.

Farmer, R.E, Jr. 1964*a*. Cottonwood flowering as related to cold requirement of flower buds. For. Sci. **10**: 296–299.

Farmer, R.E., Jr. 1964*b*. Sex ratio and sex-related characteristics in eastern cottonwood. Silvae Genet. **13**: 116–118.

Farmer, R.E., Jr. 1966*a*. Variation in time of flowering and seed dispersal of eastern cottonwood in the lower Mississippi valley. For. Sci. **12**: 343–347.

Farmer, R.E., Jr. 1966*b*. Rooting dormant cuttings of mature cottonwood. J. For. **64**: 196–197.

Farmer, R.E., Jr. 1993. Latitudinal variation in height and phenology of balsam poplar. Silvae Genet. **42**: 148–153.

Farmer, R.E., Jr., and Bonner, F.T. 1967. Germination and initial growth of eastern cottonwood as influenced by moisture stress, temperature, and storage. Bot. Gaz. **128**: 211–215.

Farmer, R.E., Jr., and Nance, W.L. 1967. Crossing eastern cottonwood in the greenhouse. *In* Proceedings of the International Plant Propagators Society, **17**: 333–338.

Farmer, R.E., Jr., and Reinholt, R.W. 1986. Genetic variation in dormancy relations of balsam poplar along a latitudinal transect in northwestern Ontario. Silvae Genet. **35**: 38–42.

Farmer, R.E., Jr., Freitag, M., and Garlick, K. 1989. Genetic variance and "C" effects in balsam poplar rooting. Silvae Genet. **38**: 62–65.

Faust, M.E. 1936. Germination of *Populus grandidentata* and *P. tremuloides* with particular reference to oxygen consumption. Bot. Gaz. **97**: 808–821.

Fechner, G.H. 1972. Development of the pistillate flower of *Populus tremuloides* following controlled pollination. Can. J. Bot. **50**: 2503–2509.

Fechner, G.H. 1976. Development of unpollinated ovules of quaking aspen. *In* Proceedings of the Northeast. For. Tree Improve. Conf. **23**: 150–157.

Fechner, G.H., Burr, K.E., and Myers, J.F. 1981. Effects of storage, temperature, and moisture stress on seed germination and early seedling development of trembling aspen. Can. J. For. Res. **11**: 718–722.

Fung, M.Y.P., and Hamel, B.A. 1993. Aspen seed collection and extraction. U.S. For. Serv. Tree Plant. Notes, **44**: 98–100.

Gallo, L.A. 1985. Uber genetisch und unweltbedingte variation bei aspen. I. Keimung und gewicht der samen. [Genetic and environmental variation in aspen. I. Germination and seed weight, from author's English summary.] Silvae Genet. **34**: 171–181.

Gammage, J.L., and Maisenhelder, L.C. 1962. Easy way to sow cottonwood nursery beds. U.S. For. Serv. Tree Plant. Notes, **51**: 19–20.

Gladysz, A. 1983. Jakosc nasion osiki produkowanych w warunkach szklarniowych. [Quality of aspen seeds produced under greenhouse conditions, translated from Polish.] Sylwan, **127**: 9–20.

Grant, M.C., and Mitton, J.B. 1979. Elevational gradients in adult sex ratios and sexual differentiation in vegetative growth rates of *Populus tremuloides* Michx. Evolution, **33**: 914–918.

Gras, M.A., Mughini, G., Cabizzosu, A., Bisoffi, S., and Girino, P. 1987. Tecnica di impollinazione controllata per i pioppi della sezione *Aigeiros*. [Technique of controlled pollination for poplars of the *Aigeiros* section, translated from Italian.] Agricoltura Ricerca, **66**: 7–10.

Hall, R.B., Hart, E.R., McNabb, H.S., and Schultz, R.C. 1991. Selection and breeding of pest-resistant clones of *Populus* for biomasss energy production in the north central region. 1991–1996 Department of the Environment Project Proposal — Program growth and yield (1988–1995 growing seasons). Iowa State University, Ames, IA. 88 p.

Hamilton, T.H. 1966. Seed storage of northern black cottonwood (*Populus trichocarpa* Torr. & Gray). B.C. For. Serv. For. Res. Rev. pp. 83–84.

Hardin, E.D. 1984. Variation in seed weight, number per capsule and germination in *Populus deltoides* Bartr. trees in southeastern Ohio. Am. Mid. Nat. **112**: 29–34.

Hartman, H.T., and Kester, D.E. 1975. Plant propagation, principles and practices. 3rd edition. Prentice-Hall, Inc., Englewood Cliffs, NJ. 662 p.

Hellum, A.K. 1973. Seed storage and germination of black poplar. Can. J. Plant Sci. **53**: 227–228.

Henry, A. 1914. The artificial production of vigorous trees. J. Dep. Agric. and Tech. Inst., (Ireland), **15**: 34–52.

Herrmann, S. 1976. Verfahren zur konservierung und erhaltung der befruchtungsfahigkeit von waldbaumpollen uber mehrere jahre. [Methods for the conservation and maintenance of the fertility of forest tree pollen over several years, from author's English summary.] Silvae Genet. **25**: 223-229.

Heslop-Harrison, J., and Heslop-Harrison, Y. 1970. Evaluation of pollen viability using induced fluorescence, intracellular hydrolysis of fluorescein diacetate. Stain Technol. **45**: 115–120.

Houle, G., and Babeux, P. 1993. Temporal variations in the rooting ability of cuttings of *Populus balsamifera* and *Salix planifolia* from natural clones-populations of subarctic Quebec. Can. J. For. Res. **23**: 2603–2608.

Howe, G.T., Hackett, W.P., Furnier, G.R., and Klevorn, R.E. 1995. Photoperiodic responses of a northern and southern ecotype of black cottonwood. Physiol. Plant. **93**: 695–708.

Il'in, A.M. 1974. Sex ratio in the aspen under different conditions of growth. Sov. J. Ecol. **4**: 162–163.

Joennoz, R., and Vallee, G. 1972 Recherche et developpement sur le peuplier dans la region de L'Est-du-Québec. II—Resultats d'hybridations artificielles chez les peupliers [Research and development on poplar in the region of eastern Quebec. II — Results of artificial hybridizations with poplars, translated from French.] Service de la recherche, Direction generale des forêts, Ministere des Terres et Forêts, Memoire No. 13. 36 p.

Johnson, L.P.V. 1945. Development of sexual and vegetative organs on detached forest tree branches cultured in the greenhouse. For. Chron. **21**: 130–136.

Johnson, L.P.V. 1946. Effect of humidity on the longevity of *Populus* and *Ulmus* seeds in storage. Can. J. Res. **24**: 298–302.

Kaul, R.B., and Kaul, M.N. 1984. Sex ratios of *Populus deltoides* and *Salix amygdaloides* (*Salicaeae*) in Nebraska. Southwest. Nat. **29**: 265–269.

Khosla, P.K., Dhall, S.P., and Khurana, D.K. 1979. Studies in *Populus ciliata* Wall. ex Royle 1. Correlation of phenotypic observations with sex of trees. Silvae Genet. **28**: 21–23.

Khurana, D.K., and Khosla, P.K. 1982. Studies in *Populus ciliata* Wall. ex Royle III. Phenotypic variation in relation to ecological blocks. J. Tree Sci. **1**: 35–45.

Klaehn, F.U., and Neu, R.L. 1960. Hardwood pollen study. Silvae Genet. **9**: 44–48.

Knox, R.B., Willing, R.R., and Pryor, L.D. 1972. Interspecific hybridization in poplars using recognition pollen. Silvae Genet. **21**: 65–69.

Koster, R. 1968. Poplar breeding in the Netherlands. International Poplar Commission, 13th Session. Food and Agriculture Organization of the United Nations, Montreal, QC. 18 p.

Kouider, M., Skirvin, R.M., Saladin, K.P., Dawson, J.O., and Jokela, J.J. 1984. A method to culture immature embryos of *Populus deltoides in vitro*. Can. J. For. Res. **14**: 956–958.

Laidly, P.R. 1990. *Populus grandidentata* Michx. bigtooth aspen. *Salicaceae* Willow family. *In* Silvics of North America. Vol. 2, Hardwoods. *Technical coordinators*: R.M. Burns and B.H. Honkala. U.S. Dep. Agric. Agric. Handb. 654. pp. 544–550.

Larson, P.R. 1958. Effect of gibberellic acid on forcing hardwood cuttings for pollen collection. U.S. For. Serv. Lake States For. Exp. Stn. Tech. Note 538. 2 p.

Larsson, H.C. 1976. Technique of mass producing *jackii* poplar seed under greenhouse conditions. Proceedings of Northeast. For. Tree Improve. Conf. **23**: 158–165.

Lester, D.T. 1961. Observations on flowering in the aspens. Proceedings of Northeast. For. Tree Improve. Conf. **8**: 35–38.

Lester, D.T. 1963*a*. Floral initiation and development in quaking aspen. For. Sci. **9**: 323–329.

Lester, D.T. 1963*b*. Variation in sex expression in *Populus tremuloides* Michx. Silvae Genet. **12**: 141–151.

Li, S.V. 1960. Use of repeated pollinations in the breeding of poplars. [Translated from Russian.] Vestn. Sel'skokhoz. Nauk., Moskova, **5**: 140–143.

Li, W. and Li, J. 1985. *In vitro* culture of hybrid ovules in *Populus* [from author's English abstract]. Sci. Silvae Sin. **21**: 339–346.

Maisenhelder, L.C. 1960. Cottonwood plantations for southern bottom lands. U.S. For. Serv. South. For. Exp. Stn. Occas. Pap. 179. 24 p.

McComb, A.L., and Lovestead, H.S. 1954. Viability of cottonwood seeds in relation to storage temperatures and humidities. U.S. For. Serv. Tree Plant. Notes, **17**: 9–11.

McDonough, W.T. 1979. Quaking aspen—seed germination and early seedling growth. U.S. For. Serv. Intermt. For. and Range Exp. Stn. Res. Pap. INT-234. 13 p.

Meilan, R., and Strauss, S.H. 1996. Stimulating precocious flowering in woody angiosperms. New Forests. In press.

Melchior, G.H., and Seitz, F.W. 1968. Interspezifische kreuzungssterilitat innerhalb der pappelsektion *Aigeiros*. [Interspecific cross sterility within the poplar section Aigerios, from author's English summary.] Silvae Genet. **17**: 88–93.

Miller, L.G. 1972. The controlled pollination of eastern cottonwood in the greenhouse. Oklahoma State University, M.Sc. thesis, Oklahoma State University, Stillwater, OK. 31 p.

Moss, E.H. 1938. Longevity of seed and establishment of seedlings in species of *Populus*. Bot. Gaz. **99**: 529–542.

Muhle Larsen, C. 1960. L'amelioration du peuplier par voie genetique. [Poplar improvement through genetic means, translated from French.] Extrait du Bulletin de la Societe Royale Forestiere de Belgique 67, No. 4. pp. 149–155.

Myers, J.F., and Fechner, G.H. 1980. Seed hairs and seed germination in *Populus*. U.S. For. Serv. Tree Plant. Notes, **31**: 3–4.

Nagaraj, M. 1952. Floral morphology of *Populus deltoides* and *P. tremuloides*. Bot. Gaz. **114**: 222–243.

Nelson, C.D., and Tauer, C.G. 1987. Genetic variation in juvenile characters of *Populus deltoides* Bartr. from the southern great plains. Silvae Genet. **36**: 216–221.

Noh, E.R., Koo, Y.B., and Lee, S.K. 1986. Hybridization between incompatible poplar species through ovary and embryo culture [from author's English summary]. Res. Rep. Inst. For. Genet. (Korea), **22**: 9–14.

Pauley, S.S. 1949. Forest-tree genetics research: *Populus* L. Econ. Bot. **3**: 299–330.

Pauley, S.S. 1950. Flowering habits in *Populus*. Genetics, **35**: 684.

Pauley, S.S., and Mennel, G.F. 1957. Sex ratio and hermaphroditism in a natural population of quaking aspen. Minn. Agr. Exp. Stn., Minn. For. Notes, **55**. 2 p.

Pauley, S.S., and Perry, T.O. 1954. Ecotypic variation of the photoperiodic response in *Populus*. J. Arn. Arb. **35**: 167–188.

Perala, D.A. 1990. *Populus tremuloides* Michx. quaking aspen. *Salicaceae* Willow family. *In* Silvics of North America. Vol. 2, Hardwoods. *Technical coordinators*: R.M. Burns and B.H. Honkala. U.S. Dep. Agric. Agric. Handb. 654. pp. 555–569.

Pohl, Z. 1962. Studia nad wzrostem I morfologia kornickich mieszancow *Populus maximowiczii* Henry. [Studies on growth and morphology of *Populus maximowiczii* Henry hybrids at Kornik Arboretum, translated from Polish.] U.S. Dep. Commer., Arbor. Kornickie, **7**: 115–184.

Pregitzer, K.S., and Barnes, B.V. 1980. Flowering phenology of *Populus tremuloides* and *P. grandidentata* and the potential for hybridization. Can. J. For. Res. **10**: 218–223.

Rajora, O.P., and Zsuffa, L. 1986. Pollen viability of some *Populus* species as indicated by *in vitro* pollen germination and tetrazolium chloride staining. Can. J. Bot. **64**: 1086–1088.

Raquin, C., Troussard, L., and Villar, M. 1993. In-ovary embryo culture as a tool for poplar hybridization. Can. J. Bot. **71**: 1271–1275.

Riemenschneider, D.E., and McMahon, B.G. 1993. Genetic variation among lake states balsam poplar populations is associated with geographic origin. For. Sci. **39**: 130–136.

Roe, E.I., and McCain, D.P. 1962. A quick method of collecting and cleaning aspen seed. U.S. For. Serv. Tree Plant. Notes, **51**: 17–18.

Ronald, W.G. 1982. Intersectional hybridization of *Populus* sections, *Leuce-Aigeiros* and *Leuce-Tacamahaca*. Silvae Genet. **31**: 94–99.

Savka, M.A., Dawson, J.O., Jokela, J.J., and Skirvin, R.M. 1987. A liquid culture method for rescuing immature embryos of eastern cottonwood. Plant Cell Tissue Organ Cult. **10**: 221–226.

Schier, G.A., and R.B. Campbell. 1976. Differences among *Populus* species in ability to form adventitious shoots and roots. Can. J. For. Res. **6**: 253–261.

Schreiner, E.J. 1974. *Populus* L. Poplar. *In* Seeds of Woody Plants in the United States. *Technical coordinator*: C.S. Schopmeyer. U.S. Dep. Agric. Agric. Handb. 450. pp. 645–655.

Seitz, F.W. 1958. Fruhtreibversuche mit bluhreisern der aspe. [Forcing tests with flowering shoots of aspen, translated from German.] Silvae Genet. **7**: 102–105.

Shea, K.R., and Kuntz, J.E. 1956. Prevention of damping-off of poplar seedlings. For. Sci. **2**: 54–57.

Simak, M. 1980. Germination and storage of *Salix caprea* L. and *Populus tremula* L. seeds. *In* Proceedings of the International Symposium on Forest Tree Seed Storage. *Edited by* B.S.P. Wang and J.A. Pitel. Canadian Forestry Service, Chalk River, ON. pp. 142–160.

Singh, V., and Arya, S.R. 1987. *Populus ciliata*—Effect of parts of catkins on seed germination. Society of Indian Foresters, **25**: 26–28.

Singh, R.V., and Gupta, K.C. 1981. Preliminary studies on germination of *Populus ciliata* seed. *In* Proceedings of Silvilculture, Management, and Utilization of Poplars. *Edited by* R.V. Singh. pp. 35–37.

Singh, R.V., Sharma, K.C., and Singh, V. 1985. Germination of *P. ciliata* seed as affected by depth of sowing. Indian For. **111**: 245–249.

Stanton, B.J. 1990. *Populus* controlled breeding procedures. [Available from Northwest Fiber Supply Div., James River Corp., Camas, WA.] 15 p.

Stettler, R.F., and Guries, R.P. 1976. The mentor pollen phenomenon in black cottonwood. Can. J. Bot. **54**: 820–830.

Stettler, R.F., Koster, R., and Steenackers, V. 1980. Interspecific crossability studies in poplars. Theor. Appl. Genet. **58**: 273–282.

Stout, A.B., and Schreiner, E.J. 1933. Results of a project in hybridizing poplars. J. Hered. **24**: 217–229.

Strobl, S. 1992. Hybrid poplar tree improvement plan for southern Ontario. [Available from Ontario Ministry of Natural Resources, Brockville, ON.] 68 p.

Tauer, C.G. 1979. Seed tree, vacuum, and temperature effects on eastern cottonwood seed viability during extended storage. For. Sci. **25**: 112–114.

Tkachenko, B.V. 1984. Raising promising poplar hybrids at the Trostyanets selection station. [Translated from Russian.] Lesovod. Agrolesomelior. Resp. Mezhved. Temat. Sb. **69**: 17–20.

Vaartaja, O. 1960. Ecotypic variation of photoperiodic response in trees especially in two *Populus* species. For. Sci. **6**: 200–206.

Valentine, F.A. 1975. Genetic control of sex ratio, earliness and frequency of flowering in *Populus tremuloides*. *In* Proceedings of Northeast. For. Tree Improve. Conf. **22**: 111–129.

Villar, M., Gaget, M., Rougier, M., and Dumas, C. 1993. Pollen-pistil interactions in *Populus*: B-galactosidase activity associated with pollen tube growth during the crosses *Populus nigra* × *P. nigra* and *P. nigra* × *P. alba*. Sex Plant Reprod. **6**: 249–256.

Walters, G.S., and Bruckmann, G. 1965. Variation in specific gravity of cottonwood as affected by tree sex and stand location. J. For. **63**: 182–185.

Weber, J.C., Stettler, R.F., and Heilman, P.E. 1985. Genetic variation and productivity of *Populus trichocarpa* and its hybrids. I. Morphology and phenology of 50 native clones. Can. J. For. Res. **15**: 376–383.

Wettstein-Westersheim, W. 1933. Die kreuzungsmethode und die beschreibung von F_1 - bastarden bei Populus. [The method of hybridization and the description of F_1 hybrids of poplars.] Z. Zuchtung. **18**: 597–626. (Plant. Breed. Abst. **4**: 163–164.)

Wilcox, J.R., and Farmer, R.E., Jr. 1968. Heritability and C effects in early root growth of eastern cottonwood cuttings. Heredity, **23**: 239–245.

Williamson, A.W. 1913. Cottonwood in the Mississippi Valley. U.S. Dep. Agric. Agric. Inf. Bull. 24. 62 p.

Winton, L.L. 1968. Fertilization in forced quaking aspen and cottonwood. Silvae Genet. **17**: 20–21.

Worsley, R.G.F. 1959. The processing of pollen. Silvae Genet. **8**: 143–148.

Wyckoff, G.W. 1975. Procedures for crossing bigtooth and quaking aspen. [Available from Institute of Paper Chemistry, Appleton, WI.] 13 p.

Wyckoff, G.W. 1980. Air-pollination chamber for use in *Populus* breeding. U.S. For. Serv. Tree Plant. Notes, **31**: 5.

Wycoff, H.B. 1960. Cottonwood seeding at the Mason State tree nursery. U.S. For. Serv. Tree Plant. Notes, **41**: 13.

Yang, Z.M., and Villar, M. 1989. Influence of thermo-hygrometric conditions on poplar pollen viability during its collection and storage: preliminary study. National Research Institute of Agronomy, INRA, Orleans, France. 13 p.

Ying, C.C., and Bagley, W.T. 1976. Genetic variation of eastern cottonwood in an eastern Nebraska provenance study. Silvae Genet. **25**: 67–73.

Ying, C.C., and Bagley, W.T. 1977. Variation in rooting capability of *Populus deltoides*. Silvae Genet. **26**: 204–207.

Zasada, J.C., and Densmore, R. 1980. Alaskan willow and balsam poplar seed viability after 3 years' storage. U.S. For. Serv. Tree Plant. Notes, **31**: 9–10.

CHAPTER 6
Poplar breeding and selection strategies

Stefano Bisoffi and Urban Gullberg

Introduction

Poplars and aspens (*Populus* spp.) are a primary source of wood in many countries of the temperate regions with current expansion to subtropical areas. Ease of vegetative propagation, fast growth, wide interspecific crossability, and plasticity with regards to end uses are the key factors of their attractiveness.

Although natural forests still represent a large source of poplar wood for some parts of the world the most common form of poplar silviculture is in artificial plantations, more or less intensively cultivated, and the most remarkable progress in breeding and selection has been made with artificial plantations as a target.

Poplar breeding is now in an expansive phase. Belgium, Canada, France, Italy, and the USA are in the process of restructuring their breeding programs and new breeding activities are being started in many more countries. Basic knowledge about poplar genetics is also rapidly increasing, since many scientific laboratories use it as a model species for a basic understanding of tree properties.

It seems therefore timely to discuss poplar breeding and selection strategies. The major strategic problems to consider are the balance between population breeding and gene technology and the integration of long-term and short-term activities. We suggest considering a systematic breeding approach where long-term and short-term breeding are integrated (Gullberg and Kang 1985; Kang and Nienstaedt 1987; Kang et al. 1996). In long-term breeding the focus is on maintaining genetic variation and creating options for the future, while in short-term

S. Bisoffi. Istituto di Sperimentazione per la Pioppicoltura, Box 116, Via di Frassineto 35, I-15033 Casale Monferrato AL, Italy.
U. Gullberg. Swedish University of Agricultural Sciences, Department of Plant Breeding Research, Box 7003, S-750 07 Uppsala, Sweden.
Correct citation: Bisoffi, S., and Gullberg, U. 1996. Poplar breeding and selection strategies. *In* Biology of *Populus* and its implications for management and conservation. Part I, Chapter 6. *Edited by* R.F. Stettler, H.D. Bradshaw, Jr., P.E. Heilman, and T.M. Hinckley. NRC Research Press, National Research Council of Canada, Ottawa, ON. pp. 139–158.

breeding all effort is focused on reaching economic goals, e.g., by clonal selection of commercial cultivars.

In the following we will justify multiple generation breeding, present the basic options for maintaining multiple generation populations and give ideas on how to use a subdivided breeding population. We shall also address some key aspects of clonal selection that are worth considering in order to maximize its efficiency.

Features of poplar breeding

Two main features have characterized the history of poplar breeding: interspecific breeding and clonal selection. Interspecific hybrids are quite often found in nature where the ranges of two species overlap (Eckenwalder 1984) and many more have appeared as a consequence of deliberate transfers of germplasm; many such spontaneous hybrids came into use in poplar culture on both sides of the Atlantic already in the 18th century (Houtzagers 1937; Garrett 1976) although the first deliberate crosses were made only at the beginning of the 20th century (Henry 1914). Interspecific hybrids were also the objective of the first breeding programs conducted with modern criteria (Jacometti 1937; Schreiner 1949, 1970).

The superiority of F_1 hybrids has been more often assumed than proved, improperly calling heterosis the superiority of selected clones in F_1 families with respect to the high parent (Muhle Larsen 1970). Recent rigorous studies, however, confirm that hybrid vigor exists (Stettler et al. 1988; Bradshaw and Stettler 1995). Superiority of F_1 hybrids seems to be generally due to complementation by dominant or partially dominant genes carried separately by two parent species at different *loci*, a general phenomenon observed also in agricultural crops (Jinks 1983; see also Stettler et al., Chapter 4).

Interspecific hybridization has also proved to be a promising starting point for the genetic dissection of many traits related to poplar growth. Indeed poplars, thanks to a relatively small genome, the availability of suitable pedigrees, and also the efficient collaboration among researchers, were among the first forest trees for which a dense linkage map was produced (Bradshaw et al. 1994). This characterization has already led to remarkable and largely unexpected findings about the inheritance of quantitative characters (Bradshaw and Stettler 1995) casting some doubts on the validity of fundamental assumptions of quantitative genetics (see also Bradshaw, Chapter 8).

The second aspect of most improvement programs has been their strong committment to clonal selection (Mohrdiek 1983; Thielges 1985; Bisoffi 1990). This was partly dictated by limited resources that imposed the maximization of short term benefits but also by the relative ease and precision of clonal evaluation with trees that are easier to propagate vegetatively than sexually. Recently, marker-assisted selection (Lander and Botstein 1989; Lande and Thompson 1990), although of debatable value in trees (Strauss et al. 1992), has also been suggested for clonal selection in poplars (Bradshaw et al. 1994; Bradshaw and Stettler 1995).

The power of clonal selection has been demonstrated by an impressive number of clonal cultivars produced by breeders and cultivated worldwide: the latest, but still incomplete, version of the Catalogue of Poplar Cultivars (Viart 1992) includes 280 cultivars that are or have been cultivated throughout the world, mostly represented by clones.

Nonconventional approaches to poplar improvement have also been tested, but so far they have had a minor impact on cultivar production. Thus triploid clones of hybrid aspens were selected for commercial use (Sekawin 1963; Mohrdieck and Melchior 1976; Mohrdieck 1983), but interest in polyploids declined in the 1970s, with the exception of some recent work in the field of molecular genetics (Bradshaw and Stettler 1993). Haploids met with more attention by researchers (Ho and Raj 1985) but found even less practical application than triploids (Wu and Nagarajan 1990). Interesting results were indeed obtained through somaclonal variation, especially in the field of *in vitro* resistance to adverse factors, such as salt, toxins, and herbicides. Poplar somaclones showing tolerance to glyphosate (Michler and Haissig 1988) and resistance to *Septoria musiva* Peck. (Ostry and Skilling 1988) were raised but the time of practical applications seems still rather far in the future. Related techniques, such as protoplast culture as a basis for somatic hybridization or other applications, although already proved viable at an experimental level (Youn et al. 1985; Russell and McCown 1986*a*, 1986*b*, 1988; Lee et al. 1987) appear even more distant from application in genetic improvement.

Poplar amenability to *in vitro* manipulation has prompted research on applications of recombinant DNA technology for its ability to add useful DNA sequences to a well-defined genotype with minimal disruption of its genomic organization. Indeed, poplars were the first forest trees to be transformed with an agronomically important gene (Fillatti et al. 1988; Sellmer and McCown 1989).

Today there are several scientific groups using poplars and molecular genetics to understand the mechanisms of gene expression in trees (see also Han et al., Chapter 9). The identification of genes of importance to cultivar performance, e.g., genes involved in resistance to herbicides, insects, diseases (viruses *in primis*), and abiotic stress factors or in biochemical pathways such as in lignin synthesis (Boerjan et al. 1995), is also under way, and one can expect that such genes will have a growing importance in breeding.

141

It is thus evident that poplar breeding has created and tested solutions that not only are justified from a short-term perspective but give options for the future. Typical long-term activities are the work on haploids and polyploids, the utilization of somaclonal variation, and the application of recombinant DNA techniques. However, a necessary component of systematic breeding, namely multiple generation breeding, has only been superficially explored.

Multiple generation breeding

Since the genetic potential of poplars for interspecific breeding and clonal selection has been well demonstrated, it is logical to pursue this line considering the new tools for hybridization and selection that gene technology has made available. Indeed it has been tempting to put most of the resources in to these activities: the variation obtained in the second generation of such breeding is profuse and recombinants that were not even imagined, considering the variation within species, can be produced. As a consequence the prevailing view among poplar breeders is to adopt breeding schemes that focus on gains to be made in one or two generations, although advanced-generation breeding has been advocated by some (Garrett 1976; Mohrdiek 1983). However, this short-term breeding approach does not necessarily mean that breeders must stop their work and restart from scratch after one cycle of breeding nor that they discard from further breeding the individuals that have been selected. Rather, they continue with the material selected with the implicit assumption that multiple generation breeding will take care of itself as long as the genetic gain is maximized.

Rapidly expanding possibilities in gene technology can further promote disregard for multiple generation breeding. Recent success of recombinant DNA technique in poplars allows rapid construction of new genotypes by gene transfer instead of selection over several generations. Improved techniques for high precision selection of valuabe genes or gene combinations through molecular markers could also detract attention from long-term breeding.

We argue, however, that the repetition of short-term breeding or the application of gene technology are unlikely to provide the best long-term results because environments in which poplars grow tend to be extremely heterogeneous temporally and spatially and our knowledge of poplar biology is poor compared with most agricultural crops. On the other hand, it has been repeatedly shown that, despite poor knowledge of the genomic structure, it is possible to drastically change the function or form of traits through multiple generation breeding.

The occurrence of environmental heterogeneity causes a more fundamental problem, namely genotype × environment interactions whereby causes and effects of genotypes and environments become intermingled. Consequently, the extrapolation of phenotype from one environment to another can be difficult, since the genes involved and their expression can change. Single-gene manipulation, by means of gene technology, is therefore not a sufficient solution for such characters, and multiple generation breeding specifically oriented towards several target environments can be at least a partial solution. Therefore, poplar breeders need to consider long-term breeding as an integral part of their breeding strategies, no matter how remote multiple generation tree breeding might appear to be: decisions made or not made now will limit the future options if the long-term aspect of tree breeding is ignored.

Here we present ideas on how population breeding might be developed for poplars. Since much international research has focused on three species that dominate poplar plantations throughout the world, *P. deltoides*, *P. nigra*, and *P. trichocarpa*, we adopt a global perspective and place emphasis on problems related to the basic structure of poplar breeding instead of on local problems of an individual species.

Maintenance of breeding material

There are at least two approaches to maintain poplar breeding populations in an evolving fashion. Fig. 1a shows the flow for single-population breeding. The selection can be through progeny testing or clonal/genotypic selection, where the selection proportion is denoted by 'b' and may differ between the sexes. Mating includes all, or a subset, of the parents. The simplest case is mass selection. The commercial varieties can be the whole selected population or, more likely, a subset of the families or selected clones.

Multiple-population breeding is illustrated in Fig. 1b. According to Kang and Nienstaedt (1987), the subdivided populations can consist of multiple populations, a collection of populations originating from different subsets of the base populations and/or copies, sharing the same ancestors. The two or more sub-populations can be either completely isolated from each other or connected through migration. The commercial material can either be drawn from a single population or from hybrids between the populations. The Italian program (Fig. 2) is an example of such multiple-population breeding.

Single-population breeding, in the most rigorous sense, is unlikely to exist in poplar breeding. Maintenance of a divided population is often motivated by the fact that most long-term breeding populations are too large to be managed as a single population. Regardless of the main emphasis, the global network of poplar breeders is likely to make replicates of the breeding population, divide it into multiple populations, or both.

Possibilities with subdivided population breeding

There are several disadvantages in maintaining a single large breeding population. The overall selection efficiency is low. Trees from the breeding population

143

Fig. 1. Single (a) and multiple (b) population breeding (adapted from fig. 3.2 of Namkoong et al. 1988).

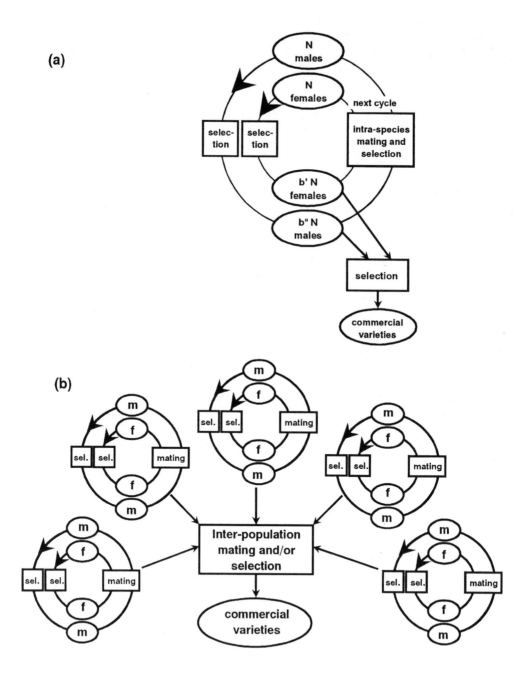

Fig. 2. The Italian breeding program for poplars (adapted from fig. 4.4 of Namkoong et al. 1988). (1) 1958–1981: collection, provenance and progeny testing, and scoring for growth, phenology, and *Melampsora* resistance that resulted in 300 *P. deltoides* selected clones; (2) 1982–1984: collection of 300 *P. nigra* clones covering the whole of Italy; (3) 1987: common tester progeny trial of *P. nigra* males: 6 *P. deltoides* × 147 *P. nigra* males; (4) 1988: polycross test of *P. deltoides* females: 95 *P. deltoides* × *P. nigra* pollen mix; (5) 1989–1991: common tester progeny trial of *P. deltoides* males: 6 *P. deltoides* females × 148 *P. deltoides* males. (6) 1990: polycross test of *P. nigra* females: 97 *P. nigra* females × *P. nigra* pollen mix.

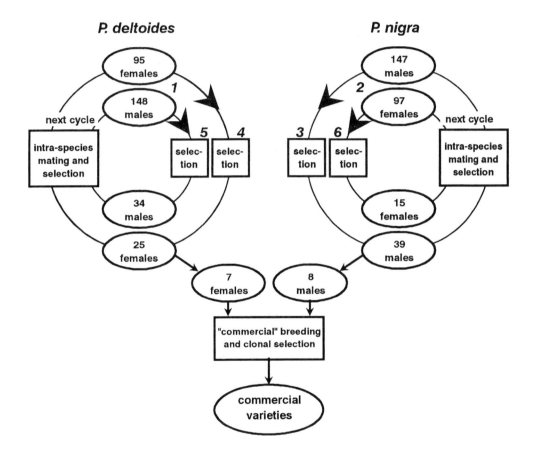

must be planted over a wide area which makes environmental control difficult as compared with agricultural crops. A single breeding population should be broadly adaptable which is only possible at the sacrifice of heritability and selection intensity. This deficiency is corrected by maintaining a subdivided breeding population with each subpopulation in a different target environment.

Further, a single population will experience one sequence of environments only and will, irrespective of population size, likely eliminate genes that would be valuable in other environments. By using a subdivided breeding population, on the other hand, the breeder has a greater chance to conserve and create genetic variation for future generations. Establishing multiple populations, i.e., subpopulations not sharing ancestors, will enable the breeder to conserve today's variability. Creating new adaptations in some of the subpopulations through selection can further improve the possibilities to cope with an unpredictable environment (Gullberg 1987; Namkoong et al. 1988, pp. 111–115).

The Italian breeding program is an example, at the species level, of how a subdivided breeding population gives flexibility in maintaining genetic variation. As can be seen in Fig. 2, the separate development of *P. deltoides* still allowed a much later change in renewal of the *P. nigra* material.

A multiple population scheme can also make full use of the highly developed, but genetically narrow, Belgian material without disrupting the structure of established breeding populations based on nonrelated material. The degree and kind of integration can be regulated by choosing the subpopulations with which the Belgian material is hybridized. In general terms, the above examples show how one can regulate the integration of any new genetic resource that is considered worth adding to a highly structured breeding population.

Besides giving flexibility to maintain genetic variation, the multiple breeding population offers possibilities to test physiological hypotheses related to poplar productivity. One example is the ideotype approach, a well-known concept (Dickmann et al. 1994; see also Dickmann and Keathley, Chapter 19) that can be tested in a subdivided population scheme. A breeder can develop multiple ideotypes for creating qualitatively different options for poplar culture. The study by Lefèvre et al. (1994) on components of resistance to *Melampsora* rust in poplars exemplifies the first step in an ideotype description relating to disease resistance. Generally, the result from developing a new ideotype through multi-generation selection will be a subpopulation with new forms, for example narrow leaves or long latency period, that can be used in future breeding and research.

In our efforts in Sweden to improve resistance to *Melampsora* in willows, this concept allowed us to use parts of the multiple breeding population to select for different types of plants (Table 1). We plan to use this material for studying components in relation to partial resistance in commercial varieties and for further studies on the mechanism of partial resistance. Since this study involves only a minor part of the breeding population, it can be done without seriously constraining our breeding goals.

The multiple breeding population can also be used to explore new genetic constructs by testing separate subpopulations. One of the most drastic would be to

Table 1. Partitioning of the Swedish breeding population of willows according to components of resistance to rust; different groups of lines (ten lines each) are aimed at increasing or decreasing the levels of expression of each component.

	Latent period		Size of uredia		No. of spores/uredia	
Group 1	Up	Down				
Group 2			Up	Down		
Group 3					Up	Down

create new species by making species hybrids that function as ordinary diploid species. This could be worth testing in Italian poplar breeding, since we know (Fig. 2) that the two species hybridize and that the hybrids are fertile. It is therefore tempting to skip the within-species element of the reciprocal recurrent selection scheme and just work on a hybrid population, single or multiple. The segregation in the F_2 is so wide as to suggest advanced-generation breeding of synthetic hybrid populations as a viable option. However, the risk of encountering problems with vitality or fertility due to recombination, as a consequence of karyotypic differences between the parent species, makes this venture risky. Perhaps the simplest way to reduce the risk is to use one or several of the subpopulations for creating the new hybrid species.

The multiple breeding population approach also enables the breeder to simplify genetic studies of poplar traits. Inbreeding can be used more efficiently than when maintaining a single population, since problems with inbreeding depression are easier to avoid. The obstacles to gene identification with heterozygosity in the parent population, as in wild or conventionally maintained poplar populations, could be overcome by creating inbred lines. For the Swedish multiple breeding population of willows we have adopted the extreme approach of letting every line turn over generations through a single brother × sister mating in order to get homozygosity quickly (Gullberg 1993). The lines are grouped and in each group they have the same breeding goal, but between groups the goals can differ. If the inbreeding depression becomes too serious (but we have had negligible problems in the first two generations, and only minor problems in the third), we will merge lines belonging to the same group and thereby recover viability and, hopefully, maintain the population structure with respect to trait composition. A further advantage of this scheme is that it facilitates the discovery of rare recessive alleles, and thereby useful genetic variants, since they will constitute up to 25% of the F_2 generation from an originally rare mutation.

Clonal selection

Clonal selection is a branch of any breeding program, its size with respect to the main "tree" depending on priorities and resources. In its simplest form clonal selection consists in starting with an abundant and genetically varied material and ending up with a limited number of clonal cultivars to be grown commercially on a large scale. The origin of the material is of course important for success but immaterial in the mechanics of the process. Seed from open or controlled pollination of selected or unselected parents belonging to wild or domesticated populations is the common starting point, although even collections of already established clones can be the starting material, if the goal is to select clones for a new environment that were selected elsewhere.

Although selected clones may also be valuable parents for advanced generation breeding, it must be remembered that immediate economic goals are the main motive for clonal selection. It may happen for instance, that clonal selection for commercial purposes is made among F_1 hybrids, whereas long-term population breeding makes use of parent species (e.g., the Italian breeding program, Fig. 2). Therefore, although an integration of long-term and short-term activities has been advocated throughout this chapter, possible conflicts or contradictions with the above discussion on population breeding must be considered in this perspective.

The following criteria for poplar selection, although over sixty years old, are still largely applicable: "fast growth (especially juvenile), wide adaptability, resistance to diseases and to frost, straight and cylindrical stems, homogeneous, white and resistant wood, suitable for pulp, boards, beams and peeling" (Jacometti 1934), even though it should possibly be further trimmed to the bare essentials. The common denominator of all breeding is the increase in production by means of high growth rates and reduced losses. Growth rate and resistance to adverse factors are thus the most important targets of breeding and selection (Thielges 1985). A disturbing factor in deciding about priorities is the inherent inertia of breeding programs, even with a relatively short-cycled tree such as poplar. Future shifts or even abrupt changes of goals (especially as a consequence of technological innovations) should be considered in the formulation of selection criteria (Namkoong 1969). A small, consistent, and relevant list of traits to be selected is essential for efficiency (Namkoong et al. 1971).

For long-lived organisms that are normally harvested only at the end of rotation (in contrast to fruit trees) and that, due to their size, need a lot of space for each tree, the choice of an optimal selection method, the exploitation of juvenile–mature correlations (JMC) and the refinement of field trial efficiency are even more crucial than for agricultural crops.

In a typically stepwise process such as clonal selection, breeders usually try to focus their attention on traits with either a high degree of stability across age (e.g., resistance to diseases) or commercial importance in juvenile material. Selection in the nursery is particularly desirable due to the limited space needed by plants at this stage.

Methods of selection

Breeders are confronted with the need to select for a number of traits either simultaneously or in sequence. Several methods exist for coping with multitrait selection: index selection, tandem selection, independent culling levels, and principal component analysis.

The theoretical superiority of index selection (Smith 1936; Hazel and Lush 1942), clearly demonstrated on the basis of the theory of linear models, is often diminished by difficulties encountered in providing reliable information on economic values to be attributed to unit improvement in any single trait and on phenotypic and genetic variance-covariance structures. Further complications arise from nonlinearity and nonindependence of economic values and on uncertainties about their future variations. Several variations of index selection have been proposed to obviate the lack of reliable information (see Lin 1978 for a review). Cotterill and Jackson (1985) describe three methods of attributing economic values that limit the subjectivity of decisions; Cotterill (1985), on the other hand, exemplifies with real data the application of two reduced methods of index selection that do not require the estimation of genetic parameters. However, due to general lack of dependable information on the variance-covariance matrices and to difficulties in attributing reliable economic values to unit trait improvements (a hard process indeed for typically multipurpose trees), there is, to our knowledge, no systematic application of index selection in the poplar world.

Tandem selection, that is the improvement of one trait at a time, is not usually considered applicable to plants with long reproductive cycles. On the other hand, the method of independent culling levels (Hazel and Lush 1942) is consciously or unconsciously applied to poplar selection because the evaluation process is a temporal sequence of tests, each providing information on a different set of traits. At each stage the population is truncated and the selected fraction is further tested. It has been proved that, especially if selective traits are correlated, the efficiency of independent culling levels is not far from that of index selection (Muir and Xu 1991) and methods for the optimization of independent culling levels have been proposed (Namkoong 1970; Smith and Quaas 1982). On the other hand, sequential evaluation in practice does not imply that only one trait is considered at each stage, so that the the problem of simultaneous multitrait selection, however simplified, is unavoidable. Xu and Muir (1992) propose a method that considers sequential selection within the framework of index selection, but the above-mentioned intrinsic problems of index selection are not removed.

A viable alternative to index selection is the use of principal component analysis (Godshalk and Timothy 1988). This was found useful in poplar selection as a way to reduce the complexity of multidimensional data sets (Bisoffi 1990). The actual selection process can be made on individual scores either by applying

149

independent culling at empirically established levels or by deriving a linear combination of them. This last case is equivalent to an index selection with arbitrary weights, with the advantage of having null covariances among the score vectors.

Juvenile/mature correlations (JMC) of growth-related traits

Apart from the problems posed by JMC in long-term breeding as mentioned earlier in this chapter, such correlations can be of great importance in clonal selection.

The cumulative nature of the commonly used size variables (e.g., height, diameter) has some interesting mathematical implications that shed light on the reliability of early selection. It can be demonstrated that if the annual increments are normally and independently distributed, the correlation between age t and age T is $(t/T)^{1/2}$ (Bisoffi 1993). Under the above assumption, age-age correlation for a given T increases monotonically with age even in the absence of any genetic correlation. This fact certainly contributes to the common observation that very reliable evaluations of rotation performances can be obtained with poplars (as well as with other forest trees) at ¼ to ½ of rotation age (Mohn and Randall 1971; Chiba and Nagata 1972; Mohrdieck 1979). This does not necessarily mean that genetic correlations between juvenile and mature performances are high, but at least that a realistic picture of the situation at the end of the rotation can be produced rather early.

Genotype × environment interactions (GEI)

The occurrence of environmental heterogeneity will, as stated before, cause problems with the extrapolation of the genetics over environments. We have discussed how this could be handled in multiple generation breeding. However, this is not sufficient, as GEI has to be dealt with also in clonal selection.

GEI is a tricky concept, first of all for the way σ_{GE}^2 is calculated, as the average squared deviation of observed values from those expected on the basis of main effect levels; secondly, because the genetic causes are far from being clear; and thirdly because there is uncertainty on how to deal with it once it has been detected. Certainly it is a factor of disturbance that diminishes the generalization of experimental results and dictates important decisions about selection strategies: shall one look for stable clones with good average performance over a range of environments or for locally specialized clones? There is no general answer to this question as it depends on the actual range of environments for which selected genotypes are intended, on the choice of test environments (typical environments or random sample: Burdon and Shelbourne 1977), on the objectives of the breeding institution, and finally, on the very nature of GEI in any specific case.

There is a rich literature on the analysis of stability (see Lin et al. 1986 for a review) that reveals a variety of meanings attached to this word (Lin and Binns 1988, 1991a, b). In fact, from a strictly mathematical point of view, any difference of slope in regressions of clonal performances on environmental means contributes to σ_{GE}^2, although it is shifts in ranking that matter from a practical point of view. Indeed, Avanzo (1978), in field trials with *P. deltoides*, found that GEI was the consequence of a higher regression slope of the clones with higher average performance, without remarkable rank changes.

It seems therefore logical to adopt estimation methods that give an insight into the very nature of GEI before making decisions that affect selection. An effective technique in most cases where a common set of genetic entries (e.g., clones) is tested in several environments is based on a mixed linear and multiplicative model (Zobel et al. 1988; Gauch 1988; Gauch and Zobel 1988) that makes use of the analysis of variance (ANOVA, linear model) for the estimation of main effects, genetic and environmental, and of principal component analysis (PCA, multiplicative model) for an interpretation of interactions.

The two-way table of interaction deviations from expected values can then be examined by rows and by columns and the PCA may be followed by a cluster analysis (with scores as input data) that gives a pictorial representation of similarities among clones or among environments with respect to the principal components. This idea represents a modification of Kempton's (1984) biplots.

PCA itself does not provide significance tests, but a validation of the model is possible for example by splitting the data set into one part used to build the model and another to validate it (Gauch 1988). Besides, we feel that significance tests are of little value in the analysis of GEI and that an insight into the causes of interactions is much more useful in actual breeding and selection work.

Experimental test design

The control of the local test environment may prove to be a crucial factor in the efficiency of evaluation trials; typical poplar trials are conducted on alluvial soils that are often heterogeneous, and spatial correlations may both decrease the power of tests and may bias the estimates of variance. Conventional experimental designs (e.g., complete randomized blocks) often prove inefficient in the early phases of clonal selection and in large progeny trials, where the number of genetic entries is high and the material available for replication is scarce (Lin and Poushinsky 1983; Besag and Kempton 1986). Designs that ensure a better statistical control of environmental variation (e.g., incomplete blocks) are rigid in their requirements (number of treatments/replications) and weak with respect to missing values.

A range of field layouts and of analytical methods that account for spatial variation in a continuous way, have been successfully employed both in nursery and stand tests. They are based on the analysis of covariance (ANCOVA) of plot values with a local fertility index as covariate. This index is either derived from check plots that are systematically distributed over the test field (Bisoffi 1991*a*), or from residues of neighbouring plot values after the elimination of treatment (clone or family) average main effects (Bisoffi 1991*a*; Pichot 1993*b*). The latter method, which was introduced by Papadakis (1937) and adopted by Wright (1978) for experiments on pines, is not sound from a purely mathematical point of view as it violates some assumptions for the validity of ANCOVA. But simulations indicate that in the presence of spatial correlation without a clearly defined trend, the estimates of treatment effects are both more accurate (closer to the true values) and more precise (smaller variance) (Pichot 1993*b*). Indeed, when Papadakis' analysis was applied to real data of an experiment with *P. trichocarpa* set out in two locations, the level of genotypic correlation was higher than with a conventional ANOVA-based analysis (Pichot 1993*a*).

The main drawback of such spatial methods, however, is the lack of specific software that is necessary for the processing of large amounts of data, often with iterative computer-intensive procedures.

Another topic that has received much attention in the past is plot size, especially for stand trials. Here again, no general rule can be given. The "rule of thumb" of adopting "the smallest size consistent with the application of the treatments, the edge effects due to neighbouring plots, and the size to which the trees are to be left to grow" (Jeffers 1958) is still valid but it does not help much in making decisions. An application of H. Fairfield Smith's empirical law[1] (Smith 1938) to eight large, mature poplar stands used as uniformity trials, revealed values of λ between 0.32 and 0.75 (Bisoffi 1991*b*) and no reason to depart from single-tree plots. Given the above considerations on age–age correlations that permit reliable evaluations quite early in the rotation, biases due to interplant competition may be minimal in the early phases of selection.

As selection proceeds and decisions are based on small differences in growth rates, larger plots may be needed. However, the common practice of providing each plot with a border row that is later eliminated from the analysis, should be seriously reconsidered. In a field trial with eight clones at an advanced stage of selection arranged in plots of 25 trees (5 × 5, with 4 replications in complete randomized blocks) Bisoffi (unpublished data) found that the gain in accuracy obtained by analysing only the 9 interior trees of each plot was largely offset by the loss in precision due to a lower number of trees for the calculation of plot means. Therefore, it is probably wiser to accept a moderate bias for a

---

[1]H.F. Smith's empirical law states, "that the variance of the mean of a plot made of n contiguous trees (σ_n^2) is related to the variance of single trees (σ_1^2) by the relationship: $\sigma_n^2 = \sigma_1^2 \cdot n^{-\lambda}$. In the case of perfect independence of neighbours $\lambda = 1$; with positive spatial correlation $\lambda < 1$; with competition $\lambda > 1$." (Smith 1938)

substantial gain in precision (the so-called "Stein effect," James and Stein 1960) especially if one considers that field trials always compare relative rather than absolute performances.

Concluding remarks

A balanced poplar breeding program should allocate resources to both population breeding and clonal selection, as well as to gene technology. Paradoxically, a major threat to population breeding comes from gene technology itself as it competes for limited research funds. Molecular tools are presented as techniques to support breeding programs but there seems to be a general tendency to spend more money on such advanced techniques than on the breeding programs they are supposed to support. For example, within the 'cluster' of forestry projects financed by the European Commission within the AIR program (Agriculture and Agro-industry, including Fisheries, 1991–1994), four out of ten projects that had a genetic component dealt solely with advanced techniques, two with conventional breeding and four had both aspects (European Commission 1995).

To some extent multigeneration selection, especially if based on a set of breeding populations, could solve the same problems as gene technology (e.g., by targeting each subpopulation as one facet of an ideotype). In comparison with the gene technology solution, the subpopulation solution is also closer to application since the wanted genotype is already created, but the knowledge it generates is less precise and can be generalized only with caution.

The practical and theoretical advantages of a subdivided breeding population compared with a single population were also discussed, but the partitioning of resources between and within populations is also worth considering. A multiple population approach may need more resources than a program relying on one or a few populations, since some of the subpopulations might be of little practical use, but in the long run the cost need not be much higher. Having in mind the complexity of large breeding programs, one should carefully consider the operational advantages of a multiple population strategy. To name just a few: (a) the size of each subpopulation can be adapted to the relative importance of the target environment/trait; (b) the generation turnovers in subpopulations are independent of each other and can be adapted to the time needed to reach maturity (variable among species and environments) or to available resources; (c) maintenance costs, if not globally reduced, are at least split into smaller portions and may thus be afforded separately by a number of coordinated units. Added flexibility would probably prove to be economically advantageous in the long run, too. Short-term or mid-term goals may be pursued within each subpopulation according to the interests and needs of each unit, while long-term benefits are ensured by the comprehensive set of all the subpopulations. Individuals or groups from different subpopulations can be mated at any time to produce offspring for genetic analysis, advanced breeding, clonal selection, or the three of them.

Both population breeding and clonal selection should be based upon quantitatively determined genetic parameters. Unfortunately the literature is not particularly rich in information on variance and covariance components and derived parameters, such as heritabilities, potential genetic gains, and genetic correlations among traits or across time. The information is so scattered among species, genetic materials, environments, cultural treatments and ages that an organic treatment of the subject at the genus level is an arduous job. Moreover, such parameters as General Combining Ability (GCA), Specific Combining Ability (SCA), broad-sense (h_{BS}^2) and narrow-sense (h_{NS}^2) heritability, GEI, and so on, are all variances or differences of variances or ratios of variances and, as such, they are affected by all the actual sources of variation in genetic experiments: sample population, test environments, cultural practices, spacing, plot size and shape, and experimental design. A change in any of these factors may substantially change the value of the estimates. Appropriate experiments in the actual or potential range of environments, with representative samples of actual species and populations and with silvicultural models as close as possible to commercial practice, are therefore needed as a background for sound decisions on breeding strategies, tactics, and techniques.

It was not our intention to give "recipes," if any exist. We only wished to point out aspects that must be considered in the actual implementation of breeding and selection, which is typically a lengthy and expensive process and therefore calls for a rational and balanced use of resources.

Finally, we advocate a closer international cooperation among breeding institutions worldwide, in much the same way as is occurring in the field of advanced genetic techniques. Breeders who have tended to view breeding programs as closed systems, impervious to "migration," should start thinking of cooperative breeding programs in regions with similar environment and institutions pursuing similar breeding goals. A multiple population breeding strategy is particularly suited to cooperative programs, as gene pools can be physically maintained and selected separately with different target traits and shared or pooled at any time. In this way, more environments and traits could be simultaneously analyzed than by any stand-alone program with a comparable individual effort.

Many questions are still unresolved and deserve further scientific attention. The following list is an attempt to identify areas in which new knowledge would represent a significant progress towards a unified view to poplar breeding, in which long-term activities, clonal selection, and new technologies can be efficiently and effectively integrated:

- Test of ideotypes and comparison of predicted vs. actual gain of quantitatively inherited characters.

- Inbreeding: evaluation of its potential as an alternative or as a preparatory step to outcrossing and interspecific breeding.

- Study of the genetic basis of resistance and virulence with combined genetic and physiological analysis of hosts and parasites.

- Juvenile-mature correlations and genotype × environment interactions: progress from a descriptive approach to the interpretation of the genetic mechanisms, with particular regard to those involved in the adaptation to a variable environment.

- Intertree competition as a function of spacing, age, and genotype in order to refine the design and analysis of field trials.

- Development of a network of poplar breeders worldwide for the exchange of information, the avoidance of duplicated efforts and, possibly, the establishment of international cooperative breeding programs.

The last point is not a research topic in itself, but it is perhaps the most crucial need to achieve a productive synthesis of scientific work with practice-oriented application. This should be the ultimate goal of research, especially in a tree whose widespread existence is so tightly linked to human use.

References

Avanzo, E. 1978. Influence de l'interaction clone milieu sur la production des peupliers cultivés. Proceedings 8th World Forestry Congress, Jakarta, 16–28 October, 1978. FID-I/17-5. 3 p.

Besag, J., and Kempton, R. 1986. Statistical analysis of field experiments using neighbouring plots. Biometrics, **42**: 231–251.

Bisoffi, S. 1990. The development of a breeding strategy for Poplars. International Poplar Commission (FAO), 35th Executive Committee Meeting, Buenos Aires 19–23 March, 1990. FO:CIP:BR/90/9. 21 p.

Bisoffi, S. 1991*a*. Nearest-neighbour and check-plots in Poplar nursery trials. Proceedings IUFRO S4.11 Conference on Optimal Design of Forest Experiments and Surveys, London, 10–14 September, 1991. pp. 49–59.

Bisoffi, S., 1991*b*. La legge empirica di H.F.Smith applicata alla sperimentazione di campagna del pioppo. Monti e Boschi, **42** (2): 49–53.

Bisoffi, S. 1993. Age-age correlations for the evaluation of forest reproductive material. *In* Proceedings of a conference held in Paris, 9–10 December, 1993 under the auspices of the European Commission. Quality of forest reproductive material in the field of the application of European Community rules. *Edited by* D. Terrasson. CEMAGREF Editions. pp. 75–91.

Boerjan, W., Baucher, M., van Doorsselaere, J., Christensen, J.-H., Meyermans, H., Chen, C., Leplé, J.-C., Chognot, E., Chabbert, B., Petit-Conil, M., Tollier, M.-T., Pilate, G., Cornu, D., Monties, B., Inzé, D., Jouanin, L., and van Montagu, M. 1995. Genetic engineering of lignin biosynthesis in poplar. Joint Meeting of the IUFRO Working Parties S.04.07 and S.04.06 on Somatic Cell Genetics and Molecular Genetics of Trees, 26–30 September, 1995. Ghent, Belgium.

Bradshaw, H.D., Jr., and Stettler, R.F. 1993. Molecular genetics of growth and development in *Populus*. I. Triploidy in hybrid poplars. Theor. Appl. Genet. **89**: 301–307.

Bradshaw, H.D., Jr., and Stettler, R.F. 1995. Molecular genetics of growth and development in *Populus*. IV. Mapping QTLs with large effects on growth, form, and phenology traits in a forest tree. Genetics, **139**: 963–973.

Bradshaw, H.D. Jr., Villar, M., Watson, D.B., Otto, K.G., Stewart, S., Stettler, R.F. 1994. Molecular genetics of growth and development in *Populus*. III. A genetic linkage map of a hybrid poplar composed of RFLP, STS, and RAPD markers. Theor. Appl. Genet. **89**: 551–558.

Burdon, R.D., and Shelbourne, C.J.A. 1977. Advanced selection strategies. Proceedings Third World Consultation on Forest Tree Breeding, 21–26 March, 1977, Canberra, Australia. FO-FTB-77-6/2. pp. 1133–1147.

Chiba, S., and Nagata, Y. 1972. Rust resistance and growth of *Populus maximowiczii* clones selected from the progenies of intraspecific hybridization. Proceedings IUFRO Genetics — SABRAO Joint Symposia, Tokyo, 1972. C-6(V). pp. 1–7.

Cotterill, P.P. 1985. On index selection II. Simple indices which require no genetic parameters or special expertise to construct. Silvae Genet. **34**: 64–69.

Cotterill, P.P., and Jackson, N., 1985. On index selection I. Methods of determining economic weight. Silvae Genet. **34**: 56–63.

Dickmann, D.I., Gold, M.A., and Flore, J.A. 1994. The ideotype concept and the genetic improvement of tree crops. Plant Breed. Rev. **12**: 163–193.

Eckenwalder, J.E. 1984. Natural intersectional hybridization between North American species of *Populus* (Salicaceae) in sections *Aigeiros* and *Tacamahaca*. II. Taxonomy. Can. J. Bot. **62**: 325–335.

European Commission. 1995. AIR Agriculture and Agro-industry, including Fisheries, 1991–1994; Non-food, Bio-energy and Forestry: Catalogue of Contracts. Office for Official Publications of the European Commission, Luxembourg.

Fillatti, J.J., Haissig, B., McCown, B., Comai, L., and Riemenschneider, D. 1988. Development of glyphosate-tolerant *Populus* plants through expression of a mutant *aroA* gene from *Salmonella typhimurium*. *In* Genetic Manipulation of Woody Plants. *Edited by* J.W. Hanover and D.E. Keathley. Plenum Press, New York and London. pp. 243–249.

Garrett, P.W. 1976. Interspecific hybridisation: the American experience. *In* Proceedings of Symposium on Eastern Cottonwood and Related Species. 28 September – 2 October, 1976, Greenville, MS. pp. 156–164.

Gauch H.G. Jr., 1988. Model selection and validation for yield trials with interaction. Biometrics, **44**: 705–715.

Gauch, H.G., and Zobel, R.W. 1988. Predictive and postdictive success of statistical analyses of yield trials. Theor. Appl. Genet. **76**: 1–10.

Godshalk, E.B., and Timothy, D.H. 1988. Factor and principal component analyses as alternatives to index selection. Theor. Appl. Genet. **76**: 352–360.

Gullberg, U. 1987. Studies on the utilization of adapted forms in the breeding of *Pinus sylvestris*. SLU, Inst.f.skogsgenetik. Res. Notes, (Uppsala, Sweden), **38**. 76 p.

Gullberg, U. 1993. Towards making willows pilot species for coppicing production. For. Chron. **69**: 721–726.

Gullberg, U., and Kang, H. 1985. A model for tree breeding. Stud. For. Suec. (Uppsala, Sweden), **169**. 8 p.

Hazel, L.N., and Lush, J.L. 1942. The efficiency of three methods of selection. J. Hered. **33**: 393–399.

Henry, A. 1914. Note on *P. generosa*. Gardeners Chronicle. pp. 257–258.

Ho, R.H., and Raj, Y. 1985. Haploid plant production through anther culture in poplars. For. Ecol. Manage. **13**: 133–142.

Houtzagers, G., 1937. Het Geslacht *Populus* in verband met zijn Beteeknis voor de Houtteelt. H. Veenman and Zonen, Wageningen, The Netherlands.

Jacometti, G. 1934. Ricerche e studi sul pioppo. *In* La 3° Mostra Forestale e Montana Arnaldo Mussolini. Comitato Nazionale Forestale, Rome, Italy. pp. 3–15.

156

Jacometti, G. 1937. I nuovi pioppi italiani. Atti del Convegno di Pioppicoltori, Comitato Nazionale Forestale, Casale Monferrato. 7 November, 1937. pp. 39–50.

James, W., and Stein, C. 1960. Estimation with quadratic loss. *In* Proceedings of 4th Berkeley Symposium, Math. Stat. Prob. University of California Press, Berkeley, CA. pp. 361–380.

Jeffers, J.N.R. 1958. Methods employed in poplar experiments. International Poplar Commission (FAO), Standing Executive Committee 14th Session, 1–4 July, 1958, Rome, Italy. FAO/CIP/CP/18 Add. 1. 21 p.

Jinks, J.L. 1983. Biometrical genetics of heterosis. *In* Heterosis. *Edited by* R. Frankel. Springer-Verlag, Berlin, Heidelberg. pp. 1–46.

Kang, H., and Nienstaedt, H. 1987. Managing long-term tree breeding stock. Silvae Genet. **36**: 30–90.

Kang, H., Lascoux, M., and Gullberg, U. 1996. Systematic tree breeding. *In* Tree Breeding. *Edited by* A.K. Mandal. Indian Council of Forestry Education. Jabalpur, India. In press.

Kempton, R.A. 1984. The use of biplots in interpreting variety by environment interactions. J. Agric. Sci. (Cambridge), **103**: 123–135.

Lande, R., and Thompson, R. 1990. Efficiency of marker-assisted selection in the improvement of quantitative traits. Genetics, **124**: 743–756.

Lander, E.S., and Botstein, D. 1989. Mapping Mendelian factors underlying quantitative traits using RFLP linkage maps. Genetics, **121**: 185–199.

Lee, J.S., Lee, S.K., Jang, S.S., and Lee, J.J. 1987. Plantlet regeneration from callus protoplasts of *Populus nigra*. Res. Rep. Inst. For. Gen. (Korea), **23**: 143–148.

Lefèvre, F., Pichot C., and Pinon, J. 1994. Intra- and interspecific inheritance of some components of the resistance to leaf rust (*Melampsora larici-populina* Kleb.) in poplars. Theor. Appl. Genet. **88**: 501–507.

Lin, C.S., and Binns, M.R. 1988. A method of analyzing cultivar location year experiments: a new stability parameter. Theor. Appl. Genet. **76**: 425–430.

Lin, C.S., and Binns, M.R. 1991*a*. Assessment of a method for cultivar selection based on regional trial data. Theor. Appl. Genet. **82**: 379–388.

Lin, C.S., and Binns, M.R. 1991*b*. Genetic properties of four types of stability parameter. Theor. Appl. Genet. **82**: 505–509.

Lin, C.S., Binns, M.R., and Lefkovitch, L.P. 1986. Stability analysis: where do we stand? Crop Sci. **26**: 894–900.

Lin, C.S., and Poushinsky, G. 1983. A modified augmented design for an early stage of plant selection involving a large number of test lines without replication. Biometrics, **39**: 553–561.

Lin, C.Y. 1978. Index selection for genetic improvement of quantitative characters. Theor. Appl. Genet. **52**: 49–56.

Michler, C.H., and Haissig, B.E. 1988. Increased herbicide tolerance of *in vitro* selected hybrid poplar. *In* Somatic cell genetics of woody plants. *Edited by* M.R. Ahuja. Kluwer Academic Publishers, Dordrecht, The Netherlands. pp. 183–189.

Mohn, C.A., and Randall, W.K. 1971. Inheritance and correlation of growth characters in *Populus deltoides*. Silvae Genet. **20**: 182–184.

Mohrdiek, O. 1979. Juvenile-mature and trait correlations in some aspen and poplar trials. Silvae Genet. **28**: 107–111.

Mohrdiek, O. 1983. Discussion: future possibilities for poplar breeding. Can. J. For. Res. **13**: 465–471.

Mohrdiek, O., and Melchior, G.H. 1976. Die Kombination von Hybrid- und Polyploidzüchtung als aussichtsreiche Methode bei Leuce-Pappeln: Vergleich zwischen diploiden und triploiden Graupappeln aus gleichen Nackommenschaften. Die Holzzucht, **30**(1): 7–10.

Muhle Larsen, C. 1970. Recent advances in poplar breeding. Int. Rev. For. Res. **3**: 1–67.

Muir, W.M., and Xu, S. 1991. An approximate method for optimum independent culling level selection for n stages of selection with explicit solutions. Theor. Appl. Genet. **82**: 457–465.

Namkoong, G. 1969. Problems of multiple-trait breeding. FAO/IUFRO Second World Consultation on Forest Tree Breeding, 7–16 August, 1969, Washington, DC. FO-FTB-69-7/4. 5 p.

Namkoong, G. 1970. Optimum allocation of selection intensity in two stages of truncation selection. Biometrics, **26**: 465–476.

Namkoong, G., Biesterfeldt, R.C., and Barber, J.C., 1971. Tree breeding and management decisions. J. For. **49**: 138–142.

Namkoong, G., Kang, H.C., and Brouard, J.S. 1988. Tree breeding: principles and strategies. Springer-Verlag, New York, Berlin, Heidelberg, London, Paris, and Tokyo. 180 p.

Ostry, M.E., and Skilling, D.D. 1988. Somatic variation in resistance of *Populus* to *Septoria musiva*. Plant Dis. **72**: 724–727.

Papadakis, J.S. 1937. Méthode statistique pour des expériences sur champ. Institut d'amélioration des plantes a Salonique, Bulletin scientifique N° 23. pp. 1–22.

Pichot, C. 1993*a*. Variabilité au stade adulte chez *P. trichocarpa* Torr. & Gray et prédiction juvenile-adulte chez *P. trichocarpa* et *P. deltoides* Bartr. (Doctoral thesis). Institut National Agronomique Paris-Grignon & INRA, Orléans.

Pichot, C. 1993*b*. Analyse de dispositifs par approches itératives prenant en compte les performances des plus proches voisins. Agronomie, **13**: 109–119.

Russell, J.A., and McCown, B.H. 1986*a*. Culture and regeneration of *Populus* leaf protoplasts isolated from non-seedling tissue. Plant Sci. **46**: 133–142.

Russell, J.A., and McCown, B.H. 1986*b*. Techniques for enhanced release of leaf protoplasts in *Populus*. Plant Cell Rep. **5**: 284–287.

Russell, J.A., and McCown, B.H., 1988. Recovery of plants from leaf protoplasts of hybrid-poplar and aspen clones. Plant Cell Rep. **7**: 59–62.

Schreiner, E.J. 1949. Poplars can be bred to order. Yearbook of Agriculture 1949. pp. 153–157.

Schreiner, E.J. 1970. Genetics of eastern cottonwood. U.S. Dep. Agric. For. Serv. Res. Pap. WO-11. 24 p.

Sekawin, M. 1963. Etude d'un peuplier tetraploïde obtenu artificiellement et de sa descendance. FAO World Consultation on Forest Genetics and Forest Tree Breeding, Stockholm, 23–30 August, 1963. FAO/FORGEN 63-1/4. 11 p.

Smith, H.F, 1936. A discriminant function for plant selection. Annals of Eugenics, **7**: 240–250.

Smith, H.F. 1938. An empirical law describing heterogeneity in the yields of agricultural crops. J. Agric. Sci. (Cambridge), **28**: 1–23.

Smith, S.P., and Quaas, R.L. 1982. Optimal truncation points for independent culling-level selection involving two traits. Biometrics, **38**: 975–980.

Stettler, R.F., Fenn, R.C., Heilman, P.E., and Stanton, B.J., 1988. *Populus trichocarpa Populus deltoides* hybrids for short rotation culture: variation patterns and 4-year field performance. Can. J. For. Res. **18**: 745–753.

Strauss, S.H., Bousquet, J., Hipkins, V.D., and Hong, Y.-P. 1992. Biochemical and molecular genetic markers in biosystematic studies of forest trees. New For. **6**: 125–158.

Thielges, B.A. 1985. Breeding poplars for disease resistance. FAO Forestry Paper No. 56. Rome, Italy. 66 p.

Viart, M. 1992. International register of the cultivars of *Populus*. *In* Proceedings of the 19th Session of the International Poplar Commission (FAO), 22–25 September 1992, Saragosa, TX. pp. 57–105.

Wright, J.W. 1978. An analysis method to improve statistical efficiency of a randomized complete block design. Silvae Genet. **27**: 12–14.

Wu, K., and Nagarajan, P. 1990. Poplars (*Populus* spp.). In vitro production of haploids. *In* Biotechnology in agriculture 12. Haploids in crop improvement 1. *Edited by* Y.P.S. Bajaj. Springer-Verlag, Berlin and Heidelberg. pp. 237–249.

Xu, S., and Muir, W.M. 1992. Selection index updating. Theor. Appl. Genet. **83**: 451–458.

Youn, Y., Lee, J.S., and Lee, S.K. 1985. Isolation and culture of protoplasts from suspension-cultured cells of *Populus alba P. glandulosa* F_1. Res. Rep. Inst. For. Gen. (Korea), **21**: 109–113.

Zobel, R.W., Wright, M.J., and Gauch, H.G., Jr. 1988. Statistical analysis of a yield trial. Agron. J. **80**: 388–393.

CHAPTER 7
Quantitative genetics of poplars and poplar hybrids

D.E. Riemenschneider, H.E. Stelzer, and G.S. Foster

Introduction

Quantitative genetics has become so closely associated with plant breeding that the two terms are often used synonomously, which is incorrect. Plant breeding is the selection of plant types that have increased utility compared to their wild progenitors. Selection is followed by the intermating of improved types to generate production populations and new breeding populations or, in the case of vegetatively propagated crops, by direct commercial deployment. The technology of plant breeding has probably been practiced for thousands of years and is one of the most important bases of modern civilization. In contrast, quantitative, or biometrical, genetics is not a technology but rather a science that seeks to explain how variability in a plant trait, or traits, is influenced by heritable and nonheritable factors. Such an explanation is achieved through the estimation of variance and covariance components attributable to experimentally established elements of genetic and environmental population structure. Quantitative genetics also seeks to explain how selection acts and how selection can be made most efficient. Evaluation of alternative selection strategies is achieved by using experimentally determined variance and covariance components to predict response to selection using appropriate quantitative genetic theory. Thus, estimation and prediction combine in the evaluation of possible alternative selection strategies allowing, a priori, the most efficient selection strategies to be identified. Quantitative genetic studies often require breeding as a prerequisite because genetically structured experimental populations are necessary for the estimation of additive and other kinds of genetic variance. Breeding, on the other hand, does

D.E. Riemenschneider. USDA Forest Service, Forestry Sciences Laboratory, 5985 Highway K, Rhinelander, WI 54501, USA.
H.E. Stelzer and G.S. Foster. USDA Forest Service, Southern Forest Experiment Station, Box 1387, Huntsville, AL 35762, USA.
Correct citation: Riemenschneider, D.E., Stelzer, H.E., and Foster, G.S. 1996. Quantitative genetics of poplars and poplar hybrids. *In* Biology of Populus and its implications for management and conservation. Part I, Chapter 7. *Edited by* R.F. Stettler, H.D. Bradshaw, Jr., P.E. Heilman, and T.M. Hinckley. NRC Research Press, National Research Council of Canada, Ottawa, ON. pp. 159–181.

not always require the use of quantitative genetics, although a lack of quantitative knowledge of breeding populations limits the efficiency with which selection and genetic gain can be accomplished. Quantitative genetics is properly applied to the study of traits that are under the control of many (i.e., >5) genes, each of which occurs in multiple forms (alleles) in a population.

In this chapter we will summarize briefly how genetic variances and covariances are estimated and how the response to artificial selection can be predicted. Then, we will discuss how quantitative genetics has been used to gain an understanding of variation in traits that are important in poplars insofar as they affect the suitability of poplars for commercial exploitation. The sections on estimation and prediction are necessarily short in relation to the volume of existing literature. For additional background the reader is directed to the works of Namkoong (1979), Cotterill and Dean (1990), Nyquist (1991), or White and Hodge (1989).

A short summary of estimation methods

Genetic and environmental variances and covariances are estimated using Analysis of Variance, an analytical method developed by R.A. Fisher in the 1920s. Analyses vary in complexity depending on the genetic structure of the experimental population and on the experimental design (i.e., the model), but always have certain features in common. First, the model must include effects due to genotype. Second, the model must include an error effect that allows: (1) tests of the hypothesis that differences among genotypes are significantly different from zero; (2) estimation of genetic variance and covariance components, and (3) effects are usually assumed random because the analytical objective is to estimate effects as variances and covariances. In the simplest possible form an appropriate analysis of variance contains only two lines and is based on the model:

[1] $\quad X_{ij} = \mu + \gamma_i + \varepsilon_i$

where X_{ij} is an experimental observation (i.e., a selection criterion) on a selection unit, is the effect of the ith genotype and is the environmental and within-genotype departure of the ijth observation from its genotypic mean. The analyses of variance and covariance can be simple, as in the case of single location tests (Table 1) or complex (see Foster and Shaw (1988) for a good example).

Genetic variance and covariance components are estimated by equating genetic expectations (expected mean squares) to observed mean squares (Table 1). Estimation can be accomplished by simple subtraction and division ([MS$_1$ – MS$_2$]/n) if the experiment is completely balanced (Table 1). Imbalanced experiments require complex Analysis of Variance computations, derivation of complex expected mean squares, and the solution of a set of simultaneous linear equations. Estimation of variance and covariance components from unbalanced

Table 1. Sample analysis of variance that allows for estimation of genetic ($\sigma_G{}^2$) and environmental ($\sigma_E{}^2$) sources of variation.

Source of variation	Degrees of freedom	Mean squares	Expected mean squares
Between genotypes	$(g - 1)$	MS_1	$\sigma_E{}^2 + n\sigma_G{}^2$
Within genotypes	$g(n - 1)$	MS_2	$\sigma_E{}^2$

Note: g = number of genotypes, n = number of replications per genotype.

experiments has, in the past, been problematic because of inadequate computer hardware, software, or both. However, recent years have seen the development of modestly priced yet powerful desktop computer systems. Operating systems are available that support unlimited virtual memory given sufficient (and inexpensive) disk storage. Statistical software packages are available that utilize virtual memory to fit large linear models and provide variance and covariance estimates from large and unbalanced experiments. Thus, there are currently no limits to the rigorous estimation of all needed variances and covariances, regardless of experimental complexity.

The array of components of variance and covariance that can be estimated depend on the genetic structure of the experimental population and on the experimental design. Single location experiments generally allow for estimation of genetic variances and one or more components of environmental variance. Estimation of variance and covariance due to genotype × environment interaction, which can be important in the design of breeding programs, requires that similar experiments be established at multiple locations.

A short summary of prediction methods

Response to artificial selection for a single trait in a population is given by the relationship:

$$[2] \quad \Delta_G = i\,h^2\,\sigma$$

or, with expansion and rearrangement by:

$$[3] \quad \Delta_G = i\,(\sigma^2/\sigma)$$

where i = selection intensity in standard deviations from the mean, h^2 = heritability, $\sigma_G{}^2$ = genetic variation among selection units, and $\sigma_P{}^2$ = phenotypic variation among selection units. Heritability is probably the most common statistic encountered in the quantitative genetic literature. Heritability can be calulated on several bases, but is always the ratio of the genetic variance among selection units to the phenotypic variance among selection units. Heritability is thus a measure of the extent to which variability in any trait is attributable to genetic variation among genotypes. The origin of Eqs. 2 and 3 can be traced to the earliest days of population genetics, and the discovery that response to natural selection is proportional to genetic variance (Fisher 1930). Proof that the same proportionality holds in the case of artificial selection, and that the proportionality constant is the intensity of artificial selection (i), can be attributed to J.B.S. Haldane (1931). Most of the development of the theory of artificial selection including multiple trait cases and the utilization of information from relatives was accomplished in the 15-year period after Fisher's (1930) and Haldane's (1931) landmark publications.

It is important to keep in mind the nature of the selection units to determine proper estimates of σ_G^2 and σ_P^2. For example, σ_G^2 is equal to the sum of all kinds of genetic variation (additive, dominance, and epistatic) when the selection units are individuals (or clones). Heritability calculated on the basis all genetic variation is referred to as a broad sense estimate (H^2). In comparison, σ_G^2 is equal to 1/4 of additive genetic variation only, or 1/2 of additive genetic variation, and 1/4 of dominance variation, when the selection units are half-sib or full-sib families, respectively. Heritability calculated on the basis of additive genetic variation is referred to as a narrow sense estimate (h^2). The most common selection unit used in quantitative genetic analyses of poplars is the clone mean, that is, the average performance of several trees that are genetically identical (ramets). Thus, broad sense heritability estimates are common within the poplar quantitative genetic literature. In such a case, σ_G^2 is the sum of all modes of genetic variation, including additive, dominance, and epistatic components. In contrast, σ_P^2 is the phenotypic variance of a clone mean. The phenotypic variance of a clone mean can be reduced in comparison to the phenotypic variance among unreplicated individuals by careful experimental design. Selection based on clone means can be highly effective because the use of all forms of genetic variation in combination with an experimentally controlled and reduced phenotypic variance leads, usually, to very high heritability. As a tradeoff, experiments with highly replicated clonal entries may contain fewer clones than more modestly replicated experiments which can lead to reduced selection intensity (i).

Response to selection can also be estimated for multiple selection criteria. When multiple criteria are utilized, selection is accomplished based on an index of the form:

[4] $I = b_1X_1 + b_2X_2 + ... b_nX_n$

where I is the index value for the selection unit, b_i are weights applied to each trait in the index and X_i is the value of the ith trait. The b_i are estimated to maximize the correlation between the index and the aggregate estimated breeding values of the selection units according to the relationship (Smith 1936; Hazel and Lush 1942; Hazel 1943):

[5] $b = P^{-1} Ga$

where b is a vector of index weights (one for each trait), P is the phenotypic covariance matrix, G is the genetic covariance matrix, and a is a vector of economic weights giving the relative value of each trait in the same measurement scale as used for covariance estimation. As in the single trait case, covariances should be appropriate to the selection units under consideration, i.e., total genotypic variation for selection among clones, phenotypic variance of clone means for selection among replicated clonal entries. Methods to estimate predicted gain in the aggregate and in individual traits are thoroughly described by Cotterill and Dean (1990).

Various modifications of the Smith selection index have been developed to meet special selection objectives. One of the most useful is the restricted selection index (Kempthorne and Nordskog 1959; Tallis 1962) that allows genetic change in one or more traits to be held to zero. The restricted selection index is calculated using the same genetic and phenotypic covariance matrices as in the unrestricted index except that they are estimated according to the matrix equation:

[6] $b = [I - P^{-1}G_r'(G_rP^{-1}G_r')]P^{-1}$

where b, P, G, and a are as before and G_r' is a modified genetic covariance matrix where rows and columns corresponding to restricted traits have been deleted.

The preceding review of estimation and prediction methods has been brief but necessary. The geneticist will find the sections lacking in detail whereas the nongeneticist may find the discussion overly complex. The conclusion of overall importance is that a body of estimation and prediction theory exists that is adequate to bring biometrical rigor to the comparison of alternative breeding strategies for virtually any possible poplar breeding problem.

Quantitative genetics and poplars

The genus *Populus* is diverse, represented by 29 species in 6 sections (or more depending on taxonomic system) distributed throughout the northern hemisphere (Dickmann and Stuart 1983; see also Eckenwalder, Chapter 1). Such a rich source of genetic variability can give rise to a multitude of breeding strategies including development of intraspecific populations of clones or families suitable for recurrent selection. In addition, many species are interfertile providing the opportunity to produce interspecific populations of diverse origin and genetic structure (Dickmann and Stuart 1983) including F_1 hybrids, advanced generation F_2 hybrids, and backcross populations suitable for direct commercial deployment. It is the role of quantitative genetic analysis to determine which of the many breeding opportunities and selection strategies are potentially the most fruitful. As in the previous discussion of quantitative genetic methods, the following summary of applications of quantitative genetics to breeding is potentially lengthy. In the interest of brevity the following discussion is organized by those selection criteria that are, in these authors' opinion, the most important according to current commercialization strategies. Thus, we will focus on growth and yield; on rooting ability because commercial deployment is most often by dormant, unrooted hardwood cutting; and on disease resistance. Our discussion of growth and yield of poplar trees in single tree or small plot experiments holds to a quantitative genetic focus. We depart from that focus to address the subject of plot configuration (i.e., the effects of competition) because empirical quantitative genetic research has not been well developed on that subject. Discussion of disease resistance will focus on *Melampsora* rust. Poplars are certainly subject to a wider range of pathogens but the bulk of the quantitative genetic literature focuses on *Melampsora* rust. *Septoria* canker, which is arguably the most limiting disease in eastern North America, is not discussed because there have been virtually no quantitative genetic studies of that disease. Poplars are ideal subjects for quantitative genetic studies because the ability to clonally replicate individuals provides the ability to estimate complex modes of gene action (Foster and Shaw 1988; Mullin and Park 1992), an ability that has not, unfortunately, been used to great advantage.

Growth and yield

Quantitative genetic studies of the genus *Populus* have emphasized the estimation of genetic and environmental variation in the height, diameter, volume, or mass of the main stem, which is reasonable considering that stem volume or mass are economically predominant selection criteria. Past research can be classified mostly into two broad areas: (1) determining how genetic variance is structured in natural populations by estimating variances due to populations, parents within populations (in narrow sense studies), and clones within populations (broad sense studies), and (2) estimating genetic covariances between stem growth and other traits, usually for the purpose of predicting indirect effects of selection for stem growth or for predicting the efficiency of various indirect selection strategies.

Initial quantitative genetic studies to determine broad-sense heritabilities for growth were conducted using clonal populations of *Populus deltoides* Bartr. from the southern United States. For example, broad sense heritabilities ranging from 0.20 to 0.30 for height and diameter after one and two growing seasons were estimated in a population of 49 clones growing in a test in Mississippi (Wilcox and Farmer 1967). Subsequent research in the same region was expanded to include more clones with attention to wood quality traits (Farmer and Wilcox 1968), evaluation of height and diameter at later ages (Mohn and Randall 1971), and additional experimental sites to estimate variance attributable to genotype × environment interaction (Mohn and Randall 1973; Randall and Cooper

1973). Studies of more broadly-based experimental populations have also confirmed significant broad-sense heritability estimates for stem growth (Foster 1986). Overall, studies demonstrated consistently significant broad-sense heritabilites for growth at all ages (up to five years) $(0.21 < H^2 < 0.50)$, significant genotypic age–age correlations among measures of tree height and diameter (especially after year 1), and genotype $\times$ environment interactions that were significant but usually less than half as large as main effects due to genotypes. Nonadditive components of variation, i.e., dominance and epistasis, may contribute significantly to broad sense genetic variation among *P. deltoides* clones (Foster 1985; Foster and Shaw 1988). Additional studies conducted with *P. deltoides* and *P. balsamifera* L. tested whether physiological preconditioning of the stock plant ("C-effects") influenced subsequent growth of the vegetatively propagated tree during the first year (Wilcox and Farmer 1968; Farmer et al. 1989) or up to three years after field planting (Farmer et al. 1988). Results demonstrated that C-effects (physiological preconditioning due to differences among stock plants of the same genotype (Lerner 1958)) could be important during the first year of growth when variance in shoot growth attributable to primary ramet effects was often as large as variance due to clones (Wilcox and Farmer 1968; Farmer et al. 1989). However, C-effects did not appear to be significant after field planting (Farmer et al. 1988). Heritability for stem height in the *P. balsmifera* field test was about 0.50 (Farmer et al. 1988). Quantitative methods have also been applied to studies of seedling populations, with the resultant narrow-sense stem growth heritability estimates being similar to broad sense estimates (*P. deltoides*, Farmer 1970; Ying and Bagley 1976; Nelson and Tauer 1987) or much lower (*P. trichocarpa* Torr. & Gray, Rogers et al. 1989).

Quantitative genetic methods have recently been used to test whether the crop ideotype concept has utility in the selection of *P. balsamifera* (Riemenschneider et al. 1992), *P. trichocarpa* (Riemenschneider et al. 1994), or hybrids between *P. deltoides* and *P. nigra* L. or *P. simonii* Carr. (Wu et al. 1992; Wu 1994a, b). Such studies typically include detailed measures of leaf, branch, and phenological characteristics, which are known to affect tree productivity (see Dickman and Keathley, Chapter 19; Ceulemans and Isebrands, Chapter 15; and Heilman et al., Chapter 18). The potential utility of the various measures as indirect selection criteria are then tested using single trait (Wu 1994b) or multiple trait selection index (Riemenschneider et al. 1992, 1994) theory. Results have demonstrated that variation in most morphological traits is heritable. However, the results concerning the utility of various traits as selection aids has been inconsistent. Studies of two-year-old *P. balsamifera* and *P. trichocarpa* (Riemenschneider et al. 1992, 1994, respectively) failed to identify selection indices that were more efficient than selection for tree height alone. Studies of hybrid populations suggest that the indirect effect of selection based on leaf size and area, branching capacity, and branch angle at age four years on stem volume at age six years could be more efficient than direct selection (Wu 1994b). However, that result depended on correcting the relative efficiencies of the two selection strategies for evaluation time (i.e., gain per year as opposed to gain per generation).

One predominant limitation of all the previously mentioned quantitative genetic studies of the growth of poplar trees has been the use of single tree or small plot studies. Commercially-grown poplar plantings tend not to be mixtures of large numbers of clones but rather mosaics of monoclonal blocks. Such dissimilarity between the experimental design used in genetic studies and the common commercial deployment method could significantly impact the effectiveness of breeding for tree growth. We address that subject in the following sections.

From individual tree to the stand

Intergenotypic competition, where trees within a species compete for limited resources (e.g., light, water, and nutrients), is a major concern in varietal deployment, and thus the development of new varieties (Foster 1992; Libby 1987). When two or more varieties are mixed together in the same plot, their yield can equal the weighted mean (complementary yield), exceed the weighted mean (overyielding), or be less than the weighted mean (underyielding). Results have been mixed In agronomic crop research on this subject. Trenbath (1974) and Marshall (1977) cited some cases where the mixed variety plot yielded even more than the best pure component variety; however, the mixed plot was usually intermediate between the two pure plots of the component varieties. Similar complexities can be expected in tree selection programs in which plot designs (often rows or single-tree plots) do not reflect the true competitive conditions within a forest stand (Adams 1980; Foster 1989, 1992, 1993; Foster and Bertolucci 1994; Foster and Knowe 1995). Cannell (1978) expressed concern over selection for competitive ideotypes rather than crop ideotypes and suggested that realized gain may be lower than expected.

The potential effects of intergenotypic competition depend on the exact deployment strategy used (Libby 1987; Foster 1992). Most deployment strategies regarding forest tree species are based upon little more than empirical observations. The simplest strategy is to collect seed from several stands within a seed source, or seed from select trees in a seed orchard, and mix it; a strategy that probably would not be readily applicable to poplars. Likewise, clones in a clonal program can also be mixed at random. An intermediate approach is to divide the select trees or clones into subsets which are known to perform particularly well on certain sites or be resistant to a particular pest. Finally, some forest managers plant mosaics of single varieties in pure patches, but with several different varieties per stand (Libby 1987; Durzan and Williams 1988). Unfortunately, there is little evidence to support the use of one strategy over another.

Genetic test designs to predict varietal yield

Considerable effort has been spent over the years to develop the most efficient progeny test design for ranking varieties of forest trees for their mean performance. Experiments that compared the different plot designs (i.e., different number of trees per plot and different plot configurations) generally showed a similar ranking of the varieties tested (Foster 1989). However, the variances among varieties decreased as the design changed from single-tree or noncontiguous plots to row plots, and finally to large block plots. Therefore, the typical design has favored the smaller plot size in order to increase the ability to detect significant differences among varieties and also to test large numbers of varieties.

Optimally, plot size should reflect stand dynamics when the experimental goal is to estimate volume yield per unit area. But, once intergenotypic competition initiates the process of crown class development, small plots are no longer an accurate representation of large plot, or stand, dynamics. This is because the size distributions of trees within the plots may be abnormal. For instance, all trees in the family plot could be either all dominants or all suppressed.

Mensurationally, the average height of the dominant-codominant trees is one of the variables needed to accurately predict stand volume (Clutter et al. 1983). The average number of trees per plot that would be needed to include the tallest 100 trees per acre would depend upon the initial spacing and the number of trees per plot. For instance, an 8×8 ft initial spacing would yield 680 trees per acre; therefore, the tallest 100 would represent 14.7 percent of the original trees. With 50-tree plots, seven dominant–codominant trees on average would be included and could be used to calculate the dominant–codominant height of the forest stand. Diameter distributions can be predicted from the dominant–codominant height, stand age, and number of trees per unit area. The number of sample trees required to estimate dominant–codominant height can vary, but the 100 tallest trees per acre is a common number of sample trees (Buford and Burkhart 1985, 1987).

It is most likely that a globally optimum experimental design does not exist. Testing objectives differ among regions and species, and are affected by anticipated commercial deployment strategies. To date, the best solution seems to include the use of sequential tests (Libby 1987). For screening large numbers of varieties, an initial group of tests is installed using small plots, and the trees are measured prior to crown closure. Then, new tests using large plots are initiated for evaluating long-term yield potential of only the better varieties.

Replacement series and yield prediction

The replacement series design has been commonly used in both agronomic and forest tree species to test the effect of intergenotypic competition on yield. In a replacement series, the varieties are grown in a series of treatments with a fixed number of plants per plot and one or more (usually two) varieties in different proportions per treatment (Harper 1977). In forestry literature, replacement series are often called mixing studies. Results from several mixing studies have been published for trees species, but most have either used seedlings or very young clonal stock. Studies have demonstrated complementary, overyielding, and underyielding responses on growth (Adams et al. 1973; Tauer 1975; Adams 1980; Tuskan and van Buijtenen 1986; von Euler et al. 1993). Whether the results from these studies can be extended to closed forest stands is yet to be determined.

Mixing studies have been conducted with *P. deltoides* (Knowe et al. 1994; Foster and Knowe 1995). Seven clones were included in pure treatment plots and also in seven two-clone mixtures at two different ratios per mixture. The plot size was 64 trees with the inner 36 being measurement trees. A 3.7 × 3.7 m spacing was used. A randomized complete block design was employed with four replications per site and two sites. Data from ages two, three, and four were analyzed. Although the trees were young, they were quite large by age four (13.23 m average height), hence stand development was well underway. Based on this experiment, Knowe et al. (1994) modeled diameter distributions in order to develop prediction equations for diameter distributions of mixed clone stands based on monoclonal stand distributions. They found that the general shape of the distribution shifted depending on the specific clone or mixture. Foster et al. (1995) noted that mixture yields for most of the two-clone groups (a group included the two monoclonal plots plus the two mixtures with those two clones) increased linearly with an increase in the proportion of the highest yielding clone. For a few two-clone groups, the trend was either overyielding or underyielding. In some cases, different trends were found at the two sites.

Results from the above mixing study were also used to simulate the stand structure and dynamics of *P. deltoides* between the ages of two and ten years by using a diameter distribution approach (Foster and Knowe 1995). Components of this approach include a dominant height growth function (which provides information about stand dynamics and is important in predicting other stand-level components), a basal area prediction function, diameter distribution prediction functions, and a height-dbh function. Simulations were performed for mixtures and monocultures of underyielding, overyielding, and complementary clones. Simulations revealed that the performance of a mixture was predictable, however, actual yields varied by environment.

In addition to the replacement series, another general type of mixing study includes a single treatment with all varieties mixed in equal proportions compared with separate single variety treatment plots. The mean of the single mixed treatment is then compared with the mean of the single variety treatments. Results from these studies are informative, but they have little predictive value, as do the replacement series studies. Most field studies with forest tree species fall into this category. A significant difference may occur between the yields of the mixed versus single variety treatments (Williams et al. 1983; Lundkvist et al. 1992) or it may not occur (Markovic and Herpka 1986; DeBell and Harrington 1993; von Euler 1993).

Two characteristics of such studies may explain the predominance of inconclusive results. Plot sizes must be relatively large to give realistic estimates of per area yields and to simulate stand development, and large plots cause the experimental errors to increase thereby requiring larger treatment differences in order to demonstrate statistical significance. A related problem is that larger plots allow fewer varieties to be tested compared with a regular progeny test, and test results then become more sensitive to the specific varieties being tested.

The number of varieties used in quantitative genetic studies and breeding tests is typically in the hundreds. Therefore, it is impractical to test all varieties in pure treatments and in all possible combinations. Some form of modeling will be necessary to help reduce the amount of testing required. For agronomic crops, Wright (1982, 1983) among others has developed models to predict yields of mixtures based on their components as well as the efficiencies of various methods of selection of components for intergenotypic mixtures. Little work has been done in this area for forest tree species. Nance (1982) used a growth and yield model to simulate the results of single family versus mixed family plantations of *Pinus taeda* L., and Knowe et al. (1994) modeled the effects of mixing clones of *P. deltoides* on diameter distributions, which can then be used in growth and yield models (Foster and Knowe 1995). Undoubtedly, factors like crown and root architecture and efficiency of use of growing space will be important in this modeling effort (Cannell 1978; Nance et al. 1983).

The value of replacement, or mixing, studies obviously depends on deployment strategy. Mixing studies have future value if select poplar clones are deployed as mixtures, which they may be for reasons of pest resistance and genetic diversity. But, experience seems to indicate that deployment will most likely be in the form of mosaics of monoclonal blocks. That being the case, it seems that the effects of intergenotypic competition on stand development lacks importance. However, it may be as difficult to define a global deployment strategy as it is to define a globally optimum testing strategy. Therefore, the wisest course of action should include at least minimal investment in studies designed to estimate and, hopefully, predict the effects of intergenotypic competition.

Rooting ability

The ability of a dormant poplar hardwood cutting to form adventitious roots is a critical biological characteristic because the dominant commercial deployment strategy is to plant unrooted cuttings. Given the importance of the trait, it is difficult to understand the shortage of quantitative genetic understanding of rooting ability compared to other commercially important traits. Estimates of genetic variances for rooting ability would provide guidelines for genetic selection for rooting. Perhaps more importantly, estimates of covariances between rooting and other commonly selected characteristics, such as growth and disease resistance, could provide estimates of (1) indirect effects of selection for stem growth on the ability of clones to form roots and (2) strategies for improving rooting ability without the need to observe the trait directly. Indirect selection strategies would be particularly useful because the direct observation of root formation and development can be laborious.

Early studies of root number, root length, and root weight produced by *P. deltoides* cuttings demonstrated significant broad sense heritability ($0.33 < H^2 < 0.58$) for all characteristics (Wilcox and Farmer 1968). Subsequent research resulted in higher broad sense heritability estimates for root number ($0.85 < H^2 < 0.91$) in the same species (Ying and Bagley 1977). Discrepancies between the two tests could have been attributed to a number of factors including: (1) local vs. regional experimental populations, (2) method of accounting for C effects, if any, and (3) computational formulations. Separation of genetic and environmental variation from C effects can be especially important because variation due to C effects can equal or exceed variation due to genetic effects (Wilcox and Farmer 1968; Farmer et al. 1989). Variation due to C effects can be confounded with genetic or environmental variation, depending on the experimental design (Wilcox and Farmer 1968). Estimates of genetic and environmental covariances between rooting and aerial characteristics are more lacking. However, past research has suggested that a genetic correlation exists between attributes of root formation and date of bud break (Wilcox and Farmer 1968; Farmer et al. 1989) suggesting that a potential indirect selection criterion might exist.

We have also conducted recent studies of rooting ability in *P. trichocarpa*. The objective of these experiments was to test whether 50 clones of *P. trichocarpa* differed in the development of shoots, leaves, and roots from dormant hardwood cuttings. We also tested whether shoot, leaf, and root dry weights measured in the current experiment were correlated with various measures of growth, morphology, and pest resistance made on the same clones during a previous clonal nursery test (Riemenschneider et al. 1992). The experimental design in the original nursery test was two replications of a randomized complete block design with three-tree plots. Trees were grown for 5 years during which time various measurements of stem growth, branch and leaf morphology, and pest resistance were obtained. Trees were cut in January after the fifth growing season to a 10 cm stump and allowed to resprout. One-year-old coppice stems from a subsample of two trees per replication from each of 50 clones were harvested in March. The top and bottom-most 20 cm of each stem were discarded and the remaining section subdivided into 10 cm cuttings which were used for a greenhouse rooting experiment. Cuttings were soaked for 1 wk in water at 21°C then planted in coarse sand in individual containers (folding book planters) in the greenhouse (22°C days, 16°C nights). The experimental design in the greenhouse was randomized complete blocks with two replications of 12 cuttings from each nursery tree (ortet). Four developing cuttings from each replication and ortet were excavated at 1, 2, and 3 weeks after planting and dissected into the original cutting and the developing shoot, leaves and roots (8 cuttings per ortet per sample time). Dry weights were determined to the nearest mg and analyzed according to the all random effects model:

[7] $X_{ijkl} = m + R_i + C_j + RC_{ij} + T_{k(ij)} + e_{l(ijk)}$

where X_{ijkl} is an experimental observation, m is the experiment mean, R_i is the effect if the ith replication, C_j is the effect if the jth clone, RC_{ij} is the clone $\times$ replication interaction effect, $T_{k(ij)}$ is the effect if the kth ortet within the jth clone and ith replication, and $e_{l(ijk)}$ is experimental error (variation among cuttings within ortets). Clone mean correlations among cutting, shoot, leaf, and root weights at each sampling time were also estimated. Further, clone mean correlations among variables measured in the current experiment and various measures of the growth, morphology, and pest resistance of the parent ortets were also estimated.

Several results of the analyses of variance were notable (Table 2). First, replication effects on cutting, shoot, and leaf weight were significant only at week one while replication effects on root weight were significant only at weeks one and two. Apparently, environmental gradients in the greenhouse were large enough to affect flushing and early growth of new shoots and roots, but not large enough to affect subsequent development. Likewise, replication $\times$ clone effects were mostly nonsignificant. Overall, results indicate that complete block (or more complex) designs may not be needed to achieve experimental efficiency in our greenhouse rooting experiments.

Second, clone effects were always significant regardless of trait or sampling time. Heritability for cutting dry weight was comparable to the heritability of other traits we have measured in the nursery in the same population. However, heritability for shoot, leaf, and root dry weight tended to be low, especially for dry weights measured 3 weeks after planting. Root dry weight measurements were characterized by very high coefficients of variation (83.1, 78.2, and 82.6% at 1, 2, and 3 weeks after planting, respectively) (Table 2). It has been our general experience in the conduct of rooting experiments that root weights are usually associated with high experimental errors. Such errors are commonly attributed to loss of roots during destructive excavation. But, the root systems observed in the current system were simple and we are confident that they were excavated without loss. Therefore, we conclude that high errors were an inherent attribute of root systems in this study and not attributable to incautious handling.

Third, variance due to ortet within clone (i.e., C-effects) was consistently significant (Table 2). The component of variation due to ortet within clone in the

Table 2. Analyses of variance of cutting, shoot, leaf, and root dry weight in an experiment to test differences among 50 clones of *Populus trichocarpa* in their ability to develop new shoots and adventitious roots from dormant hardwood cuttings.

Source of variation	df	Week 1				Week 2				Week 3				Expected mean square
		Cutting (×10³)	Shoot (×10³)	Leaves (×10³)	Roots (×10⁵)	Cutting (×10⁴)	Shoot (×10⁴)	Leaves (×10³)	Roots (×10⁴)	Cutting (×10⁴)	Shoot (×10⁴)	Leaves (×10³)	Roots (×10⁴)	
Rep	1.00	2.05ns	12.68**	8.26**	16.82*	1.57ns	10.81ns	0.26ns	4.69**	1.30ns	0.26ns	5.76ns	1.05ns	$\sigma_E^2+4\sigma_T^2+8\sigma_{RC}^2+400\sigma_R^2$
Clone	49.00	5.27**	3.01**	3.41**	8.16**	4.30**	7.85**	6.79**	1.44**	4.56**	18.91**	11.78**	7.12**	$\sigma_E^2+4\sigma_T^2+8\sigma_{RC}^2+16\sigma_C^2$
Rep× clone	49.00	1.85*	0.89ns	0.71ns	2.74ns	1.17ns	2.70ns	2.45ns	0.43ns	1.29ns	7.66ns	5.14ns	1.30ns	$\sigma_E^2+4\sigma_T^2+8\sigma_{RC}^2$
Ramet/ clone	100.00	1.07**	0.70**	0.90**	3.18**	0.93**	2.41**	2.41**	0.34**	1.23**	8.18**	5.06**	1.42**	$\sigma_E^2+4\sigma_T^2$
Cutting/ ramet	600.00	0.34	0.25	0.33	2.12	0.30	1.34	0.93	0.19	0.28	4.40	1.92	0.76	σ_E^2
Variance components														
Rep		0.00	0.03	0.02	0.04	0.00	0.02	0.00	0.01	0.00	0.00	0.00	0.00	
Clone		0.21	0.13	0.17	0.34	0.20	0.32	0.27	0.06	0.20	0.70	0.42	0.36	
Rep× clone		0.10	0.02	0.00	0.05	0.03	0.04	0.06	0.01	0.01	0.00	0.01	0.01	
Ramet/ clone		0.18	0.11	0.14	0.26	0.16	0.27	0.37	0.04	0.24	0.95	0.78	0.16	
Cutting/ ramet		0.34	0.25	0.33	2.12	0.30	1.34	0.93	0.19	0.28	4.40	1.92	0.76	

Note: * = mean square significant at $0.01 < p < 0.05$; ** = mean square significant at $0.01 < p$, ns = not significant. Units are mg corrected as indicated for readability.

current experiment was high in relation to variation due to clones (Table 2) and sometimes exceeded that due to clones (i.e., shoot and leaf dry weights at 3 weeks after planting). The consistently large ortet effects in the current experiment support two conclusions. First, our genetics trial experiments should incorporate a strategy to estimate or eliminate ortet effects because, if ignored, those effects can be confounded with variance due to clone or error (depending on experimental design) which can lead to faulty predictions of selection efficiency. Second, significant ortet effects suggest that some experimental attention should be given to treatments, apart from genetic selection, that might enhance the expression of desired characteristics.

Correlation analyses also revealed several interesting results. First, cutting dry weights were correlated among sample times (Table 3) as would be expected when sampling at random from a genetically variable population. But, cutting dry weights were uncorrelated (except for a single case) with the dry weight of the developing shoots, leaves, or roots (Table 3). Likewise, incorporating cutting dry weight as a covariate during analyses of variance of shoot, leaf, or root dry weight had little outcome on the results of the analyses, including the proportion of variation due to ortets within clones. Thus, an understanding of the nature of C effects on shoot and root development will probably not be gained by simple measurement of cutting mass. Shoot and leaf weights tended to be strongly correlated both within and between sample times (Table 3). Correlations between shoot and leaf weights and root weight were nonsignificant at week one but increased thereafter. Overall, correlation analyses suggested that shoots and leaves were more tightly linked allometrically than were shoots and roots during very early development, but that an allometric relation was established among all developing plant parts by the third week after planting. The third week after planting was also associated with a decline in extraordinarily high shoot-root ratios of 18.0, 18.8, and 15.4 for weeks 1, 2, and 3, respectively.

We also completed correlation analyses between variables measured in the current experiment and previous measures of growth and morphology on the clones. Cutting dry weight was significantly correlated with several measures of leaf size made on 2-year-old clones. Correlations were strongest with leaf area on LPI 10 and 15 leaves when leaf measurements were made near the end of the growing season on September 1. Cutting dry weight was also correlated with clone height at age 1 year. Shoot and leaf weight were also correlated with age 2 year budset date and with date of bud-break at the start of the third growing season. Root dry weight was negatively correlated with bud-break date indicating that some increase in rooting ability might be achieved by selecting for clones that flush rapidly at the beginning of the growing season, an observation also made in past studies (Wilcox and Farmer 1968; Farmer et al. 1989). Overall, current results suggest that genetic selection for rooting ability could increase rooting at reasonable selection intensities, and that additional increases could be achieved through a more complete knowledge of the effects of ortet preconditioning.

Table 3. Clone mean correlations among cutting, shoot, leaf, and root dry weights (g) at 1, 2, and 3 wks after planting.

	Week 1				Week 2				Week 3			
	Cutting	Shoot	Leaves	Roots	Cutting	Shoot	Leaves	Roots	Cutting	Shoot	Leaves	Roots
Week 1 cutting weight		0.18	-0.23	-0.06	0.93**	0.18	0.08	-0.16	0.90**	0.00	0.06	-0.24
Week 1 shoot weight			-0.52**	-0.03	0.22	0.64**	0.15	-0.09	0.18	0.34*	0.14	-0.22
Week 1 leaf weight				0.11	-0.28	-0.03	0.46**	0.31*	-0.25	0.26	0.30*	0.54**
Week 1 root weight					-0.06	0.06	0.23	0.56**	-0.08	0.08	0.42**	0.51**
Week 2 cutting weight						0.25	0.06	-0.19	0.91**	0.04	0.03	-0.32*
Week 2 shoot weight							0.47**	0.17	0.24	0.70**	0.42**	0.07
Week 2 leaf weight								0.56**	0.05	0.65**	0.77**	0.64**
Week 2 root weight									-0.13	0.38**	0.61**	0.81**
Week 3 cutting weight										0.11	0.14	-0.24
Week 3 shoot weight											0.66**	0.35*
Week 3 leaf weight												0.64**

Note: * = correlation significant at $0.01 < p < 0.05$; ** = correlation significant at $p < 0.01$.

Disease resistance

Melampsora rust is an ideal example for the application of quantitative genetics to breeding for disease resistance in poplars for two reasons. First, the *Melampsora*-poplar pathosystem is biologically complex and therefore of significant fundamental interest. Four species of *Melampsora* are indigenous to North America and show host specificity (Stack and Ostry 1990; Hsiang and Chastagner 1990; Pinon 1992) or mixed infections on individual hosts and host tissues (Stack and Ostry 1990). In addition to pathogen species complexity, *Melampsora*-poplar interactions may depend on pathogen race (Prakash and Thielges 1987), host geographic origin (Hsiang et al. 1993) and environmental conditions (Singh and Heather 1982; Prakash and Thielges 1988). Second, *Melampsora* rust can cause both growth loss (Widin and Schipper 1981; Wang and van der Kamp 1992) and mortality (Newcombe et al. 1994). Thus, estimation of genetic and phenotypic variances for incidence and severity of *Melampsora* rust, and covariances with other selection criteria, should be of significant practical interest.

The quantitative inheritance of *Melampsora* rust incidence on poplars was established at least as early as 1966 with estimation of narrow ($0.38 < h^2 < 0.66$) and broad ($0.66 < H^2 < 0.88$) sense heritabilities for rust incidence in seedling and clonal populations of *P. deltoides* (Jokela 1966) . Subsequent studies have added heritability estimates of rust incidence or severity in *P. deltoides* (Farmer and Wilcox 1968; Thielges and Adams 1975), *P. tremula* L., and *P. tremuloides* Michx. (Gallo et al. 1985), open-pollinated progeny of *P. deltoides* (Nelson and Tauer 1987), *P. balsamifera* (Riemenschneider et al. 1992), *P. trichocarpa* and its hybrids (Hsiang et al. 1993; Riemenschneider et al. 1994) and hybrids among *P. deltoides, P. nigra*, and *P. maximowiczii* Henry (Rajora et al. 1994). Sufficient research was accomplished to subdivide symptom development into several subcategories (latent period, number of uredia, size of uredia) (Pichot and Tessier du Cros 1993) in a factorial mating among *P. deltoides* parents. Reports of nonzero heritability estimates are characteristic of all the aforementioned studies. However, some reports of narrow sense (additive genetic variance) heritability estimates need to be carefully interpreted because populations of interspecific hybrids may not adhere to all the assumptions, i.e., random mating and linkage equilibrium, required for genetic variance component estimation. Nonetheless, estimates of narrow and broad sense heritabilities suggest that significant population improvement in resistance to *Melampsora* rust can be accomplished in many, if not most, important species and species hybrids.

Covariances between the incidence and severity of *Melampsora* rust and growth of the host plant have been less commonly estimated and results have been inconsistent. For example, correlations between rust and tree height or diameter have been reported zero, positive, or negative depending on host species, age at which observations were made, and experimental location (Farmer and Wilcox 1968; Gallo et al. 1985; Nelson and Tauer 1987; Riemenschneider et al. 1992; Pichot and Tessier du Cros 1993; Riemenschneider et al. 1994; Rajora et al. 1994). Inconsistency among estimates of the covariance between rust and growth are troubling because rust is important, in a crop development context, only insofar as rust affects growth of the host plant. It has been suggested that poplar genotypes may differ in the rate with which physiological debilitation occurs in the face of rust infection and that common once-per-year observation strategies (Riemenschneider et al. 1994) may preclude a quantitative genetic understanding of rust-growth relations. It has been demonstrated that the temporal phases of rust infection can be described quantitatively (Pichot and Teissier du Cros 1993). It has also been very clearly demonstrated by Wang and van der Kamp (1992) that poplar genotypes differ in both rust disease severity (frequency and distribution of uredia) and tolerance (relative loss of yield among clones exhibiting the same level of disease severity). We suggest here that consideration of temporal dynamics of rust infection along with careful analysis of growth impacts is prerequisite to a quantitative genetic understanding of rust-growth dynamics.

Conclusions and recommendations

Several conclusions can be reached based on 30 years of quantitative genetics research in poplars. First, and most important, virtually every trait that has ever been studied has turned out to be heritable, at least to some extent. This conclusion holds true over a very broad range of experimental poplar populations including pure species (both clonal and seedling) and interspecific hybrids. As a corollary, it can be concluded that response to direct selection for most traits can be expected in most poplar populations. Second, covariance structures describing relationships among various above-ground plant traits (branching, leaf morphology, phenology, etc.) suggest that strong genetic relationships between those traits and stem growth are common. However, efficient strategies for using auxiliary traits as indirect selection criteria for stem growth, if they exist at all, have been difficult to discover. This conclusion raises the question of how, or whether, a quantitative genetic basis can be established for the crop ideotype concept, including the use of physiological and physiologically based selection criteria, as it is applied to the selection of fast growing poplars. Third, knowledge of covariances between stem growth and rooting ability or disease resistance is rudimentary, although new information is continually emerging. Fourth, knowledge of covariances between stem growth in different competitive environments (i.e., single tree plot genetic experiments vs. large plot yield trials) is mostly lacking and limits the confidence with which quantitative genetic inferences can be made between different competitive settings. In general, it is possible to predict that direct selection for stem growth will result in significant response. But, the indirect effects of that selection on other important characteristics (i.e., disease resistance, large plot growth, and yield) cannot be predicted with confidence.

Overall, we have recommendations for future research on estimation of variance and covariance components and on prediction of selection response. First, in regard to estimation needs, our knowledge of covariances between tree growth and other important selection criteria is inadequate. We believe rooting ability has been largely ignored in quantitative genetic studies because a kind of *de facto* selection is always applied for rooting ability. Serial propagation of test clones occurs as testing progresses from large populations with low replication to small, select, populations with high replication. Thus, clones that root poorly are effectively eliminated from clonal selection programs. Yet, this strategy ignores possible correlated effects of *de facto* selection for rooting on growth potential, wood quality, and other factors. A more reasonable strategy would be to evaluate rooting ability in quick (1–3 weeks) experiments as we have described. Then, covariances among selection criteria could be estimated and selection could be practiced in a more deterministic fashion. Second, our knowledge of covariances between tree growth and disease resistance is also inadequate. *Melampsora* leaf rust is the most studied of all poplar diseases, yet results regarding the covariance between rust and growth are inconsistent. Consideration of seasonal dynamics in disease progression coupled with detailed growth measurements is recommended. Studies designed to achieve quantitative genetic links between growth and *Septoria* canker are also critically needed.

Two additional needs in the area of prediction are also apparent. First, we know of no quantitative genetic basis for predicting the performance of hybrid progeny derived from wide interspecific crosses, and have not discussed the subject in this chapter. Gains in achieving such an understanding seem to be more quickly forthcoming from analyses of quantitative trait loci and from physiological studies of tree growth and development in advanced generation hybrid poplar populations (see Stettler et al., Chapter 4; Bradshaw, Chapter 8). The application of quantitative genetics to the prediction of hybrid performance might be difficult because wide crosses probably violate assumptions regarding random mating and linkage equilibrium, which precludes all but broad-sense heritability estimates. Thus, some theoretical treatment of the subject may be prerequisite to empirical research. Second, additional knowledge is needed to determine how small-plot selection relates to deployment strategies, genetic diversity, and predictability of growth and yield in large plots. Reasons for this knowledge gap are well known, i.e., large, long-term, relatively expensive studies are needed. Data from replacement series experiments, as exemplified by Knowe et al. (1994), may be useful, depending on deployment strategy. We further suggest that existing and future knowledge of the physiology of poplar growth and poplar crown architecture (i.e., the crop ideotype) (see Dickman and Keathley, Chapter 19; Ceulemans and Isebrands, Chapter 15; and Heilman et al., Chapter 18), could lead to the development of new, informational hypotheses about stand development and the effects of clonal mixtures and mosaics. Such hypotheses could have great potential value to poplar breeding and selection, but discrepancies between physiological and quantitative genetic experimentation need to be reconciled.

Poor integration of genetic and physiological approaches to the study of the growth and development of poplar trees and stands has, thus far, limited progress.

References

Adams, W.T. 1980. Intergenotypic competition in forest trees. *In* Proceedings of the 6th North American Forest Biology Workshop. *Edited by* B.P. Dancik and K.O. Higginbotham. University of Alberta. Edmonton, AB. pp. 1–14.

Adams, W.T., Roberds J.H., and Zobel, B.J. 1973. Intergenotypic interactions among families of loblolly pine (*Pinus taeda* L.). Theor. Appl. Genet. **43**: 319–322.

Buford, M.A., and Burkhart, H.E. 1985. Dynamics of improved loblolly pine plantations and the implications for modeling growth of improved stands. *In* Proceedings of the 18th Southern Forest Tree Improvement Conference. Long Beach, MS. pp. 170–177.

Buford, M.A., and Burkhart, H.E. 1987. Genetic improvement effects on growth and yield of loblolly pine plantations. For. Sci. **33**: 707–724.

Cannell, M.G.R. 1978. Improving per hectare forest productivity. *In* Proceedings of the 5th North American Forest Biology Workshop. University of Florida, Gainesville, FL. pp. 120–148.

Clutter, J.L., Forston, J.C., Pienaar, L.V., Brister, G.H., and Bailey, R.L. 1983. Timber management: a quantitative approach. John Wiley & Sons, New York, NY.

Cotterill, P.P., and Dean, C.A. 1990. Successful tree breeding with index selection. Commonwealth Scientific and Industrial Research Organization, Melbourne, Australia. 79 p.

DeBell, D.S., and Harrington, C.A. 1993. Deploying genotypes in short-rotation plantations: mixtures and pure cultures of clones and species. For. Chron. **69**: 705–713.

Dickmann, D.I., and Stuart, K.W. 1983. The culture of poplars in eastern North America. University Publications, Michigan State University, East Lansing, MI. 168 p.

Durzan, H.W., and Williams, C.G. 1988. Matching loblolly pine families to regeneration sites. South. J. Appl. For. **12**: 166–169.

Farmer, R.E., Jr. 1970. Genetic variation among open-pollinated progeny of eastern cottonwood. Silv. Genet. **19**: 149–151.

Farmer, R.E., Jr., and Wilcox, J.R. 1968. Preliminary testing of eastern cottonwood clones. Theor. Appl. Genet. **38**: 197–201.

Farmer, R.E., Jr., Garlick, K., and Watson, S.R. 1988. Heritability and C effects in a 3-year-old balsam poplar clonal test. Can. J. For. Res. **18**: 1059–1062.

Farmer, R.E., Jr., Freitag, M., and Garlick, K. 1989. Genetic variance and "C" effects in balsam poplar rooting. Silv. Genet. **38**: 62–65.

Fisher, R.A. 1930. The genetical theory of natural selection. Clarendon Press, Oxford, UK.

Foster, G.S. 1985. Genetic parameters for two eastern cottonwood populations in the lower Mississippi valley. *In* Proceedings of the 18th Southern Forest Tree Improvement Conference. Long Beach, MS. pp. 258–266.

Foster, G.S. 1986. Provenance variation of Eastern cottonwood in the lower Mississippi valley. Silv. Genet. **35**: 32–38.

Foster, G.S. 1989. Inter-genotypic competition in forest trees and its impact on realized gain from family selection. *In* Proceedings of the 20th Southern Forest Tree Improvement Conference. Clemson University, Charleston, SC. pp. 21–35.

Foster, G.S. 1992. Estimating yield: beyond breeding values. *In* Handbook of quantitative forest genetics. *Edited by* L. Fins, S.T. Friedman, and J.V. Brotschol. Kluwer Academic, The Netherlands. pp. 229–269.

Foster, G.S. 1993. Varietal deployment strategies and genetic diversity: balancing productivity and stability of forest stands. *In* Proceedings of the 24th Canadian Tree Improvement Association Meeting, Part 2. *Edited by* J. Lavereau. Natural Resources Canada. Fredericton, NB. pp. 15–19.

Foster, G.S., and Bertolucci, F.L.G. 1994. Clonal development and deployment strategies to enhance gain while minimizing risk. *In* ?? *Edited by* R.R.B. Leakey and A.C. Newton. HMSO Publishers, London, UK. pp. 103–111.

Foster, G.S., and Knowe, S.A. 1995. Deployment and genetic gains. *In* Proceedings of CRCTHF-IUFRO Conference Eucalypt Plantations: Improving fibre yield and quality. *Edited by* B.M. Potts, N.M.G. Borralho, J.B. Reid, R.N. Comer, W.N. Tibbits, and C.A. Raymond. Hobart, Australia. pp. 19–24.

Foster, G.S., and Shaw, D.V. 1988. Using clonal replicates to explore genetic variation in a perennial plant species. Theor. Appl. Genet. **76**: 788–794.

Gallo, L.A., Stephan, B.R., and Krusche, D. 1985. Genetic variation of *Melampsora* leaf rust resistance in progenies of crossings between and within *Populus tremula* and *P. tremuloides* clones. Silv. Genet. **34**: 208–214.

Haldane, J.B.S. 1931. The causes of evolution. Harper and Brothers, New York, NY.

Harper, J.L. 1977. Population biology of plants. Academic Press, New York, NY.

Hazel, L.N. 1943. The genetic basis for constructing selection indexes. Genetics, **28**: 476–490.

Hazel, L.N., and Lush, J.L. 1942. The efficiency of three methods of selection. J. Hered. **33**: 393–399.

Hsiang, T., and Chastagner, G.A. 1990. Incidence of *Melampsora* rust in a seedling plantation of hybrid poplar. *In* Recent research on foliage diseases. *Edited by* W. Merrill and M.E. Ostry. U.S. For. Serv. Gen. Tech. Rep. WO-56, Washington, DC. pp. 125–129.

Hsiang, T., Ghastagner, G.A., Dunlap, J.M., and Stettler, R.F. 1993. Genetic variation and productivity of *Populus trichocarpa* and its hybrids. VI. Field susceptibility of seedlings to *Melampsora occidentalis* leaf rust. Can. J. For. Res. **23**: 436–441.

Jokela, J.J. 1966. Incidence and heritability of *Mempampsora* rust in *Populus deltoides*. *In* Breeding pest resistant trees. *Edited by* H.D. Gerhold, E.J. Schriener, R.E. McDermott, and J.A. Winieski. Pergamon Press, Oxford, UK. pp. 111–117.

Kempthorne, O., and Nordskog, A.W. 1959. Restricted selection indices. Biometrics, **16**: 10–19.

Knowe, S.A., Foster, G.S., Rousseau, R.J., and Nance, W.L. 1994. Eastern cottonwood clonal mixing study: predicted diameter distributions. Can. J. For. Res. **24**: 405–414.

Lerner, I.M. 1958. The genetic basis of selection. John Wiley and Sons, New York, NY. 298 p.

Libby, W.J. 1987. Testing for clonal forestry. Ann. Forest. **13**: 69–75.

Lundkvist, K., Eriksson, G., and Norell, L. 1992. Performance of clonal mixtures and single-clone plots in young *Picea abies* trials. Scand. J. For. Res. **7**: 53–62.

Markovic, J., and Herpka, I. 1986. Plantations in short rotations. *In* Poplars and Willows in Yugoslavia. Poplar Research Institute. Novi Sad. pp. 182–198.

Marshall, D.R. 1977. The advantages and hazards of genetic homogeneity. *In* The genetic basis of epidemics in agriculture. *Edited by* P.R. Dey. Academy of Sciences. New York, NY. pp. 1–20.

Mohn, C.A., and Randall, W.K. 1971. Inheritance and correlation of growth characters in *Populus deltoides*. Silv. Genet. **20**: 182–184.

Mohn, C.A., and Randall, W.K. 1973. Interaction of cottonwood clones with site and planting year. Can. J. For. Res. **3**: 329–332.

Mullin, T.J., and Park, Y.S. 1992. Estimating genetic gains from alternative breeding strategies for clonal forestry. Can. J. For. Res. **22**: 14–23.

Namkoong, G. 1979. Introduction to quantitative genetics in forestry. U.S. Dep. Agric., Tech. Bull. No. 1588. 342 p.

Nance, W.L. 1982. Simulated growth and yield of single-family versus multi-family loblolly pine plantations. *In* Proceedings of the 2nd Biennial Southern Siviculture Conference. U.S. For. Serv., South. For. Exp. Stn. New Orleans, LA. pp. 446–453.

Nance, W.L., Land, S.B., Jr., and Daniels, R.F. 1983. Concepts for analysis of intergenotypic competition in forest trees. *In* Proceedings of the 17th Southern Forest Tree Improvement Conference. University of Georgia. Athens, GA. pp. 131–145.

Nelson, C.D., and Tauer, C.G. 1987. Genetic variation in juvenile characters of *Populus deltoides* Bartr. from the southern great plains. Silv. Genet. **36**: 216–221.

Newcombe, G., Chastagner, G.A., Schuette, W., and Stanton, B.J. 1994. Mortaility among hybrid poplar clones in a stool bed following leaf rust caused by *Melampsora medusae* f.sp. *deltoidae*. Can. J. For. Res. **24**: 1984–1987.

Nyquist, W.E. 1991. Estimation of heritability and prediction of selection response in plant populations. Crit. Rev. Plant Sci.ences. **10**: 235–322.

Pichot, C., and Tessier du Cros, E. 1993. Susceptibility of *P. deltoides* Bartr. to *Melampsora larici-populina* and *M. allii-populina*. II. Quantitative analysis of a 6×6 factorial mating design. Silv. Genet. **42**: 188–199.

Pinon, J. 1992. Variability in the genus *Populus* in sensitivity to *Melampsora* rusts. Silv. Genet. **41**: 25–34.

Prakash, C.S., and Thielges, B.A. 1987. Pathogenic variation in *melampsora medusae* leaf rust of poplars. Euphytica, **36**: 563–570.

Prakash, C.S., and Thielges, B.A. 1988. Interaction of geographic isolates of *Melampsora medusae* and *Populus*:effect of temperature. Can. J. Bot. **67**: 486–490.

Rajora, O.P., Zsuffa, L., and Yeh, F.C. 1994. Variation, inheritance and correlations of growth characters and *Melampsora* leaf rust resistance in full-sib families of *Populus*. Silv. Genet. **43**: 219–226.

Randall, W.K., and Cooper, D.T. 1973. Predicted genotypic gain from cottonwood clonal tests. Silv. Genet. **22**: 165–167.

Riemenschneider, D.E., McMahon, B.E., and Ostry, M.E. 1992. Use of selection indices to increase tree height and to control damaging agents in 2-year-old balsam poplar. Can. J. For. Res. **22**: 561–567.

Riemenschneider, D.E., McMahon, B.E., and Ostry, M.E. 1994. Population-dependent selection strategies needed for 2-year-old black cottonwood clones. Can. J. For. Res. **24**: 1704–1710.

Rogers, D.L., Stettler, R.F., and Heilman, P.E. 1989. Genetic variation and productivity of *Populus trichocarpa* and its hybrids. III. Structure and pattern of variation in a 3-year field test. Can. J. For. Res. **19**: 372–377.

Singh, S.J., and Heather, W.A. 1982. Temperature sensitivity of qualitative race-cultivar interactions in *Melampsora medusae* Thum. and *Populus* species. Eur. J. For. Path. **12**: 123–127.

Smith, H.F. 1936. A discriminant function for plant selection. Ann. Eugen. **7**: 240–250.

Stack, R.W., and Ostry, M.E. 1990. Melampsora leaf rust on *Populus* in the North Central United States. *In* Recent research on foliage diseases. *Edited* by W. Merrill and M.E. Ostry. U.S. Dep. Agric. For. Serv. Gen. Tech. Rep. WO-56, Washington, DC. pp. 119–124.

Tallis, G.M. 1962. A selection index for optimum genotype. Biometrics, **18**: 120–122.

Tauer, C.G. 1975. Competition between selected black cottonwood genotypes. Silv. Genet. **24**: 44–49.

Thielges, B.A., and Adams, J.C. 1975. Genetic variation and heritability of *Melampsora* leaf rust resistance in Eastern cottonwood. For. Sci. **21**: 278–282.

Trenbath, B.R. 1974. Biomass productivity of mixtures. Adv. Agron. **26**: 177–210.

Tuskan, G.A., and van Buijtenen, J.P. 1986. Inherent differences in family response to inter-family competition in loblolly pine. Silv. Genet. **35**: 163–173.

von Euler, F. 1993. Plot design effects on phenotypic variability and dynamics of field competition between six *Pinus sylvestris* L. varieties. Scand. J. For. Res. **8**: 163–173.

von Euler, F., Ekberg, I., and Eriksson. G. 1993. Pairwise competition among progenies derived from matings within and between three origins of *Picea abies* (L.) Karst. in a close-spaced nursery trial. *In* Competition in the progeny testing of forest trees. *Edited by* F. von Euler. Department of Forest Genet., Swedish Univ. Agric. Sci., Uppsalla, Sweden, Research Notes. **48**: 1–13.

Wang, J., and van der Kamp, B.J. 1992. Resistance, tolerance, and yield of western black cottonwood infected by *Melampsora* rust. Can. J. For. Res. **22**: 183–192.

White, T.L., and Hodge, G.R. 1989. Predicting breeding values with applications in forest tree improvement. Kluwer, Dordrecht, The Netherlands.

Widin, K.D., and Schipper, A.L., Jr. 1981. Effect of Melampsora medusae leaf rust infection on yield of hybrid poplars in the north-central United States. Eur. J. For. Path. **11**: 438–448.

Wilcox, J.R., and Farmer, R.E., Jr. 1968. Variation and inheritance of juvenile characters of eastern cottonwood. Silv. Genet. **16**: 162–165.

Wilcox, J.R., and Farmer, R.E., Jr. 1968. Heritability and C effects in early root growth of eastern cottonwood cuttings. Heredity, **23**: 239–245.

Williams, C.G., Bridgwater, F.E., and Lambeth, C.C. 1983. Performance of single family versus mixed family plantation blocks of loblolly pine. *In* Proceedings of the 17th Southern Forest Tree Improvement Conference. University of Georgia, Athens, GA. pp. 194–202.

Wright, A.J. 1982. Some implications of a first-order model of interplant competition for the means and variances of complex mixtures. Theor. Appl. Genet. **69**: 91–96.

Wright, A.J. 1983. The expected efficiencies of some methods of selection of components for intergenotypic mixtures. Theor. Appl. Genet. **67**: 45–52.

Wu, R.-L., Wang, M.-X., and Huang, M.-R. 1992. Quantitative genetics of yield breeding for *Populus* short rotation culture. I. Dynamics of genetic control and selection model of yield traits. Can. J. For. Res. **22**: 175–182.

Wu, R.-L. 1994a. Quantitative genetics of yield breeding for *Populus* short-rotation culture. II. Genetic determination and expected selection response of tree geometry. Can. J. For. Res. **24**: 155–165.

Wu, R.-L. 1994b. Quantitative genetics of yield breeding for *Populus* short rotation culture. III. Efficiency of indirect selection on tree geometry. Theor. Appl. Genet. **88**: 803–811.

Ying, C.C., and Bagley, W.T. 1976. Genetic variation of Eastern cottonwood in an Eastern Nebraska provenance study. Silv. Genet. **25**: 67–73.

Ying, C.C., and Bagley, W.T. 1977. Variation in rooting capability of *Populus deltoides*. Silv. Genet. **26**: 204–207.

CHAPTER 8
Molecular genetics of *Populus*

H.D. Bradshaw, Jr.

Introduction

The elucidation of the structure and function of DNA is one of the greatest achievements of basic research. This discovery culminates a trend of reductionism in biology, which progressed from the study of individual organisms, to organs within organisms, cells within organs, nuclei within cells, and chromosomes within nuclei before arriving at the molecular level of DNA structure. The "central dogma of molecular biology," which describes the flow of biological information from DNA to RNA to protein, stands with Darwinian natural selection as one of the great organizing principles in biology.

Our detailed understanding of the physical, chemical, and biological properties of the DNA molecule has three important consequences. First, we can now *explain* genetic phenomena that had been recognized for centuries or decades, such as mutation, variation (polymorphism) within populations (see Whitham et al., Chapter 11), and resistance to disease (see Newcombe, Chapter 10), in terms of precisely-defined changes in the sequence of nucleotide bases in DNA. For example, the difference between a diseased and a disease-resistant plant can be traced to an alteration in just one of the hundreds of millions of nucleotides in the genome. Second, we are able to *test* the effects of direct or indirect manipulation of the DNA sequence of an organism. This test may involve breeding and selection for opposite states of a genetically-determined trait, e.g., high and low wood specific gravity, or could make use of genetic engineering to introduce a novel gene to measure its effect (see Han et al., Chapter 9). Third, we can *predict* the outcome of genetic experiments. For instance, the physical structure of DNA gives clues as to which forms of radiation are most likely to result in genetic damage, DNA chemistry permits the identification of potential carcinogens, and knowledge of the biological function of DNA allows purposeful

Abbreviations: QTL, quantitative trait locus; RFLP, restriction fragment length polymorphism; RAPD, random amplified polymorphic DNA; MAS, marker-aided selection.
H.D. Bradshaw, Jr. Center for Urban Horticulture, Box 354115, University of Washington, Seattle, WA 98195-4115, USA. e-mail address: toby@u.washington.edu
Correct citation: Bradshaw, H.D., Jr. 1996. Molecular genetics of *Populus*. *In* Biology of *Populus* and its implications for management and conservation. Part I, Chapter 8. *Edited by* R.F. Stettler, H.D. Bradshaw, Jr., P.E. Heilman, and T.M. Hinckley. NRC Research Press, National Research Council of Canada, Ottawa, ON. pp. 183–199.

modifications to be made to existing genes, with precise expectations about the phenotype of transgenic organisms (see Han et al., Chapter 9).

The earliest work in molecular genetics was focused on mutations of genes with large effects on the appearance (i.e., phenotype) of the organism. Frequently these studies were, by necessity, carried out in only a few model species, such as the fruit fly *Drosophila*. Until recently, the molecular biology juggernaut has had little impact on quantitative genetics, which is the study of continuously variable traits (see Riemenschneider et al., Chapter 7) and the scientific basis of most plant and animal breeding programs. Yet, molecular methods have the same potential for increasing our ability to explain, test, and predict quantitative genetic phenomena as they have shown for "classical" genetics, and to extend these investigations to any sexually-reproducing organism, including forest trees. Further, molecular tools can be used not as a replacement for traditional quantitative genetics in plant breeding, but rather as an adjunct to the powerful statistical framework which defines contemporary quantitative genetics. The merging of statistical and molecular genetics will have at least two beneficial effects. First, quantitative geneticists will appreciate the precision afforded by molecular techniques; i.e., differences among individual trees, families, and environments will be described not just in abstract terms of means and variances, but in concrete terms of genes and biochemical pathways. Second, molecular geneticists will benefit by dealing with large populations of highly-variable tree genotypes growing in highly-variable environments, thus receiving a much-needed dose of reality to temper the extravagant claims sometimes made from within the cloistered laboratory environment!

In forest trees, where long generation intervals, high genetic load, and other factors (see Bisoffi and Gullberg, Chapter 6) impede progress by traditional quantitative and classical genetic approaches, molecular genetics has much to offer. In exploring the biology of *Populus*, there are two general goals to be attained by the application of molecular tools to questions in quantitative genetics.

- Acquire a detailed understanding of the genetic architecture and evolution of complex processes such as tree growth, development, and adaptation, and their integration with anatomy and physiology.

- Determine the genetic contributions of various *Populus* species and provenances to hybrid superiority, identify new gene and trait combinations with potential commercial importance, and devise long-term breeding and selection strategies to produce these desired genotypes.

The first of these goals is largely a matter of basic research, while the second goal requires the application of basic research results to solve practical problems in the genetic improvement of *Populus*. Progress has been made toward both goals, with many questions remaining to be answered.

Quantitative genetics at the molecular level

The contribution of statistical genetics to plant and animal breeding has been strongly positive, and quantitative geneticists have set a high standard in experimental design and analysis. Traditionally, quantitative geneticists have worked at the level of species, populations within species, families within breeding populations, and, where clonal propagation is possible, at the level of individual organisms within families. Sources of phenotypic variation can be identified as either genetic, environmental, or a combination of the two (see Riemenschneider et al., Chapter 7). The genetic variance may be further subdivided into additive and nonadditive components. This level of resolution is adequate for predicting the magnitude of genetic gain in each generation of breeding, as long as the assumptions of the analysis are met, but it falls short of identifying the proximate cause of genetic variation: individual genes of differing magnitudes of effect and modes of action/interaction. The resolution of conventional quantitative genetics is limited by the dimensions of the experimental unit; hence, parents and their offspring are ranked by the *average* performance of the progeny rather than by the presence or absence of specific alleles of genes responsible for the observed variation in clonal performance and breeding value. It is the aggregate effect of individual genes that determines the genetic variation we use as raw material for plant breeding, and the reductionist approach to understanding this aggregate effect is first to understand the role of each gene. At the molecular level, we wish to know the following for every quantitative trait:

- The chromosomal location of each quantitative trait locus (QTL) affecting the trait.

- The magnitude of effect of each QTL on the observed phenotype.

- The mode of gene action at each QTL (additive, dominant/recessive, over-dominant).

- The effect of interactions among different QTLs (epistasis).

- The parental source of beneficial QTL alleles.

None of these parameters are estimated well by quantitative genetic analysis at the level of families or clones. Even the "additive" component of genetic variance determined by conventional quantitative methods can be due entirely to individual genes (i.e., QTLs) with dominant/recessive modes of action. An understanding of genetic variance at the molecular level will permit us to study the genes involved in adaptive variation in natural populations and to manipulate with greater precision the genes necessary for genetic improvement of commercial clones. The term "molecular breeding" has been coined to describe the practical application of the union of quantitative and molecular genetics.

The resolving power of the quantitative genetic method is increased by observing inheritance at the molecular level, in much the same way that using clones represents an improvement over using families as the basic experimental unit. The extension of quantitative genetics to the molecular level is straightforward in principle, and has become widely investigated in animal and plant systems as the necessary biotechnology has been simplified (Tanksley 1993). There are two prerequisites for a molecular-level investigation of quantitative traits. First, a pedigree must be found, or produced by new breeding, in which the trait(s) of interests are segregating. A simple example would be a full-sib family (i.e., the offspring of a controlled mating where both the seed and pollen parents are known) containing some trees susceptible to infection by a pathogen while the remainder of their "brothers" and "sisters" are resistant to infection. A more complex trait, such as stem volume growth or water use efficiency, may also be found segregating within such a full-sib family. The power to detect QTLs is proportional to the number of offspring sampled and the total amount of genetic variation in the trait(s) being measured. This generally limits QTL mapping to those traits that can be assessed in at least 100 offspring, and it is not unusual to have thousands of offspring in a single mapping experiment. The second requirement for a QTL mapping project is that the pedigree has segregating variation for a collection of genetic markers, and that there are enough of these markers to construct a genetic map of all the chromosomes. A genetic marker identifies a unique location on a single chromosome, and at the same time tells us which parent provided this piece of its chromosome to the offspring. Genetic markers based on differences in DNA sequence among individuals are most often used for genome map construction. These markers are detected by a variety of techniques and have been given acronyms like RFLP (i.e., restriction fragment length polymorphism; Botstein et al. 1980) and RAPD (random amplified polymorphic DNA; Williams et al. 1990). DNA markers have achieved some public recognition by virtue of their use in human forensic "DNA fingerprinting," and have been used in *Populus* to study gene flow in natural hybrid zones (Keim et al. 1989), reproductive biology and polyploidy (Bradshaw and Stettler 1993), inbreeding depression (Bradshaw and Stettler 1994), and for verification of clonal identity (i.e., the poplar version of "fingerprinting"; Castiglione et al. 1993). A typical pattern of genetic marker data in a *Populus* pedigree is shown in Fig. 1. By collecting genetic marker data from several hundred genetic loci in up to 379 F_2 offspring of a 3-generation *Populus* pedigree, we have constructed linkage maps of the 19 poplar chromosomes (Bradshaw et al. 1994; Fig. 2). Similar maps have been made for aspen (*Populus tremuloides*; Liu and Furnier 1993), *Eucalyptus* (*E. grandis* and *E. urophylla*; Grattapaglia and Sederoff 1994), loblolly pine (Devey et al. 1994), slash pine (Nelson et al. 1993), maritime pine (Plomion et al. 1995), and other forest trees, as well as in many crop plants and animal species.

With pedigree and linkage map in hand, how do we make a systematic search for QTLs governing expression of a phenotypic trait? If the trait is a simple one,

Fig. 1. Genetic marker (RFLP) segregation in a backcross between a TD female F_1 hybrid and a pure *P. deltoides* (DD) male. Lane 1 shows the marker genotype at locus P201 for a female *P. trichocarpa* (TT), in lane 2 is a male *P. deltoides* (DD), and in lane 3 is their F_1 female hybrid offspring (TD). The female *P. trichocarpa* is homozygous for the T allele at this locus (i.e., its genotype is TT), the male *P. deltoides* is homozygous for the D allele (i.e., its genotype is DD), and the F_1 hybrid is heterozygous (i.e., its genotype is TD, with one allele inherited from each of its parents). The TD hybrid was backcrossed to the DD male parent, yielding the B_1 progeny shown in lanes 4-30. Each backcross offspring inherits a D allele from its male parent, but may inherit either a T allele or a D allele from its F_1 hybrid female parent. Approximately half the B_1 offspring will have the DD genotype (one dark band) at this marker locus, and the other half will have the TD genotype (two dark bands).

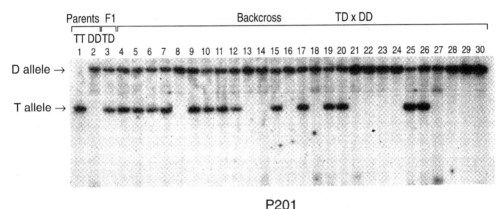

P201

such as a disease resistance provided by a single gene (see Newcombe, Chapter 10), the trait itself can be treated as a genetic marker since there are discrete phenotypes — "resistant" and "susceptible." In this way, RFLP, AFLP, or RAPD markers with the same pattern of inheritance as the disease resistance are found readily on the genome map (Devey et al. 1995; Newcombe et al. 1996; Villar et al. 1996; Cervera et al. 1996). For most traits, including adaptive traits such as date of bud burst (a measure of tolerance to late spring frost) or commercial traits such as height growth, the offspring in the pedigree do not fall into discrete classes such as "early bud burst" and "late bud burst," or "short" and "tall," but rather have a continuous distribution of phenotypic values from "early" to "late" and "short" to "tall." This lack of discrete phenotypic classes is due either to control of the trait more than one gene, or to environmental variation, or to some combination of the two (see Riemenschneider et al., Chapter 7). Despite the lack of discrete classes for these phenotypes, QTLs can be detected by statistical methods, and their chromosome position, magnitude of effect, and mode of action estimated. Consider a simplified analysis of the backcross marker information shown in Fig. 1. The backcross offspring are the result of mating a *Populus trichocarpa* × *P. deltoides* hybrid (TD) to a pure *P. deltoides* (DD). The TD diploid hybrid can produce haploid gametes (i.e., pollen and ovules) of two

Fig. 2. Linkage map of the *Populus* genome. By scoring hundreds of DNA markers in hundreds of F$_2$ (TD × TD) offspring in Family 331, a map of the poplar chromosomes can be constructed and used to search for QTLs affecting growth and development. Each vertical bar represents a linkage group (roughly equivalent to a chromosome), and each number represents a different genetic marker along the length of the chromosome.

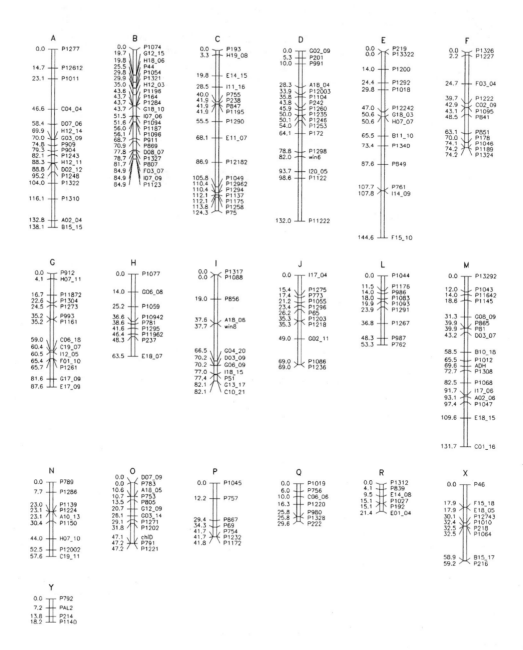

types at any genetic locus: "T" and "D." The pure *P. deltoides* can only produce gametes with a "D" allele at any position in the genome. The union of "T" or "D" gametes from the hybrid with "D" gametes from the pure *P. deltoides* to form the backcross offspring yields two possible diploid genotypes at any place (i.e., locus) along a chromosome: TD or DD. As expected from Mendel's principle of independent assortment, a particular backcross tree may be TD at one chromosomal location and DD at another. At each locus along the chromosome, the pedigreed offspring are divided into genotypic classes based on genetic marker information (Fig. 1). Using a *t*-test statistic, it is possible to determine if the *average* height growth of the TD genotypic class is significantly different from the *average* of the DD genotypes. If the mean phenotypes of the two genotypic classes are different, we can conclude that there must be one or more QTLs near this genetic marker that have an effect on height growth. The magnitude of the difference between the phenotypic average of each genotypic class is an estimate of the contribution of this QTL to the total variation in height within the pedigree. If the TD class of offspring is taller (on average) than the DD class at this QTL, we can further deduce that the *P. trichocarpa* allele has a positive effect on height growth (Fig. 3).

The QTL mapping process can be extended by considering more complex mating designs, such as a TD × TD F_2 cross with 3 genotypic classes (TT, TD, DD), or by treating the F_2 as a cross between two F_1 hybrid trees descended from heterozygous parents (e.g., $T_1D_1 \times T_2D_2$) which can generate 4 F_2 genotypic classes (T_1T_2, T_1D_2, T_2D_1, D_1D_2) at any locus. The statistical test for QTLs may use more powerful detection methods such as interval mapping (Lander and Botstein 1989) and refinements thereof (Jansen 1993; Jansen and Stam 1994; Zeng 1994; Kruglyak and Lander 1995).

The genetic architecture of growth, development, and adaptation

Historically, there have been two approaches to the study of genetics. The first approach is "classical" genetics, which makes use of mutants to study processes, such as growth and development, one gene at a time. While classical genetics has proven very powerful because it is focused on the proximate cause of genetic variation, i.e., mutations in single genes, it is not easy to apply in forest trees because of the difficulty of producing near-isogenic lines or clones differing only in the mutant gene. Many of the mutants of classical genetics are recessive to the normal allele, and so require two generations and close inbreeding to reveal. The long generation interval and high genetic load of many tree species make classical genetics impractical, although biotechnological techniques are showing an increasing promise in this arena (see Han et al., Chapter 9). The second approach to investigating genetics is the quantitative method (see Riemenschneider et al., Chapter 7). The statistical treatment of quantitative genetic

Fig. 3. Principle of QTL mapping in a backcross. To locate QTLs controlling a continuously variable trait, such as height growth, the backcross progeny are divided into two groups based on their marker genotype (TD or DD) at each locus along the chromosome. At the locus defined by Marker 1, both the TD and DD offspring have the same average height growth, suggesting that there is no QTL nearby. Moving down Chromosome 1 to Marker 2, it is clear that the TD offpsring are, on average, taller than the DD offspring. There must be a QTL near Marker 2 which affects height growth in the backcross, and trees with the T allele will be taller.

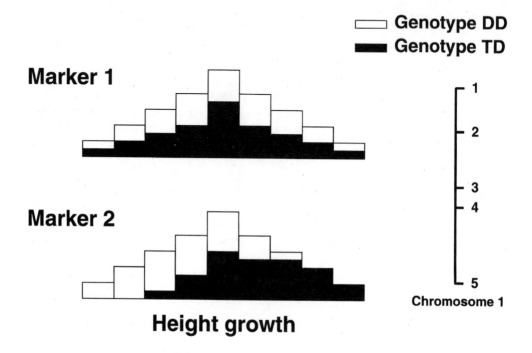

variation is simplified if the preconditions of Fisher's "infinitesimal model" are met. The infinitesimal model asserts that continuously variable traits are the result of the action of a very large number of unlinked genes (i.e., "polygenes"), each gene contributing only a small amount of the variation in the trait. These assumptions do not reflect biological reality very well, since we know that some genes do have a large effect on the phenotype, and that these genes may be physically linked to each other on the same chromosome. It should be pointed out that the existence of continuous variation for a trait does not provide any positive evidence that a large number of genes governs the trait; it is equally plausible that the absence of distinct phenotypic classes is due wholly or partly to the environmental "noise" in the phenotypic data (Falconer 1989). Genome mapping experiments are designed to merge the best features of the classical and quantitative methods to explain, test, and predict, gene-by-gene, continuous variability in tree growth, development, and adaptation across environments.

In *Populus*, the first steps have been taken to search the genome for QTLs affecting growth and adaptation. Genetic markers have been developed and used to make linkage maps of the *Populus* genome (Liu and Furnier 1993; Bradshaw et al. 1994; Fig. 2). A 3-generation pedigree involving interspecific hybrids of *P. trichocarpa* and *P. deltoides* has been produced and tested in three different replicated clonal trials (Wu and Stettler 1994; see Stettler et al., Chapter 4). The first pilot-scale was installed at Washington State University Farm 5 in Puyallup, Washington in 1991, and marker genotypes determined at 343 loci for 54 diploid F_2 trees in Family 331 (Bradshaw and Stettler 1995). Some examples of QTL discovery in Family 331 are given below, to indicate the types of traits amenable to molecular genetic analysis.

One of the most readily-mapped quantitative traits in *Populus* is the date of spring bud burst. Bud flush is highly heritable, with 98% or more of the phenotypic variance being under genetic control (Bradshaw and Stettler 1995). Spring phenology is also an adaptive trait that shows latitudinal and elevational trends within species, since early leaf flush prolongs the growing season but makes the tree more susceptible to late frosts. In the F_2 Family 331, more than a third of the phenotypic variance in the date of spring bud burst can be explained by the genotypic data from a single RFLP marker (P1308), suggesting that QTLs of large effect play a role in adaptation of hybrid poplars (Fig. 4). A genetic model incorporating this marker and information from four additional markers found near bud-burst QTLs is sufficient to account for essentially all of the variation in this trait. The difference in average date of spring bud flush between F_2 trees with either all 5 "early" QTL alleles or all 5 "late" alleles is 28 days, giving the earliest genotypes an extra month of growing season while simultaneously exposing them to an additional month of potential frost damage. The contributions of the *P. trichocarpa* and *P. deltoides* parents to "early" and "late" bud burst were determined, along with the mode of action of each QTL. In every instance of nonadditive QTL action, "early" is dominant to "late" (Bradshaw and Stettler 1995). Thus, from this QTL mapping experiment we were able to meet the goals set forth earlier in this chapter: determination of the number of QTLs and their chromosomal position, their magnitude of effect, mode of action/interaction, and source of alleles for "earliness" and "lateness." The most important conclusion drawn from this work is that a very small proportion of the genome controls a large proportion of the quantitative variation in an adaptive trait.

A similar, if less striking, pattern of inheritance was found when stem volume was examined after 2 years of growth in the replicated trial, with 45% of the genetic variance in stem volume explained by just two of the 343 genetic markers on the map. It seems that growth, a continuously variable trait, may also be affected by QTLs of large magnitude. Environmental variation in growth generally prevents the detection of these QTLs unless molecular genetic methods are brought to bear on the problem. The heterosis observed in F_1 hybrids may be ascribed to the complementary action of dominant alleles for increased height

Fig. 4. QTL affecting date of bud burst in spring in F_2 Family 331. By scanning Linkage Group M using the molecular markers along its length, a QTL is found near the RFLP marker P1308. There is a high probability (LOD = 4.33; roughly equivalent to a *P* value of 0.0002) that the QTL exists, and it explains 36.3% of the total observed variation in the trait. F_2 trees homozygous for the *P. trichocarpa* allele (TT) flush their leaves 18 days before those homozygous for the *P. deltoides* allele (DD). The heterozygotes (TD) are intermediate in date of bud burst, suggesting additive QTL action.

Linkage group M

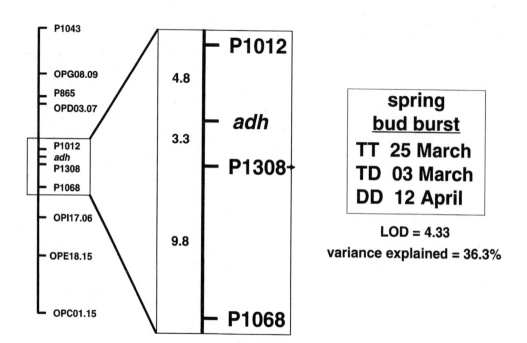

and diameter growth from the *P. trichocarpa* and *P. deltoides* parents, respectively (Bradshaw and Stettler 1995; see Stettler et al., Chapter 4).

This pattern of quantitative trait control by a few loci of large effect has been reported in herbaceous plants (e.g., Paterson et al. 1988; Stuber et al. 1992) and trees (e.g., Groover et al. 1994; Grattapaglia et al. 1995), although the relatively small progeny sizes used may have biased the estimates of QTL magnitude upward (Beavis 1994). There is now little doubt that QTLs of significant magnitude exist as statistical entities in the many organisms subjected to mapping experiments. QTL mapping is able to *explain* continuous phenotypic variation in terms of Mendelian inheritance of subchromosomal regions of the genome, although the number of genes in these chromosomal regions could be more than one.

Because the most commonly-used *Populus* QTL mapping pedigree is the F_2 of an interspecific hybrid, it segregates widely for many other traits that differ between the parental *P. trichocarpa* and *P. deltoides* species. The entire 3-generation pedigree, including more than 350 F_2 offspring, is "portable" because it has been vegetatively propagated, making it attractive to collaborators interested in the genetic control of morphological, physiological, and pathological traits. The adoption of Family 331 and its progenitors as a genetic "standard" for *Populus* research allows cumulative, detailed knowledge to develop, and fosters interactions among scientists in different disciplines. For example, the coincidence of map position for QTLs affecting both leaf area and disease resistance might suggest that the combined talents of a physiologist and a pathologist could shed light on the impact of leaf rust on tree growth. To give a sense of the range of traits being studied using molecular genetic analysis as a tool, a (necessarily incomplete) list of current QTL mapping efforts in this pedigree is shown in Table 1. The most current information about the pedigree and its genome is available on the World Wide Web at http://poplar1.cfr.washington.edu.

The future of molecular genetics in *Populus*

To move beyond *explanation* to the *testing* and *prediction* phases of QTL mapping requires multiple generations of breeding or biotechnological intervention. The initial attempts to validate the existence of QTLs by using markers linked to them to guide indirect selection in herbaceous plant breeding appear encouraging (deVicente and Tanksley 1993; Azanza et al. 1994). Genome maps and markers may also be used as tools to clone directly the genes responsible for important phenotypes such as disease resistance (Martin et al. 1993; Wing et al. 1994; Song et al. 1995), both to study the biology of these genes and to use them as transgenes for the genetic improvement of agricultural plants. Similar efforts to validate QTLs and clone important genes based on map position are needed in forest trees. In what way will these objectives be met with *Populus*?

First, the QTLs identified in one mapping experiment must be *tested* in other experiments, with more F_2 offspring and in different growing environments. We have begun to do this using the F_2 Family 331, along with its F_1 and parental generations, planted in two additional replicated clonal trials in strongly contrasting growing environments east (Boardman) and west (Clatskanie) of the Cascade Range in Oregon (see Stettler et al., Chapter 4). The pattern of QTL expression in these two new environments can be compared with that found in the first trial in Puyallup, Washington, west of the Cascades. The expectation is that some QTLs will be important in all environments, while others will be important in only one or two of the three environments tested. It is further possible that the favorable genotype at a QTL in one environment may be unfavorable in another, giving a gene-by-gene dissection of genotype-by-environment interaction (see Riemenschneider et al., Chapter 7; Stettler et al.,

Table 1. Traits segregating and mapped (or being mapped) in the F_2 Family 331.

Growth and form	Physiology/Anatomy	Pathology	Wood quality
Height, diameter, and volume growth	Osmotic potential and adjustment	*Melampsora* spp. leaf rusts	Specific gravity
Stem dry weight and harvest index	Adventitious rooting from cuttings	*Septoria populicola* leaf spot	Fiber length
Stem taper	Organogenesis in tissue culture	*Septoria musiva* stem canker	Cell wall thickness
Branch number and length	Water use efficiency	Eriophyid mite leaf damage	Cellulose/hemicellulose proportion
Branch angle	Nutrient use efficiency	*Venturia* shoot blight	Lignin concentration and composition
Leaf angle	Leaf nutrient levels	*Phratora purpurea* leaf beetle damage	Bark thickness
	Isoprene emissions		
	Hydraulic conductivity		
	Wind stress response		
	Growth regulator ("hormone") concentration		
	Single leaf area		
	Leaf cell size and number		
	Stomatal density and distribution		
	Leaf area on sylleptic and proleptic branches		

Chapter 4). Preliminary results from such a comparative QTL mapping experiment are shown in Fig. 5. Using stem height, basal area, and volume traits after 1 and 2 years of growth as the subject of QTL mapping, it seems clear that some QTLs, such as those on linkage groups E and M governing stem height, are important in all three environments (Fig. 5). In every environment the TT homozygotes at the QTL on linkage group E perform best, while the TD (i.e., heterozygous) genotypes are superior when linkage group M is considered (Fig. 5). However, not all QTLs significantly affect growth in all environments; some are specific to a single environment, e.g., the stem volume QTL on linkage group C at Boardman. Other QTLs, such as that controlling basal area and stem volume on linkage group J, are detected as significant in two different environments, but the better genotype is TT on the west side of the Cascades and DD on the east side (Fig. 5).

Second, from QTL mapping experiments designed to sample the major growing environments, we can produce a "molecular ideotype" describing the genetic

Fig. 5. QTL conservation across environments. The pattern of QTLs controlling height growth after 1 and 2 yrs in the field (HT1, HT2), stem basal area (BA1, BA2), and stem volume index (VOL1, VOL2) is shown for the F_2 Family 331 in three environments: Puyallup (western Washington), Clatskanie (western Oregon), and Boardman (eastern Oregon). The best-performing genotype, i.e., TT (homozygous for the *P. trichocarpa* allele), TD (heterozygous), or DD (homozygous for the *P. deltoides* allele), is shown below each linkage group. For example, on Linkage Group E, trees homozygous for the *P. trichocarpa* allele (TT) are taller in year 1 and have greater stem volume after 2 yrs than either TD or DD marker genotypes.

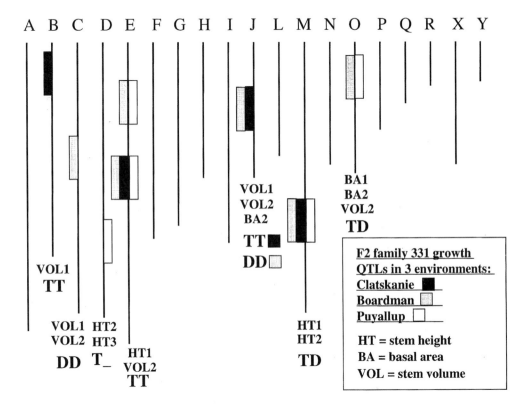

constitution of the most favorable allele combinations from any two parental species, such as *P. trichocarpa* and *P. deltoides*, involved in hybrid poplar breeding. Development of molecular ideotypes moves QTL mapping into the *predictive* phase, since breeding and selection of ideotypes has the goal of producing superior recombinant genotypes (see Dickmann and Keathley, Chapter 19). Using Fig. 5 to illustrate the construction of a molecular ideotype, we would predict that F_2 trees homozygous for the *P. trichocarpa* allele (i.e., with marker genotype TT) on linkage group E and heterozygous (TD) on linkage group M would be superior to the alternative genotypes in all environments. Mendelian genetics tells us that only 1/8 of F_2 offspring will have the desired genotype at both of these loci. Using molecular markers linked to these two QTLs, the DNA from each offspring can be screened for the presence of the favorable genotype and the best 1/8 of F_2 offspring can be selected for additional evaluation in the field. The remaining 7/8 of their siblings can be rejected without further testing, since the genetic markers have shown them to lack crucial alleles needed for maximum growth. This is the essence of marker-aided selection (MAS); the prediction of the future phenotype of a tree based only upon genetic marker data collected early in the seedling stage. As more and more QTLs are mapped, and their preferred allelic configuration determined, MAS becomes increasingly powerful (Lande and Thompson 1990). For example, the disease resistance of a seedling could be predicted without ever exposing the nursery to the pathogen, as long as linkage between genetic markers and disease resistance has been established by QTL mapping experiments. Since resources for seedling testing are limited, a prescreening for disease susceptibility using genetic markers saves the trouble and expense of extensive testing of clones that are destined to become diseased. Although MAS has the potential to be an important component of tree improvement (Williams and Neale 1992), there remains considerable debate about its long-term merits (Strauss et al. 1992). Experiments designed to resolve this issue should be a high priority.

Third, QTL mapping should be extended to all important *Populus* species and generations (F_1, F_2, backcross, intraspecific crosses) to pinpoint sources of valuable alleles for natural and managed populations, and to steer the development of rational strategies for the domestication of wild *Populus* while preserving the natural genetic variation upon which all long-term breeding efforts must ultimately depend. While most operational breeding of *Populus* now employs either pure species or F_1 interspecific hybrids, it is not difficult to imagine a *Populus* ideotype which "borrows" wood quality QTLs from *P. deltoides*, stem form from *P. trichocarpa*, early leaf flush from *P. maximowiczii*, and bacterial canker resistance from *P. nigra*. There are certain to be beneficial alleles found at high frequency in some *Populus* species that are completely lacking or in low frequency in other species. QTL mapping should play a prominent part in the identification of the patterns of genetic variation within and among species, and in the design of multispecies hybrids with unique trait combinations. Even where it may be preferable to plant a pure (or nearly-pure) species, such as in restoration of degraded natural habitats, genome maps and genetic markers may be used to recruit single genes from a donor species to improve the recipient species. To illustrate this point, consider the possibility that an exotic disease becomes established and the native *Populus* species is uniformly susceptible. Resistance to the disease could be obtained by hybridization to another species, and backcrossing to the native *Populus* used to introgress the resistance allele into the "native" genetic background. Genetic maps and markers are used to select among the backcross progeny for those trees most like the recurrent parent, yet still having the introgressed disease resistance allele from the donor species. MAS is already used for this purpose in crop breeding (e.g., Azanza et al. 1994).

Fourth, it will be necessary to integrate the QTL statistical maps with physical maps of the *Populus* chromosomes, and to take the final step in reductionist genetics by cloning the genes that are the QTLs. This will probably be approached from two directions: first, by sequencing all the expressed genes in *Populus*; and, second, by constructing physical maps of the *Populus* chromosomes with overlapping fragments of cloned DNA. Both of these strategies are being employed in the human genome project and the genome projects of other model organisms in biology (e.g., Adams et al. 1991; Schmidt and Dean 1993; Olson 1995). *Populus* is an attractive model forest tree for this kind of effort, since it has a small genome (Bradshaw and Stettler 1993), an acceptable genome map (Liu and Furnier 1993; Bradshaw et al. 1994) with numerous QTLs and other genes of basic or commercial interest located (Bradshaw and Stettler 1995; Newcombe et al. 1996; Villar et al. 1996; Cervera et al. 1996), and an excellent transformation/regeneration system suitable for confirming the activity of cloned genes (see Han et al., Chapter 9). The sooner we get started, the sooner we'll be describing quantitative traits in the refreshingly simple four-letter AGCT alphabet of molecular biology, rather than using all the letters of the Greek alphabet (and then some) that the statisticians seem to need!

Perhaps the most crucial factor for long-term success in understanding the biology of *Populus* at the molecular level is the continued collaboration of anatomists, biochemists, physiologists, pathologists, and ecologists with breeders and geneticists. Interdisciplinary research over the past 30 yrs has put *Populus* in a preeminent position as a model for forest tree biology. The use of "standard" pedigrees with extensive molecular genetic information is a sensible way to maintain this integrated approach.

Acknowledgements

I am indebted to Reini Stettler for his collaboration in mapping QTLs in *Populus* and for his unflagging enthusiasm and encouragement. Many of our colleagues have contributed to this work by collecting genetic and phenotypic data, establishing and maintaining field plantations, and providing valuable insights into forest tree biology. This work was supported by the USDA National Research Initiative Competitive Grants Program, the U.S. Department of Energy Biofuels Feedstock Development Program, the Washington Technology Centers, the Consortium for Plant Biotechnology Research, Boise Cascade Corporation, James River Corporation, and the Poplar Molecular Genetics Cooperative.

References

Adams, M.D., Kelley J.M., Gocayne, J.D., Dubnick, M., Polymeropoulos, M.H., Xiao, H., Merril, C.R., Wu, A., Olde, B., Moreno, R.F., Kerlavage, A.R., McCombie, W.R., and Venter, J.C.

1991. Complementary DNA sequencing: expressed sequence tags and human genome project. Science (Washington, DC), **252**: 1651–1655.

Azanza, F., Young, T.E., Kim, D., Tanksley, S.D., and Juvik, J.A. 1994. Characterization of the effect of introgressed segments of chromosome 7 and 10 from *Lycopersicon chmielewskii* on tomato soluble solids, pH, and yield. Theor. Appl. Genet. **87**: 965–972.

Beavis, W.D. 1994. The power and deceit of QTL experiments: Lessons from comparative QTL studies. *In* Proceedings of the Corn and Sorghum Industry Research Conference, American Seed Trade Association, Washington, DC. pp. 250–266.

Botstein, D., White, R.L., Skolnick, M., and Davis, R.W. 1980. Construction of a genetic linkage map in man using restriction fragment length polymorphisms. Am. J. Hum. Genet. **32**: 314–331.

Bradshaw, H.D., Jr., and Stettler, R.F. 1993. Molecular genetics of growth and development in *Populus*. 1. Triploidy in hybrid poplars. Theor. Appl. Genet. **86**: 301–307.

Bradshaw, H.D., Jr., and Stettler, R.F. 1994. Molecular genetics of growth and development in *Populus*. 2. Segregation distortion due to genetic load. Theor. Appl. Genet. **89**: 551–558.

Bradshaw H.D., Jr., and Stettler, R.F. 1995. Molecular genetics of growth and development in *Populus*. 4. Mapping QTLs with large effects on growth, form, and phenology in a forest tree. Genetics, **139**: 963–973.

Bradshaw, H.D., Jr., Villar, M., Watson, B.D., Otto, K.G., Stewart, S., and Stettler, R.F. 1994. Molecular genetics of growth and development in *Populus*. 3. A genetic linkage map of a hybrid poplar composed of RFLP, STS, and RAPD markers. Theor. Appl. Genet. **89**: 167–178.

Castiglione, S., Wang, G., Damiani, G., Bandi, C., Bisoffi, S., and Sala, F. 1993. RAPD fingerprints for identification and for taxonomic studies of elite poplar (*Populus* spp) clones. Theor. Appl. Genet. **87**: 54–59.

Cervera, M.T., Gusmao, J., Steenackers, M., Peleman, J., Storme, V., Vanden Broeck, A., Van Montagu, M., and Boerjan, W. 1996. Identification of AFLP molecular markers for resistance against *Melampsora larici-populina* in *Populus*. Theor. Appl. Genet. In press.

Devey, M.E., Fiddler, T.A., Liu, B.H., Knapp, S.J., and Neale, D.B. 1994. An RFLP linkage map for loblolly pine based on a three-generation outbred pedigree. Theor. Appl. Genet. **88**: 273–278.

Devey, M.E., Delfino-Mix, A., Kinloch, B.B., and Neale, D.B. 1995. Random amplified polymorphic DNA markers tightly linked to a gene for resistance to white pine blister rust in sugar pine. Proc. Natl. Acad. Sci. U.S.A. **92**: 2066–2070.

deVicente, M.C., and Tanksley, S.D. 1993. QTL analysis of transgressive segregation in an interspecific tomato cross. Genetics, **134**: 585–596.

Falconer, D.S. 1989. Introduction to quantitative genetics. Longman Group UK Ltd., Essex, UK. 438 p.

Grattapaglia, D., and Sederoff, R. 1994. Genetic linkage maps of *Eucalyptus grandis* and *Eucalyptus urophylla* using a pseudo-testcross mapping strategy and RAPD markers. Genetics, **137**: 1121–1137.

Grattapaglia, D., Bertolucci, F.L., and Sederoff, R.R. 1995. Genetic mapping of QTLs controlling vegetative propagation in *Eucalyptus grandis* and *E. urophylla* using a pseudo-testcross strategy. Theor. Appl. Genet. **90**: 933–947.

Groover, A., Devey, M., Fiddler, T., Lee, J., Megraw, R., Mitchell-Olds, T., Sherman, B., Vujcic, S., Williams, C., and Neale, D. 1994. Identification of quantitative trait loci influencing wood specific gravity in an outbred pedigree of loblolly pine. Genetics, **138**: 1293–1300.

Jansen, R.C. 1993. Interval mapping of multiple quantitative trait loci. Genetics, **135**: 205–211.

Jansen, R.C., and Stam, P. 1994. High resolution of quantitative traits into multiple loci via interval mapping. Genetics, **136**: 1447–1455.

Keim, P., Paige, K.N., Whitham, T.G., and Lark, K.G. 1989. Genetic analysis of an interspecific hybrid swarm of *Populus*: Occurrence of unidirectional introgression. Genetics, **123**: 557–565.

Kruglyak, L., and Lander, E.S. 1995. A nonparametric approach for mapping quantitative trait loci. Genetics, **139**: 1421–1428.

Lande, R., and Thompson, R. 1990. Efficiency of marker-assisted selection in the improvement of quantitative traits. Genetics, **124**: 743–756.

Lander, E.S., and Botstein, D. 1989. Mapping Mendelian factors underlying quantitative traits using RFLP linkage maps. Genetics, **121**: 185–199.

Liu, Z., and Furnier, G.R. 1993. Inheritance and linkage of allozymes and RFLPs in trembling aspen. J. Hered. **84**: 419–424.

Martin, G.B., Brommonschenkel, S.H., Chunwongse, J., Frary, A., Ganal, M.W., Spivey, R., Wu, T.Y., Earle, E.D., and Tanksley, S.D. 1993. Map-cased cloning of a protein kinase gene conferring disease resistance in tomato. Science (Washington, DC), **262**: 1432–1436.

Nelson, C.D., Nance, W.L., and Doudrick, R.L. 1993. A partial genetic linkage map of slash pine (*Pinus elliottii* Engelm. var. *elliottii*) based on random amplified polymorphic DNAs. Theor. Appl. Genet. **87**: 145–151.

Newcombe, G., Bradshaw, H.D., Jr., Chastagner, G.A., and Stettler, R.F. 1996. A major gene for resistance to *Melampsora medusae* f.sp. *deltoidae* in a hybrid poplar pedigree. Phytopathology, **86**: 87–94.

Olson, M.V. 1995. A time to sequence. Science (Washington, DC), **270**: 394–396.

Paterson, A.H., Lander, E.S., Hewitt, J.D., Peterson, S., Lincoln, S.E., and Tanksley, S.D. 1988. Resolution of quantitative traits into Mendelian factors by using a complete linkage map of restriction fragment length polymorphisms. Nature (London), **335**: 721–726.

Plomion, C., O'Malley, D.M., and Durel, C.E. 1995. Genomic analysis in maritime pine (*Pinus pinaster*) — comparison of two RAPD maps using selfed and open-pollinated seeds of the same individual. Theor. Appl. Genet. **90**: 1028–1034.

Schmidt, R., and Dean, C. 1993. Towards construction of an overlapping YAC library of the *Arabidopsis thaliana* genome. BioEssays, **15**: 63–69.

Song, W.Y., Wang, G.L., Chen, L.L., Kim, H.S., Pi, L.Y., Holsten, T., Gardner J., Wang, B., Zhai, W.X., Zhu, L.H., Fauquet, C., and Ronald, P. 1995. A receptor kinase like protein encoded by the rice disease resistance gene, Xa21. Science (Washington, DC), **270**: 1804–1806.

Strauss, S.H., Lande, R., and Namkoong, G. 1992. Limitations of molecular-marker-aided selection in forest tree breeding. Can. J. For. Res. **22**: 1050–1061.

Stuber, C.W., Lincoln, S.E., Wolff, D.W., Helentjaris, T., and Lander, E.S. 1992. Identification of genetic factors contributing to heterosis in a hybrid from two elite maize inbred lines using molecular markers. Genetics, **132**: 823–839.

Tanksley, S.D. 1993. Mapping polygenes. Annu. Rev. Genet. **27**: 205–233.

Villar, M., Lefevre, F., Bradshaw, H.D., and Teissier du Cros, E. 1996. Molecular genetics of rust resistance in poplars (*Melampsora larici-populina* Kleb./*Populus* sp.) by bulked segregant analysis in a 2 × 2 factorial mating design. Genetics, **143**: 531–536.

Williams, C.G., and Neale, D.B. 1992. Conifer wood quality and marker-aided selection: A case study. Can. J. For. Res. **22**: 1009–1017.

Williams, J.G.K., Kubelik, A.R., Livak, K.J., Rafalski, J.A., and Tingey, S.V. 1990. DNA polymorphisms amplified by arbitrary primers are useful as genetic markers. Nucleic Acids Res. **18**: 6531–6535.

Wing, R.A., Zhang, H.B., and Tanksley, S.D. 1994. Map-based cloning in crop plants — Tomato as a model system. 1. Genetic and physical mapping of *jointless*. Mol. Gen. Genet. **242**: 681–688.

Wu, R., and Stettler, R.F. 1994. Quantitative genetics of growth and development in *Populus*. 1. A three-generation comparison of tree architecture during the first 2 years of growth. Theor. Appl. Genet. **89**: 1046–1054.

Zeng, Z.B. 1994. Precision mapping of quantitative trait loci. Genetics, **136**: 1457–1468.

CHAPTER 9
Cellular and molecular biology of *Agrobacterium*-mediated transformation of plants and its application to genetic transformation of *Populus*

Kyung-Hwan Han, Milton P. Gordon, and Steven H. Strauss

Introduction

Genetic engineering is a means for insertion of genes that confer desired traits not readily available in sexually accessible gene pools. In contrast to sexual breeding, it allows new genes to be added while the genotypes of elite clones are preserved, and can therefore reduce the time required to produce an elite line. Rather than treat poplars in isolation, we organize our review around principles of plant transformation. However, we also discuss their relevance to poplar transformation and review progress in the field.

Tissue culture and regeneration

Summary of work on Populus

Since Winton (1970) obtained shoot regeneration from callus tissue of *P. tremuloides*, many authors successfully regenerated shoots from *in vitro* cultures of several species and hybrids of *Populus* (for review; see Lubrano 1992; Chun 1993; Ernst 1993). Explant sources initiated *in vitro* include shoot and root tips, stem nodes and internodes, gametophytic and sporophytic tissues, leaf and petiole

K.-H. Han and S.H. Strauss. Department of Forest Science, College of Forestry, Oregon State University, Corvallis, OR 97331, USA. Fax: (541) 737-1390; e-mail address: hanky@fsl.orst.edu, strauss@fsl.orst.edu
M.P. Gordon. Department of Biochemistry, SJ-70, University of Washington, Seattle, WA 98195, USA. Fax: (206) 685-8279.
Correct citation: Han, K.-H., Gordon, M.P., and Strauss, S.H. 1996. Cellular and molecular biology of *Agrobacterium*-mediated transformation of plants and its application to genetic transformation of *Populus*. *In* Biology of *Populus* and its implications for management and conservation. Part I, Chapter 9. *Edited by* R.F. Stettler, H.D. Bradshaw, Jr., P.E. Heilman, and T.M. Hinckley. NRC Research Press, National Research Council of Canada, Ottawa, ON. pp. 201–222.

segments, micro-cross sections of leaf midveins, and cambial tissues. Somatic embryogenesis has been demonstrated with *Populus* species (Cheema 1989; Michler and Bauer 1991; Park and Son 1988*a*; reviewed in Michler 1994), and may provide a regeneration system for both micropropagation and genetic transformation. Poplar protoplasts have been isolated and cultured from leaf tissues (Park and Han 1986; Park and Son 1988*b*; Russell and McCown 1988) and cell cultures (Cheema 1988; Youn et al. 1985), and plants were regenerated (Park and Son 1988*b*; Russell and McCown 1988).

Pathways for regeneration and somaclonal variation

There are several ways to achieve shoot regeneration from unorganized tissues in culture. Shoots obtained by direct organogenesis are less likely to have somaclonal variation than those produced by other methods such as somatic embryogenesis and callus-organogenesis. Organogenesis from callus involves a dedifferentiation-redifferentiation process, and, therefore, appears to promote somaclonal variation. However, this pathway also increases cell division, and may enhance opportunities for *Agrobacterium* transformation.

Somaclonal variation in poplar is induced largely as a result of prolonged culture in the presence of synthetic growth regulators (e.g., more than 16 months: Saieed et al. 1994). The frequency of morphologically detectable somaclonal variation in poplar ranged from 1% (Antonetti and Pinon 1993) up to 24% (Son et al. 1993) of regenerated plants.

Factors affecting regeneration

Plant genotype

Adventitious root and shoot regeneration *in vitro* is under strong genetic control and can be affected by major genes (Han et al. 1995*a*). In a test of 16 clones of *Populus deltoides*, four clones reacted significantly better than the others and six clones did not respond at all (Coleman and Ernst 1989). Thus, when not precluded by commercial goals, genotypes should be preselected for regenerability.

Physiological condition of explants

The physiological state of explants is influenced by the environmental conditions under which the source material is grown. Shading and fertilization (consequently increasing the NO_3^-/NH_4^+ ratio) of mother plants improved shoot regeneration *in vitro* in *Matthiola incana* (Mizozoe et al. 1993). Preconditioning source tissues on MS medium (Murashige and Skoog 1962) containing BAP (N_6-benzylaminopurine) was important for shoot regeneration in *Lathyrus* species (Malik et al. 1992).

In vitro materials are usually preferred over field-grown or greenhouse materials for adventitious shoot regeneration. There is no need to sterilize *in vitro* explants, and they usually retain greater juvenility and thus superior regeneration capacity. Field grown explants are also commonly more lignified, rich in phenolics, and physiologically dormant (Civinova and Sladky 1990). However, unlignified stem portions of greenhouse-grown hybrid cottonwood have been efficiently transformed with *Agrobacterium rhizogenes* (Han et al. 1995*b*).

Tissue culture conditions

Tissue culture components, particularly nutrient composition, growth regulators, light regime, temperature, and gelling agents, are important factors for regeneration. For example, auxin pulses of specific durations induced competence for shoot regeneration from callus of *Populus deltoides* (Coleman and Ernst 1990). By adapting the NO_3^-/NH_4^+ ratio of the medium, De Block (1990) was able to reduce explant necrosis and increase callus formation in *Populus trichocarpa* × *P. deltoides*.

Medium pH can strongly affect regenerability (Minocha 1987). Use of an acid pH blocked organogenesis from callus culture of *Populus nigra*. Repeated subculture of the calli on fresh medium containing BAP at pH 5.8 allowed organogenesis to resume (Hrib 1993).

Gene transfer methods

Transformation techniques

Agrobacterium-mediated transformation

This method uses the soil bacteria *Agrobacterium tumefaciens* and *A. rhizogenes*. After the bacteria bind to cells, they transfer a segment of DNA from their large endogenous Ti (tumor-inducing) or Ri (root-inducing) plasmids. The transferred DNA (T-DNA) is covalently incorporated into plant genomes. Under optimal conditions the transformation frequency achieved by this method is high, but efficiency is highly dependent on bacterial and plant genotype (Binns 1990).

Direct DNA transfer

Although *Agrobacterium*-mediated gene transfer systems have been developed to such an extent that the genetic transformation of many plant species has become routine, a number of agronomically important crop species and genotypes remain recalcitrant to transformation and regeneration using this system. In the search for species- and genotype-independent transformation methods, several techniques based on the direct delivery of naked DNA to the plant cells were developed. These methods include biolistics, electroporation, microinjection, PEG- or liposome-mediated DNA uptake, and silicon carbide whisker-mediated transformation.

In biolistics, high velocity microprojectiles carry DNA past cell walls and membranes into plant cells. This technique has generated many transgenic organisms, including microorganisms, mammalian cells, and a large number of herbaceous and woody plant species (Christou 1994) including poplar (McCown et al. 1991; Devantier et al. 1993). It has also been used to transform cellular organelles. Electroporation has been used for transformation of protoplasts from several species (Fromm et al. 1985) including poplar (Chupeau et al. 1994), and electroporation conditions have been found that deliver DNA molecules directly into plant tissues (D'Halluin et al. 1992). In silicon carbide whisker-mediated transformation, gene transfer is achieved by vortexing plant cells with a premixed naked DNA/silicon carbide (SiC) whiskers. Kaeppler et al. (1990) first demonstrated GUS gene expression in maize and tobacco suspension cells after vortexing for 60 seconds with SiC fibers. The same group later recovered stably transformed colonies using this technique (Kaeppler et al. 1992). It has also been used to transform *Chlamydomonas* (Dunahay 1993) and maize (Frame et al. 1994).

Meristem transformation

In recent years, direct introduction of DNA into meristems has been explored for several species. Because of the multicellular structure of meristems, the resulting plants are initially chimeric for the introduced genes. Homogeneously transformed individuals can subsequently be obtained by selfing the treated generation and selecting for the added trait, or through extensive vegetative propagation and selection. Meristematic cells of seed-derived embryos have been transformed by *Agrobacterium* (Gould et al. 1991), electroporation (Akella and Lurquin 1993), and microparticle acceleration (Schnall and Weissinger 1993). The advantage in this approach is that shoots develop directly from the primary and secondary meristems without an intervening explant-organogenesis phase. This minimizes the necessity for treatment with phytohormones and thus opportunity for somaclonal variation. Its limitations are extremely low transformation frequency, difficult manipulation of tissue from small-seeded species, and extensive chimerism.

In planta transformation

Feldmann and Marks (1987) reported that germinating *Arabidopsis thaliana* seeds exposed to *Agrobacterium* gave rise to transformed progeny without tissue culture. Chang et al. (1994) transformed *A. thaliana* by inoculating decapitated plantlets with *Agrobacterium*. This technique was used successfully to introduce resistance to the herbicide Basta® (Hoechst AG, Germany) into *A. thaliana* (Bouchez et al. 1993). In a mature *A. thaliana* plant, vacuum infiltration of *Agrobacterium* cells containing a binary vector resulted in both transformed and untransformed vegetative sectors (Bechtold et al. 1993). Transformants were recovered from the progeny of the infiltrated plants. A new technique based on electroporation of DNA into intact nodal meristems *in planta* was successful in generating transgenic leguminous plants (Chowrira et al. 1995). There have been a number of reports of pollen transformation over the years (Nishihara et al.

1993); however, to our knowledge no reproducible systems have been established.

Poplar transformation

Biolistics, electroporation, and *Agrobacterium* have all been used to transform poplars (Table 1). However, *Agrobacterium* has accounted for the large majority of published reports. We have also chosen to focus on *Agrobacterium* in our lab because although efficiency varies widely, it transfers T-DNA to nearly all poplar clones that have been studied (reviewed in Jouanin et al. 1993). The main obstacle is targeting and isolating regenerable cells. Because the main advantage to biolistics is facile insertion into cells, it appears to provide little advantage over *Agrobacterium*. Moreover, recent studies have shown that *Agrobacterium* can be used to transform organized tissues, such as meristems (Gould et al. 1991). Nonetheless, direct DNA delivery techniques may be useful for transformation of elite clones that are highly resistant to *Agrobacterium* infection.

Major components of *Agrobacterium*-mediated transformation system

Biology of Agrobacterium-*mediated transformation*

Components of T-DNA transfer system

There are three genetic elements necessary for an *Agrobacterium* cell to transfer T-DNA to plants (Zupan and Zambryski 1995). The first are the T-DNA border-repeat-sequences. These 25 bp, highly conserved DNA sequences flank and delimit the T-DNA. Only those sequences between the borders are transferred to the plant cells. There are no genes within the T-DNA that are necessary for T-DNA transfer. Therefore, the naturally occurring T-DNA-born oncogenes (tumor-inducing genes) can be replaced with the genes of interest, thus 'disarming' the wild type T-DNA.

The second element necessary for T-DNA transfer is the set of virulence genes (*vir*A, B, C, D, E, and G), which reside in a region outside the T-DNA of the Ti- or Ri-plasmid (reviewed in Birot et al. 1987). The product of *vir*A is a periplasmic membrane protein that senses specific phenolic signals (e.g., acetosyringone and related compounds) released from the wounded plant cells. The *vir*A proteins also interact with a periplasmic galactose binding protein (encoded by a chromosomal gene *chv*E) important for the binding of a sugar 'co-inducer,' also released from the wounded plant cells. Upon receiving the phenolic and sugar inducers, the *vir*A protein is autophosphorylated, then transfers the phosphate group to the *vir*G protein, activating it. Activated *vir*G protein acts as a transcriptional activator of the remaining *vir* genes involved in the processing and transfer of the T-DNA from Ti-plasmid to plant cells.

Table 1. Genetic transformation of poplars.

Species/hybrid	Method	Transgenes*	Result**	Reference
P. trichocarpa×P. deltoides	A. tumefaciens	Ti T-DNA	T, S, SE	Parsons et al. 1986
P. tremuloides	A. tumefaciens	CAT, nptII	T, R, S, SE	Minocha et al. 1987
P. alba×P. grandidentata	A. tumefaciens	aroA, nptII	T, R, S, SE, GT	Fillatti et al. 1987
P. trichocarpa×P. deltoides	A. tumefaciens	aroA, nptII	T, R, S, SE, GT	Riemenschneider et al. 1988
P. alba×tremula	A. rhizogenes	Ri T-DNA	T, R, S, SE	Phytoud et al. 1987
P. trichocarpa×P. deltoides	A. tumefaciens	bar, nptII	T, R, S, SE, GT	De Block 1990
P. deltoides	A. tumefaciens	Ti T-DNA	T, SE	Riemenschneider 1990
P. tomentosa	A. rhizogenes	Ri T-DNA	T, R, SE, GT	Bu et al. 1991
P. tomentosa	A. tumefaciens	Ti T-DNA	T	Bu et al. 1991
P. alba×P. grandidentata	A. tumefaciens	nptII, pin2-CAT	T, R, S, SE, GT	Klopfenstein et al. 1991
P. alba×P. grandidentata	Biolistics	BT, GUS, nptII	T, R, S, SE, GT	McCown et al. 1991
P. nigra×P. trichocarpa	Biolistics	BT, GUS, nptII	T, R, S, SE, GT	McCown et al. 1991
P. tremula×P. alba	A. tumefaciens	GUS, nptII, Ti T-DNA	T, R, S, SE, GT	Brasileiro et al. 1991
P. deltoides×P. nigra / P.nigra×P. maximowiczii	A. rhizogenes/ A. tumefaciens	Ri T-DNA/Ti T-DNA	T, R, GT	Charest et al. 1992
P. tremula×P. alba	A. tumefaciens	GUS, nptII	T, R, S, SE	Leple et al. 1992
P. tremula×P. alba	A. tumefaciens	crs1-1, nptII	T, R, S, SE, GT	Brasileiro et al. 1992
P. tremula×P. tremuloides	A. tumefaciens	hptII, luxab, nptII	T, R, S, SE	Nilsson et al. 1992
P. deltoides×P. nigra	Biolistics	GUS, nptII	T	Devantier et al. 1993
P. nigra×P. maximowiczii	Biolistics	GUS, nptII	T	Devantier et al. 1993
P. tremula×P. alba	Electroporation	crs1-1, nptII, PAT	T, R, S	Chupeau et al. 1993
P. nigra	A. tumefaciens	GUS, nptII, Ti T-DNA	T, S	Confalonieri et al. 1994
P. alba×P. grandidentata	A. tumefaciens	Ac, BT, hptII, nptII	T, S, SE	Howe et al. 1994
P. tremula×P. alba	A. tumefaciens	ipt, nptII	T, R, S, SE	Schwartzenberg et al. 1994
P. trichocarpa×P. deltoides	A. rhizogenes	nptII, Ri T-DNA	T, R, S, SE, GT	Han, Gordon, and Strauss, in review
P. tremula×P. alba	A. tumefaciens	CPsyn, GME, GUS, nptII	T, R, S, SE, GT	Han, Ma, and Strauss, in prep.
P. trichocarpa×P. deltoides	A. tumefaciens	CPsyn, GME, GUS, nptII	T, R, S, SE	Han, Ma, and Strauss, in prep.
P. tremula×P. alba	A. tumefaciens	DTA, GUS, nptII, PtFL	T, R, S, SE	Rottmann and Strauss, in prep.

***Ac**: Ac transposable element; **ALS**: acetolactate synthase (Sulfonylurea^r); *aroA*: mutant EPSP synthase (glyphosate^r); *bar*: phosphinothricin acetyltransferase; **BT**: *Bacillus truringiensis* toxin; **CAT**: chloramphenicol acetyltransferase; **Cpsyn**: EPSP synthase (glyphosate^r); *crs1–1*: mutant acetolactate synthase (chlorsulfuron^r); **DTA**: diphtheria toxin A; **GME**: Degrade glyphosate; **GUS**: β-glucuronidase; *hptII*: hygromycin phosphotransferase; *luxab*: luciferase; *nptII*: neomycin phosphotransferase; **PAT**: phosphinothricin synthase (Basta^r); **PtFL**: poplar leafy.

T: transformation; **R: regeneration; **S**: selection; **SE**: stable expression; **GT**: greenhouse test.

The final element consists of the set of chromosomal genes such as *chv*A, *chv*B, *chv*D, *chv*E, and *psc*A (*exo*C). Some of these genes are responsible for exopolysaccharide production by *Agrobacterium*, which are essential for the attachment of bacterium to plant cells (Douglas et al. 1985). Many other chromosomal gene products probably also contribute to virulence.

Choice of species

The soil bacteria *Agrobacterium tumefaciens* and *A. rhizogenes* are the two species widely used to transform higher plants. Both systems rely on a similar transformation mechanism. One of the differences between the two species is that genetically transformed plants can be regenerated from roots induced by wild-type *A. rhizogenes*, while transformed plant cells with wild-type Ti plasmids can not be regenerated into plants. *A. rhizogenes* has been successfully used to transform many recalcitrant species including a leguminous tree, black locust (Han et al. 1993), and hybrid cottonwood (Han et al. 1995*b*). Plants regenerated from Ri-transformed roots, however, display the 'hairy-root syndrome' caused by the Ri T-DNA born genes (Schmulling et al. 1988). Transformation with a disarmed *A. tumefaciens* strain produces transgenic plants without such abnormal phenotypes. However, with disarmed *A. tumefaciens* species a precise regeneration/selection system has to be developed for each plant species or genotype.

Choice of strains

Plant species and/or genotypes usually exhibit strain specificity in their susceptibility to infection by *Agrobacterium*. Since virulence of specific strains also varies depending on plant species, the choice of bacterial strains is important for transformation of recalcitrant species. The recent success in rice transformation, a species once thought beyond the host-range of *Agrobacterium*, can be attributed partly to the use of a supervirulent strain (Bo542) and a 'super-binary' vector which carries the virulence region of supervirulent plasmid pTiBo542 (Hiei et al. 1994). Other workers have also emphasized the importance of 'supervirulent' strains of *A. tumefaciens* for transformation of recalcitrant species, including rice (Hiei et al. 1994) and soybean (Kovacs and Pueppke 1993).

Strain specificity is important in transformation of tree species. The type of *Agrobacterium* strain was the most important factor for transformation of *Taxus* species (Han et al. 1994). Most transgenic poplars have been produced using nopaline strains of *Agrobacterium* (Table 2). Among three strains (Ach5, A281, and C58), C58 was the most virulent on *Populus nigra* leaf discs (Confalonieri et al. 1994). Transformation with an octopine strain such as LBA4404 was not successful in poplar (Fillatti et al. 1987; Pythoud et al. 1987). Our recent results showed that EHA105 was better than C58 and LBA4404 when assaying expression of GUS-intron transgene in an aspen and two hybrid cottonwood clones (unpublished data).

Table 2. Strains of *Agrobacterium* spp. used for genetic transformation of woody plants.

	Strains	Opine	Characteristics	Wood plant species/reference
A. tumefaciens	Ach5		Wild type, carrying pTiAch5	*Alnus* & *Betula*/Mackay et al. 1988; *Populus*/Fillatti et al. 1987
	LBA4404		Ach5 Ti-cured, Rif, Strep^r	*Vitis*/Mullins et al. 1990; *Populus*/Leple et al. 1992; *Populus*/Confalonieri et al. 1994
	A6	Octopine	Wild type, carrying PtiA6	*Robinia*/Davis & Keathley 1989
	A348		Wild type, C58 chromosome + pTiA6	*Populus*/Parsons et al. 1986
	C58		Wild type, carrying pTiC58	*Salix*/Vahala et al 1989; *Populus*/Riemenschneider 1990; *Populus*/Brasileiro et al. 1991; *Populus*/Charest et al. 1992; *Taxus*/Han et al. 1994; *Populus*/Confalonieri et al. 1994
	C58/pMP90		C58 Ti-cured, Rif, Gm^r	*Populus*/De Block 1990; *Populus*/Brasileiro et al. 1992; *Populus*/Leple et al. 1992; *Populus*/Schwartzenberg et al. 1994; *Populus*/current work in our lab
	C58SZ707	Nopaline	C58 Ti-cured, Strep^r	*Populus*/Howe et al. 1994; *Liquidambar*/Stomp and Chen, in prep.
	GV2260		C58 Ti-cured	*Populus*/Confalonieri et al. 1994
	GV3101		C58 Ti-cured	*Salix*/Vahala et al. 1989
	GV3850		C58 Ti-cured carrying pGV3850::pKU3	*Populus*/Confalonieri et al. 1994; *Populus*/Howe et al. 1994
	A281		Supervirulent wild type, C58 chromosome+pTiBo542, Rif^r	*Populus*/Parsons et al. 1986; *Pinus*/Loopstra et al. 1990; *Prunus*/Scorza et al. 1990; *Populus*/Klopfenstein et al. 1991; *Populus*/Confalonieri et al. 1994
	EHA101	Agropine	A281 Ti-cured, Rif, Strep^r, Km^r	*Populus*/Howe et al. 1994
	EHA105		A281 Ti-cured, Rif, Gm^r	*Populus*/Current work in our lab
	Bo542		Supervirulent w.t., carrying pTiBo542	*Pinus*/Loopstra et al. 1990; *Populus*/Brasileiro et al. 1991; *Taxus*/Han et al. 1994

Table 2 (*concluded*).

	Strains	Opine	Characteristics	Wood plant species/reference
	A4		Wild type, carrying pRiA4b	*Malus*/Lambert & Tepfer 1992; *Populus*/Brasileiro et al. 1991; *Populus*/Charest et al. 1992
A. Rhizogenes	R1000	Agropine	C58 chromosome+pRiA4b	*Robinia*/Han, unpublished; *Populus*/Bu et al. 1991
	R1601		R1000+pTVK291, Km^r, Cb^r	*Populus*/Pythoud et al. 1987; *Robinia*/Han et al. 1993; *Eucalyptus*/Macrae & Van Staden 1993; *Populus*/Han et al., in prep.
	11325		Wild type, carrying pRi11325	*Larix*/Huang et al. 1991; *Larix*/Shin et al. 1994

Induction conditions

Agrobacterium transformation requires that plant cells produce appropriate signal molecules for induction of *vir* genes and subsequent T-DNA transfer into plant cells. The virulence (*vir*) region in Ti- or Ri-plasmid is the master switch for the transformation process. Wounded plant cells excrete low molecular weight *vir* gene-inducing molecules. These molecules include acetosyringone (AS) and hydroxy-acetosyringone (OH-AS) (Stachel et al. 1985). Cangelosi et al. (1990) found certain sugars induced *vir* genes synergistically with phenolic inducers. Multiple copies (Liu et al. 1992) or constitutive expression (Hansen et al. 1994) of *vir*G improved transformation efficiency. Wound response is likely to vary depending on cell type, physiological conditions of explants, explant source, genotype, and the manner in which explants are manipulated. In culture, however, inducers can be added directly, theoretically reducing genotype and explant specificity. *Agrobacterium* virulence is usually improved by the addition of inducers, such as acetosyringone and sugars, to the co-cultivation medium (Mathis and Hinchee 1994).

Inoculation and co-cultivation

The manner in which explants are inoculated with *Agrobacterium* can significantly influence transformation efficiency. Vacuum infiltration during *Agrobacterium* inoculation enhances transformation frequency in rice (Rout et al. 1995). McGranahan et al. (1988) tested three different inoculation methods (sonication, prick, and cut) to infect walnut embryos and found that cutting embryos was the most successful in obtaining infection.

Our laboratory has found that slicing the explants in bacterial suspension results in higher transformation frequency than does cutting in water and subsequently dipping in bacteria. Wounding poplar leaf discs with nontraumatic forceps also improved transformation frequency based on transient GUS expression (unpublished data). Enzymatic digestion of cell walls did not improve transformation efficiency in rice (Hiei et al. 1994).

Transformation efficiency is also affected by co-cultivation conditions such as medium composition and type (solid or liquid), growth regulators, presence or absence of a nurse culture, and light regime (Mathis and Hinchee 1994). Tobacco suspension cells were successfully used as a nurse culture during co-cultivation in poplar transformation (Fillatti et al. 1987). The use of feeder layers during co-cultivation was critical in tomato transformation (Van Roekel et al. 1993). Transformation efficiency in peanut (Mansur et al. 1993) and in poplar (Confalonieri et al. 1994) was greatest when the explants were co-cultivated on a solid medium rather than a liquid medium. The presence of growth regulators, glucose, and acetosyringone in the co-cultivation medium enhanced the transformation efficiency for clover (Khan et al. 1994). Co-cultivation of *Populus nigra* explants for 48 h significantly improved transformation efficiency over 24 h co-cultivation (Confalonieri et al. 1994).

Titer of the *Agrobacterium* inoculum

A high *Agrobacterium* titer is desirable for transformation. However, in many species this results in tissue necrosis or inhibition of regeneration. For example, we found that an *Agrobacterium* concentration of higher than 1.0 OD_{620} ($\approx$ 5×10^8 cells/mL) caused tissue necrosis on poplar stem explants after 2 days of co-cultivation. Nevertheless, a concentration of 7×10^8 cells/mL gave the best transformation in *Populus nigra* based on transient GUS expression (Confalonieri et al. 1994). In other cases transformation efficiency continued to increase as the *Agrobacterium* cell concentration was raised from 10^6 to 10^{10} cells/mL (Lin et al, 1994). Reported cell titers in successful poplar transformation range from 2×10^8 to 9×10^8 cells/mL.

Co-inoculation and co-transformation

Combinations of disarmed plasmids in separate hosts, or combinations of wild type and disarmed strains, can be used to insert multiple transgenes and to stimulate transformation frequency, respectively. When two nopaline type *Agrobacterium* strains, each carrying a distinct disarmed T-DNA, were used to infect *Brassica napus*, high transformation frequencies (38–85%) of both T-DNAs were obtained (De Block and Debrouwer 1991). An *Agrobacterium* suspension containing both a wild-type strain and a disarmed strain was successfully used to transform poplar (Brasileiro et al. 1991).

Co-transformation with two separate DNAs contained within a single bacterium can facilitate transformation of recalcitrant species. Under some conditions, different T-DNAs can be integrated into unlinked sites at high frequencies. Binary vectors derived from *A. tumefaciens* can be efficiently transferred into plant cells when they are placed in a virulent strain of *A. rhizogenes*. The plant tissues produce hairy roots most of whose cells are transgenic. The hairy roots also are often capable of regenerating into whole plants, and a high frequency (39–85%) of co-transfer of wild type and binary vector T-DNAs is generally obtained (De Block and Debrouwer 1991; Hamamoto et al. 1990). When the co-transformed tomato tissues were cultured on a medium supplemented with appropriate growth regulators, they produced callus instead of hairy roots (Shahin et al. 1986), and phenotypically-normal transgenic plants could be obtained from this callus. *A. rhizogenes* has been the only vector able to give rise to transformed plants in several recalcitrant tree species (Han et al. 1993, 1995*b*; Shin et al, 1994).

Biology of plant materials in relation to competence for transformation

Plant genotype

Plant genotype influences transformation frequency due to variation in DNA transfer efficiency from *Agrobacterium* and regenerability into plants. We have observed clone-specific responses of poplar tissues to *Agrobacterium* DNA transfer from transient expression studies of introduced GUS-intron genes (K.-H. Han, unpublished results). Susceptibility of seedlings to *Agrobacterium* showed familial differences in intra- and inter-specific F_1 and F_2 hybrid *Populus* families that was primarily associated with the female parent (Riemenschneider 1990).

Physiological conditions of explants

The physiological state of explants is influenced by the environmental conditions under which the source material is grown. Unlignified stem portions of greenhouse-grown hybrid cottonwood have been efficiently transformed with *Agrobacterium rhizogenes* (Han et al. 1995*b*). Explant age significantly affected *Agrobacterium*-mediated transformation of walnut embryos. Inoculations of young embryos resulted in a significantly higher transformation frequency than did inoculations of old embryos (McGranahan et al. 1988).

Explant type

Explant type (e.g., leaves, stems, petioles, cotyledons, roots, and hypocotyls) can have a significant influence on regenerability and transformation frequency. Strong tissue-specific binding affinity of *Agrobacterium* was observed in *Zea mays* (Graves et al. 1988); bacteria primarily adhered to cells in the vascular bundles of young internodes. Internodes and petioles of poplars often show the greatest regenerative capacity (K.-H. Han and C. Ma, unpublished results). Explant type also affects the type and copy number of T-DNA integration. For example, single-copy T-DNA insertions were the predominant form of integration in root-derived *Arabidopsis* transgenics, whereas multiple insertions were found in leaf discs (Grevelding et al. 1993). Agroinfection, the *Agrobacterium*-mediated transfer of viral sequences to plant cells, was used to study the competence of the shoot apical meristem of immature maize embryos at different developmental stages (Schlappi and Hohn 1992). Results suggested that competence for agroinfection was developmentally regulated and that *Agrobacterium*-mediated transformation may require differentiation of tissue in the maize shoot apical meristem prior to wounding.

Explant manipulation

Transformation is influenced by explant preparation prior to *Agrobacterium* inoculation. Important factors are wounding during dissection (which stimulates plant defense mechanisms and release of phenolics), cell division frequency, and the regeneration competence of cells at the wound sites. We found that keeping the explants wet while they were dissected, (which may reduce ethylene action and oxidative browning), minimized tissue browning in cottonwood hybrids. Cutting cotyledons in half longitudinally improved shoot regeneration in watermelon (Compton and Gray 1993). This type of explant manipulation also gives improved transformation frequency in hybrid poplar, based on transient GUS expression assay (K.-H. Han and C. Ma, unpublished results). Microprojectile bombardment of plant tissue prior to *Agrobacterium* inoculation increased transformation frequency in tobacco (Bidney et al. 1992).

Cell division

Cell division is necessary for successful *Agrobacterium*-mediated transformation. In *Arabidopsis*, GUS gene expression increased as the period of preincubation on callus induction medium was extended, suggesting that actively-dividing cells are more susceptible to *Agrobacterium* infection (Akama et al. 1992). The highest competence for transformation occurred in cotyledons of tomato during the first 24-h interval after wounding. Use of a virulence-inducing factor such as acetosyringone could reverse the decline in transformation competence up to 96 h after wounding (Davis et al. 1991). A window of competence for transformation was related to the timing of cell division at the wound site (Braun and Mandle 1948). Only during active division did transformation occur. In cotyledons and leaf explants of *Arabidopsis*, dedifferentiating mesophyll cells were competent for transformation, a result of wounding and phytohormone treatments (Sangwan et al. 1992).

A preculture period prior to *Agrobacterium* infection can enhance transformation frequencies dramatically. Some recalcitrant crops show an increase in transformation efficiency if they are conditioned on a preculture medium (e.g., callus induction medium) designed to promote cell division (Mathis and Hinchee 1994). The preculture treatment increased the number of competent cells for *Agrobacterium* transformation (Sangwan et al. 1992). In recalcitrant species, cell activation and division following wounding may not be sufficient for transformation without exogenous application of growth regulators.

DNA transcription and replication

Both activation of transcription and DNA replication appear to be important for T-DNA integration, which may explain the strong dependence of transformation on cell division. There is evidence that the T-DNA integrates preferentially into regions that can potentially be transcribed (Kertbundit et al. 1991). Tobacco cells bombarded at the M- and G_2-phases yielded 4–6 times higher transformation efficiencies than those at S- and G_1-phases (Iida et al. 1991). Studies on the structure of integrated T-DNA in plant genomes strongly suggest that host DNA synthesis is required (Gheysen et al. 1987).

Selection

Use of dominant selectable markers is an integral part of transformation strategies. Markers available include genes encoding antibiotic-, antimetabolite-, and herbicide-resistance, and those conferring resistance to toxic levels of amino acids or analogs. If necessary, these marker genes can be eliminated from transgenic plants after selection via site-specific recombination (Ow and Medberry 1995). However, in contrast to species where recombinases are introduced by sexual crosses, vegetatively propagated woody species such as poplar would require an efficient inducible promoter for stimulating recombinase activity.

Important factors for efficient selection of transgenic cells are type of selectable markers, their expression level, and timing and intensity of selection after transformation (for review see Angenon et al. 1994). The sensitivity of plant cells to the selection agent depends on the genotype, the physiological condition, size, and type of explant, and tissue culture conditions. Therefore, the minimum level of a selection agent that can fully inhibit the growth of untransformed cells should be determined for each transformation and regeneration system. Depending on the type of selectable marker, it may be necessary to adjust the tissue culture system to avoid unnecessary inhibitory effects on transformed cells caused by the selection agent. For example, phosphinothricin causes an accumulation of NH_4^+ in the plant cells by inhibiting glutamine synthetase (De Block et al. 1987). Therefore, when phosphinothricin is used as a selection agent, the use of tissue culture conditions which lower the metabolic activity of the explants is important (De Block et al. 1991). Howe et al. (1994) found that hygromycin, which led to far fewer escapes from selection than kanamycin, was preferable for selection of transgenic poplar suspension cultures.

Management of transgene expression

In most plant transformation systems, one or two selectable markers (i.e., antibiotic resistance genes) are coinserted along with the gene of interest to distinguish transgenic from nontransgenic cells. Subsequently, transgenic shoots are selected based on their ability to grow in the presence of the antibiotics, against which the selectable marker genes impart resistance. Because of the low levels of selectable marker gene expression, it is likely that many successfully transformed plant cells are killed during the initial selection period, resulting in low recovery of transgenic shoots. Moreover, the expression of many recovered transformants is too low or erratic for commercial value. This recalcitrance may be overcome by increasing the uniformity and expression level of selectable marker genes.

Position effects and cosuppression

The level and specificity of expression of introduced genes can vary widely among individual transformants (Hobbs et al. 1990; Hollick and Gordon 1993). These "position effects" are observed in both animals and plants. Although several novel gene transfer methods have been developed for plants, a majority of detectable transformants express the introduced gene at low levels (Peach and Velten 1991). This low level expression is attributed to variation in transcriptional activity of the genomic insertion site. Gene silencing has also emerged in recent years as a substantial problem for the application of genetic engineering to plant improvement (Finnegan and McElroy 1994). Several different genetic mechanisms appear to be involved, and usually are associated with multiple copies of transgenes and DNA methylation.

Matrix attachment regions (MARs)

MARs are DNA fragments that appear to both increase and stabilize expression of flanked transgenes (reviewed in Breyne et al. 1994). MARs help to organize chromatin into loop domains by interacting with the proteinaceous scaffold. The loops are thought to delimit differing zones of transcriptional activity, isolating flanked genes from influences of adjacent DNA. Because about 80% of plant chromatin is in an inactive conformation at any given time, blocking such influences may result in enhanced gene expression. MARs have been identified in many organisms, including humans, chickens, potato, soybean, tobacco, and yeast (Breyne et al. 1994). We have found that tobacco-derived MARs can enhance transformation frequency and transgene expression in poplars (K.-H. Han and S.H. Strauss, unpublished results).

DNA methylation and demethylating agents

DNA methylation has an important role in the control of gene expression, and is usually associated with gene silencing (Matzke et al. 1989). *In vitro* DNA methylation inhibited gene expression in transgenic tobacco (Weber et al. 1990). The demethylating agent 5-azacytidine (azaC) activated silent T-DNA genes in transgenic tobacco transformed by *Agrobacterium tumefaciens* (Bochardt et al. 1992). Treatment of *Agrobacterium* or leaf disks with azaC generated more transformants than did controls (Palmgren et al. 1993).

Transgene copy number

Inserting multiple copies of a transgene may lead to cosuppression and consequent loss of gene expression. *Agrobacterium*-mediated transformation appears to result in fewer transgene copies than do direct DNA transformation methods (De Block 1993). The stage of the cell cycle also influences transgene insertion patterns. PEG/Ca^{2+}-mediated transformation of nonsynchronized protoplasts generally resulted in simple integration patterns (single copy insertions). Transformation of synchronized protoplasts in the synthetic (S) and mitotic (M) phase led to more complex patterns (Kartzke et al. 1990). As discussed above, root explants tend to give rise to fewer transgene copies than do leaves in *Arabidopsis* (Grevelding et al. 1993).

Transformation booster sequences (TBSs)

Genomic DNA segments that enhance transformation efficiencies have been reported for *Aspergillus* (Cullen et al. 1987), *Petunia* (Meyer et al. 1988), and tobacco (Marchesi et al. 1989). These are known as "transformation enhancing sequences" or "transformation booster sequences." Use of the petunia TBS resulted in a several-fold increase in transformation frequencies using biolistics (Buising and Benbow 1994). However, no increase in transgene expression was observed in these experiments. The mechanism(s) by which these sequences may stimulate homologous recombination remains unclear. TBS contains sequences for DNA unwinding elements (DUEs), matrix attachment regions (MARs), and topoisomerase binding sites (Buising and Benbow 1994). Its usefulness for *Agrobacterium*-mediated transformation has not been reported.

Future prospects

We are facing the very real prospect of a world in the 21st century with a population at least twice as large as todays. This world will need increasing quantities of fiber, fuel, chemical feedstocks, and other products that can be provided by silviculture. In addition, population pressure will reduce the amount of land on which to grow these materials. Woody crops will therefore need to be bred and engineered to be as productive as possible.

Genes currently under study for genetic engineering of poplar, including those for herbicide and insect resistance, reproductive sterility (Strauss et al. 1995), and lignin modification (Feuillet et al. 1995), can all enhance wood production and reduce environmental impacts of intensive tree culture. However, large increases in transformation efficiency are needed to accommodate the numerous genotypes and rapid turnaround in poplar breeding programs. Improvements to transformation will therefore remain a priority area for genetic engineering research for the foreseeable future.

Acknowledgements

We thank Marc De Block and an anonymous reviewer for their helpful comments.

References

Akama, K., Shiraishi, H., Ohta, S., Nakamura, K., Okada, K., and Shimura, Y. 1992. Efficient transformation of *Arabidopsis thaliana*: Comparison of the efficiencies with various organs, plant ecotypes and *Agrobacterium* strains. Plant Cell Rep. **12**: 7–11.

Akella, V. and Lurquin, P.F. 1993. Expression in cowpea seedlings of chimeric transgenes after electroporation into seed-derived embryos. Plant Cell Rep. **12**: 110–117.

Angenon, G., Dillen, W., and van Montagu, M. 1994. Antibiotic-resistance markers for plant transformation. *In* Plant Molecular Biology Manual. *Edited by* S.B. Gelvin and R.A. Schilperoort. Kluwer Academic Publishers, Boston, MA. pp. C1: 1–13.

Antonetti, P.L.E. and Pinon, J. 1993. Somaclonal variation within poplar. Plant Cell Tissue Organ Cult. **35**: 99–106.

Bechtold, N., Ellis, J., and Pelletier, G. 1993. *In planta Agrobacterium* mediated gene transfer by infiltration of adult *Arabidopsis thaliana* plants. C. R. Acad. Sci. Paris, Science de la vie/Life Sci. **316**: 1194–1199.

Bidney, D., Scelonge, C., Martich, J., Burrus, M., Sims, L., and Huffman, G. 1992. Microprojectile bombardment of plant tissues increases transformation frequency by *Agrobacterium tumefaciens*. Plant Mol. Biol. **18**: 301–313.

Binns, A.N. 1990. *Agrobacterium*-mediated gene delivery and the biology of host range limitations. Physiol. Plant. **79**: 135–139.

Birot, A.-M., Bouchez, D., Casse-Delbart, F., Durand-Tardif, M., Jouanin, L., Pautot, V., Robaglia, C., Tepfer, D., Tepfer, M., Tourneur, J., and Vilane, F. 1987. Studies and uses of the Ri plasmids of *Agrobacterium*. Plant Physiol. Biochem. (Paris), **25**(3): 323–335.

Bochardt, A., Hodal, L., Palmgreen, G., Mattsson, O., and Okkels, F.T. 1992. DNA methylation is involved in maintenance of an unusual expression pattern of an introduced gene. Plant Physiol. **99**: 409–414.

Bouchez, D., Camilleri, C., and Caboche, M. 1993. A binary vector based on Basta resistance for *in planta* transformation of *Arabidopsis thaliana*. C. R. Acad. Sci. Paris, Science de la vie/Life Sci. **316**: 1188–1193.

Brasileiro, A.C.M., Leple, J.-C., Muzzin, J., Ounnoughi, D., Michel, M.-F., and Jouanin, L. 1991. An alternative approach for gene transfer in trees using wild-type *Agrobacterium* strains. Plant Mol. Biol. **17**: 441–452.

Brasileiro, A.C.M., Tourneur, C., Leple, J.-C., Combes, V., and Jouanin, L. 1992. Expression of the mutant *Arabidopsis thaliana* acetolactate synthase confers chlorsulfuron resistance to poplar. Transgenic Research **1**: 133–141.

Braun, A.C. and Mandle, R.J. 1948. Studies on the inactivation of the tumor inducing principle in crown gall. Growth **12**: 255–269.

Breyne, P., Montagu, M.V., and Gheysen, G. 1994. The role of scaffold attachment regions in the structural and functional organization of plant chromatin. Transgenic Res. **3**: 195–202.

Bu, X.-X, Lin, Z.-P., and Chen, W.-L. 1991. Transformation of *Populus tomentosa* by *Agrobacterium* and regeneration of transformed plantlets. Acta Bot. Sin. **33**(3): 206–213.

Buising, C.M. and Benbow, R.M. 1994. Molecular analysis of transgenic plants generated by microprojectile bombardment: effect of petunia transformation booster sequence. Mol. Gen. Genet. **243**: 71–81.

Cangelosi, G.A., Ankenbauer, R.G., and Nester, N.W. 1990. Sugars induce the virulence genes through a periplasmic binding protein and a transmembrane signal protein. Proc. Natl. Acad. Sci. U.S.A. **87**: 6708–6712.

Chang, S.S., Park, S.K., Kim, B.C., Kang, B.J., Kim, D.U., and Nam, H.G. 1994. Stable genetic transformation of *Arabidopsis thaliana* by *Agrobacterium* inoculation *in planta*. The Plant J. **5**(4): 551–558.

Charest, P.J., Stewart, D., and Budicky, P.L. 1992. Root induction in hybrid poplar by *Agrobacterium* genetic transformation. Can. J. For. Res. **22**: 1832–1837.

Cheema, G.S. 1988. Isolation and culture of protoplasts from totipotent cell cultures of *Populus ciliata*. Progr. Plant Protoplast Res. pp. 107–108.

Cheema, G.S. 1989. Somatic embryogenesis and plant regeneration from cell suspension and tissue cultures of mature Himalayan poplar (*Populus ciliata*). Plant Cell Rep. **8**: 124–127.

Chowrira, G.M., Akella, V., and Lurquin, P.F. 1995. Electroporation-mediated gene transfer into intact nodal meristems *in planta*. Mol. Biotechnol. **3**: 17–23.

Christou, P. 1994. Gene transfer to plants *via* particle bombardment. *In* Plant Molecular Biology Manual. *Edited by* S.B. Gelvin and R.A. Schilperoort. Kluwer Academic Publishers, Boston, MA. pp. A2: 1–15.

Chun, Y.W. 1993. Clonal propagation in non-aspen poplar hybrids. *In* Micropropagation of woody plants. *Edited by* M.R. Ahuja. Kluwer Academic Publisher, Dordrecht, The Netherlands. pp. 209–222.

Chupeau, M.C., Pautot, V., and Chupeau, Y. 1994. Recovery of transgenic trees after electroporation of poplar protoplasts. Transgenic Res. **3**(1): 13–19.

Civinova, B., and Sladky, Z. 1990. Stimulation of the regeneration capacity of tree shoot segment explants *in vitro*. Biol. Plant. **32**(6): 407–413.

Coleman, G.D., and Ernst, S.G. 1989. *In vitro* shoot regeneration in *Populus deltoides*: effect of cytokinin and genotype. Plant Cell Rep. **8**: 459–462.

Coleman, G.D., and Ernst, S.G. 1990. Shoot induction competence and callus determination in *Populus deltoides*. Plant Sci. **71**: 83–92.

Compton, M.E., and Gray, D.J. 1993. Shoot organogenesis and plant regeneration from cotyledons of diploid, triploid, and tetraploid watermelon. J. Am. Soc. Hortic. Sci. **118**(1): 151–157.

Confalonieri, M., Balestrazzi, A., and Bisoffi, S. 1994. Genetic transformation of *Populus nigra* by *Agrobacterium tumefaciens*. Plant Cell Rep. **13**: 256–261.

Cullen, D., Wilson, L.J., Grey, G.L., Henner, D.J., Turner, G., and Ballance, D.J. 1987. Sequence and centromere proximal location of a transformation-enhancing fragment *ans*1 from *Aspergillus nidulans*. Nucleic Acids Res. **15**: 9163–9175.

Davis, J.M., and Keathley, D.E. 1989. Detection and analysis of T-DNA in crown gall tumors and kanamycin-resistant callus of *Robinia pseudoacacia*. Can. J. For. Res. **19**: 1118–1123.

Davis, M.E., Miller, A.R., and Linebeger, R.D. 1991. Temporal competence for transformation of *Lycopersicon esculentum* (L. Mill.) cotyledons by *Agrobacterium tumefaciens*: relation to wound-healing and soluble plant factors. J. Exp. Bot. **42**(236): 359–364.

De Block, M. 1990. Factors influencing the tissue culture and the *Agrobacterium*-mediated transformation of hybrid aspen and poplar clones. Plant Physiol. **93**: 1110–1116.

De Block, M. 1993. The cell biology of plant transformation: current state, problems, prospects and the implications for plant breeding. Euphytica, **71**: 1–14.

De Block, M., and Debrouwer, D. 1991. Two T-DNAs co-transformed into *Brassica napus* by a double *Agrobacterium tumefaciens* infection are mainly integrated at the same locus. Theor. App. Genet. **82**: 257–263.

De Block, M., Botterman, J., Vandewiele, M., Dochx, J., Thoen, C., Gossele, V., Movva, N.R., Thompson, C., Van Montagu, M., and Leemans, J. 1987. Engineering herbicide resistance in plants by expressing of a detoxifying enzyme. EMBO J. **6**: 2513–2518.

De Block, M., Debrouwer, D., and Tenning, P. 1989. Transformation of *Brassica napus* and *Brassica oleracea* using *Agrobacterium tumefaciens* and the expression of the *bar* and *neo* genes in the transgenic plants. Plant Physiol. **91**: 694–701.

De Block, M., De Sonville, A., and Debrouwer, D. 1995. The selection mechanism of phosphinothrien is influenced by the metabolic status of the tissue. Planta, **197**: 619–626.

Devantier, Y.A., Moffat, B., Jones, C., and Charest, P.J. 1993. Microprojectile-mediated DNA delivery to the *Salicaceae* family. Can. J. Bot. **71**(11): 1458–1466.

D'Halluin, K., Bonne, E., Bossut, M., De Beuckeleer, M., and Leemans, J. 1992. Transgenic maize plants by tissue electroporation. Plant Cell, **4**: 1495–1505.

Douglas, C.J., Staneloni, R.J., Rubin, R.A., and Nester, E.W. 1985. Identification and genetic analysis of an *Agrobacterium tumefaciens* chromosomal virulence region. J. Bacteriol. **161**: 850–860.

Dunahay, T.G. 1993. Transformation of *Chlamydomonas reinhardtii* with silicon carbide whiskers. Biotechniques, **15**: 452–460.

Ernst, S.G. 1993. In vitro culture of pure species non-aspen poplars. *In* Micropropagation of woody plants. *Edited by* M.R. Ahuja. Kluwer Academic Publisher, Dordrecht, The Netherlands. pp. 195–207.

Feldmann, K.A., and Marks, M.D. 1987. *Agrobacterium*-mediated transformation of germinating seeds of *Arabidopsis thaliana*: a non-tissue culture approach. Mol. Gen. Genet. **208**: 1–9.

Feuillet, C., Lauvergeat, V., Deswarte, C., Pilate, G., Boudet, A., and Grima-Pettenati, J. 1995. Tissue- and cell-specific expression of a cinnamyl alcohol dehydrogenase promoter in transgenic poplar plants. Plant Mol. Biol. **27**: 651–667.

Fillatti, J., Sellmer, J., McCown, B., Haissig, B., and Comai, L. 1987. *Agrobacterium* mediated transformation and regeneration of *Populus*. Mol. Gen. Genet. **206**: 192–199.

Finnegan, J., and McElroy, D. 1994. Transgene inactivation: plants fight back! Bio/Technology, **12**: 883–888.

Frame, B.R., Draton, P.R., and Bagnall, S.V. 1994. Production of fertile transgenic maize plants by silicon carbide whisker-mediated transformation. Plant J. **6**: 941–948.

Fromm, M.E., Tayler, L.P., and Walbot, V. 1985. Expression of genes transferred into monocot and dicot plant cells by electroporation. Proc. Natl. Acad. Sci. U.S.A. **82**: 5824–5828.

Gheysen, G., Van Montagu, M., and Zambryski, P. 1987. Integration of *Agrobacterium tumefaciens* T-DNA involves rearrangements of target plant DNA sequences. Proc. Natl. Acad. Sci. U.S.A. **84**: 9006–9010.

Gould, J., Devey, M., Hasegawa, O., Ulian, E.C., Peterson, G., and Smith, R.H. 1991. Transformation of *Zea mays* L. using *Agrobacterium tumefaciens* and the shoot apex. Plant Physiol. **95**: 426–434.

Graves, A.E., Goldman, S.L., Banks, A.W., and Graves, A.C.F. 1988. Scanning electron microscope studies of *Agrobacterium tumefaciens* attachment to *Zea mays*, *Gladiolus* sp., and *Triticum aestivum*. J. Bacteriol. **170**: 2395–2400.

Grevelding, C., Fantes, V., Kemper, E., Schell, J., and Masterson, R. 1993. Single-copy T-DNA insertions in *Arabidopsis* are the predominant form of insertion in root-derived transgenics, whereas multiple insertions are found in leaf discs. Plant Mol. Biol. **23**: 847–860.

Hamamoto, H., Boulter, M.E., Shirsat, A.H., Croy, E.J., and Ellis, J.R. 1990. Recovery of morphologically normal transgenic tobacco from hairy roots co-transformed with *Agrobacterium rhizogenes* and a binary vector plasmid. Plant Cell Rep. **9**: 88–92.

Han, K.-H., Keathley, D.E., Davis, J.M., and Gordon, M.P. 1993. Regeneration of a transgenic woody legume (*Robinia pseudoacacia* L., black locust) and morphological alterations induced by *Agrobacterium rhizogenes*-mediated transformation. Plant Sci. **88**: 149–157.

Han, K.-H, Fleming, P., Walker, K., Loper, M., Chilton, W.S., Mocek, U., Gordon, M.P., and Floss, H.G. 1994. Genetic transformation of mature *Taxus*: an approach to genetically control the *in vitro* production of the anticancer drug, taxol. Plant Sci. **95**: 187–196.

Han, K.-H., Bradshaw, H.D., Jr., and Gordon, M.P. 1995a. Adventitious root and shoot regeneration *in vitro* is under major gene control in an F$_2$ family of hybrid poplar (*Populus trichocarpa* × *P. deltoides*). For. Genet. **1**(3): 139–146.

Han, K.-H., Gordon, M.P., and Strauss, S.H. 1995b. *Agrobacterium rhizogenes*-induced phenotypic changes in hybrid poplar. In press.

Hansen, G., Das, A., and Chilton, M.-D. 1994. Constitutive expression of the virulence genes improves the efficiency of plant transformation by *Agrobacterium*. Proc. Natl. Acad. Sci. U.S.A. **91**: 7603–7607.

Hiei, Y., Ohta, S., Komari, T., and Kumashiro, T. 1994. Efficient transformation of rice (*Oryza sativa* L.) mediated by *Agrobacterium* and sequence analysis of the boundaries of the T-DNA. Plant J. **6**(2): 271–282.

Hobbs, S.L.A., Kpodar, P., and DeLong, C.M.O. 1990. The effect of T-DNA copy number, position and methylation on reporter gene expression in tobacco transformants. Plant Mol. Biol. **15**: 851–864.

Hollick, J.B., and Gordon, M.P. 1993. A poplar tree proteinase inhibitor-like gene promoter is responsive to wounding in transgenic tobacco. Plant Mol. Biol. **22**: 561–572.

Howe, G.T., Goldfarb, B., and Strauss, S.H. 1994. *Agrobacterium*-mediated transformation of hybrid poplar suspension cultures and regeneration of transformed plants. Plant Cell Tissue Organ Cult. **36**: 59–71.

Hrib, J. 1993. Effect of acid pH on organogenesis of poplar *in vitro*. Biologia, **48**(1): 89–92.

Huang, Y., Diner, A., and Karnosky, D. 1991. *Agrobacterium rhizogenes*-mediated genetic transformation and regeneration of a conifer: *Larix decidua*. In Vitro Cell. Dev. Biol. **27P**: 201–207.

Iida, A., Yamashida, T., Yamada, Y., and Morikawa, H. 1991. Efficiency of particle-bombardment-mediated transformation is influenced by cell stage in synchronized cultured cells of tobacco. Plant Physiol. **97**: 1585–1587.

Jouanin, L., Brasileiro, A.C.M., Leple, J.C., Pilate, G., and Cornu, D. 1993. Genetic transformation: a short review of methods and their applications, results and perspectives for forest trees. Ann. Sci. For. **50**: 325–336.

Kaeppler, H.F., Gu, W., and Somers, D.A. 1990. Silicon carbide fiber-mediated DNA delivery into plant cells. Plant Cell Rep. **9**: 415–418.

Kaeppler, H.F., Somers, D.A., and Rines, H.W. 1992. Silicon carbide fiber-mediated stable transformation of plant cells. Theoret. Appl. Genet. **84**: 560–566.

Kartzke, S., Saedler, H., and Meyer, P. 1990. Molecular analysis of transgenic plants derived from transformations of protoplasts at various stages of the cell cycle. Plant Sci. **67**: 63–72.

Kertbundit, S., De Greve, H., Deboeck, F., Van Montagu, M., and Hernalsteens, J.-P. 1991. *In vivo* random β-glucuronidase gene fusions in *Arabidopsis thaliana*. Proc. Natl. Acad. Sci. U.S.A. **88**: 5212–5216.

Khan, M.R., Tabe, L.M., Heath, L.C., Spencer, D., and Higgins, T.J.V. 1994. *Agrobacterium*-mediated transformation of subterranean clover (*Trifolium subterraneum* L.). Plant Physiol. **105**: 81–88.

Klopfenstein, N.B., Shi, N.Q., Kernan, A., McNabb, H.S., Jr., Hall, R.B., Hart, E.R., and Thornburg, R.W. 1991. Transgenic *Populus* hybrid expresses a wound-inducible potato proteinase inhibitor II–CAT gene fusion. Can. J. For. Res. **21**: 1321–1328.

Kovacs, L.G., and Pueppke, S.G. 1993. The chromosomal background of *Agrobacterium tumefaciens* Chry5 conditions high virulence on soybean. Mol. Plant-Microbe Interact. **6**(5): 601–608.

Lambert, C., and Tepfer, D. 1992. Use of Agrobacterium rhizogenes to create transgenic apple trees having an altered organogenic response to hormones. Theor. Appl. Genet. **85**: 105–109.

Leple, J.C., Brasileiro, A.C.M., Michel, M.-F., Delmotte, F., and Jouanin, L. 1992. Transgenic poplars: expression of chimeric genes using four different constructs. Plant Cell Rep. **11**: 137–141.

Lin, J.-J., Assad-Garcia, N., and Kuo, J. 1994. Effects of *Agrobacterium* cell concentration on the transformation efficiency of tobacco and *Arabidopsis thaliana*. Focus, **16**(3): 72–77.

Liu, C.-N., Li, X.-Q., and Gelvin, S.B. 1992. Multiple copies of *vir*G enhance the transient transformation of celery, carrot and rice tissues by *Agrobacterium tumefaciens*. Plant Mol. Biol. **20**: 1071–1087.

Loopstra, C.A., Stomp, A.-M., and Sederoff, R.R. 1990. *Agrobacterium*-mediated DNA transfer in sugar pine. Plant Mol. Biol. **15**: 1–9.

Lubrano, L. 1992. Micropropagation of pooplars (*Populus* spp.). *In* Biotechnology in Agriculture and Forestry, vol. 18. High-tech and Micropropagation II. *Edited by* Y.P.S. Bajaj. Springer-Verlag, Berlin Heidelberg. pp. 151–178.

Mackay, J., Seguin, A., and Lalonde, M. 1988. Genetic transformation of 9 *in vitro* clones of *Alnus* and *Betula* by *Agrobacterium tumefaciens*. Plant Cell Rep. **7**: 229–232.

Macrae, S., and Van Staden, J. 1993. *Agrobacterium rhizogenes*-mediated transformation to improve rooting ability of eucalyptus. Tree Physiol. **12**: 411–418.

Malik, K.A., Khan, S.T.A., and Saxena, P.K. 1992. Direct organogenesis and plant regeneration in preconditioned tissue cultures of *Lathyrus ciera* L., *L. ochrus* (L.) DC. and *L. sativus* L. Ann. Bot. **70**: 301–304.

Mansur, E.A., Lacorte, C., Freitas, V.G.D., Oliveira, D.E.D., Timmerman, B., and Cordeiro, A.R. 1993. Regulation of transformation efficiency of peanut (*Arachis hypogaea L.*) explants by *Agrobacterium tumefaciens*. Plant Sci. **89**(1): 93–99.

Marchesi, M.L., Castiglione, S., and Sala, F. 1989. Effect of repeated DNA sequences on direct gene transfer in protoplast of *Nicotiana plumbaginifolia*. Theo. Appl. Genet. **78**: 113–118.

Mathis, N.L., and Hinchee, M.A.W. 1994. *Agrobacterium* inoculation techniques for plant tissues. *In* Plant Molecular Biology Manual. *Edited by* S.B. Gelvin and R.A. Schilperoort. Kluwer Academic Publishers, Boston, MA. pp. B6: 1–9.

Matzke, M.A., Primig, M., Trnovsky, J., and Matzke, A.J.M. 1989. Reversible methylation and inactivation of marker genes in sequentially transformed tobacco plants. EMBO J. **8**: 643–649.

McCown, B.H., McCabe, D.E., Russell, D.R., Robison, D.J., Barton, K.A., and Raffa, K.F. 1991. Stable transformation of *Populus* and incorporation of pest resistance by electric discharge particle acceleration. Plant Cell Rep. **9**: 590–594.

McGranahan, G.H., Leslie, C.A., Uratsu, S.L., Martin, L.A., and Dandekar, A.M. 1988. *Agrobacterium*-mediated transformation of walnut somatic embryos and regeneration of transgenic plants. Bio/Technology, **6**: 800–804.

Meyer, P., Kartzke, S., Niedenhof, I., Heidmann, I., Bussman, K., and Saedler, H. 1988. A genomic DNA segment from *Petunia hybrida* leads to increased transformation frequencies and simple integration patterns. Proc. Natl. Acad. Sci. U.S.A. **85**: 8568–8572.

Michler, C.H. 1994. Somatic embryogenesis in *Populus* spp. *In* Somatic embryogenesis in woody polants. *Edited by* S. Jain, P. Gupta, and R. Newton. Kluwer Academic Publisher, Dordrecht, The Netherlands. pp. 89–97.

Michler, C.H., and Bauer, E.O. 1991. High frequency somatic embryogenesis from leaf tissue of *Populus* spp. Plant Sci. **77**: 111–118.

Minocha, S.C. 1987. pH of the culture medium and the growth and metabolism of cells in culture. *In* Cell and Tissue Culture in Forestry, Vol. 1, General Principles and Biotechnology. *Edited by* J.M. Bonga and D.D. Durzan. Martinus Nijhof Publishers, Dordrecht, The Netherlands. pp. 125–141.

Minocha, S.C., Noh, E.W., and Kausch, A.P. 1986. Tissue culture and genetic transformation in *Betula papyrifera* and *Populus tremuloides*. TAPPI J. **68**: 116–119.

Mizozoe, M., Inagaki, N., Okano, M., Kanechi, M., and Maekawa, S. 1993. Effects of shading and fertilization of mother plants, and ammonium:nitrate ratio in the medium on *in vitro* organogenesis of stock (*Matthiola incana* R. Br.). J. Jpn. Soc. Hortic. Sci. **61**(3): 625–633.

Mullins, M.G., Tang, F.C.A., and Facciotti, D. 1990. *Agrobacterium*-mediated genetic transformation of grapevines: transgenic plants of *Vitis rupestris* Scheele and buds of *Vitis vinifera* L. Bio/Technology, **8**: 1041–1045.

Murashige, T., and Skoog, F. 1962. A revised medium for rapid growth and bioassay with tobacco tissue cultures. Plant Physiol. **15**: 473–479.

Nilsson, O., Torsen, A., Sitbon, F., Little, C.H.A., Chalupa, V., Sandberg, G., and Olsson, O. 1992. Spatial pattern of cauliflower mosaic 35S promoter-luciferase expression in transgenic hybrid aspen trees monitored by enzymatic assay and non-destructive imaging. Transgenic Res. **1**: 209–220.

Nishihara, M., Ito, M., Tanaka, I., Kyo, M., Ono, K., Irifune, K., and Morikawa, H. 1993. Expression of the β-glucuronidase gene in pollen of lily (*Lilium longifloum*), tobacco (*Nicotiana tabacum*), *Nicotiana rustica*, and peony (*Paeonia lactiflora*) by particle bombardment. Plant Physiol. **102**: 357–361.

Ow, D.W., and Medberry, S.L. 1995. Genome manipulation through site-specific recombination. Crit. Rev. Plant Sci. **14**(3): 239–261.

Palmgren, G., Mattson, O., and Okkels, F.T. 1993. Treatment of *Agrobacterium* or leaf disks with 5-azacytidine increases transgene expression in tobacco. Plant Mol. Biol. **21**: 429–435.

Park, Y.G., and Han, K.-H. 1986. Isolation and culture of mesophyll protoplasts from in vitro cultured *Populus alba* × *P. glandulosa*. J. Kor. For. Soc. **73**: 33–42.

Park, Y.G. and Son, S.H. 1988*a*. *In vitro* organogenesis and somatic embryogenesis from punctured leaf of *Populus nigra* × *P. maximowiczii*. Plant Cell Tissue Organ Cult. **15**(2): 95–105.

Park, Y.G. and Son, S.H. 1988*b*. Culture and regeneration of *Populus alba* × *P. glandulosa* leaf protoplasts isolated from *in vitro* cultured explant. J. Kor. For. Soc. **77**: 208–215.

Parsons, J.F., Sinkar, V.P., Stettler, R.F., Nester, E.W., and Gordon, M.P. 1986. Transformation of poplar by *Agrobacterium tumefaciens*. Bio/Technology, **4**: 533–536.

Peach, C., and Velten, J. 1991. Transgene expression variability (position effect) of CAT and GUS reporter genes driven by linked divergent T-DNA promoters. Plant Mol. Biol. **17**: 49–60.

Phythoud, F., Sinkar, V.P., Nester, E.W., and Gordon, M.P. 1987. Increased virulence of *Agrobacterium rhizogenes* conferred by the *vir* region of pTiBo542: application to genetic engineering of poplar. Bio/Technology, **5**: 1323–1327.

Riemenschneider, D., Haissig, B.E., Sellmer, J., and Fillatti, J. 1988. Expression of an herbicide tolerance gene in young plants of a transgenic hybrid poplar clone. *In* Somatic Cell Genetics of Woody Plants. *Edited by* M.R. Ahuja. Kluwer Academic Publishers, Boston, MA. pp. 73–80.

Riemenschneider, D. 1990. Susceptibility of intra- and inter-specific hybrid poplars to *Agrobacterium tumefaciens* strain C58. Phytopathology, **80**: 1099–1102.

Rout, J.R., Gordon, M.P., Lucas, W.J., and Nester, E.W. 1995. *Agrobacterium*-mediated gene transfer to rice (*Oriza sativa* L.). In Vitro Cell. & Dev. Biol. **31**(3): 28A.

Russell, J.A., and McCown, B.H. 1988. Recovery of plants from leaf protoplasts of hybrid-poplar and aspen clones. Plant Cell Rep. **7**: 59–62.

Saieed, N.T., Douglas, G.C., and Fry, D.J. 1994. Induction and stability of somaclonal variation in growth, leaf phenotype and gas exchange characteristics of poplar regenerated from callus culture. Tree Physiol. **14**: 1–16.

Sangwan, R.S., Bourgeois, Y., Brown, S., Vasseur, G., and Sangwan-Norreel, B. 1992. Characterization of competent cells and early events of *Agrobacterium* mediated genetic transformation in *Arabidopsis thaliana*. Planta, **188**(3): 439–456.

Schlappi, M., and Hohn, B. 1992. Competence of immature maize embryos for *Agrobacterum*-mediated gene transfer. Plant Cell, **4**: 7–16.

Schmulling, T., Schell, J., and Spena, A. 1988. Single genes from *Agrobacterium rhizogenes* influence plant development. EMBO J. **7**: 2621–2629.

Schnall, J.A., and Weissinger, A.K. 1993. Culturing peanut (*Arachis hypogae* L.) zygotic embryos for transformation *via* microprojectile bombardment. Plant Cell Rep. **12**: 316–319.

Schwartzenberg, K.V., Doumas, P., Jouanin, L., and Pilate, G. 1994. Enhancement of the endogenous cytokinin concentration in poplar by transformation with *Agrobacetrium* T-DNA gene *ipt*. Tree Physiol. **14**: 27–35.

Scorza, R., Morgens, P.H., Cordts, J.M., Mante, S., and Callahan, A.M. 1990. *Agrobacterium*-mediated transformation of peach (*Prunus persica* L. batch) leaf segments, immature embryos, and long-term embryogenic callus. In Vitro Cell. & Dev. Biol. **26**: 829–834.

Shahin, E.A., Sukhapinda, K., Simpson, R.B., and Spivey, R. 1986. Transformation of cultivated tomato by a binary vector in *Agrobacterium rhizogenes*: transgenic plants with normal phenotype harbor binary vector T-DNA, but no Ri-plasmid T-DNA. Theor. Appl. Genet. **72**: 770–777.

Shin, D.-L., Podila, G.K., Huang, Y., and Karnosky, D.F. 1994. Transgenic larch expressing genes for herbicide and insect resistance. Can. J. For. Res. **24**: 2059–2067.

Son, S.H., Moon, H.K., and Hall, R.B. 1993. Somaclonal variation in plants regenerated from callus culture of hybrid aspen (*Populus alba* L. × *P. grandidentata* Michx.). Plant Sci. **90**: 89–94.

Stachel, S.E., Messens, E., Van Montagu, M., and Zambryski, P. 1985. Identification of the signal molecules produced by wounded plant cells which activate the T-DNA transfer process in *Agrobacterium tumefaciens*. Nature (London), **318**: 624–629.

Strauss, S.H., W.H. Rottmann, A.M. Brunner, and Sheppard, L.A. 1995. Genetic engineering of reproductive sterility in forest trees. Mol. Breed. **1**: 5–26.

Vahala, T., Stabel, P., and Eriksson, T. 1989. Genetic transformation of willows (*Salix* spp) by *Agrobacterium tumefaciens*. Plant Cell Rep. **8**: 55–58.

Van Roekel, J.S.C., Damm, B., Melchers, L.S., and Hoekema, A. 1993. Factors influencing transformation frequency of tomato (*Lycopersicon esculentum*). Plant Cell Rep. **12**(11): 644–647.

Weber, H., Ziechmann, C., and Graessmann, A. 1990. *In vitro* DNA methylation inhibits gene expression in transgenic tobacco. EMBO J. **9**: 4409–4415.

Winton, L.L. 1970. Shoot and tree production from aspen tissue cultures. Am. J. Bot. **57**: 904–909.

Youn, Y., Lee, J.S., and Lee, S.K. 1985. Isolation and culture of protoplasts from suspension-cultured cells of *Populus alba* × *P. glandulosa* F$_1$. Res. Rep. Inst. For. Genet. **21**: 109–111.

Zupan, J.R., and Zambryski, P. 1995. Transfer of T-DNA from *Agrobacterium* to the plant cell. Plant Physiol. **107**: 1041–1047.

CHAPTER 10
The specificity of fungal pathogens of *Populus*

George Newcombe

Introduction

Species of *Populus* L. are widely distributed and abundant in many different environments of the northern hemisphere. *Populus* hybrids are cultivated in suitable environments of both northern and southern hemispheres. The fungi that affect *Populus* species are extremely diverse; for example, more than 250 species of fungi are known to be associated with decay of *P. tremuloides* Michx. alone in North America (Lindsey and Gilbertson 1978). New host–fungus associations continue to be discovered; in the last 4 years Callan (1995) listed 41 fungal taxa on *Populus* that had not been reported before in British Columbia. Some fungal pathogens of *Populus* are widely distributed, whereas others are only known from a single region. Some have become widely distributed in association with the expansion of hybrid poplar culture, whereas others have remained confined to original distributions on native *Populus* and hybrid poplar introduced to the area. Many pathogens occur primarily in mature stands, whereas others are problematic in young hybrid poplar in intensively managed short rotations.

The negative effects of pathogens on poplar growth have inspired poplar breeders to make disease resistance the major selection criterion in poplar tree improvement for the past 50 years (R. Stettler, personal communication). In spite of these efforts, in most poplar-growing areas, few clones combine necessary disease resistance with desirable form and growth characteristics. The probability of selecting such clones in recurrent F_1 breeding is low given the diversity of pathogens attacking *Populus* in any given region. Furthermore, when planted extensively, such clones may in time succumb to the evolution of new pathotypes and pathogen races unless a way is found to breed for durable resistance.

G. Newcombe. Washington State University, Puyallup Research and Extension Center, Puyallup, WA 98371-4998, USA.
Correct citation: Newcombe, G. 1996. The specificity of fungal pathogens of *Populus*. *In* Biology of Populus and its implications for management and conservation. Part I, Chapter 10. *Edited by* R.F. Stettler, H.D. Bradshaw, Jr., P.E. Heilman, and T.M. Hinckley. NRC Research Press, National Research Council of Canada, Ottawa, ON. pp. 223–246.

The primary focus of this chapter is on the host specificity of fungal pathogens of *Populus*. "Host specificity" is used here in preference to the less felicitous term "host range"; a pathogen will specifically attack host taxa in its host range, but may also exhibit more exclusive specificity within that range. "Host range" is a phytopathological term that should not be confused with geographic range or distribution. Pathogens may be specific to a single susceptible species of *Populus* or may exhibit specificity toward a subsection, section, or the entire genus. In some cases, pathogens of *Populus* also affect species of *Salix* L. or even other genera of woody plants. In most cases, "susceptible species" actually vary in response to the pathogen, in that some individuals are susceptible and others are resistant.[1]

Pathogen taxonomy must also be host-independent at the species level for there to be any noncircular discussion of the host specificity of the pathogen. At the subspecific level, a pathogen species may be divided into *formae speciales*, or pathotypes, or races, and these designations do depend on considerations of host specificity. Pathogen taxonomy integrates knowledge of the organism's morphological, ecological, chemical, and molecular variation together with studies of mating behaviour and reproductive isolation from congeneric species. Hopefully, this review will point out the many areas in which studies of the taxonomy of pathogens of *Populus* would benefit poplar research.

Intraspecific resistance within a susceptible host species has been the subject of a number of genetic studies. There are also cases to be discussed in which the resistance of an interspecific hybrid has been analyzed. In a few cases, resistance genes of major effect have been described and placed on the *Populus* linkage map (see Bradshaw, Chapter 8). Studies on disease resistance of hybrids are of great interest to poplar breeders and growers. Such studies, when coupled with knowledge of original or native distributions of both host and pathogen, point to resistance often being of "exapted" nature in *Populus* (Gould and Vrba 1982). Exaptations are characters evolved for other usages (or for no function at all), that later prove to have current adaptive value, unlike adapations that were favored by selection for their current role.

It can safely be said that our knowledge of the specificity of fungal pathogens of *Populus* is fragmentary, even for the best-studied pathogens. For example, no pathogen thought to be genus-specific is known to cause disease on all species of the 6 sections of *Populus*. In fact, virtually nothing is known of the pathogens of sections *Turanga* Bunge, *Abaso* Ecken., and *Leucoides* Spach, and this chapter will, by necessity, only deal with sections *Populus* Ecken. (= sect. *Leuce* Duby), *Aigeiros,* and *Tacamahaca* (Eckenwalder 1977). Jean Pinon (1992) has summarized the few records of the occurrence of leaf rust on sections *Turanga* and *Leucoides*. Thus, pathogens thought to be restricted to one of the latter 3 sections

---

[1]Resistant plants allow little or no reproduction of pathogens infecting them.

might also occur on any or all of the 3 unstudied sections. The fact that *Aigeiros* and *Tacamahaca* are closely related sections differing only in vegetative characters (Eckenwalder 1977) is reflected by the relative rarity of pathogens discriminating between the two (i.e., specific to one or the other). More commonly, fungal pathogens are specific either to section *Populus* or to sections *Aigeiros* and *Tacamahaca*. Of course, knowledge of the host specificity of fungal pathogens of *Populus* depends on the still-evolving taxonomy of the pathogens themselves, that occasionally is compromised by incorrect disease records of the past in which both host and pathogen may have been misidentified.

Taxonomic authorities for fungal names are as per Farr et al. (1989), except for Eurasian fungi not yet recorded in North America for which authorities are as stated by the authors cited. For common names of poplar diseases, I have complied with the system of first naming the host, and secondly describing symptoms, followed by mention of the causal organism (Holliday 1989; Kommedahl 1995).

Pathogens might hybridize, particularly in natural hybridization zones of the host. Natural hybridization of sympatric *Populus* species is quite common (Eckenwalder 1984), but evidence of hybridization among pathogens is scant, although little effort has been directed toward its detection. Any such efforts at the phylogenetic level for a given pathogen may have to take into consideration ancient hybridization of the pathogen parallelling that of the host (Barnes 1967).

The choice of fungal pathogens of *Populus* reviewed here reflects the North American bias of the author, working in the Pacific Northwest of the United States. Research on poplar pathogens in North America, where extensive natural populations of native *Populus* and relatively little poplar culture prevail, differs from that in Europe or Asia, where cultivated poplars are the focus of research. Another difference lies in the major pathogens studied. A recent review of poplar pathogens attacking poplar cultivars registered in Europe focused on bacterial canker (*Xanthomonas populi*), leaf spot [*Marssonina brunnea* (Ell. and Ev.) Magn.], leaf rust (*Melampsora* Castagne), poplar mosaic virus, and bark necrosis [*Discosporium populeum* (Sacc.) Sutton] (Pinon and Valadon 1996). Three of these five (i.e., bacterial canker, poplar mosaic virus, and bark necrosis) are of little or no importance in North America, whereas *Septoria musiva* Peck, the cause of the most serious poplar stem canker in North America, receives extensive review here even though it does not occur in Europe. In Europe the emphasis has been on diseases of registered cultivars (Pinon and Valadon 1996). Bacterial pathogens of *Populus* are not reviewed here even though they cause important diseases in Europe. Readers interested in a recent review of bacterial pathogens of *Populus* should consult Nesme and Ménard (1995).

In summary, this review will focus on the host specificity of fungal pathogens of *Populus*, from the generic level to what little is known at the genetic level. The pathogens chosen for review should illustrate the problems that poplar breeders and pathologists jointly face in attempting to incorporate disease resistance into commercial poplar clones and cultivars. Finally, the idea is explored that there might be a link between the susceptibility of native *Populus* species to a large number of different indigenous pathogens, and the resistance of these species to devastating epidemics associated with introduced pathogens. This link might be mediated by pathogen competition or by interactions with the saprophytic and hyperparasitic microflora. In spite of the movement of poplar cuttings around the world and the known introductions of pathogens, *Populus* has not yet had its "chestnut blight" or "Dutch elm disease."

Foliage diseases

Poplar leaf rust caused by Melampsora *species*

Undoubtedly, more study has been devoted to leaf rust of *Populus* than to any other poplar disease. Poplar leaf rust has been researched primarily in relatively young, plantations. Leaf rust can be very damaging to poplar (Widin and Schipper 1981; Wang and van der Kamp 1992), and can even be linked to the mortality of young trees (Newcombe et al. 1994). The *Melampsora* species causing poplar leaf rust are generally thought of as heteroecious, alternating between the *Populus* host on which uredinial and telial states are formed in summer and fall, and the aecial host in spring. Aecial hosts are primarily conifers. Damage to some aecial hosts may also be considered important, as in pine twist rust caused by *M. pinitorqua* Rostrup.

There has been much speculation regarding the mode of overwintering of *Melampsora* species affecting *Populus*. *Melampsora* forms telia of contiguous teliospores on leaves of their poplar host in the fall. Telia overwinter and undergo vernal meiosis resulting in haploid basidiospores that infect the alternate host. It is on the alternate host, frequently a conifer, where mating occurs to restore the dikaryotic condition of the uredinial-telial state. Some species apparently perennate exclusively in the telial state and therefore must infect a suitable aecial host in the subsequent spring (e.g., *M. occidentalis* H. Jacks. on the aecial host *Pseudotsuga menziesii* (Mirb.) Franco that commonly co-occurs with the telial host *P. trichocarpa* Torr. & Gray in the Pacific Northwest). However, another little-appreciated perennation mode, uredinial-state overwintering, is known to occur in coastal Italy where uredinia of *M. pinitorqua* appear on young shoots of aspen before signs of infection appear on local *Pinus* L. which act as the aecial host (Longo et al. 1975). Uredinial-state overwintering allows successful genets to become perennial and also gives them a selective advantage (i.e., earlier infection of the telial host) each spring over the sexual recombinants occurring on *Pinus*. Uredinial-state overwintering is difficult to prove but is suspected to occur to varying extents in various *Melampsora* species on both *Populus* and *Salix* (Gremmen 1980; Royle and Ostry 1995).

A number of fungal pathogens of *Populus* have been introduced with poplar culture into regions and continents where they are exotic. The most striking examples are of *Melampsora larici-populina* Kleb. and *M. medusae* Thuem. f.sp. *deltoidae* Shain. Originally Eurasian and North American, respectively, these two species now occur in almost all parts of the world where hybrid poplar is grown (Newcombe and Chastagner 1993*a*, 1993*b*; Walker et al. 1974). Expansion within North America of the geographical range of *M. occidentalis* was noticed recently (Moltzan et al. 1993). Similar intracontinental expansions in the geographical ranges of *Melampsora* species with poplar culture probably occurred in Europe and Asia prior to the development of means of identifying leaf-rust fungi. Thus, original distributions of Eurasian *Melampsora* species may by now have been obscured.

Knowledge of the host specificity exhibited by *Melampsora* species toward *Populus* is dependent on the taxonomy of *Melampsora* itself. In spite of being best-studied among genera of fungal pathogens that affect *Populus*, *Melampsora* taxonomy is beset by controversy. There are differing opinions regarding: the separation of *M. medusae* into two *formae speciales* (Shain 1988); the existence of *formae speciales* (Viennot-Bourgin 1937) or races of *M. allii-populina* Kleb. (J. Pinon, personal communication); the reduction to synonymy of *M. medusae* and *M. albertensis* Arth. by Ziller (1965); the *M. populnea* (Pers.) P. Karst. "species complex" (Pinon 1973); and the relatively recent description of new species of *Melampsora* (Bagyanarayana and Ramachar 1984; Shang et al. 1986) that have not yet been integrated into the body of knowledge of longer-standing species.

The host specificity of *Melampsora* on *Populus* has been the subject of recent review (Pinon 1992) and, to avoid redundancy, apart from general points, only those more recent papers dealing with this subject will be discussed here. *Melampsora* taxa are known or suspected to be specific either to sections *Aigeiros* and *Tacamahaca* or to section *Populus*. Interestingly, of the 6 taxa specific to sections *Aigeiros* and *Tacamahaca*, there are 2 indigenous in North America (*M. medusae* f.sp. *deltoidae* and *M. occidentalis*), 2 in Europe (*M. larici-populina* and *M. allii-populina*), and 3 in Asia (*M. multa* Shang, Pei et Z.W. Yuan, *M. ciliata* Wall., and *M. larici-populina*). As already mentioned, the native distributions of *M. larici-populina* and *M. medusae* f.sp. *deltoidae* have been expanded by introductions with poplar culture. Additional *Melampsora* species specific to section *Populus* and indigenous to either North America or Eurasia have also been described: *M. albertensis*, *M. abietis-canadensis* C.A. Ludw. ex Arth., and *M. medusae* f.sp. *tremuloidae* Shain in North America; *M. pulcherrima* Maire in Europe, and the *M. populnea* species complex comprised of *M. pinitorqua*, *M. rostrupii* Wagner, *M. magnusiana* Wagner, and *M. larici-tremulae* Kleb. in Europe and Asia. The first group of 6 taxa, although sharing host section specificity, can easily be distinguished by inoculating differential hosts from within the 2 sections (Pinon 1992). Further subspecific distinctions have been readily made at the level of races in *M. larici-populina* (Pinon et al. 1987; Pinon and Peulon 1989; Pinon et al. 1994; Pinon 1995). Pathogenic variation has been investigated in *M. medusae* f.sp. *deltoidae* (Prakash and Thielges 1987), *M. occidentalis* (Hsiang and van der Kamp 1985; Hsiang and Chastagner 1993), and *M. allii-populina* (J. Pinon, personal communication), without the description of races. *Melampsora larici-tremulae* and *M. pinitorqua* have been distinguished on a telial host set of *Populus* clones (Gremmen 1980).

Estimates of the broad-sense heritability of resistance to poplar leaf rust have generally been high, indicating strong genetic control (Thielges and Adams 1975; Newcombe et al. 1996). Even in studies where H^2 was not calculated, it was evident that most of the variation in rust severity was among clones, rather than within them (Eldredge et al. 1973). Inheritance of resistance to *M. medusae* f.sp. *deltoidae* has also been studied in intraspecific progenies of *P. deltoides* (Prakash and Heather 1986, 1989). Inheritance of resistant infection type, a "qualitative" trait, depended upon the isolate of *M. medusae* f.sp. *deltoidae* used for inoculations and varied from control by a single gene to that of 3 or 4 genes. Inheritance of latent period[2], infection frequency, and sporulation capacity was thought to be polygenic. No F_2 or backcross progenies were available for study, however, and interpretation of results was thus tentative.

With a three-generation, *Populus trichocarpa* × *P. deltoides* (T × D) mapping pedigree (Bradshaw et al. 1994), there has been an opportunity to analyze genetic control of resistance to leaf rust and other pathogens in a more conclusive manner than before. Analyses of rust severity in the field and in growth-room experiments indicated that a single gene, *Mmd1*, has played a major role in resistance to *M. medusae* f.sp. *deltoidae* in the Pacific Northwest since the introduction of the latter into the region in 1991 (Newcombe et al. 1996). A dominant allele at *Mmd1*, inherited from the *P. trichocarpa* parent, controls necrotic flecking that is strongly associated with resistance. The position of the *Mmd1* flecking locus on the linkage map is on group Q (see Bradshaw, Chapter 8). It is not yet known whether the *Mmd1* gene is race-specific or not. Associations between molecular markers and resistance to *M. larici-populina* are also being investigated in Europe (Villar et al. 1996).

Poplar leaf and shoot blight caused by Venturia *species*

Poplar leaf and shoot blight is a disease that occurs in the spring across Europe, Asia, and North America. It has been researched primarily in relatively young plantations. Symptoms usually appear first on young leaves. Black, necrotic lesions often spread down the midvein and the affected leaf lamina curls or cups. Conidia form on the leaf lesions, and petioles and young shoots also can become necrotic. Infected shoot tissue is black and reflexed so that the shoot tip

---

[2]The period of time between inoculation and first appearance of rust pustules or uredinia.

resembles a "shepherd's crook." Susceptible clones can lose all of their young foliage if spring weather is wet and thus conducive to infection. Cankers sometimes form.

Venturia species on *Populus* appear to be at least section-specific. Morelet (1985) recognized three varieties of *Venturia tremulae* Aderhold, that cause leaf and shoot blight of aspens. The three varieties also appear to be exclusive to either Europe or North America. *Venturia tremulae* var. *tremulae* occurs only in Europe on *P. tremula* and its hybrids, and sometimes on *P. alba* L. Morelet (1986) recognized two races of *V. tremulae* var. *tremulae* following inoculation of a set of host clones. *Venturia tremulae* var. *populi-albae* occurs in Europe on *P. alba* only, and thus represents the only known species-specific *Venturia* on *Populus*. *Venturia tremulae* var. *grandidentatae* occurs only in North America on *P. tremuloides*, *P. grandidentata* Michx., introduced *P. alba*, and intrasectional hybrids. Dance (1959, 1961*b*) also regarded *V. tremulae* to be section-specific. *Venturia macularis* (Fr.:Fr.) Müller et von Arx, prior to Morelet's work (1985, 1987), was regarded as the cause of leaf and shoot blight of aspens, and the only species of *Venturia* on section *Populus* (Sivanesan 1974). Morelet (1983) proved via inoculation studies that *V. macularis* and *V. viennotii* Morelet are not pathogenic although they may still be specific to section *Populus*. In North America, in addition to *V. tremulae* var. *grandidentatae*, *V. borealis* Funk also occurs on *P. tremuloides*, but only in the Yukon and central and northwest British Columbia (Funk 1989). *Venturia* on *Populus* section *Populus* in Asia appears to be unstudied.

Venturia populina (Vuill.) L. Fabricius, as it is currently understood, is genus-specific but not section-specific as it occurs on species of closely related sections *Tacamahaca* and *Aigeiros* and their intersectional hybrids (Dance 1961*a*). Dance (1961*a*) recognized however that there did appear to be a difference in host specificity between *V. populina* in Europe and North America. In Europe, where no species of *Tacamahaca* occur naturally (Pauley 1949), *V. populina* causes leaf and shoot blight primarily on species of *Aigeiros*, especially the native *P. nigra* L., and intrasectional hybrids (Giorcelli and Vietto 1992). Species of *Tacamahaca*, its intrasectional hybrids, and even intersectional hybrids with *Aigeiros*, tend to be resistant (Giorcelli and Vietto 1992). In contrast, in North America, *V. populina* affects species of *Tacamahaca* (Dance 1961*a*) and intersectional hybrids with *Aigeiros* in the Pacific Northwest (G. Newcombe, unpublished data). It is probable, but not yet proven, that the resistance of some clones of intersectional hybrids in the Pacific Northwest is contributed by the *Aigeiros* parent. Given that *V. populina* in the Pacific Northwest attacks some intersectional hybrid clones, this population of the taxon, as currently defined, is not strictly section-specific. The population of this taxon in India also occurs both on *Tacamahaca* (*P. ciliata* Wall.) and *Aigeiros* (*P. nigra*), although only the imperfect state, *Pollaccia elegans* Serv., was reported (Khan and Misra 1989).

In addition to *V. populina*, *V. mandshurica* Morelet has recently been described in China (Morelet 1993; Wu and Sutton 1995), causing a disease known as poplar "grey spot." It may be specific to sections *Aigeiros* and *Tacamahaca* since it has been reported mainly on their intersectional hybrid, *P. simonii* Carr. × *P. nigra*, and reports of it on section *Populus* are disputed (M. Morelet, personal communication).

Poplar leaf spot caused by Marssonina species

Several *Marssonina* species cause leaf-spot diseases of *Populus*. Petioles and young shoots may also be affected. *Marssonina* diseases have been researched primarily in relatively young plantations. Distinctive symptom expression was thought to characterize different *Marssonina* species, and thus have diagnostic value, but Spiers (1984) showed that leaf-spot symptoms varied with host. The only exception was the host-specific "dendritic" symptom caused by *M. populi* (Lib.) Magnus on *P. tremuloides* and *P. angustifolia* James, that consisted of "large, circular, dendritic, chestnut brown blotches and threads" forming a tree-like, branching pattern (Spiers 1984). In general, even if without diagnostic value, the most economically important *Marssonina* species, *M. brunnea* f.sp. *brunnea* Spiers, causes "small (1–2 mm) discrete, circular to angular, amphigenous punctiform black spots on most hosts" (Spiers 1984).

There has been controversy about host specificity, given the confusion concerning the taxonomy of *Marssonina* species. A recent comparative study of type and herbarium specimens of *Marssonina* recognized four species on the basis of conidial morphometric characters: *M. populi*, *M. castagnei* (Desmaz. and Mont.) Magn., *M. balsamiferae* Y. Hiratsuka, and *M. brunnea* (Spiers 1988). Although variation in conidial characters can be induced by environmental, host, and cultural factors, taxonomic differences among species were not obscured (Spiers 1990).

Marssonina brunnea has been divided into two *formae speciales*: (*i*) *M. brunnea* f.sp. *brunnea*, that is pathogenic to *P. deltoides* Bartr. Ex. Marsh. and its hybrids with *P. nigra* (i.e., *P. ×euramericana*) but not pathogenic to *P. tremula* or *P. tremuloides* in subsection *Trepidae* of section *Populus*, and (*ii*) *M. brunnea* f.sp. *trepidae* Spiers, that is pathogenic to *P. tremula* and/or *P. tremuloides* but not pathogenic to *P. deltoides* or its hybrids with *P. nigra* (Spiers 1984). *Marssonina populi* is specific to sections *Aigeiros* and *Tacamahaca*, and is not pathogenic to section *Populus*. *Marssonina castagnei*, long thought to be specific to *P. alba* only (Pirozynski 1974*b*), was shown by Spiers (1984) to infect and sporulate on representatives of *Aigeiros* and *Tacamahaca*; curiously, *M. castagnei* did not infect *Trepidae*, even though it is most aggressive on the other subsection, *Albidae*, of section *Populus*.

Gremmen (1965) described the sexual, *Drepanopeziza* states of 3 of the 4 currently recognized *Marssonina* species, although only one description, that of

D. punctiformis, was original. Hiratsuka (1984) described the 4th *Marssonina* species, *M. balsamiferae*, that occurs on *P. balsamifera* L. in Manitoba and northern Ontario, but did not observe its sexual state. The sexual, *Drepanopeziza* states of *M. castagnei,* and *M. populi* have been the subject of less study in North America than in Europe (Boyer 1961; Pirozynski 1974*a, b,* and *c*). Pirozynski (1974*c*) apparently ignored or discounted the report by Boyer (1961) of the common occurrence of *Drepanopeziza populorum* (Desmaz.) Höhn., the sexual state of *M. populi*, in Quebec. *Drepanopeziza punctiformis*, the sexual state of *M. brunnea*, has been the subject of only one research study in North America (Ostry 1987). It may be that in some regions of North America (e.g., the Pacific Northwest), *Drepanopeziza* states are rarely, or never, produced.

Apart from *M. balsamiferae* which is native to the Canadian prairies, the original, indigenous distributions of *Marssonina* species on *Populus* are not clear. Pinon (1984) has suggested that *M. brunnea* was introduced into Europe from North America, since disease-free, susceptible clones had long been grown in Europe before the epidemics there of the early 1960s. One would suppose that *M. castagnei* is native to Europe, given its relative specificity toward the European white poplar, *P. alba*. It is difficult even to speculate on the original distribution of *M. populi*.

Although there have been no studies of the genetic control of resistance to *Marssonina*, it is apparent that *Tacamahaca* species may, in general, be a source of resistance to leaf spot caused by any *Marssonina* species pathogenic to poplar (Spiers 1984). However, F_1 hybrids between *Aigeiros* and *Tacamahaca* are generally susceptible to *M. brunnea* f.sp. *brunnea*, even though Spiers (1984) showed that they are resistant to the other 4 *Marssonina* taxa. Advanced generations (F_2s, backcrosses, etc.) may have to be bred to capture resistance to *M. brunnea* f.sp. *brunnea*.

Any generalizations about host range of *Marssonina* species affecting *Populus* must be tempered by the fact that, with the exception of *M. brunnea* f.sp. *brunnea*, only a single isolate of each taxon was used in Spiers' inoculation studies (1984). Isolates of *M. brunnea* f.sp. *brunnea* collected in various countries varied in aggressiveness (Spiers 1983). As is generally the case for foliar pathogens of *Populus*, leaf age significantly affects resistance to *Marssonina*. Young, soft, half-to three-quarter-expanded leaves are more susceptible than slightly older, soft, fully-expanded leaves which, in turn, were more susceptible than old, mature, hard leaves (Cellerino et al. 1978; Spiers and Hopcroft 1984). Spiers and Hopcroft (1984) suggest that mature leaves be used in a leaf–disk assay for resistance.

Poplar leaf spot and stem canker caused by Septoria *species*

Septoria musiva and *S. populicola* Peck were thought of as leaf-spotting parasites of native North American poplars until 1939 when it was shown that *S. musiva* causes cankers as well as leaf spots on certain introduced hybrid poplars (Bier 1939). Since then, these pathogens have been researched mainly in young plantations. Bier linked *S. musiva* to a *Mycosphaerella* sexual state and described infection of stems of the host via mechanical wounds, uninjured lenticels, leaf petioles, or stipules. He supposed that leaf spotting alone would reduce annual growth and that advanced cankers might result from the combined attack of *S. musiva* and *Cytospora chrysosperma* (Pers.) Fries. Bier stated that in Canada, *S. musiva* was distributed from Quebec to Alberta on many *Populus* species and hybrids. He was first to report variation in resistance to stem canker. It has since become clear that canker caused by *S. musiva* is a major disease that limits the growth of hybrid poplars in much of eastern North America (Ostry et al. 1989).

Bier's work was followed closely by that of Thompson (1941). The teleomorphic state of *S. musiva* was described and named *Mycosphaerella populorum* G.E. Thompson and that of *S. populicola, M. populicola* G.E. Thompson. Apart from morphometric differences (i.e., sizes of fruiting bodies and spores), a significant difference in host range was reported from inoculation studies: *S. musiva* caused disease on 26 taxa of *Populus* in sections *Aigeiros, Tacamahaca,* and *Populus,* whereas *S. populicola* caused disease on only 3 taxa (including *P. trichocarpa*) in section *Tacamahaca.* Thompson also mentioned that his inoculation studies revealed that pycnidia and conidia varied in size depending on the host species.

Septoria populi Desmaz. causes a poplar leaf spot disease in Europe. It is apparently of periodic importance (J. Pinon, personal communication), not unlike *S. populicola* in North America.

A number of studies have indicated that *S. musiva* may take years to establish itself in a new plantation. Hybrid poplar plantations in southern Ontario were first free of, and then severely affected by, canker (Spielman et al. 1986). In a hybrid poplar planting near Rhinelander, Wisconsin, there also was a lag of a few years before stem canker became prevalent (Ostry and McNabb 1985). Initial inoculum for new hybrid poplar plantings is thought to come from native local poplars (i.e., *P. balsamifera* and *P. deltoides,* for example) on which *S. musiva* is endemic (Spielman et al. 1986; Ostry and McNabb 1986).

Populus species and hybrids vary in their resistance to leaf spot and canker caused by *S. musiva.* Native local poplars have generally been thought of as canker-resistant, though susceptible to leaf spots. Somaclonal variants[3] derived from susceptible hybrid poplar clones have, in some cases, proven to be resistant regenerants (Ostry and Skilling 1988). Hybrids with parental species from the section *Tacamahaca* have generally been highly susceptible to leaf spot and canker (Ostry 1987), although some hybrids with *P. maximowiczii* Henry parentage

---

[3]Somaclonal variation is variability found among plants regenerated from tissue culture of a single genotype.

may be exceptions. *P. trichocarpa* hybrids have not done well in trials in the midwestern U.S., owing to their susceptibility to Septoria canker, although the same clones have never been affected by stem canker in the Pacific Northwest. *Septoria populicola* is prevalent in the Pacific Northwest but *S. musiva* and the cankers it causes have not been detected, even in plantings of susceptible clones established 15 years ago (Newcombe et al. 1995).

Pathogenic variation in *S. musiva* and *S. populicola* is even less well documented than variation in host resistance. Isolates from southern Ontario, the American midwest, and Mississippi proved to be similar in aggressiveness when inoculated on 3 susceptible hybrid poplar clones (Spielman et al. 1986). The authors admitted, however, that a different inoculation procedure might have allowed detection of differences in aggressiveness. On the other hand, Krupinsky (1989) did find differences in aggressiveness among local and regional isolates of *S. musiva*. More recent work has also shown that pathogenic variation exists in *S. musiva* and *S. populicola* (Royle and Ostry 1995) using laboratory screening procedures (Ostry et al. 1988).

Evidence for 3 quantitative trait loci (QTL) of major effect in resistance to *S. populicola* has recently been obtained (Newcombe and Bradshaw 1996). In each of 2 years, 2 complementary QTLs accounted for roughly 70% of the genetic variance in resistance to leaf spot. The 3 QTLs are on linkage groups X, A, and M (see Bradshaw, Chapter 8). The resistance alleles are contributed by the *P. deltoides* parent, and are dominant to the *P. trichocarpa* alleles.

By contrast, recessive resistance or dominant susceptibility might explain the prevalence of susceptibility among F_1 hybrids to *S. musiva*. Although the classical quadratic check for a gene-for-gene interaction is based on dominant alleles for both resistance and avirulence, recessive resistance genes are not uncommon (Kesseli et al. 1993). The best known example is probably that of resistance of oats to *Bipolaris victoriae* (F. Meehan & Murphy) Shoemaker, the causal agent of victoria blight, where both susceptibility to *C. victoriae* and sensitivity to the toxin, victorin, are controlled by the dominant allele at the Vb locus.

Stem diseases

Aspen canker caused by Hypoxylon mammatum

Canker fungi are often a disease problem only in certain regions, even if their species distribution is much wider. Hypoxylon canker of aspen is a good example. Research on this disease has been conducted in variously aged, natural or managed stands of the host. *Hypoxylon mammatum* (Whal.) Miller (= *H. pruinatum* (Klotz.) Cke.), has been described as the most destructive disease of *P. tremuloides* in southern Michigan (Graham and Harrison 1954) and generally in the Great Lakes region, northeastern North America, and the northwestern prairies (Hinds 1985; Manion and Griffin 1986), even though its distribution extends over the entire continent (Manion and Griffin 1986; Farr et al. 1989). In an extensive survey of aspen cankers in Colorado, Juzwik et al. (1978) were only able to find Hypoxylon canker in 1 of 9 national forests. Regional importance may have to do with population differentiation in fungal virulence or aggressiveness, or in host resistance. It might also be a function of climate, or of interactions with other organisms. Insects have been shown in some studies in the Great Lakes region to be almost entirely responsible for the initial injury allowing infection of *H. mammatum* to take place (Graham and Harrison 1954; Anderson et al. 1979; Ostry and Anderson 1983). Downy woodpeckers (*Picoides pubescens* L.) feeding on insects also play a role in inoculating trees (Ostry et al. 1982). Many unsuccessful attempts to infect trees with ascospores of *H. mammatum* were followed by successful inoculation of galls via oviposition wounds made by *Saperda inornata* (Ostry and Anderson 1995). Other studies indicated a minor role for insects and suggested that infection takes place at the base of first- and second-season-old dead branches (Manion 1975). The absence of aspen canker in parts of North America might have to do with the absence of the organisms facilitating infection in the Great Lakes region. Alternatively, it might be a function of competitive canker fungi or other bark inhabitants (Bier 1964).

After observing that drought stress increases the susceptibility of aspen to Hypoxylon canker, researchers have tried to determine the mechanism of such a change in the host–pathogen relationship. Canker fungi, including *H. mammatum*, commonly occur as endophytes in unstressed aspen (Bier 1964; Chapela 1989) but opportunistically exploit drought-stressed trees to cause cankers. The opportunity may derive from the effect of drought stress in reducing concentrations of antifungal compounds such as salicin, salicortin, and catechol that normally hold the endophyte in check (Kruger and Manion 1994). The low incidence of Hypoxylon canker on aspen in the Rocky Mountains, Alaska, and the Pacific coast might be explained by genetically controlled response to moisture stress, distinguishing these populations from susceptible ones in the Great Lakes region.

Hypoxylon mammatum, as currently defined, has a broad host range encompassing many hardwood genera (Manion and Griffin 1986). An inoculation study employing isolates from *P. tremuloides* demonstrated however a much narrower host range restricted to *P. tremuloides* and its hybrids, and *P. tremula* (Berbee and Rogers 1964). In a field study of *H. mammatum* on *P. tremuloides* in Michigan, moderate genetic control of resistance was inferred from much greater variation in incidence of Hypoxylon canker among clones than within clones (Copony and Barnes 1974). In Europe, *H. mammatum* is known to occur on *P. tremula*, and other species and intrasectional hybrids in section *Populus* (Pinon 1979). However, it also causes cankers on *P. trichocarpa* in Europe (Terrasson et al. 1988), and various hybrid poplar clones in North America (Ostry and McNabb 1986). *Hypoxylon mammatum* produces toxins, and assays have been based on the toxin (Schipper 1978). On the other hand, toxin assays have been shown to be poorly correlated with various measures of disease in the field (Bélanger et al. 1989). A genetic analysis of sensitivity of *P. tremuloides* to culture filtrates containing toxins of *H. mammatum* suggested control by a small number of genes (Kruger and Manion 1993).

Decay of aspen caused by Phellinus *tremulae*

The most important decay of aspen is caused by the white-rotting fungus *Phellinus tremulae* (Bond.) Bond. & Borisov. (Thomas et al. 1960). Research has focused on the incidence of decay in older, natural stands. *Phellinus tremulae* in Europe is regarded as species-specific, occurring only on *Populus tremula* (Niemelä 1974). In North America, its host specificity is uncertain, in the absence of modern systematic investigations. However, in both Europe and North America only aspens seem to have serious problems with decay.

Artificial inoculations of mycelial cultures of the fungus into sapwood and heartwood of living aspens have resulted in typical decay from which *Ph. tremulae* could be reisolated (Riley 1952). Basidiospores have also been successfully used to inoculate sapwood wounds of aspen (Manion and French 1968). Highly significant differences in amount of decay and position of decay columns among clones of *P. tremuloides* have been reported (Wall 1971). Recent studies of *Ph. tremulae* in Europe have focused on the fine-scale genetic variability of the fungus in individual trees (Holmer et al. 1994).

Much recent work has been done in Alberta, Canada, on black stem galls on aspen and their relationship to decay caused by *Ph. tremulae*. Trees with galls tend to be free of conks of *Ph. tremulae* more often than trees without galls (Crane et al. 1994). Conks of *Ph. tremulae* are a visible external indicator of decay. Given the relationship between galls and conks, the presence of galls should help attempts to estimate decay and merchantable volume (Crane et al. 1994). To understand why decay would be reduced in galled trees, the microflora of galls has been described, paying particular attention to organisms that might cause the galls or that might demonstrate antifungal activity toward *Ph. tremulae*. Of approximately 100 wood-inhabiting fungi isolated, *Peniophora polygonia* (Pers.:Fr.) Bourdot & Galzin and *Phoma etheridgei* Hutchison et Y. Hiratsuka were the only species that were antagonistic to *Ph. tremulae* (Hutchison et al. 1994). Work is in progress to characterize the inhibitory compounds produced by these fungi and to determine whether they cause galls correlated with reduced decay.

The basis of the relative resistance of species of *Populus* in sections *Aigeiros* and *Tacamahaca* to decay is unknown. Although 70 species of fungi were associated with decay of *P. trichocarpa* near Quesnel, British Columbia, only 6 species caused significant loss in living trees and *Polyporus delectans* Peck and *Pholiota populnea* (Pers.:Fr.) Kuyper & Tjall.-Beukers caused 92% of this loss (Thomas and Podmore 1953). Decay volume was negligible in trees less than 80 years old. In older trees there was a high incidence of decay but the average volume of decay per infected tree was low. Decay resistance in *P. trichocarpa* has been attributed to its near-anaerobic wetwood (van der Kamp et al. 1979).

Poplar canker and dieback caused by Discosporium populeum

What had been known as Dothichiza canker is a more serious disease in Europe than in North America (Waterman 1957). It is not clear whether the difference has to do with differential aggressiveness of *Discosporium populeum* or with environmental differences. In Europe, trees are more susceptible to this canker if they have first been attacked by leaf rust. It is not even entirely clear to taxonomists that the fungus is the same on the two continents (B.E. Callan, personal communication). The host range is reportedly in *Aigeiros* and *Tacamahaca*.

Poplar canker caused by Cytospora chrysosperma (Pers.:Fr.) Fr.

Even today, debate lingers as to whether this fungus is a pathogen capable of causing cankers or a secondary colonizer of cankers instigated by another agent. Schreiner (1931) thought *Valsa sordida* Nitschke (anamorph: *Cytospora chrysosperma*) to be a fairly "vigorous parasite." His inoculations resulted commonly in disease if inoculum was placed on wounds and if the trees were then grown in poor soil conditions. Christensen (1940) disputed the pathogenic ability of *C. chrysosperma*. Finding that *C. chrysosperma* is a common endophyte in many apparently healthy *Populus* spp., he dismissed the notion that *C. chrysosperma* is pathogenic even when others had shown that inoculation of stressed trees can result in cankers on which pycnidia of *C. chrysosperma* develop. The common occurrence of the fungus as an endophyte in *Populus* has been confirmed in more recent studies; *C. chrysosperma* is one of a guild of fungi persisting in a latent, or at least symptomless, form in healthy aspen wood (Chapela 1989). Christensen (1940) saw no point in trying to select clones resistant to "Cytospora canker." In general, it cannot currently be said that even genus-specificity has been proven for *C. chrysosperma* since woody genera outside the *Salicaceae* are listed as hosts (Farr et al. 1989).

Host–pathogen interactions

Assays to determine host–pathogen relationships

Precise knowledge of host specificity is typically acquired through use of an inoculation assay. Detached leaves and leaf disks have been used extensively by poplar pathologists, often with results that reflect those in the field. In particular, Shain and Cornelius (1979) established the utility of the leaf–disk assay in studies of poplar leaf rust. However, in some cases, the leaf–disk assay may be inappropriate. For example, the *Mmd1* gene for resistance to *Melampsora medusae* f.sp. *deltoidae* is expressed by necrotic flecking in response to inoculation of attached leaves of growing plants, but not consistently manifested when leaf disks are inoculated (Newcombe et al. 1996).

Similarly, appropriate methods of evaluation of disease in the field require further study. This has rarely been done with poplar pathogens, but a comparison of methods for evaluating foliar infection by *Marssonina brunnea* f.sp. *brunnea* furnishes an example of what should more often be done (Joannes and Pinon 1982). In general, there should be more work to determine the best inoculation assays and methods of field evaluation of disease with the most important poplar pathogens.

Breeding and selection for disease resistance

There has long been interest in knowing whether disease resistance might be linked with sex in the dioecious poplars (Pauley 1949). However, since Pauley's statement that "disease resistance and vigor appear to be associated with the male sex," there has been neither confirmation nor refutation of his generalization. But, it seems unlikely, given the multiplicity of pathogens affecting any given *Populus* species or hybrid, that all disease resistance genes would be linked to those for sex determination. Some clustering of disease resistance genes is to be expected, given results in better-studied plants (e.g., lettuce — Kesseli et al. 1993) but, to date, the *Mmd1* locus for resistance to rust caused by *Melampsora medusae* f.sp. *deltoidae* and the 3 QTLs for resistance to *Septoria populicola* are unlinked (Newcombe and Bradshaw 1996).

Hybrid poplar clones and registered cultivars in Europe were, in many cases, selected for resistance to *Melampsora* and *Marssonina* species, or bacterial canker. However, such resistance has frequently been defeated by race or pathotype microevolution (e.g., *Melampsora larici-populina* in western Europe) or by other pathogens (e.g., *Marssonina*-resistant clones have been *Venturia*-susceptible in Italy).

By analyzing the genetic control of resistance to pathogens and insect pests in the Pacific Northwest, and coordinating the information by placing the genes and QTLs for resistance on the *Populus* linkage map, it is hoped that the best-case strategy for breeding for multiple disease resistance for the region can be devised. Similar efforts may be undertaken by poplar breeders and pathologists elsewhere. The problem to be overcome is, in part, the lack of linkage among relevant genes for disease resistance to many pathogens. Furthermore, durable genes for resistance, that would not be defeated by pathotype microevolution, are not currently selectable.

Some poplar breeder/pathologist teams may hope that by selecting for quantitative resistance traits, durable resistance under polygenic control will be obtained. However, in a recent analysis involving the same isolate of *Melampsora medusae* f.sp. *deltoidae* for which the *Mmd1* major gene has been defined, such "quantitative resistance" traits as latent period and infection efficiency were found to be affected by *Mmd1* as well as by other unlinked QTLs (G. Newcombe and H.D. Bradshaw, Jr. unpublished).

Adapted versus exapted resistance to pathogens

In many cases it is impossible to reconstruct the relationship between host and pathogen to determine the nature of resistance. Original distributions may not be known and host materials supporting appropriate genetic inference may not be available. But in some cases, host resistance has been determined to be adapted following coevolution of host and pathogen, or exapted, in the sense of Gould and Vrba (1982). The latter implies that resistance is expressed in a novel confrontation between host and pathogen. Such confrontations result from the introduction, frequently in hybrids, of exotic host germplasm, or from the inadvertent introduction of exotic pathogens. This presents a research opportunity in *Populus* quite unlike that in other crops domesticated long ago, where the "widespread movement of crops and pathogens can produce a complicated pattern of escapes, encounters and re-encounters" (Lenné and Wood 1991).

Where the relationship between pathogen and *Populus* can be reconstructed, major-gene resistance appears to be, more often than not, exapted. *Melampsora medusae* f.sp. *deltoidae* is relatively nonaggressive compared to *M. occidentalis* on *P. trichocarpa* (G. Newcombe, unpublished observations) and, by definition, is virtually nonpathogenic on *P. tremuloides* (Shain 1988; Newcombe and Chastagner 1993*a*). These two *Populus* species are the only native poplars of the maritime Pacific Northwest west of the Cascade Mountains. It is thus not surprising that there are no records of the occurrence of *M. medusae* f.sp. *deltoidae* in this region prior to 1991 (Newcombe and Chastagner 1993*a*). Genetic analysis of a T × D hybrid pedigree, as discussed above, revealed that the major gene for resistance, *Mmd1*, was inherited from the *P. trichocarpa* parent (Newcombe et al. 1996). It could not be adaptive in the latter as *M. medusae* f.sp. *deltoidae* had not been present in the region. Analysis of the same pedigree has revealed that resistance to an indigenous pathogen, *Septoria populicola*, is inherited in the form of dominant QTLs from the exotic parent, *P. deltoides*. Poplar anthracnose caused by *Glomerella cingulata* (Stoneman) Spauld. & H. Schrenk, poplar leaf bronzing caused by an undescribed eriophyid mite, poplar leaf and shoot blight caused by *Venturia populina*, and leaf spot of *P. trichocarpa* × *P. maximowiczii* hybrids caused by *S. populicola*, are other examples from the Pacific Northwest that conform to this exapted pattern. In Europe, resistance of poplar hybrids to *Melampsora larici-populina* is typically inherited from the exotic parent and not from the coevolved, *P. nigra* parent (Pinon 1992). Similarly, it appears that rust resistance is exapted in *Salix*; clones of North American origin are resistant in the U.K. and those of European origin are resistant in North American trials (Royle and Ostry 1995). A few examples of adaptive resistance are also apparent. The resistance of *P. deltoides* and susceptibility of *P. tricho-carpa* to canker caused by *Septoria musiva*, is the most notable example.

Differentiation of populations within host and pathogen

Populus species are widely distributed and significant regional and local population differentiation is common. Many studies have indicated that populations of a given *Populus* species react differently to a particular pathogen population (Prakash and Thielges 1987; Hsiang et al. 1993; Pinon 1992) and that relatively aggressive pathogen populations may be found on relatively susceptible host populations (Hsiang and Chastagner 1993). In Europe, studies of rust populations on wild *P. nigra* are just beginning (J. Pinon, personal communication). A pattern of susceptibility to local pathogens to which other, more distant host populations are more resistant would result in exapted resistance conferring disease resistance to the intraspecific, interpopulation progeny.

Other studies of natural plant pathosystems have found evidence of local population differentiation for susceptibility to a fungal pathogen (Parker 1985). Correlations between disease susceptibility and other fitness traits in a local host population may result in nonadaptive changes in disease resistance, as the frequency of resistant individuals declines in response to natural selection on another trait (Parker 1991).

With respect to poplar leaf spot, Zalasky (1978) tested inoculum of both *Septoria musiva* and *S. populicola* from one source area in Manitoba against seedlings of *P. balsamifera* from 6 source areas. Seedlings from the inoculum source area were more susceptible to both species of *Septoria* than seedlings from 4 other source areas; 2 localities within the 6th source area however were the most susceptible.

There may be local differentiation in the disease resistance of natural hybrids of the host. The natural hybrid of *P. tremuloides* and *P. grandidentata*, *P. ×smithii* Boivin, has been observed to be more susceptible to dieback, leaf rust, and stem canker than its parents in southeastern Michigan (Henry and Barnes 1977).

Parallels between Populus and its fungal pathogens

Coevolution, in the strict sense, of host and pathogen would result in parallel phylogenies. No studies of comparative phylogeny of *Populus* and a pathogen have been reported, however. *Melampsora* would be the logical first choice for such a study. For the time being, we must make do with the reflection of host taxonomy found in the host specificity of pathogens, the major theme of this review. Other reflections of the host might be expected in terms of coinciding distributions, or similar tendencies toward natural hybridization.

There are no species of *Populus* that have circumpolar native distributions spanning North America and Eurasia but many pathogens are reputed to have such (e.g., *Venturia populina*, *Hypoxylon mammatum*, *Phellinus tremulae*, *Discosporium populeum*, *Ceratocystis fimbriata*, etc.). Each fungal pathogen with a reported circumpolar distribution should be considered a candidate for critical comparisons of Eurasian and North American specimens.

Hybridization of fungal pathogens of *Populus* is virtually unknown, in spite of the proclivities of the host. Only 2 examples are apparent in the literature. First, the 2 European varieties of *Venturia tremulae* Aderh., *V. tremulae* var. *populi-albae* and *V. tremulae* var. *tremulae*, were easily hybridized in culture by Morelet (1987). Secondly, Spiers and Hopcroft (1994) reported a naturally occurring hybrid of *M. medusae* f.sp. *deltoidae* and *M. larici-populina* in New Zealand. Hybridization might explain some of the morphological variation encountered in pathogens of Populus.

Concluding remarks

Contrary to the expectation arising from spectacular examples of introduced diseases such as chestnut blight, white pine blister rust, and Dutch elm disease, exotic pathogens have not seriously affected native *Populus* species. *Melampsora medusae* f.sp. *deltoidae*, after having been inadvertently introduced into Europe, has curiously not spread to many susceptible clones occurring outside its relatively limited European range (Pinon 1986). *Melampsora larici-populina*, a Eurasian rust, after having been first found in North America in 1991 (Newcombe and Chastagner 1993*b*), has also not spread to susceptible *P. trichocarpa* and hybrid clones, but remained confined to a small region in the Pacific Northwest, although it is apparently more widespread in California. In contrast, when these same rust fungi have been introduced to poplar-growing areas of the southern hemisphere where *Populus* species do not occur naturally, epidemics have resulted.

Although much of the information on the host specificity of fungal pathogens of *Populus* is inconclusive in any general sense, and some of it is possibly incorrect, it does appear that *Populus* species are susceptible to a large number of fungal pathogens, particularly local or native ones (Callan and Ring 1994). Better information on host specificity will only be obtained as the taxonomy of pathogens becomes better understood and inoculation studies with defined pathogen isolates are performed.

Each poplar-growing region of the world will continue to face the common problem of breeding for disease resistance even though its particular pathogen spectrum may well be region-specific. Effective breeding strategies for resistance will only replace selection programs once genetic analyses of resistance have been performed. Better understanding will best be achieved if pathologists in each region analyze the genetic control of resistance of common host materials to their region-specific pathogen spectrum. In this regard, the University of Washington-bred, 3-generation hybrid pedigree is becoming prominent.

Before tinkering too much with disease resistance, we would do well to strive to understand the basis of the natural resistance of *Populus* species to devastating epidemics caused by introduced pathogens. Could it be that susceptibility to many different pathogens competing for the same niche (e.g., leaves, or stems, or roots), protects native *Populus* species from devastating epidemics? Bier (1963) showed that disease control could be achieved by microorganisms residing on healthy bark and foliage. We would do well to seek out and adapt for our needs not only genetically based host resistance but other forms of biologically based disease control that contribute to the prominence of *Populus* in the natural world.

Acknowledgements

I would like to express my sincere gratitude to Jean Pinon, Mike Ostry, Brenda Callan, and Reini Stettler for their thoughtful reviews of this chapter. It has been greatly improved by their efforts.

References

Anderson, N.A., Ostry, M.E., and Anderson, G.W. 1979. Insect wounds as infection sites for *Hypoxylon mammatum* on trembling aspen. Phytopathology, **69**: 476–479.

Bagyanarayana, G., and Ramachar, P. 1984. New rusts on the genus *Populus* — II. Curr. Sci. **53**: 863–865.

Barnes, B.V. 1967. Indications of possible mid-Cenozoic hybridization in the aspens of the Columbia Plateau. Rhodora, **69**: 70–81.

Bélanger, R.R., Falk, S.P., Manion, P.D., and Griffin, D.H. 1989. Tissue culture and leaf spot bioassays as variables in regression models explaining *Hypoxylon mammatum* incidence on *Populus tremuloides* clones in the field. Phytopathology, **79**: 318–321.

Berbee, J.G., and Rogers, J.D. 1964. Life cycle and host range of *Hypoxylon pruinatum* and its pathogenesis on poplars. Phytopathology, **54**: 257–261.

Bier, J.E. 1939. Septoria canker of introduced and native hybrid poplars. Can. J. Res. (C) **17**: 195–204.

Bier, J.E. 1963. Tissue saprophytes and the possibility of biological control of some tree diseases. For. Chron. **39**: 82–84.

Bier, J.E. 1964. The relation of some bark factors to canker susceptibility. Phytopathology, **54**: 272–275.

Boyer, M.G. 1961. Variability and hyphal anastomoses in host-specific forms of *Marssonina populi* (Lib.) Magn. Can. J. Bot. **39**: 1409–1427.

Bradshaw, H.D., Villar, M., Watson, B.D., Otto, K.G., and Stewart, S. 1994. Molecular genetics of growth and development in *Populus*. III. A genetic linkage map of a hybrid poplar composed of RFLP, STS, and RAPD markers. Theor. Appl. Genet. **89**: 551–558.

Callan, B.E. 1995. Fungal diseases and host-fungus associations of *Populus* in British Columbia. International Poplar Symposium, University of Washington, Seattle, WA, Aug. 20–25, 1995. Program abstract. pp. 71.

Callan, B.E., and Ring, F.M. 1994. An annotated host fungus index for *Populus* In British Columbia. Joint Publication of the Canadian Forest Service and the British Columbia Ministry of Forests. 50 p.

Cellerino, G.P., Anselmi, N., and Pinon, J. 1978. Influence de l'âge des feuilles de peuplier sur la sensibilité à *Marssonina brunnea*. Eur. J. For. Pathol. **8**: 273–279.

Chapela, I.H. 1989. Fungi in healthy stems and branches of American beech and aspen: a comparative study. New Phytol. **113**: 65–75.

Christensen, C.M. 1940. Studies on the biology of *Valsa sordida* and *Cytospora chrysosperma*. Phytopathology, **30**: 459–475.

Copony, J.A., and Barnes, B.V. 1974. Clonal variation in the incidence of Hypoxylon canker on trembling aspen. Can. J. Bot. **52**: 1475–1481.

Crane, P.E., Blenis, P.V., and Hiratsuka, Y. 1994. Black stem galls on aspen and their relationship to decay by *Phellinus tremulae*. Can. J. For. Res. **24**: 2240–2243.

Dance, B.W. 1959. A cultural connection between *Venturia tremulae* Aderh. and its imperfect stage in Ontario. Can. J. Bot. **37**: 1139–1140.

Dance, B.W. 1961a. Leaf and shoot blight of poplars (section *Tacamahaca* Spach) caused by *Venturia populina* (Vuill.) Fabric. Can. J. Bot. **39**: 875–890.

Dance, B.W. 1961b. Spore dispersal in *Pollaccia radiosa* (Lib.) Bald. et Cif. Can. J. Bot. **39**: 1429–1435.

Eckenwalder, J.E. 1977. North American cottonwoods (*Populus*, Salicaceae) of sections Abaso and Aigeiros. J. Arnold Arbor. Harv. Univ. **58**: 193–208.

Eckenwalder, J.E. 1984. Natural intersectional hybridization between North American species of *Populus* (Salicaceae) in sections *Aigeiros* and *Tacamahaca*. II. Taxonomy. Can. J. Bot. **62**: 325–335.

Eldredge, K.G., Matheson, A.C., and Stahl, W. 1973. Genetic variation in resistance to poplar leaf rust. Aust. For. Res. **6**: 53–59.

Farr, D.F., Bills, G.F., Chamuris, G.P., and Rossman, A.Y. 1989. Fungi on plants and plant products in the United States. APS Press, St. Paul, MN. 1252 p.

Funk, A. 1989. Observations on an aspen leaf spot disease and associated fungus, *Pollaccia borealis*. Can. J. Plant Pathol. **11**: 153–356.

Giorcelli, A., and L. Vietto. 1992. Poplar disease situation in Italy (1990–1991). Report (pp. 282–287) to the 19th Session of the International Poplar Commission Meeting in Zaragoza, Spain. 737 p.

Gould, S.J., and Vrba, E.S. 1982. Exaptation — a missing term in the science of form. Paleobiology, **8**: 4–15.

Graham, S.A., and Harrison, R.P. 1954. Insect attacks and Hypoxylon infections in aspen. J. For. **52**: 741–743.

Gremmen, J. 1965. Three poplar-inhabiting *Drepanopeziza* species and their life-history. Nova Hedwigia, **9**: 170–176.

Gremmen, J. 1980. Problems and prospects in breeding *Melampsora* rust-resistant poplars. Folia For. (Helsinki), **422**: 5–9.

Henry, R.M., and Barnes, B.V. 1977. Comparative reproductive ability of bigtooth and trembling aspen and their hybrid. Can. J. Bot. **55**: 3093–3098.

Hinds, T.E. 1985. Diseases. *In* Aspen: ecology and management in the western United States. *Edited by* N.V. DeByle and R.P. Winokur. USDA For. Serv. Gen. Tech. Rep. RM-119: 87–106.

Hiratsuka, Y. 1984. New leaf spot fungus, *Marssonina balsamiferae*, on *Populus balsamifera* in Manitoba and Ontario. Mycotaxon, **19**: 133–136.

Holliday, P. 1989. A dictionary of plant pathology. Cambridge University Press, Cambridge, England.

Holmer, L., Nitare, L., and Stenlid, J. 1994. Population structure and decay pattern of *Phellinus tremulae* in *Populus tremula* as determined by somatic incompatibility. Can. J. Bot. **72**: 1391–1396.

Hsiang, T., and Chastagner, G.A. 1993. Variation in *Melampsora occidentalis* rust on poplars in the Pacific Northwest. Can. J. Plant Pathol. **15**: 175–181.

Hsiang, T., and van der Kamp, B.J. 1985. Variation in rust virulence and host resistance of *Melampsora occidentalis* on black cottonwood. Can. J. Plant Pathol. **7**: 247–252.

Hsiang, T., Chastagner, G.A., Dunlap, J.M., and Stettler, R.F. 1993. Genetic variation and productivity of *Populus trichocarpa* and its hybrids. VI. Field susceptibility of seedlings to *Melampsora occidentalis* leaf rust. Can. J. For. Res. **23**: 436–441.

Hutchison, L.J., Chakravarty, P., Kawchuk, L.M., and Hiratsuka, Y. 1994. *Phoma etheridgei* sp. nov. from black galls and cankers of trembling aspen (*Populus tremuloides*) and its potential role as a bioprotectant against the aspen decay pathogen *Phellinus tremulae*. Can. J. Bot. **72**: 1424–1431.

Joannes, H., and Pinon, J. 1982. Comparaison de deux méthodes d'estimation de l'infection de jeunes peupliers par le *Marssonina brunnea* (Ell. & Ev.) P. Magn. Eur. J. For. Pathol. **12**: 87–96.

Juzwik, J., Nishijima, W.T., and Hinds, T.E. 1978. Survey of aspen cankers in Colorado. Plant Dis. Rep. **62**: 906–910.

Kesseli, R., Witsenboer, H., Stanghellini, M., Vandermark, G., and Michelmore, R. 1993. Recessive resistance to *Plasmopara lactucae-radicis* maps by bulked segregant analysis to a cluster of dominant disease resistance genes in lettuce. Mol. Plant-Microbe Interact. **6**: 722–728.

Khan, S.N., and Misra, B.M. 1989. Pollaccia blight of poplars in India. Eur. J. For. Pathol. **19**: 379–381.

Kommedahl, T. 1995. Disease names and binomial nomenclature. Phytopath. News **29**: 160.

Kruger, B.M., and Manion, P.D. 1993. Genetic control of *Populus tremuloides* sensitivity to metabolites of *Hypoxylon mammatum*. Can. J. Bot. **71**: 1276–1279.

Kruger, B.M., and Manion, P.D. 1994. Antifungal compounds in aspen: effect of water stress. Can. J. Bot. **72**: 454–460.

Krupinsky, J.M. 1989. Variability in *Septoria musiva* in aggressiveness. Phytopathology, **79**: 413–416.

Lenné, J.M. and Wood, D. 1991. Plant diseases and the use of wild germplasm. Annu. Rev. Phytopathol. **29**: 35–63.

Lindsey, J.P., and Gilbertson, R.L. 1978. Basidiomycetes that decay aspen in North America. Bibl. Mycol. **63**: 1–406.

Longo, N., Moriondo, F., and Naldini Longo, B. 1975. The status of *Melampsora pinitorqua* Rostr. in Italy. Eur. J. For. Pathol. **5**: 147–152.

Manion, P.D. 1975. Two infection sites of *Hypoxylon mammatum* in trembling aspen (*Populus tremuloides*). Can. J. Bot. **53**: 2621–2624.

Manion, P.D., and French, D.W. 1968. Inoculation of living aspen trees with basidiospores of *Fomes igniarius* var. *populinus*. Phytopathology, **58**: 1302–1304.

Manion, P.D., and Griffin, D.H. 1986. Sixty-five years of research on Hypoxylon canker of aspen. Plant Dis. **70**: 803–808.

Moltzan, B.D., Stack, R.W., and Mason, P.A. 1993. First report of *Melampsora occidentalis* on *Populus trichocarpa* in the central United States. Plant Dis. **77**: 953.

Morelet, M. 1983. Systématique et biologie des Venturia inféodés aux peupliers de la section Leuce. Thèse Doct. Fac. Sciences, Univ. Nancy I. 118 p.

Morelet, M. 1985. Les *Venturia* des peupliers de la section Leuce. I. Taxinomie. Cryptogamie, Mycol. **6**: 101–117.

Morelet, M. 1986. Les risques d'adaptation de *Venturia tremulae* aux trembles sélectionnés. Bull. OEPP (Organ. Eur. Mediterr. Prot. Plant.)/EPPO (Eur. Mediterr. Plant Prot. Organ.) Bull. **16**: 589–592.

Morelet, M. 1987. Les *Venturia* des peupliers de la section Leuce. II. Biologie culturale. Eur. J. For. Pathol. **17**: 85–93.

Nesme, X., and Ménard, M. 1995. Intérêt de l'étude de la diversité intraspécifique des bactéries dans la lutte contre les bactérioses du peuplier. C.R. Acad. Agric. Fr. **81**: 125–136.

243

Newcombe, G., and Bradshaw, H.D., Jr. 1996. Quantitative trait loci conferring resistance in hybrid poplar to leaf spot caused by *Septoria populicola*. Can. J. For. Res. **26**: 1943–1950.

Newcombe, G., and Chastagner, G.A. 1993*a*. A leaf rust epidemic of hybrid poplar along the Lower Columbia River caused by *Melampsora medusae*. Plant Dis. **77**: 528–531.

Newcombe, G., and Chastagner, G.A. 1993*b*. First report of the Eurasian poplar leaf rust fungus, *Melampsora larici-populina*, in North America. Plant Dis. **77**: 532–535.

Newcombe, G., Chastagner, G.A., Schuette, W., and Stanton, B.J. 1994. Mortality among hybrid poplar clones in a stool bed following leaf rust caused by *Melampsora medusae* f.sp. *deltoidae*. Can. J. For. Res. **24**: 1984–1987.

Newcombe, G., Chastagner, G.A, Callan, B.E., and Ostry, M.E. 1995. An epidemic of Septoria leaf spot on *Populus trichocarpa* in the Pacific Northwest in 1993. Plant Dis. **79**: 212.

Newcombe, G., Bradshaw, H.D., Jr., Chastagner, G.A, and Stettler, R.F. 1996. A major gene for resistance to *Melampsora medusae* f.sp. *deltoidae* in a hybrid poplar pedigree. Phytopathology, **86**: 87–94.

Niemelä, T. 1974. On fennoscandian polypores. III. *Phellinus tremulae* (Bond.) Bond. & Borisov. Ann. Bot. Fenn. **11**: 202–215.

Ostry, M.E. 1987. Biology of *Septoria musiva* and *Marssonina brunnea* in hybrid *Populus* plantations and control of Septoria canker in nurseries. Eur. J. For. Pathol. **17**: 158–165.

Ostry, M.E., and Anderson, N.A. 1983. Infection of trembling aspen by *Hypoxylon mammatum* through cicada oviposition wounds. Phytopathology, **73**: 1092–1096.

Ostry, M.E., and Anderson, N.A. 1995. Infection of *Populus tremuloides* by *Hypoxylon mammatum* ascospores through *Saperda inornata* galls. Can. J. For. Res. **25**: 813–816.

Ostry, M.E., and McNabb, H.S., Jr. 1985. Susceptibility of *Populus* species and hybrids to disease in the north central United States. Plant Dis. **69**: 755–757.

Ostry, M.E. and McNabb, H.S., Jr. 1986. *Populus* species and hybrid clones resistant to *Melampsora, Marssonina,* and *Septoria*. U.S. For. Serv. Res. Pap. NC-272. 7 p.

Ostry, M.E., and Skilling, D.D. 1988. Somatic variation in resistance of *Populus* to *Septoria musiva*. Plant Dis. **72**: 724–727.

Ostry, M.E., Daniels, K., and Anderson, N.A. 1982. Downy woodpeckers — a missing link in a forest disease cycle? Loon, **54**: 170–175.

Ostry, M.E., McRoberts, R.E., Ward, K.T., and Resendez, R. 1988. Screening hybrid poplars in vitro for resistance to leaf spot caused by *Septoria musiva*. Plant Dis. **72**: 497–499.

Ostry, M.E., Wilson, L.F., and McNabb, H.S., Jr. 1989. Impact and control of *Septoria musiva* on hybrid poplars. U.S. For. Serv. Gen. Tech. Rep. NC-133.

Parker, M.A. 1985. Local population differentiation for compatibility in an annual legume and its host-specific fungal pathogen. Evolution, **39**: 713–723.

Parker, M.A. 1991. Nonadaptive evolution of disease resistance in an annual legume. Evolution, **45**: 1209–1217.

Pauley, S.S. 1949. Forest-tree genetics research: *Populus* L. Econ. Bot. **3**: 299–330.

Pinon, J. 1973. Les rouilles du Peuplier en France. Eur. J. For. Pathol. **3**: 221–228.

Pinon, J. 1979. Origine et principaux caractères des souches françaises d'*Hypoxylon mammatum* (Wahl.) Miller. Eur. J. For. Pathol. **9**: 129–142.

Pinon, J.D. 1984. Management of diseases of poplars. Eur. J. For. Pathol. **14**: 415–425.

Pinon, J. 1986. Status of *Melampsora medusae* in Europe. Bull. OEPP (Organ. Eur. Mediterr. Prot. Plant.)/EPPO (Eur. Mediterr. Plant Prot. Organ.) Bull. **16**: 547–551.

Pinon, J. 1992. Variability in the genus *Populus* in sensitivity to Melampsora rusts. Silvae Genet. **41**: 25–34.

Pinon, J. 1995. Variabilité des rouilles du peuplier et évolution de leurs populations. Conséquences sur les stratégies de lutte. C.R. Acad. Agric. Fr. **81**: 99–109.

Pinon, J., and Peulon, V. 1989. Mise en évidence d'une troisième race physiologique de *Melampsora larici-populina* Klebahn en Europe. Cryptogam. Mycol. **10**: 95–106.

Pinon, J., and Valadon, A. 1996. Comportement des cultivars de peupliers commercialisables dans l'Union Européenne vis à vis de quelques parasites majeurs. Ann. Sci. For. In press.

Pinon, J., Van Dam, B.C., Genetet, I., and De Kam, M. 1987. Two pathogenic races of *Melampsora larici-populina* in north-western Europe. Eur. J. For. Pathol. **17**: 47–53.

Pinon, J., Newcombe, G., and Chastagner, G.A. 1994. Identification of races of *Melampsora larici-populina*, the Eurasian poplar leaf rust fungus, in California and Washington. Plant Dis. **78**: 101.

Pirozynski, K.A. 1974*a*. *Marssonina brunnea*. Fungi Canadenses No. 13. National Mycol. Herb. Biosystematic Res. Inst., Agric. Can., Ottawa, ON. 2 p.

Pirozynski, K.A. 1974*b*. *Marssonina castagnei*. Fungi Canadenses No. 14. National Mycol. Herb. Biosystematic Res. Inst., Agric. Can., Ottawa, ON. 2 p.

Pirozynski, K.A. 1974*c*. *Marssonina populi*. Fungi Canadenses No. 15. National Mycol. Herb. Biosystematic Res. Inst., Agric. Can., Ottawa, ON. 2 p.

Prakash, C.S., and Heather, W.A. 1986. Inheritance of resistance to races of *Melampsora medusae* in *Populus deltoides*. Silvae Genet. **35**: 74–77.

Prakash, C.S., and Heather, W.A. 1989. Inheritance of partial resistance to two races of leaf rust, *Melampsora medusae* in eastern cottonwood, *Populus deltoides*. Silvae Genet. **38**: 90–94.

Prakash, C.S., and Thielges, B.A. 1987. Pathogenic variation in *Melampsora medusae* leaf rust of poplars. Euphytica, **36**: 563–570.

Riley, C.G. 1952. Studies in forest pathology. IX. *Fomes igniarius* decay of poplar. Can. J. Bot. **30**: 710–734.

Royle, D.J., and Ostry, M.E. 1995. Disease and pest control in the bioenergy crops poplar and willow. Biomass Bioenergy, **9**: 69–79.

Schipper, A.L., Jr. 1978. A *Hypoxylon mammatum* pathotoxin responsible for canker formation in quaking aspen. Phytopathology, **68**: 866–872.

Schreiner, E.J. 1931. Two species of *Valsa* causing disease in *Populus*. Am. J. Bot. **18**: 1–29.

Shain, L. 1988. Evidence for *formae speciales* in the poplar leaf rust fungus, *Melampsora medusae*. Mycologia, **80**: 729–732.

Shain, L., and Cornelius, P.L. 1979. Quantitative inoculation of eastern cottonwood leaf tissue with *Melampsora medusae* under controlled conditions. Phytopathology, **69**: 301–304.

Shang, Y., Pei, M., and Yuan, Z. 1986. A new rust fungus on poplars. Acta Mycol. Sinica Suppl. I: 180–184.

Sivanesan, A. 1974. *Venturia macularis*. CMI Descriptions of Pathogenic Fungi and Bacteria No. 403. Commonwealth Mycological Institute, Surrey, England.

Spielman, L.J., Hubbes, M., and Lin, D. 1986. *Septoria musiva* on hybrid poplar in southern Ontario. Plant Dis. **70**: 968–971.

Spiers, A.G. 1983. Host range and pathogenicity studies of *Marssonina brunnea* to poplars. Eur. J. For. Pathol. **13**: 181–196.

Spiers, A.G. 1984. Comparative studies of host specificity and symptoms exhibited by poplars infected with *Marssonina brunnea*, *Marssonina castagnei* and *Marssonina populi*. Eur. J. For. Pathol. **14**: 202–218.

Spiers, A.G. 1988. Comparative studies of type and herbarium specimens of *Marssonina* species pathogenic to poplars. Eur. J. For. Pathol. **18**: 140–156.

Spiers, A.G. 1990. Influence of environmental, host and cultural factors on conidium morphology of *Marssonina* species pathogenic to poplars. Eur. J. For. Pathol. **20**: 154–166.

Spiers, A.G., and Hopcroft, D.H. 1984. Influence of leaf age, leaf surface and frequency of stomata on the susceptibility of poplar cultivars to *Marssonina brunnea*. Eur. J. For. Pathol. **14**: 270–282.

Spiers, A.G., and Hopcroft, D.H. 1994. Comparative studies of the poplar rusts *Melampsora medusae*, *M. larici-populina* and their interspecific hybrid *M. medusae-populina*. Mycol. Res. **98**: 889–903.

Terrasson, D., Pinon, J., and Ridé, M. 1988. L'infection naturelle de *Populus trichocarpa* par *Hypoxylon mammatum*. Rev. For. Fr. **40**: 126–130.

Thielges, B.A., and Adams, J.C. 1975. Genetic variation and heritability of Melampsora leaf rust resistance in eastern cottonwood. For. Sci. **22**: 278–282.

Thomas, G.P., and Podmore, D.G. 1953. Studies in forest pathology: XI. Decay in black cotton-wood in the middle Fraser region, British Columbia. Can. J. Bot. **31**: 675–692.

Thomas, G.P., Etheridge, D.E., and Paul, G. 1960. Fungi and decay in aspen and balsam poplar in the boreal forest region, Alberta. Can. J. Bot. **38**: 459–466.

Thompson, G.E. 1941. Leaf-spot diseases of poplars caused by *Septoria musiva* and *S. populicola*. Phytopathology, **31**: 241–254.

Van der Kamp, B.J., Gokhale, A.A., and Smith, R.S. 1979. Decay resistance owing to near-an-aerobic conditions in black cottonwood wetwood. Can. J. For. Res. **9**: 39–44.

Viennot-Bourgin, G. 1937. Contribution à l'étude de la flore cryptogamique du bassin de la Seine (11è note). Deux urédinales nouvelles. Rev. Pathol. Vég. **24**: 78–85.

Villar, M., Lefèvre, F., Bradshaw, H.D., Jr., and Teissier du Cros, E. 1996. Molecular genetics of rust resistance in poplars (*Melampsora larici-populina* Kleb./*Populus* sp.) by bulked segregant analysis in a 2×2 factorial mating design. Genetics, **143**: 531–536.

Walker, J., Hartigan, D., and Bertus, A.L. 1974. Poplar rusts in Australia with comments on potential conifer rusts. Eur. J. For. Pathol. **4**: 100–118.

Wall, R.E. 1971. Variation in decay in aspen stands as affected by their clonal growth pattern. Can. J. For. Res. **1**: 141–146.

Wang, J., and van der Kamp, B.J. 1992. Resistance, tolerance, and yield of western black cottonwood infected by *Melampsora* rust. Can. J. For. Res. **22**: 183–192.

Waterman, A.M. 1957. Canker and dieback of poplars caused by *Dothichiza populea*. For. Sci. **3**: 175–183.

Widin, K.D., and Schipper, A.L., Jr. 1981. Epidemiology and impact of *Melampsora medusae* leaf rust infection on yield of hybrid poplars in the north-central United States. Eur. J. For. Pathol. **11**: 438–448.

Zalasky, H. 1978. Stem and leaf spot infections caused by *Septoria musiva* and *S. populicola* on poplar seedlings. Phytoprotection, **59**: 43–50.

Ziller, W.G. 1965. Studies of western tree rusts. VI. The aecial host ranges of *Melampsora albertensis*, *M. medusae*, and *M. occidentalis*. Can. J. Bot. **43**: 217–230.

CHAPTER 11
Ecological and evolutionary implications of hybridization: *Populus*–herbivore interactions

Thomas G. Whitham, Kevin D. Floate,
Gregory D. Martinsen, Elizabeth M. Driebe,
and Paul Keim

Introduction

Because natural hybrid swarms possess great genetic variation, they host a potentially unique array of ecological and evolutionary interactions. Due to their propensity to naturally hybridize and an abundance of associated herbivores and pathogens, poplars (*Populus*) are ideally suited for studies of these interactions. Schreiner (1971) wrote "If poplars are not the most pest-ridden of the world's important timber trees, they certainly rank high in this respect..." At least 300 species of insects have been reported on living *P. tremuloides* (Davidson and Prentice 1968) and over 650 species of fungi have been reported on *Populus* (Ostry et al. 1988; Farr et al. 1989). Due to their economic importance and disproportionate contribution to biodiversity in threatened riparian communities, it is important to understand the ecology and evolution of this genus and its larger role in the ecosystem. These factors in combination with ease of cloning and rapid growth, make poplars ideally suited for experimental manipulation.

Zones of overlap and hybridization are a common feature of riparian *Populus* forests in North America. Native species of section *Tacamahaca* (*Populus angustifolia, P. balsamifera, P. trichocarpa*) hybridize with native species of section *Aigeiros* (*P. fremontii, P. deltoides*) wherever species in the two sections

T.G. Whitham, G.D. Martinsen, E.M. Driebe, and P. Keim. Department of Biological Sciences, Northern Arizona University, Flagstaff, AZ 86011, USA.
K.D. Floate. Agriculture and Agri-Food Canada, P.O. Box 3000, Lethbridge, AB T1J 4B1, Canada.
Correct citation: Whitham, T.G., Floate, K.D., Martinsen, G.D., Driebe, E.M., and Keim, P. 1996. Ecological and evolutionary implications of hybridization: *Populus*–herbivore interactions. *In* Biology of *Populus* and its implications for management and conservation. Part I, Chapter 11. *Edited by* R.F. Stettler, H.D. Bradshaw, Jr., P.E. Heilman, and T. M. Hinckley. NRC Research Press, National Research Council Canada, Ottawa, ON. pp. 247–275.

are sympatric (Eckenwalder 1984*a*, *b*). This occurs frequently in mountainous regions where high elevation species overlap with lower elevation species along most river systems. Hybridization also occurs within sections, as has been reported between narrowleaf cottonwood, *P. angustifolia*, and balsam poplar, *P. balsamifera* (Brayshaw 1965).

Observations and experiments have shown that these hybrid zones are often centers of insect abundance (Fig. 1). For example, the 13 km hybrid zone of *P. angustifolia* × *P. fremontii* on the Weber River (Utah, USA) represents less than 3% of the cottonwood population in this drainage system, yet supports 94% of the population of the leaf-feeding beetle, *Chrysomela confluens* (Floate et al. 1993), and 85–100% of the population of the gall producing aphid, *Pemphigus betae* (Whitham 1989). The greater concentration of *P. betae* in hybrid zones has now been observed in six other river systems in the United States and Canada (Floate et al. 1996). Similar concentrations of herbivores and/or pathogens on naturally occurring hybrids have been reported for willows in the eastern United States (Fritz et al. 1994), *Eucalyptus* in Australia (Whitham et al. 1994), pinyon

Fig. 1. A. On the Weber River, over a 3-yr period of censusing, beetles were concentrated in the hybrid zone of Fremont cottonwood, *Populus fremontii* and narrowleaf cottonwood, *P. angustifolia* (adapted from Floate et al. 1993). **B.** Over a 6-yr period of censusing, gall aphids were concentrated in the same cottonwood hybrid zone (adapted from Whitham 1989).

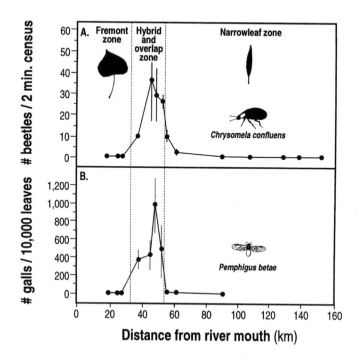

pines in the southwest United States (Christensen et al. 1995), and sedges in Scandinavia (Ericson et al. 1993).

Because the fossil record of the Tertiary shows that intersectional hybridization in *Populus* was fairly common in western North America (Eckenwalder 1984*c*), insects presumably have had ample time to colonize both the parental species and their natural hybrids, yet they have not. This raises a fundamental question, "Why are some species so concentrated on so few host plants?" We believe that the answer to this question rests on an understanding of how hybridization affects the genetics of plant resistance traits, which in turn affect the evolution and ecology of insects and other dependent organisms. In this chapter we focus on how natural hybridization between native *Populus* has influenced plant-insect interactions. First, we examine the defensive chemistry of *Populus* and review how hybridization affects resistance traits. Second, we examine how herbivores have ecologically and perhaps evolutionarily responded to natural hybrids. Third, we examine the conservation implications for both the trees and their dependent communities.

Defensive chemistry

Phenolics are the only class of secondary compounds in the Salicaceae (Palo 1984). These carbon-based compounds include phenolic glycosides, flavanoids, and tannins; no nitrogen-based defensive compounds are known to exist in *Populus*. Although the mechanism by which phenolics are defensive is still debated, at least one study supports the defensive role of phenolics in *Populus*. Zucker (1982) showed an inverse relationship between the concentration of total phenols in individual leaves of narrowleaf cottonwood and the distribution of the leaf-galling aphid, *P. betae*. Leaves without galls had twice the total phenol concentration as leaves with 2 or 3 galls. Also, when 2 neighboring trees with different aphid densities were compared, the tree with 6 times as many aphids had only one-third the phenolic content as the tree with low aphid densities.

Phenolic glycosides

This group includes salicin, salicortin, tremuloiden, and tremulacin, which are the most thoroughly studied defensive compounds in *Populus*. Phenolic gly-cosides affect insect feeding, oviposition and growth, and mammalian feeding. These compounds occur in the leaves, twigs, and bark of *Populus* (Tahvanainen et al. 1985*a*, Reichardt et al. 1990) where they may attract or deter herbivores. For example, several studies have shown that chrysomelid beetles are attracted to willow and poplar hosts with high leaf phenolic glycoside concentrations (Rowell-Rahier 1984; Smiley et al. 1985; Tahvanainen et al. 1985*b*; Denno et al. 1990; Soetens et al. 1991; Rank 1992). Salicin from the plant is then converted by these specialized beetles into salicylaldehyde, which the beetles use as a chemical defense against arthropod predators (Pasteels et al. 1983; Kearsley and Whitham 1992).

Oviposition preference is also influenced by foliar phenolic glycoside concentrations. Moderate concentrations of salicortin and salicin are necessary to stimulate feeding and oviposition of the chrysomelid beetle, *Chrysomela scripta*, but high concentrations of tremulacin may deter these behaviors (Bingaman and Hart 1992, 1993). Thus, different compounds from the same chemical class can have very different effects on the oviposition behavior of *C. scripta*.

Although phenolic glycosides do not seem to be an effective deterrent to the feeding of specialist insects, several studies show that they are effective against generalist insects (Rowell-Rahier 1984) and mammals (Edwards 1978; Tahvanainen et al. 1985a; Basey et al. 1990). For example, the mountain hare, *Lepus timidus*, preferred the mature twigs of eight salicaceous tree species over juvenile twigs in field feeding trails (Tahvanainen et al. 1985a). The juvenile twigs had greater concentrations of phenolic glycosides than mature twigs (up to twice as much) and, across all eight species, the concentration of phenolic glycosides was negatively correlated with hare feeding.

The concentration of phenolic glycosides also can influence herbivore growth. Pupal weights of the large aspen tortrix, *Choristoneura conflictana*, were reduced by approximately 30, 35, and 40%, respectively when raised on artificial diets of salicortin, tremulacin, and a combination of the two (Clausen et al. 1989). However, it may not be possible to obtain general predictions of how phenolic glycosides impact herbivores. Whereas the growth of the eastern tiger swallowtail butterfly, *Papilio glaucus glaucus*, was reduced by high concentrations of salicortin and tremulacin, the growth of the another subspecies, *P. glaucus canadensis*, was unaffected by high concentrations of any of the four phenolic glycosides found in quaking aspen (Lindroth et al. 1988).

Tannins and flavanoids

Although condensed tannins are present in poplars (Julkunen-Titto 1986), their effect on herbivores is unclear. Fifth instar larvae of the large aspen tortrix declined in weight when fed an artificial diet high in both condensed tannins and a phenolic glycoside-containing methanol extract from the leaves of *P. tremuloides* (Bryant et al. 1987). Larvae reared on a diet containing only the methanol extract did not pupate, whereas all larvae pupated that were fed a diet with only condensed tannins. Subsequent experiments with tannin only and no tannin control diets showed a significant but small effect of tannins on growth. Thus, phenolic glycosides had a major effect on insect growth, whereas tannins had only a minor effect.

Flavanoids are another class of compounds that may serve a defensive role in poplars, but their ecological functions remain largely unstudied. Bud exudates of almost every species of *Populus* have high concentrations of flavanoids (Wollenweber 1975). Their abundance and diversity has made them helpful in chemical taxonomy where they have been used to discriminate between poplar species, hybrids, and even clones (e.g., Crawford 1974).

Hybrid poplars and defensive chemistry

The secondary chemistry of hybrid plants differs from parental species in several ways. In a synthesis of 24 studies of 22 different plant genera, Rieseberg and Ellstrand (1993) showed that most hybrids accumulated the secondary compounds of both parental species even when the secondary chemistry of the parental species differed. For example, *P. fremontii* contains only flavone glycosides, *P. angustifolia* contains only flavonol glycosides, but their hybrids contain both (Crawford 1974). Similar patterns have been described by Jones and Seigler (1975), Greenaway et al. (1991, 1992), and Orians and Fritz (1995). Hybrids also may contain novel compounds that are absent in their parent species or lack compounds found in their parent species (Wollenweber 1975; Greenaway et al. 1992). These qualitative differences may result in some hybrids being better defended against herbivores than either of their parents, while other hybrids may be less defended.

Although the above studies document qualitative differences in secondary chemistry between parental species and their hybrids, secondary chemistry also may differ quantitatively. For example, leaves of the willow, *Salix sericea*, contain high concentrations of two phenolic glycosides (salicortin and 2'-cinnamoylsalicortin), but low concentrations of condensed tannins (Orians and Fritz 1995). Conversely, *S. eriocephala* leaves contain no phenolic glycosides, but their concentration of condensed tannins is about 6 times greater than *S. sericea*. In general, hybrids show an additive inheritance of these 2 classes of compounds (Fig. 2). However, hybrids exhibited greater variation in the concentrations of these compounds than their parental species.

Extensive variation in hybrid defenses is probably common, especially in hybrid swarms where both F_1 and advanced generation hybrids coexist. For example, Soetens et al. (1991) quantified the phenolic glycoside and trichome densities of hybrid willow leaves and using cluster analysis classified shrubs into 4 groups. At one end of the continuum, leaves contained high levels of phenolic glycosides and very few trichomes, whereas, at the other end, concentrations of phenolic glycosides were low and trichome densities high. Most importantly, the combination of chemical and mechanical defense traits were good predictors of insect abundance.

Many studies have pooled all classes of hybrids from F_1s to complex backcrosses, but this practice can mask important differences among hybrids that affect herbivores. For example, F_1 hybrids and some advanced generation hybrids may contain the defenses of both parent species. However, the segregation of traits

Fig. 2. A. The concentration of a phenolic glycoside (salicortin) in hybrids is intermediate between the two willow parent species (*Salix eriocephala* and *S. sericea*). **B.** The same pattern occurs with condensed tannins. The hybrids contain the defensive chemistry of both parents at intermediate concentrations (adapted from Orians and Fritz 1995).

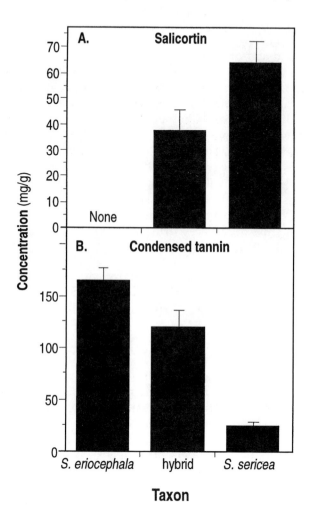

during backcrossing may produce many advanced generation hybrids that have reduced defenses. Preliminary data (L. Mota-Brava, personal communication) suggest that *Populus angustifolia* have high levels of tannins but low levels of salicin. In contrast, *P. fremontii* have low levels of tannins but high levels of salicin. F_1 hybrids and some backcross hybrids have intermediate levels of both compounds, whereas most backcrosses are highly variable (i.e., high concentrations of both to nearly none and all combinations and permutations in between). Just as mutant fruit flies have been crucial to the study of gene expression and regulation, such extreme variation in the chemistry of hybrids suggests they can be used as experimental tools to precisely examine plant–herbivore interactions.

Induced defenses

Two major patterns of induction (i.e., the triggering of plant defenses) have been documented in poplars. First, herbivory can induce an increase in chemical defenses in damaged and/or adjacent plant parts. Second, these same stimuli can trigger the abscission of whole plant parts such as leaves, shoots, or small branches.

Clausen et al. (1989) proposed a model that includes two levels of plant chemical response, both translocation and enzyme-mediated changes for short-term induction of defensive compounds in quaking aspen, *P. tremuloides*. Concentrations of 2 phenolic glycosides, salicortin and tremulacin, increased by 16 and 6%, respectively, following simulated herbivory of the large aspen tortrix. It is thought that these glycosides are translocated from internode stores to leaves. Furthermore, when leaves were crushed, these compounds were converted into salicin, tremuloiden, and 6-hydroxy–2-cyclohexenone (6-HCH). 6-HCH breaks down to the toxic compounds phenol and catechol. Experiments indicated that these conversions were enzymatically mediated. Thus, translocation and enzyme-mediated changes in these compounds, and the fact that large aspen tortrix is negatively affected by phenolic glycosides supports the hypothesis that the feeding activity of herbivores can induce chemical defenses in aspen.

The abscission of infected or damaged plant parts is a second type of induced response. Two species of leaf-galling aphids (*Pemphigus*) induce premature leaf abscission by *Populus angustifolia* and *P. fremontii* (Williams and Whitham 1986). Within a few days of bud break in early spring, trees selectively abscise the leaves on which these aphids are attempting to form galls. Because aphids on abscised leaves die, selective abscission can reduce aphid loads on *P. angustifolia* and *P. fremontii* by 25 and 53%, respectively. Considering that an 18 m tree may support 81 000 galls containing 4.5 million aphids (Whitham 1983) that extract resources from both galled and adjacent ungalled leaves (Larson and Whitham 1991), herbivore-induced abscission allows poplars to reduce their herbivore loads and selectively eliminate damaged leaves early in the season before the aphid population rises.

An induced defense to one herbivore may attract other herbivore species. After beavers cut down cottonwood trees, the stump resprouts, producing foliage preferred by the beetle, *Chrysomela confluens* (Martinsen et al. 1996). Adult beetles were 15 times more abundant on resprout growth than nearby nonresprout growth from the same clone. Leaves from resprout growth had twice the concentration of phenolic glycosides (mostly salicortin) than leaves from nonresprout growth. Also, beetle larvae raised on resprout growth grew 16–20% larger and 10% faster than larvae raised on nonresprout growth. Subsequent experiments showed that beetles fed resprout growth were better defended from their predators (i.e., *Formica* ants) than beetles fed non-resprout growth. Thus, the induction of increased defenses by one herbivore (in this case beavers) led to increased attack by another herbivore (i.e., chrysomelid beetles) that has evolved a specialized pathway for converting the plant's defense into their own defense.

Variation in defensive chemistry

We emphasize that the concentration of phenolics among and within individual poplars is highly variable. For example, leaves at earlier stages of phenological development have higher concentrations of phenolic glycosides than leaves at later stages (Rowell-Rahier 1984; Lindroth et al. 1987; Bingaman and Hart 1993). Consequently, within an individual tree or even a single branch, leaves at different phenological stages may have different concentrations of phenolic glycosides.

Abiotic factors also may change concentrations of plant secondary chemicals and the tree's palatability to herbivores. Nitrogen fertilization of *P. tremuloides* decreased the concentration of tannins and phenolic glycosides, and increased nitrogen levels in leaves, making them better food for lepidopteran larvae (Bryant et al. 1987). Similarly, *Salix dasyclados* grown in a light stressed environment produced about one-third the phenolics as plants grown in conditions of high light and low nutrients, and plants grown in optimal conditions (high light and nutrients) (Larsson et al. 1986). Increased atmospheric CO_2 also increased plant secondary compounds in *P. tremuloides* (Lindroth et al. 1993).

In summary, much of the variation in defensive chemistry examined in this section can be predictably related to plant genotype, seasonal patterns of plant phenology, developmental patterns of plant growth (e.g., juvenile versus mature), nutrient availability and environmental stress. This temporal and spatial variation may benefit plants by limiting the window of exposure to herbivory and/or by making it difficult for herbivores to track their host plants in ecological and evolutionary time (Whitham 1983). While we have only scratched the surface of understanding the defensive chemistry of *Populus*, a major goal should be to integrate plant defensive chemistry with ecological genetics. Hybrid poplars with their variable chemistry, genetics and herbivore responses to individual clones offer great promise in achieving this goal.

Resistance and introgression in hybrid swarms

Effects of hybridization on resistance

Four genetically-based mechanisms of resistance and susceptibility describe the patterns of herbivory on hybrid plants relative to their parental species (Fritz et al. 1994). These are: (*i*) additive, whereby hybrids do not differ from the average between the 2 parents, (*ii*) dominance, in which hybrid resistance differs significantly from the mean resistance of both parentals, but not from one of the parents, (*iii*) hybrid susceptibility, in which herbivores achieve higher densities and/or higher performance on hybrids than on parental species, and (*iv*) hybrid resistance, whereby hybrid plants are more resistant than either parent. Examples of these patterns for diverse systems are reviewed in Strauss (1994).

Examples of each of these genetic mechanisms can be found in natural hybrid swarms where adjacent hybrids may exceed the resistance or susceptibility traits of either parental species. Figure 3 shows the distribution of tree resistances to the leaf-galling aphid, *P. betae*, in both the hybrid zone and in the adjoining narrowleaf cottonwood zone (Whitham 1989). Tree resistance was measured as the survival rate of colonizing aphids that attempted to induce galls. Two major patterns are shown. First, the average tree in the hybrid zone is more susceptible than the average tree in the narrowleaf cottonwood zone (Fremont cottonwood is not included as it is not a host for this aphid species). Second, although tree resistance in the narrowleaf zone appears normally distributed, tree resistance in the hybrid/overlap zone is not. Although a few trees are extremely resistant, the distribution is skewed towards greater susceptibility. Whereas both zones exhibit considerable variation in resistance traits, the hybrid zone appears to exhibit the best and the worst.

Patterns of resistance likely reflect the genetic resistance mechanisms involved (e.g., monogenic or polygenic), their dominance relationships (Gallun and Khush 1980), the pattern of hybridization (e.g., F_1 sterility, unidirectional or bidirectional introgression; Keim et al. 1989; Floate and Whitham 1993), and the level of backcrossing. For example, RFLP genetic analyses of natural hybrids between Fremont and narrowleaf cottonwood revealed that F_1 and BC_1 hybrids to narrowleaf cottonwood were almost totally resistant to the aphid, *P. betae*. In contrast, complex backcrosses with narrowleaf cottonwood were very susceptible (Paige et al. 1990; Floate and Whitham 1995). The opposite pattern occurred with another herbivore; the bud-galling mite, *Aceria parapopuli*, reached outbreak proportions on F_1s, but was virtually absent on complex backcrosses to narrowleaf cottonwood (Floate and Whitham 1995). Herbivore densities also differ across categories of hybrid eucalypts (*Eucalyptus* — Morrow et al. 1994; Whitham et al. 1994), spruce (*Picea* — Manley and Fowler 1969), and oak (*Quercus* — Moorehead et al. 1993). In part, these variable distributions result from segregating traits which differentially affect the defensive chemistry of different hybrid classes (see previous section). These studies emphasize two points. First, all hybrids are not equal; identifying plants as being "hybrid" without recognition of the degree of hybridization (e.g., complex backcrosses versus F_1) may mask important differences in susceptibility traits. Second, to understand the effect of hybridization on dependent community members, the observed variation among hybrids in susceptibility traits may require the use of known crosses and molecular genetics techniques to identify the genetic composition of individual trees.

255

Fig. 3. Trees in the hybrid zone of Fremont and narrowleaf cottonwood are significantly more susceptible to the gall aphid, *Pemphigus betae*, than are trees in the narrowleaf zone. Aphid survival on Fremont cottonwood is zero, but it is not a host for this gall aphid. Aphid survival was quantified by transferring 4000 newly-hatched stem mothers onto trees in both zones and recording their survival 45 days later (adapted from Whitham 1989).

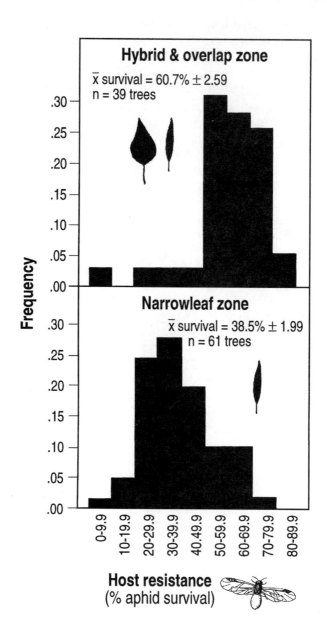

So as not to imply that herbivores respond only to genetic based factors that may be accentuated in hybrid zones, we emphasize that environmental stress also may significantly affect the susceptibility of plants in hybrid zones. Plants at the edge of their range where water, nutrients, and temperatures are limiting may be stressed and most susceptible to pest attack (Whitham et al. 1994). This may be especially important for hybrids which often occur at the edge of the ranges of both parental species. Although plant stress has often been associated with increased attack by some herbivores (White 1969), vigorous plants suffering little stress are the preferred hosts of other herbivores (Price 1991).

Environmental and genetic hypotheses are not mutually exclusive; genotype by environment interactions may also influence plant resistance (Tingey and Singh 1980). To examine clone by environment interactions, we have planted the same cottonwood clones in both stressful and benign environments, and then quantified the survival and reproduction of gall aphids transferred to these trees. Although we have found strong clonal host effects on the survival of gall aphids, no predictable site or stress impacts have been detected (T.G. Whitham and K.D. Floate, unpublished data). Clearly, additional experiments are needed to make generalizations for other members of the herbivore community.

Patterns of hybridization and introgression

There are 4 major patterns of plant hybridization that are likely to affect the distributions of herbivores and pathogens. First, the parental species may produce sterile F_1 hybrids and, hence, all hybrids will contain 50% of the genomes of each parent. Second, F_1s may backcross to only one of the parental species; i.e., unidirectional introgression. This pattern generates a continuum of hybrid genotypes between the backcross species to F_1s, but leaves a large genetic gap between F_1s and the nonbackcross species. Third, F_1 hybrids may bidirectionally introgress (i.e., backcross to both species) to form a continuum of hybrid genotypes between the parents. Fourth, hybrids may interbreed, also forming a continuum of hybrid genotypes.

RFLP genetic analyses of trees from a natural hybrid zone along the Weber River in Utah document a pattern of unidirectional introgression between Fremont and narrowleaf cottonwood (Keim et al. 1989; Paige and Capman 1993). These authors found a gradient of intermediates between narrowleaf and F_1s, but no intermediates between F_1s and Fremont cottonwood. Because all of the analyzed hybrids lacked loci homozygous for Fremont cottonwood alleles, they concluded that the hybrids did not backcross to Fremont cottonwood and did not interbreed. Thus, even though hybrids make up most of the trees in the hybrid zone, they successfully reproduce only when they backcross to narrowleaf cottonwood. Using allozyme analyses, this unidirectional pattern of introgression also has been shown with a second pair of cottonwoods (*P. angustifolia* × *P. deltoides*) that hybridize along Boulder Creek in Colorado, USA (J. Mitton, personal communication). The following section emphasizes how these patterns of hybridization and their effects on plant defenses have important impacts on herbivores, pathogens, and other dependent members of the cottonwood community.

Herbivore responses to hybrids

Host shifting and the hybrid-bridge hypothesis

Host shifting is presently explained by either preadaptation or mutation hypotheses, both of which require herbivores to make a "single jump" in shifting from one host species to another (Bush 1974, 1975; Tabashnik 1983; Futuyma 1983; Jermy 1984; Thomas et al. 1987). These hypotheses do not consider the potential role of natural plant hybrids that bridge the genetic, chemical, and morphological gaps between parental species and may allow herbivores to shift in a "series of gradual steps." We refer to the potential role of hybrid intermediates serving as stepping stones that facilitate host shifts between parental species as the "hybrid-bridge" hypothesis (Floate and Whitham 1993).

The hybrid-bridge hypothesis is supported by studies that examined the distributions of common galling insects and mites that are indigenous to either Fremont or narrowleaf cottonwood along the Weber River, Utah, USA. If the hybrid-bridge hypothesis is correct, this hybridizing system with unidirectional introgression should support three predictions: (*i*) the lack of intermediates between Fremont cottonwood and F_1 hybrids should inhibit herbivores native to Fremont cottonwood from shifting onto F_1 hybrids; (*ii*) the presence of intermediates between F_1 hybrids and narrowleaf cottonwood should promote shifting of herbivores indigenous to narrowleaf cottonwood onto F_1 hybrids; and (*iii*) herbivores that have shifted from narrowleaf cottonwood onto F_1s should be unable to shift further onto Fremont cottonwood due to the absence of intermediates between F_1s and Fremont cottonwood.

Our first prediction is supported by each of 4 leaf-galling aphid species indigenous to Fremont cottonwood. In the absence of backcross intermediates, no successful shifts from Fremont to F_1s or beyond to narrowleaf cottonwood were observed (Fig. 4). Although characteristic signs of their failed galls are found on F_1s, signifying that these aphids naturally encounter and recognize hybrid trees as hosts, each species has thus far been unsuccessful in bridging the gap between Fremont cottonwood and F_1 hybrids.

Our second and third predictions are supported by each of 4 species indigenous to narrowleaf cottonwood. In the presence of backcross intermediates, 3 species of leaf-galling aphids and a bud-galling mite have successfully shifted from narrowleaf cottonwood onto F_1 hybrids (Fig. 5). However, having shifted from narrowleaf cottonwood to F_1s via backcrosses, these herbivores have then encountered the gap between F_1s and Fremont cottonwood and have been unable

Fig. 4. The distributions of four leaf-galling aphid species indigenous to Fremont cottonwood show that these species have not bridged the morphological and genetic gap to F$_1$ hybrids. The sole exception to this pattern (asterisk) was a single *Pemphigus* sp. gall found on one F$_1$ hybrid (adapted from Floate and Whitham 1993).

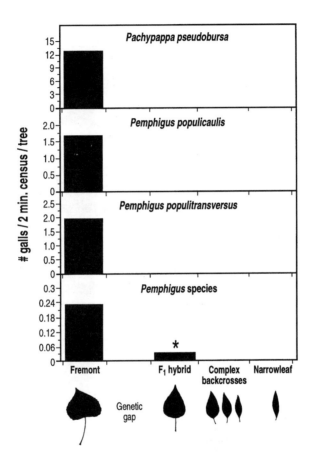

to successfully cross this barrier (Fig. 5). Hence, each of the 3 predictions tested in this cottonwood system have been supported 4 times by different galling species. The same patterns were obtained in a second year's census.

Host shifting and predicting pest problems with hybrids

The hybrid bridge hypothesis (Floate and Whitham 1993) makes predictions about the origins of potential pest species that could cause economic damage. Based upon the pattern of introgression or lack thereof (i.e., F$_1$ sterility, uni- or bidirectional introgression) and a knowledge of the pests of the parental species,

Fig. 5. The distributions of four leaf-galling species (1 mite, top panel, and 3 aphids) indigenous to narrowleaf cottonwood show that these species have shifted to F_1 hybrids via hybrid intermediates, but they have not bridged the morphological and genetic gap between F_1s and Fremont cottonwood where hybrid intermediates are absent. The asterisk indicates low densities of only 17 galls found on 30 trees (adapted from Floate and Whitham 1993).

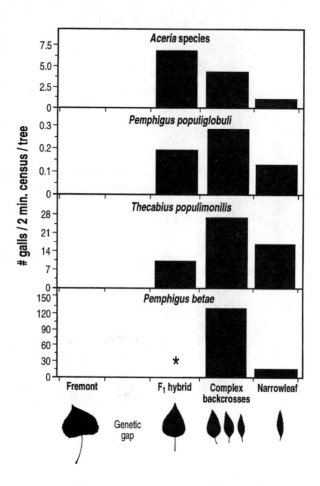

we can predict how economically-important hybrids may acquire pest species. For example, in progressing from the least genetic gaps between species (i.e., bidirectional introgression), to unidirectional introgression with one major genetic gap, to F_1 sterility with 2 major gaps, we would predict an associated decline in potential pest species on F_1 hybrids. Thus, all else being equal, we would expect F_1 hybrids between species that exhibited bidirectional introgression to accumulate the herbivores of both parental species. In contrast, we predict that sterile F_1 hybrids that are isolated by the lack of intermediate backcrosses, will accumulate the fewest pest species.

Although tests of these predictions are lacking with other poplars, hybrids of bidirectional introgressing *Eucalyptus*, accumulated all of the insects and fungi of both parental species (Whitham et al. 1994). Furthermore, the distribution of specialist insects and fungi on hybrids was highly predictable. The most preferred hybrid phenotype was the one most similar to the preferred host species. Thus, if species A and B exhibit bidirectional introgression, backcrosses to species A are far more likely to acquire species indigenous to species A, while backcrosses to species B are far more likely to acquire species indigenous to species B. We refer to the predictable use of hybrids as the "phenoypic affinity hypothesis" in which the phenotypic affinity or relatedness of a hybrid with one of its parents will predictably affect which suite of herbivores it acquires.

If herbivores discriminate among hybrid classes and their parental hosts as the above studies suggest, then the distributions of arthropods may compliment the use of plant morphological traits as a means of discriminating among hybrid classes and their parental species. To examine the abilities of herbivores as taxonomists, trees in a hybrid zone of *P. angustifolia* × *P. fremontii* were classified on the basis of traditional morphological traits and then reclassified solely on the basis of their arthropod communities (Floate and Whitham 1995). Both methods of classification resulted in distinct groupings of Fremont cottonwood, F_1 types, and complex backcrosses hybrids (the local rarity of narrowleaf cottonwood prevented their inclusion in the study). Most importantly, there was a 98% level of agreement in the classification of individual trees by both methods suggesting that, even among hybrids, arthropods can be very selective. Insect distributions also have been used to distinguish parental and hybrid categories of willow (*Salix* — Fritz et al. 1994), subspecies of rubber rabbitbrush (*Chrysothamnus nauseosus* — Floate et al. 1995), and to a lesser extent, parental and hybrid categories of oak (*Quercus* — Aguilar and Boecklen 1992).

Although there seems to be substantial predictability in determining how patterns of hybridization and introgression affect arthropod distributions on hybrids, it is difficult to predict which species could reach outbreak levels. For example, within the hybrid zone on the Weber River, the bud-galling mite, *Aceria* sp. is absent from nearly all Fremont, narrowleaf, and complex backcross cottonwoods. However, on F_1 type hybrids (i.e., F_1 and BC_1), half of all censused trees were attacked and we recorded densities of up to 60 000 golf-ball sized galls on individual trees (T.G. Whitham, unpublished data). Thus, a mite that is relatively rare throughout the river system has reached outbreak proportions on a very specific and relatively rare class of hybrids. Although these mites are frequently abundant on hybrid poplars planted as ornamentals and in shelterbelts (Ives and Wong 1988), based upon their low densities on the parental species in the wild, it would have been difficult to predict their potential to become pests.

261

Using hybrid susceptibility in pest management

Why do some organisms like the gall aphid, *P. betae* (Fig. 1), show such spectacularly high concentrations in the hybrid zone and an almost complete failure to colonize adjacent zones of either narrowleaf or Fremont cottonwood? Whitham (1989) suggested that for this aphid, susceptible hybrids represented ecological sinks in which hybrids attract aphids from other potential hosts. Furthermore, if in the process of specializing on susceptible hybrids, aphids lost the ability to utilize the more resistant parental cottonwood host species, then susceptible hybrids could be considered evolutionary sinks as well. The implications for pest management are considerable. Atsatt and O'Dowd (1976) argued that the presence of susceptible host species or varieties could lengthen the useful life of resistant hosts by slowing pest evolution. Simulation models by Gould (1986) for mixing resistant and susceptible varieties of wheat support this logic and suggest that susceptibility could be used as a management tool to conserve resistance genes.

If susceptible hybrids act as evolutionary sinks for pests, one should then expect that the presence of hybrids may affect insect evolution. Moran and Whitham (1988) suggested that life cycle evolution of *P. betae* has been affected by the pattern of resistance resulting from hybridization in cottonwoods. In the hybrid zone where susceptible hosts are common, *P. betae* annually alternates between cottonwoods and herbaceous plants. In the adjacent narrowleaf zone where trees are more resistant and aphid survival is reduced, *P. betae* has abandoned its cottonwood host and remains almost solely on its herbaceous hosts. Reciprocal aphid transfer experiments in the field and the performance of aphid clones in the lab showed that the differences between the simple and complex life cycles are, in part, genetically based (Moran and Whitham 1988; Moran 1991). Such different life cycles suggest that plant resistance and susceptibility influenced by hybridization can affect insect evolution and perhaps even speciation.

Although hybrids might act as ecological and evolutionary sinks for some pests, the reverse may be true for other organisms; i.e., susceptible hybrids could serve as sources of infection rather than as sinks. In Scandinavia, Ericson et al. (1993) found that densities of the smut, *Anthracoidea fischeri*, was 30–80 times greater on hybrid sedges (*Carex canescens* × *C. mackenziei*) than on the parental species. They concluded that the presence of smut is dependent upon the presence of the hybrid sedges. If true, then hybrids may serve as the source of smut outbreaks, not as a sink as was suggested for *P. betae* aphids on hybrid cottonwoods. The critical test is to remove the hybrids and see what happens to the smut. If hybrids were removed, one might achieve control of the smut. Conversely, removal of susceptible hybrids may simply select for smuts with adaptations to survive on the more resistant parental species. These 2 contrasting studies illustrate the dilemma of using susceptibility as a management tool and emphasize the need to further understand the evolution of plant–pest interactions to make appropriate management decisions.

Hybridization and tree development

Developmental changes within individual trees affect defensive chemistry and can determine the distribution of herbivores (Bryant et al. 1985; Tahvanainen et al. 1985a; Reichardt et al. 1990). Because F_1 type hybrids, complex backcrosses, and their parent species often differ developmentally (e.g., the expression of juvenile and mature traits, ratios of juvenile to mature branches, propensity to clone), we predict that varation in developmental processes are likely to affect dependent organisms.

For example, Kearsley and Whitham (1989) found that juvenile ramets of cottonwood clones were dominated by the leaf-feeding beetle, *C. confluens*, whereas mature ramets of the same clones were dominated by the leaf-galling aphid, *P. betae*. On mature ramets beetle larvae had 50% lower survival and took 25% longer to reach adulthood than on juvenile ramets. Not surprisingly, their densities were 400 times greater on the superior juvenile ramets than on inferior mature ramets. In contrast, aphid survivorship was 50% higher on mature ramets than on juvenile ramets, and aphids colonized the superior mature ramets at a rate 70 times greater than juvenile ramets. An examination of 42 diverse taxa (e.g., herbivores, predators, parasites) from 8 arthropod orders showed similar patterns (Waltz and Whitham 1996). This suggests that developmental changes in resistance and susceptibility as poplars age has general impacts on community structure.

Mammalian responses to hybrids

Diverse mammals consume the bark, leaves, and roots of *Populus* (e.g., hares — Bryant 1981, pocket gophers — Cantor and Whitham 1989, beavers — Basey et al. 1990, elk — Romme et al. 1995). Although little is known about mammalian responses to the patterns of hybridization in *Populus*, the abundance of mammalian herbivores (both native and domesticated) combined with the economic and conservation importance of *Populus* make it important to fill this gap in our knowledge. We expect mammalian herbivores to be sensitive to the variation arising from hybridizing poplars for three reasons. First, phenolic glycosides are abundant in the Salicaceae and are known to be defensive towards mammals (Edwards 1978; Tahvanainen et al. 1985a). Second, the patterns of hybridization are known to affect these important defenses (e.g., Fig. 2). Third, in an unintended experiment, Whitham et al. (1991) found that when elk moved into a cottonwood garden and browsed at their leisure on intermixed clones of *P. fremontii* × *P. angustifolia*, they were very selective in their browsing. During this winter trial, all replicates of each clone suffered either ≥99% branch loss or ≤2% branch loss. This apparent all or none feeding on individual clones suggests that elk feed in response to different concentrations of cottonwood defenses. If this experiment is an indication of mammalian selectivity, such studies could be important in preventing losses in large poplar plantations and restoration projects where large ungulates may browse during the winter.

Mutualisms and hybrid cottonwoods

The effects of ant and mycorrhizal mutualisms on hybrid poplar interactions with herbivores and pathogens is another area of research that is poorly understood, but likely to be important. Ants often provide trees with protection against herbivores in return for rewards they receive either directly or indirectly from the tree (e.g., Huxley and Cutler 1991). Floate and Whitham (1994) found that *Chaitophorus* sp. aphids attract aphid-tending ants which subsequently reduced herbivory 2-fold by the beetle, *Chrysomela confluens*. Because beetle defoliation of juvenile trees can average 25%, the aggressive behavior of ants towards all nonaphid arthropods likely outweighs the negative effects of aphids. Furthermore, as beetle populations are concentrated in the hybrid zone (Fig. 1), the protective services of ants are likely to be more important to hybrid cottonwoods than to either parental species.

Because mycorrhiza have a positive, if not crucial role in the survival and performance of their host plants, especially those living in nutrient poor soils (Meyer 1973), it is important to examine how the patterns of hybridization and herbivory affect these fungal mutualists. Although no studies have been done with *Populus*, Rabin and Pacovsky (1985) experimentally demonstrated that the presence of mycorrhiza negatively affected the growth and survival of two insect herbivores of soybean.

Conversely, herbivory can negatively affect mycorrhiza. Gehring and Whitham (1991) found that pinyon pine, *Pinus edulis*, susceptible to the stem-boring moth, *Dioryctria albovittella*, had mycorrhizal levels 33% less than trees resistant to moth attack. When moths were experimentally removed from susceptible trees, mycorrhiza rebounded to levels that were indistinguishable from resistant trees. Because plant defensive chemistry and herbivore distributions are affected by the patterns of hybridization (e.g., Figs. 2, 4, and 5), it seems likely that the mycorrhizal mutualists of these trees will also be affected. Furthermore, in light of the findings of Rabin and Pacovsky (1985), we should also explore the possibility that the resistance traits of some hybrids to insect attack may be derived, in part, from an ability to maintain superior mycorrhizal associates that enable plants to better resist herbivory.

Conservation of cottonwood hybrid zones

Poplars are one of the most productive and sensitive components of riparian ecosystems in western North America. The number of vertebrate species associated with poplar communities is 4 times higher than the number associated with spruce-fir, lodgepole pine, or Douglas fir communities (Finch and Ruggiero 1993). However, human activities result in the loss of over 100 000 ha of riparian habitat each year (Finch and Ruggiero 1993). Although *Populus* conservation is specifically addressed in Chapter 3 by Braatne, Heilman, and Rood, here we emphasize the importance of conserving hybrid populations. Although hybrids have no protection status under the US Endangered Species Act of 1973, there are numerous reasons for their conservation.

Hybrid zones as centers of biodiversity and unique community assemblages

If many species respond to hybrid zones as those shown in Fig. 1, then hybrid zones should be both centers of abundance and species richness. Although the issue of how hybridization affects arthropod diversity in poplar communities is unknown, it has been examined with *Eucalyptus*. Whitham et al. (1994) quantified the distributions of 40 insect and fungal taxa in a natural hybrid swarm formed by *E. amygdalina* and *E. risdonii* (both endemic to Tasmania and the latter endangered). The average hybrid tree supported 53% more insect and fungal species, and relative abundances averaged 4 times greater on hybrids than on either eucalypt species growing in pure stands. Also, 50% of the species coexisted only in the hybrid zone which produced in a unique assemblage of species. Subsequent studies in common gardens using known *Eucalyptus* crosses, demonstrated that hybridization was responsible for many of the patterns observed in nature (Dungey et al. 1994).

The genetic variability that occurs within a hybrid swarm affects the rest of the community and affects biodiversity. For example, within a cottonwood hybrid swarm, trees susceptible to the aphid, *P. betae*, had 31% greater species richness and 26% greater relative abundance than aphid-resistant trees (Dickson and Whitham 1996). Aphid removal experiments demonstrated that this effect was largely due to the presence of aphids and their galls, which attracted other herbivores and their associated predators and parasites. Even vertebrate insectivores such as Black-capped Chickadees, *Parus atricapillus*, were attracted to trees with higher insect loads.

Hybrid zones as centers of plant and insect evolution

Interspecific and intergeneric hybridization in plants is widespread and has long been thought to have played an important role in plant evolution (e.g., Stebbins 1950). Stace (1987) estimated that 50–70% of all angiosperms arose via hybridization events. Based upon comparisons of chloroplast and nuclear DNA, *Populus nigra*, may be of hybrid origin (Smith and Sytsma 1990, but see also Eckenwalder, Chapter 1). Since the late Miocene when the first sympatric occurrences of sections *Aigeiros* and *Tacamachaca* were recorded, intersectional poplar hybrids have been fairly common in the fossil record (Eckenwalder 1984c). Thus, it seems important to conserve the widespread and natural hybridization that is observed in poplars.

Natural hybrids also should be conserved for their potential economic value. Eckenwalder (1984*b*) notes that the natural intersectional hybrid, *Populus* ×*acuminata* is planted widely throughout the Great Plains of North America where its hardiness and attractive foliage make it a superior street tree. Furthermore, the hybrid *P.* ×*canadensis* (also known as *P.* ×*euramericana*), is used extensively in forestry and is the progeny of *P. deltoides* and *P. nigra* (cited in Smith and Sytsma 1990). Considering that *P. nigra* may have evolved from an ancient hybridization event (Smith and Sytsma 1990) and has subsequently been used by man to produce commercially important hybrids (e.g., *P.* ×*canadensis*), the conservation of natural hybrids is important from both evolutionary and economic perspectives (Whitham et al. 1991).

If plants speciate via hybridization, it seems likely that their dependent insect herbivores and other organisms might simultaneously speciate on hybrids and their derivative species (Floate and Whitham 1993). Because these new hosts represent an environment distinct from the old host, herbivore populations may be subject to divergent selection pressures which could then lead to reproductive isolation, race formation, and eventual speciation (e.g., see above example of *P. betae* life cycle variation in apparent response to hybridization by cottonwoods).

Although herbivores show clear ecological responses to hybrid plants and hybrid zones (i.e., density and performance), their ability to respond evolutionarily is likely dependent upon the long-term persistence of hybrid zones. Considering the presence of hybrid poplars in the fossil record (Eckenwalder 1984*c*) and the fact that hybridization has often been tied to periods of environmental disturbance such as glaciation, many hybrid zones may have originated long ago (Rattenbury 1962; Grant 1971; Kat 1985). Furthermore, Kemperman and Barnes (1976) described large clones of aspen, *P. tremuloides*, covering 81 ha and speculated that individual clones could date back to the Pleistocene and earlier in areas where no glaciation occurred. Although insects and other organisms with short generation times are clearly capable of rapid evolution, evolutionary responses to plants in small and ephemeral hybrid zones would not be expected. Thus, we predict that plant–herbivore interactions will vary from hybrid zone to hybrid zone depending on their size and age (Floate et al. 1996).

Hybrids as essential insect habitat

Cottonwood hybrid zones may represent essential habitat for some insect species. For the insects shown in Fig. 1, the cottonwood hybrid zone is the center of their distribution and may represent a refugium where they can survive when their population crashes elsewhere (Whitham 1989). Such examples are not unique to cottonwoods. Whitham et al. (1994) found that 5 of 40 insect and fungal species examined on hybrid *Eucalyptus* were nearly restricted to the hybrid zone. Furthermore, if hybrid zones support high concentrations of herbivores, they also are likely to support high densities of beneficial predators and parasites (Whitham et al. 1991). Thus, protecting hybrid zones may protect potential sources of biocontrol agents. Because insect conservationists argue that it is most important to save sites of rich diversity (e.g., Morris et al. 1990), where natural hybrid zones support an exceptionally rich fauna, their conservation is warranted.

Hybrid cottonwoods as superior avian nesting habitat

Riparian habitats are especially important to birds (Carothers et al. 1974; Szaro 1980; Knopf 1985). Over 50% of 166 species breeding in riparian areas of the Southwest are completely dependent on this habitat (Johnson et al. 1977). Several authors report that cottonwood stands contain the highest bird densities of riparian areas in western North America (Carothers et al. 1974; Johnson et al. 1977; Strong and Bock 1990).

Our studies along the Weber River suggest that hybrid cottonwoods may increase the value of riparian habitat for nesting birds (Martinsen and Whitham 1994). We found almost 3 times as many nests in the hybrid zone as in either Fremont or narrowleaf cottonwood zones. Also, within the hybrid zone, nest densities in F_1 type trees were 2 times greater than in adjacent Fremont or narrowleaf cottonwoods (Fig. 6). We suspect that the distinctive branching pattern of F_1 type hybrids (see photos in Martinsen and Whitham 1994), may enhance their attractiveness to nesting birds. In general, F_1 type hybrids have many large lateral branches that are used for nest placement. Additionally, because hybrid and overlap zones contain the distinctive architectural types of Fremont, narrowleaf and their hybrids, the greater architectural diversity of this zone may support a greater diversity of birds, each with different nest requirements.

Increased bird nest densities in the hybrid zone also may be a response to abundant food resources. A single hybrid cottonwood can support 4.5 million aphids (Whitham 1983) that attracts insectivorous birds such as Black-capped Chickadees, which selectively forage where aphid densities are highest (Martinsen and Whitham 1994; Dickson and Whitham 1996).

Threats to hybrid poplars

Riparian habitats and poplar hybrid zones are threatened in western North America by dams, diversion of river waters, cattle, housing developments, and agriculture. Because dam building prevents natural flooding, its impact on drought-intolerant poplars is especially severe (reviewed by Rood and Mahoney 1990). Dams and water channelization prevent the flooding required for seedling establishment, and increase the susceptibility of mature poplars to drought. Because poplar hybrid zones are often found at the mouths of canyons

Fig. 6. A. Significant differences were found in bird nest distributions in which trees in the hybrid zone had significantly more nests than trees in the Fremont cottonwood zone or the narrowleaf cottonwood zone. **B.** Significant differences were found in nest distributions among tree types within the hybrid zone in which F_1 type hybrids had significantly more nests than adjacent Fremont and backcross/narrowleaf cottonwoods (adapted from Martinsen and Whitham 1994).

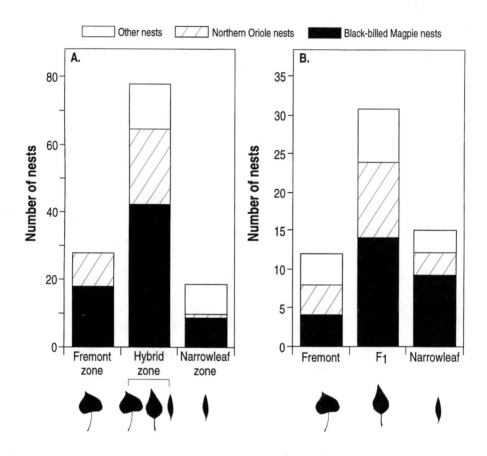

(Eckenwalder 1984b) where dams are frequently built (T.G. Whitham, personal observation), they also may be destroyed by the lakes created when rivers are impounded. Furthermore, in the arid Southwest, riparian habitats are favorite sites for housing developments that fragment gallery forests and require channelization and levees to protect buildings. For example, all the hybrid zones on rivers emptying into Great Salt Lake are threatened by the expansion of cities that are rapidly filling Salt Lake Valley. Historically, because poplars may have been the only tree available, they were harvested for use in construction and as fuel wood. Also, riparian forests have been cleared to provide land for both grazing and agriculture. Cattle grazing remains a serious threat to poplars, especially to seedlings and saplings, which are often eaten or trampled (Rood and Mahoney 1990).

Another potential threat to native stands of poplars is hybridization with exotic poplars, which are widely grown as shelter belts, ornamentals, and as a source of fiber and energy. Gene flow from a large population adapted to one environment into a small population adapted to another environment may have two consequences: (*i*) outbreeding depression and disruption of local adaptation in the small population, and (*ii*) extinction of the small population by genetic assimilation or swamping (Ellstrand 1992). Thus, declining populations of native poplars which are growing near large plantations of introduced trees may be especially vulnerable. For example, extensive plantings of numerous cultivars of *P. ×canadensis* are displacing "pure" *P. nigra* from its native habitat in Europe (cited in Smith and Sytsma 1990).

Conclusions and implications

Hybridization between dominant plant species such as poplars has important ecological and evolutionary consequences for both the hybridizing species and their dependent community. Natural hybridization between *Populus angustifolia* and *P. fremontii* in the western US affects gene flow, plant defensive chemistry, the distributions of arthropod herbivores and nesting birds. Although some hybrids are extremely resistant to herbivores (i.e., heterosis), most appear to be poorly defended (i.e., hybrid breakdown). As a consequence, some herbivores have ecologically and perhaps even evolutionarily responded to these hybrids. For the beetle, *Chrysomela confluens*, and the aphid, *Pemphigus betae*, the hybrid zone is the geographic center of their distributions. Furthermore, *P. betae* has apparently evolved different life cycles to live in hybrid and pure host zones. Even vertebrates are affected by hybridizing poplars. With the distinctive branch architecture of hybrids, we find that bird nest densities are greatest in F_1 type hybrids within a hybrid swarm, and greater in the hybrid swarm than in pure stands of the parental species. We suspect that mammalian herbivores and the mycorrhizal mutualists of poplars also may be affected by the patterns of hybridization. Because natural hybrid zones can be centers of biodiversity, their conservation may warrant special consideration as hybrids are not currently protected by laws such as the Endangered Species Act of 1973 in the USA.

Studies of natural poplar hybrids and hybrid zones also have applied implications. Hybrids may facilitate host shifts of pests and pathogens by providing intermediates that allow pests to gradually experience the defenses of another plant species rather than having to make an abrupt host shift to a novel genome. Thus, an understanding of the patterns of hybridization and its effects on defensive chemistry may allow us to predict which pests are likely to attack hybrids of known parentage. Although our studies argue that natural hybrids are part of a dynamic ecosystem and should be conserved, we caution that gene flow brought about by large scale plantings of exotic hybrid poplars could swamp native gene pools and pose a conservation hazard.

Acknowledgements

We thank R. Stettler and Y. Linhart for numerous comments on the manuscript. This research has been supported by grants from NSF, USDA, and DOE.

References

Aguilar, J.M., and Boecklen, W.J. 1992. Patterns of herbivory in the *Quercus grisea* × *Quercus gambelii* species complex. Oikos, **64**: 498–504.

Atsatt, P.R., and O'Dowd, D.J. 1976. Plant defense guilds. Science (Washington, DC), **193**: 24–29.

Basey, J.M,., Jenkins, S.H., and Miller, G.C. 1990. Food selection by beavers in relation to inducible defenses of *Populus tremuloides*. Oikos, **59**: 57–62.

Bingaman, B.R., and Hart, E.R. 1992. Feeding and oviposition preferences of adult cottonwood leaf beetles (Coleoptera: Chrysomelidae) among *Populus* clones and leaf age classes. Envir. Entomol. **21**: 508–517.

Bingaman, B.R., and Hart, E.R. 1993. Clonal and leaf age variation in *Populus* phenolic glycosides: Implications for host selection by *Chrysomela scripta* (Coleoptera: Chrysomelidae). Envir. Entomol. **22**: 397–403.

Brayshaw, T.C. 1965. Native poplars of southern Alberta and their hybrids. Can. For. Serv. Publ. 1109.

Bryant, J.P. 1981. Phytochemical deterrence of snowshoe hare browsing by adventitious shoots of four Alaskan trees. Science (Washington, DC), **313**: 889–890.

Bryant, J.B., Clausen, T.P., and Kuropat, P. 1985. Interactions of snowshoe hare and feltleaf willow in Alaska. Ecology, **66**: 1564–1573.

Bryant, J.P., Clausen, T.P., Reichardt, P.B., McCarthy, M.C., and Werner, R.A. 1987. Effect of nitrogen fertilization upon the secondary chemistry and nutritional value of quaking aspen (*Populus tremuloides* Michx.) leaves for the large aspen tortrix (*Choristoneura conflictana* (Walker)). Oecologia (Berlin), **73**: 513–517.

Bush, G.L. 1974. The mechanism of sympatric host race formation in the true fruit flies (Tephritidae). *In* Genetic mechanisms of speciation in insects. *Edited by* M.J.D. White. Australian and New Zealand Book Co., Sydney, Australia. pp. 3–23.

Bush, G.L. 1975. Sympatric speciation in phytophagous parasitic insects. *In* Evolutionary strategies of parasitic insects and mites. *Edited by* P.W. Price. Plenum Press, New York, NY. pp. 187–207.

Cantor, L., and Whitham, T.G. 1989. Importance of belowground herbivory: pocket gophers may limit aspen to rock outcrop refugia. Ecology, **70**: 962–970.

Carothers, S.W., Johnson, R.R., and Aitchison, S.W. 1974. Population and social organization of southwestern riparian birds. Am. Zool. **14**: 97–108.

Christensen, K.M., Whitham, T.G., and Keim, P. 1995. Herbivory and tree mortality across a pinyon pine hybrid zone. Oecologia (Berlin), **101**: 29–36.

Clausen, T.P., Reichardt, P.B., Bryant, J.P., Werner, R.A., Post, K., and Frisby, K. 1989. Chemical model for short-term induction in quaking aspen (*Populus tremuloides*) foliage against herbivores. J. Chem. Ecol. **15**: 2335–2346.

Crawford, D.J. 1974. A morphological and chemical study of *Populus acuminata* Rydberg. Brittonia, **26**: 74–89.

Davidson, A.G., and Prentice, R.M. 1968. Insects and diseases. *In* Growth and utilization of poplars in Canada. *Edited by* J.S. Maini and J.H. Cayford. Ministry of Forestry and Rural Development, Ottawa, ON. pp. 116–144.

Denno, R.F., Larsson, S., and Olmstead, K.L. 1990. Role of enemy-free space and plant quality in host plant selection by willow leaf beetles. Ecology, **71**: 124–137.

Dickson, L.L., and Whitham, T.G. 1996. Genetically-based plant resistance traits affect arthropods, fungi, and birds. Oecologia (Berlin), **106**: 400–406.

Dungey, H.S., Potts, B.M., Whitham, T.G., and Morrow, P.A. 1994. Plant hybrid zones as centres of biodiversity: evidence from *Eucalyptus*. *In* Proceedings of the International Forest Biodiversity Conference. Conserving biological diversity in temperate forest ecosystems — towards sustainable management. 4–9 December 1994. Canberra, Australia. pp. 67–68.

Eckenwalder, J.E. 1984*a*. Natural intersectional hybridization between North American species of *Populus* (Salicaceae) in sections *Aigeiros* and *Tacamahaca*. I. Population studies of *P.* ×*parryi*. Can. J. Bot. **62**: 317–324.

Eckenwalder, J.E. 1984*b*. Natural intersectional hybridization between North American species of *Populus* (Salicaceae) in sections *Aigeiros* and *Tacamahaca*. II. Taxonomy. Can. J. Bot. **62**: 325–335.

Eckenwalder, J.E. 1984*c*. Natural intersectional hybridization between North American species of *Populus* (Salicaceae) in sections *Aigeiros* and *Tacamahaca*. III. Paleobotany and evolution. Can. J. Bot. **62**: 336–342.

Edwards, W.R.N. 1978. Effect of salicin content on palatibility of *Populus* foliage to opossum (*Trichosurus vulpecula*). N. Z. J. Sci. **21**: 103–106.

Ellstrand, N.C. 1992. Gene flow by pollen: implications for plant conservation genetics. Oikos, **63**: 77–86.

Ericson. L., Burdon, J.J., and Wennstrom, A. 1993. Inter-specific host hyrbids and phalacrid beetles implicated in the local survival of smut pathogens. Oikos, **68**: 393–400.

Farr, D.F., Bills, G.F., Cnamuris, G.P., and Rossman, A.Y. 1989. Fungi on plants and plant products. The US APS Press, St. Paul, MN.

Finch, D.M., and Ruggiero, L.F. 1993. Wildlife habitats and biological diversity in the Rocky Mountains and northern Great Plains. Nat. Areas J. **13**: 191–203.

Floate, K.D., and Whitham, T.G. 1993. The "hybrid bridge" hypothesis: host shifting via plant hybrid swarms. Am. Nat. **141**: 651–662.

Floate, K.D., and Whitham, T.G. 1994. Aphid-ant interaction reduces chrysomelid herbivory in a cottonwood hybrid zone. Oecologia (Berlin), **97**: 215–221.

Floate, K.D., and Whitham, T.G. 1995. Insects as traits in plant systematics: their use in discriminating between hybrid cottonwoods. Can. J. Bot. **73**: 1–13.

Floate, K.D., Kearsley, M.J.C., and Whitham, T.G. 1993. Elevated herbivory in plant hybrid zones: *Chrysomela confluens*, *Populus* and phenological sinks. Ecology, **74**: 2056–2065.

Floate, K.D., Whitham, T.G., and Keim, P. 1994. Morphological versus genetic markers in classifying hybrid plants. Evolution, **48**: 929–930.

Floate, K.D., Fernandes, G.W., and Nilsson, J.A. 1995. Distinguishing intrapopulational categories of plants by their insect faunas: galls on rabbitbrush. Oecologia (Berlin), **105**: 221–229.

Floate, K.D., Martinsen, G.D., and Whitham, T.G. 1996. Cottonwood hybrid zones as centers of abundance for gall aphids in western North America: importance of relative habitat size. J. Anim. Ecol. In press.

Fritz, R.S., Nichols-Orians, C.M., and Brunsfeld, S.J. 1994. Interspecific hybridization of plants and resistance to herbivores: hypotheses, genetics, and variable responses in a diverse herbivore community. Oecologia (Berlin), **97**: 106–117.

Futuyma, D.J. 1983. Selective factors in the evolution of host choice by phytophagous insects. *In* Herbivorous insects. *Edited by* S. Ahmad. Academic Press, New York, NY. pp. 227–244.

Gallun, R.L., and Khush, G.S. 1980. Genetic factors affecting expression and stability of resistance. *In* Breeding plants resistant to insects. *Edited by* F.G. Maxwell and P.R. Jennings. Wiley, New York, NY. pp. 63–85.

Gehring, C.A., and Whitham, T.G. 1991. Herbivore-driven mycorrhizal mutualism in insect-susceptible pinyon pine. Nature (London), 353: 556–557.

Gould, F. 1986. Simulation models for predicting durabililty of insect-resistant germ plasm: Hessian fly (Diptera: Cecidomyiidae)-resistant winter wheat. Envir. Entomol. 15: 11–23.

Grant, V. 1971. Plant speciation. Columbia University Press, New York, NY.

Greenaway, W., English, S., Whatley, F.R., and Rood, S.B. 1991. Interrelationships of poplars in a hybrid swarm as studied by gas chromatography -mass spectrometry. Can. J. Bot. 69: 203–208.

Greenaway, W., English, S., and Whatley, F.R. 1992. Relationships of *Populus* ×*acuminata* and *Populus* ×*generosa* with their parental species examined by gas chromatography — mass spectrometry of bud exudates. Can. J. Bot. 70: 212–221.

Huxley, C.R., and Cutler, D.F. 1991. Ant-plant interactions. Oxford University Press, Oxford, New York, NY.

Ives, W.G.H., and Wong, H.R. 1988. Tree and shrub insects of the prairie provinces. Can. For. Serv., North. For. Res. Cent., Edmonton, AB. Inf. Rep. NOR-X-292.

Jermy, T. 1984. Evolution of insect/host plant relationships. Am. Nat. 124: 609–630.

Johnson, R.R., Haight, L.T., and Simpson, J.M. 1977. Endangered species vs. endangered habitats: a concept. *In* Importance, preservation and management of riparian habitat: a symposium. *Technical coordinators:* R.R. Johnson and D.A. Jones, Jr. U.S. Dep. Agric., For. Serv., Gen. Tech. Rep. RM-43. pp. 68–79.

Jones, A.G., and Seigler, D.S. 1975. Flavonoid data and populational observations in support of hybrid status for *Populus acuminata*. Biochem. Syst. Ecol. 2: 201–206.

Julkunen-Titto, R. 1986. A chemotaxonomic survey of of phenolics in leaves of Northern Salicaceae species. Phytochemistry, 25: 663–667.

Kat, P.W. 1985. Historical evidence for fluctuation in levels of hybridization. Evolution, 39: 1164–1169.

Kemperman, J.A., and Barnes, B.V. 1976. Clone size in American aspens. Can. J. Bot. 54: 2603–2607.

Kearsley, M.J.C., and Whitham, T.G. 1989. Developmental changes in resistance to herbivory: implications for individuals and populations. Ecology, 70: 422–434.

Kearsley, M.J.C., and Whitham, T.G. 1992. Guns and butter: a no cost defense against predation. Oecologia (Berlin), 92: 556–562.

Keim, P., Paige, K.N., Whitham, T.G., and Lark, K.G. 1989. Genetic analysis of an interspecific hybrid swarm of *Populus*: occurrence of unidirectional introgression. Genetics, 123: 557–565.

Knopf, F.L. 1985. Significance of riparian vegetation to breeding birds across an altitudinal cline. *In* Riparian ecosystems and their management: reconciling conflicting uses. *Technical coordinators*: R.R. Johnson, C.D. Ziebell, D.R. Patten, P.F. Ffolliot, and R.H. Hamre. U.S. Dep. Agric, For. Serv., Gen. Tech. Rep. RM-120. pp. 105–111.

Larson, K.C., and Whitham, T.G. 1991. Manipulation of food resources by a gall-forming aphid: the physiology of sink-source interactions. Oecologia (Berlin), 88: 15–21.

Larsson, S., Wirén, A., Lundgren, L., and Ericsson, T. 1986. Effects of light and nutrient stress on leaf phenolic chemistry in *Salix dasyclados* and susceptibility to *Galerucella lineola* (Coleoptera). Oikos, 47: 205–210.

Lindroth, R.L., Hsia, M.T.S., and Scriber, J.M. 1987. Seasonal patterns in the phytochemistry of three *Populus* species. Biochem. Syst. Ecol. 21: 535–542.

Lindroth, R.L., Scriber, J.M., and Hsia, M.T.S. 1988.. Chemical ecology of the tiger swallotail: mediation of host use by phenolic glycosides. Ecology, 69: 814–822.

Lindroth, R.L., Kinney, K.K., and Platz, C.L. 1993. Responses of deciduous trees to elevated atmospheric CO_2: Productivity, phytochemistry, and insect performance. Ecology, 74: 763–777.

Manley, S.A.M., and Fowler, D.P. 1969. Spruce budworm defoliation in relation to introgression in red and black spruce. For. Sci. **15**: 365–366.

Martinsen, G.D., and Whitham, T.G. 1994. More birds nest in hybrid cottonwoods. Wilson Bull. **106**: 474–481.

Martinsen, G.D., Driebe, E.M., and Whitham, T.G. 1996. Indirect interactions mediated by changing plant chemistry: beaver browsing benefits beetles. Ecology. In review.

Meyer, F.H. 1973. Distribution of ectomycorrhizae in native and man-made forests. *In* Ectomycorrhizae: their ecology and physiology. *Edited by* J.C. Marks and T.T. Kozlowski. Academic Press, New York, NY. pp. 79–106.

Moorehead, J.R., Taper, M.L., and Case, T.J. 1993. Utilization of hybrid oak hosts by a monophagous gall wasp: How little host character is sufficient? Oecologia (Berlin), **95**: 385–392.

Moran, N.A. 1991. Phenotype fixation and genotypic diversity in the complex life cycle of the aphid *Pemphigus betae*. Evolution, **45**: 957–970.

Moran, N.A., and Whitham, T.G. 1988. Evolutionary reduction of complex life cycles: loss of host alternation in *Pemphigus* (Homoptera: Aphididae). Evolution, **42**: 717–728.

Morris, M.G., Collins, N.M., Vane-Wright, R.I., and Waage, J. 1990. The utilization and value of non-domesticated insects. *In* The conservation of insects and their habitats. *Edited by* N.M. Collins and J.A. Thomas. Academic Press, London, UK. pp. 319–347.

Morrow, P.A., Whitham, T.G., Potts, B.M., Ladiges, P., Ashton, D.H., and Williams, J. 1994. Gall-forming insects concentrate on hybrid phenotypes of *Eucalyptus*. *In* The ecology and evolution of gall-forming insects. *Edited by* P.W. Price, W.J. Mattson, and Y.N. Baranchikov. U.S. Dep. Agric., For. Serv., North Cent. For. Exp. Stn. GTR NC-174. pp. 121–134.

Orians, C.M., and Fritz, R.S. 1995. Secondary chemistry of hybrid and paretal willows: phenolic glycosides and condensed tannins in *Salix sericea*, *S. eriocephala* and their hybrids. J. Chem. Ecol. **21**: 1245–1253.

Ostry, M.E., Wilson, L.F., McNabb, H.S., Jr., and Moore, L.M. 1988. A guide to insect, disease and animal pests of poplars. U.S. Dep. of Agric. (Washington, DC), Agric. Handb. 677.

Paige, K.N., Keim, P., Whitham, T.G., and Lark, K.G. 1990). The use of restriction fragment length polymorphisms to study the ecology and evolutionary biology of aphid-plant interactions. *In* Aphid-plant genotype interactions. *Edited by* R.K. Campbell and R.D. Eikenbary. Elsevier, Amsterdam. pp. 69–87.

Paige, K.N., and Capman, W.C. 1993. The effects of host-plant genotype, hybridization, and environment on gall-aphid attack and survival in cottonwood: the importance of genetic studies and the utility of RFLP's. Evolution, **47**: 36–45.

Palo, R.T. 1984. Distribution of birch (*Betula* spp.), willow (*Salix* spp.), and poplar (*Populus* spp.) secondary metabolites and their potential role as chemical defense against herbivores. J. Chem. Ecol. **10**: 499–520.

Pasteels, J.M., Rowell-Rahier, M., Braekman, J.C., and Dupont, A. 1983. Salicin from host plant as precursor of salicylaldehyde in defensive secretion of chrysomeline larvae. Physiol. Entomol. **8**: 307–314.

Price, P.W. 1991. The plant vigor hypothesis and herbivore attack. Oikos, **62**: 244–251.

Rabin, L.B., and Pacovsky, R.S. 1985. Reduced larva growth of two lepidoptera (Noctuidae) on excised leaves of soybean infected with a mycorrhizal fungus. J. Econ. Entomol. **78**: 1358–1363.

Rank, N.E. 1992. Host plant preference based on salicylate chemistry in a willow leaf beetle (*Chyrsomela aeneicollis*). Oecologia (Berlin), **90**: 95–101.

Rattenbury, J.A. 1962. Cyclic hybridization as a survival mechanism in the New Zealand forest flora. Evolution, **16**: 348–363.

Reichardt, P.B., Bryant, J.P., Mattes, B.R., Clausen, T.P., Chapin, F.S., III, Meyer, M. 1990. Winter chemical defense of Alaskan balsam poplar against snowshoe hares. J. Chem. Ecol. **16**: 1941–1959.

Rieseberg, L.H., and Ellstrand, N.C. 1993. What can morphological and molecular markers tell us about plant hybridization. Crit. Rev. Plant Sci. **12**: 213–241.

Romme, W.H., Turner, M.G., Wallace, L.L., and Walker, J.S. 1995. Aspen, elk, and fire in northern Yellowstone National Park. Ecology, **76**: 2097–2106.

Rood, S.B., Mahoney, J.M. 1990. Collapse of riparian poplar forests downstream from dams in western prairies: probable causes and prospects for mitigation. Environ. Manage. **14**: 451–464.

Rowell-Rahier, M. 1984. The food plant preferences of *Phratora vitellinae* (Coleoptera: Chrysomelidae). A. field observations. Oecologia (Berlin), **64**: 369–374.

Schreiner, E.J. 1971. Genetics of Eastern cottonwood. U.S. Dep. Agric., For. Serv. Res. Pap. WO-11.

Smiley, J.T., Horn, J.M., and Rank, N.E. 1985. Ecological effects of salicin at three trophic levels: new problems from old adaptations. Science (Washington, DC), **229**: 649–651.

Smith, R.L., and Sytsma, K.J. 1990. Evolution of *Populus nigra* (Sect. *Aigeiros*): Introgressive hybridization and the chloroplast contribution of *Populus alba* (Sect. *Populus*). Am. J. Bot. **77**: 1176–1187.

Soetens, P.H., Rowell-Rahier, M., and Pasteels, J.M. 1991. Influence of phenolglucosides and trichome density on the distribution of insects herbivores on willows. Entomol. Exp. Appl. **59**: 175–187.

Stace, C.A. 1987. Hybridization and the plant species. *In* Differentiation patterns in higher plants. *Edited by* K.M. Urbanska. Academic Press, New York, NY. pp. 115–127.

Stebbins, G.L. 1950. Variation and evolution in plants. Columbia University Press, New York, NY.

Strauss, S.Y. 1994. Levels of herbivory and parasitism in host hybrid zones. TREE, **9**: 209–214.

Strong, T.R., and Bock, C.E. 1990. Bird species distribution patterns in riparian habitats in southeastern Arizona. Condor, **92**: 866–885.

Szaro, R.C. 1980. Factors influencing bird populations in southwestern riparian forests. *In* Workshop proceedings: management of western forests and grasslands for nongame birds. *Technical coordinator*: R.M. DeGraff. U.S. Dep. Agric, For. Serv., Gen. Tech. Rep. INT-86. pp. 403–418.

Tabashnik, B.E. 1983. Host range evolution: the shift from native legume hosts to alfalfa by the butterfly, *Colias philodice eriphyle*. Evolution, **37**: 150–162.

Tahvanainen, J., Helle, E., Julkenen-Tiitto, R., Lavola, A. 1985*a*. Phenolic compounds of willow bark as deterrents against feeding by mountain hare. Oecologia (Berlin), **65**: 319–323.

Tahvanainen, J., Julkenen-Tiitto, R., and Kettunen, J. 1985*b*. Phenolic glycosides govern the food selection pattern of willow feeding leaf beetles. Oecologia (Berlin), **67**: 52–56.

Thomas, C.D., Ng, D., Singer, M.C., Mallet, J.L.B., Parmesan, C., and Billington, H.L. 1987. Incorporation of a European weed into the diet of a North American herbivore. Evolution, **41**: 892–901.

Tingey, W.M., and Singh, S.R. 1980. Environmental factors influencing the magnitude and expression of resistance. *In* Breeding plants resistant to insects. *Edited by* F.G. Maxwell and P.R. Jennings. Wiley, New York, NY. pp. 87–113.

Waltz, A.M., and Whitham, T.G. 1996. Plant development directly and indirectly affects arthropod community structure: opposing impacts of species removal. Ecology. In press.

White, T.C.R. 1969. An index to measure weather-induced stress of trees associated with outbreaks of psyllids in Australia. Ecology, **50**: 905–909.

Whitham, T.G. 1983. Host manipulation of parasites: within-plant variation as a defense against rapidly evolving pests. *In* Variable plants and herbivores in natural and managed systems. *Edited by* R.F. Denno and M.S. McClure. Academic Press, New York, NY. pp. 15–41.

Whitham, T.G. 1989. Plant hybrid zones as sinks for pests. Science (Washington, DC), **244**: 1490–1493.

Whitham, T.G., Maschinski, J., Larson, K.C., and Paige, K.N. 1991. Plant responses to herbivory: the continuum from negative to positive and underlying physiological mechanisms. *In* Plant-animal interactions: evolutionary ecology in tropical and temperate regions. *Edited by* P.W. Price, T.M. Lewinsohn, G.W. Fernandes, and W.W. Benson. Wiley, New York, NY. pp. 227–256.

Whitham, T.G., Morrow, P.A., and Potts, B.M. 1991. Conservation of hybrid plants. Science (Washington, DC), **254**: 779–780.

Whitham, T.G., Morrow, P.A., and Potts, B.M. 1994. Plant hybrid zones as centers of biodiversity: the herbivore community of two endemic Tasmanian eucalypts. Oecologia (Berlin), **97**: 481–490.

Williams, A.G., and Whitham, T.G. 1986. Premature leaf abscission: an induced plant defense against gall aphids. Ecology, **67**: 1619–1627.

Wollenweber, E. 1975. Flavonoidmuster als systematisches merkmal in der gattung *Populus*. Biochem. Syst. Ecol. **3**: 35–45.

Zucker, W.V. 1982. How aphids choose leaves: the role of phenolics in host selection by a galling aphid. Ecology, **63**: 972–981.

PART II. PHYSIOLOGY OF GROWTH, PRODUCTIVITY, AND STRESS RESPONSE

Overview

T.M. Hinckley

The overview of the first part of this book used four themes (*Understanding natural variation, Genetic manipulation of natural variation, Conservation and restoration of variation*, and *Poplar as a model tree*) to present the associated chapters. Certain elements of those themes are repeated in the chapters associated with this part; for example, the use of *Poplar as a model tree*. The pioneering work of Dr. Philip Larson on the study of vascularization and the development of meristematic tissues in *Populus*, described in Chapters 13 (Telewski, Aloni, and Sauter) and 15 (Ceulemans and Isebrands) set the stage for subsequent understanding of leaf development and maturation, patterns of translocation, the movement of wound signals, etc. This understanding has laid much of the foundation for conceptual models used in studying other woody plants. More recent research has shown that *Populus* offers important opportunities for understanding leaf growth, leaf senescence, the regulation of water transport both within the plant and from the plant to the atmosphere, carbon fixation and allocation, etc. Aspects of this historical work and the newly emerging findings within *Populus* physiology and morphology are developed in the chapters found in Part II. Poplar may then serve as a model tree for a number of biological disciplines. Poplar may also serve as a model by which different disciplines are bridged and integrated. Chapter 19 (Dickmann and Keathley) synthesizes much of our current understanding of Poplar as a model tree and provides the reader with a sense of how physiology, culture, breeding, and molecular biology all fit nicely together within this organism.

The other three themes from the first part deal with some aspect of *variation* (its understanding, manipulation, and preservation). Much of the material presented in the second part of this book also deals with variation, but in a context that is relatively new for physiologists, including physiological ecologists and production physiologists. First, variation is a major issue in the emerging efforts to scale or integrate information collected at one level of biological organization and to extrapolate that information to higher levels. Second, genetic and life history variation exist in the material studied. Previously efforts were made to minimize or control this variation (assuming it was recognized). Now there is a much greater emphasis on the specific use, the potential manipulation and the understanding of genetic variation in the elucidation of plant processes. We will see that these two contexts are intimately intertwined.

Scaling involves several issues (Fig. 1). First, can information collected at lower levels of biological information (e.g., at the organelle, cell, or organ level) be scaled or taken to higher levels of biological information (e.g., the whole tree, stand, ecosystem)? If the answer is yes, or even maybe, then what conditions or constraints are placed on how information collected at a lower level is taken to a higher level? For example, physiologists like to measure stomatal conductances and net photosynthesis of leaves. Someone managing a poplar plantation might like to know how much water the plantation uses or needs in a day or how much volume growth will occur over a growing season. Indeed such questions regarding water use are arising even on the moister west side of the Cascades in the Pacific Northwest. As poplar culture moves into former agricultural land, for a period there will be a mosaic of land uses; that is, poplar plantations intermixed with agricultural fields. Owners of the agricultural fields may perceive the poplar tree as a giant wick that will drain the soil of all moisture. The challenge for the plantation manager will be to allay such fears. The challenge for the physiologist is to provide definitive answers to these questions. Initially, the physiologist may only have leaf level information, but as illustrated in Chapters 15 (Ceulemans and Isebrands) and 18 (Heilman, Hinckley, Roberts, and Ceulemans), physiologists are beginning to derive whole-tree and stand answers. Finally, can one skip the lower levels and come at the issue from a top-down or a higher level? A section in Chapter 18 on remote sensing examines one aspect of this last question.

Almost every chapter in Part II directly or indirectly addresses some issue of scaling. Van Volkenburgh and Taylor (Chapter 12, Leaf growth physiology) describe the mechanisms of leaf growth in general and in *Populus* specifically. Many environmental factors, such as anoxia, soil drought, and elevated carbon dioxide concentrations, affect leaf growth by impacting cell wall extensibility. In addition, elevated carbon dioxide stimulates cell production or increases the rate of leaf primordia production in some clones. Since leaf area and stem-wood production are closely linked, understanding how environmental factors affect leaf initiation, growth, and area is clearly important. As Van Volkenburgh and Taylor note in their chapter: "These data on leaf growth and elevated CO_2 suggest that in the future, the rate of leaf expansion, leaf production, and total tree leaf area may be stimulated for a wide range of hybrid poplar clones." Two unanswered questions, however, remain: (1) will the leaf and whole-plant-level increases also be found in stands and (2) what is the nature of the genetic control over leaf production? The former is a classical scaling question whereas the latter is important for breeding and selection.

In Chapter 13, Telewski, Aloni, and Sauter review current knowledge regarding the development, structure, and function of secondary tissue in *Populus*. They

Fig. 1. Scaling diagram for leaf-to-stand levels of biological organization in *Populus*. Diagram emphasizes carbon exchange and growth. Activities and processes at one scale may be propagated to higher scales (arrow) and higher scales may set limits or determine feedback on processes at lower scales. For example, the position of a leaf on the tree determines (1) the leaf's position within the canopy and its microclimate, (2) its hydraulic connection to the tree, and (3) its source–sink regimes. The entire system illustrated here is affected by genotype and by variables at smaller and larger scales not illustrated.

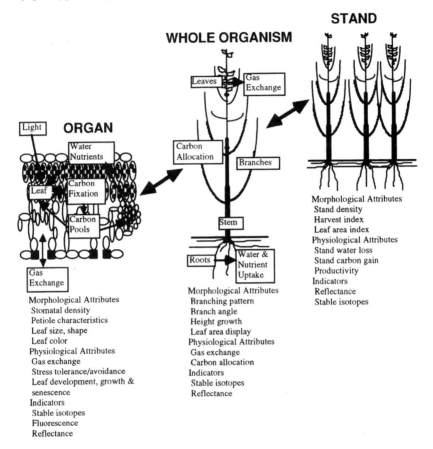

frame much of their chapter around the concept that structure and function are intimately related and that secondary cell and tissue structure vary considerably, depending on genetic predisposition and environmental and cultural conditions. It is from this variation that whole-tree- and stand-level gains in function (i.e., reduced xylem hydraulic vulnerability, see Chapter 16, Blake, Sperry, Tschaplinski, and Wang, for more detail) and wood properties (i.e., fiber size) might be realized. Because *Populus* can serve as a model system for understanding primary and secondary tissue form and function, Telewski et al. see novel as well as practical pieces of information emerging from continued studies of the economically and biologically important tissue that connects leaves to roots. Again the challenge will be to apply these cell and tissue levels of understanding to the whole organism and stand.

In addition to reviewing what is known regarding carbon assimilation and carbon allocation and partitioning in *Populus*, Ceulemans and Isebrands (Chapter 15) explain the nature and function of a number of whole-tree growth models that have been developed specifically for poplar. Modeling is an important aspect of scaling or integration. Modeling allows one to use mechanistic information collected at, for example, the leaf level to understand and predict whole-canopy carbon gain. Through the use of models, the effects of changes in canopy architecture or stand density can be evaluated. Modeling probably provides the best current tool for scaling information collected at smaller scales to higher scales; however, this cannot be undertaken without an understanding of the scaling process. The close agreement between seasonal carbon fixation estimated from ECOPHYS and actual individual plant productivity suggests that these poplar models have predictive power and at the same time offer experimental opportunities.

Chapters 18 (Heilman et al.) and 19 (Dickmann and Keathley) both try to make the leap to plantations of poplar. Clearly, many gains have been made in our whole-tree to stand-level understanding of this genus, but the insights gained at the organ scale in carbon, water, and stress physiology (Chapters 15, 16, and 17) are still elusive at these higher levels. A similar discrepancy is evident when we compare our level of understanding of aboveground growth and development (e.g., Chapters 12, 13, 15) to belowground processes (Chapter 14, Pregitzer and Friend). Root systems and root function, particularly in a field context, are much more difficult to study than leaves and canopies.

The second issue that all of the chapters in Part II directly or indirectly address is the issue of genetic variation. Figure 2 illustrates the nature of this genetic variation among closely related F_2 individuals of hybrid Family 331, a pedigree formed from the hybridization of *P. trichocarpa* and *P. deltoides* parents and the subsequent breeding of two F_1 offspring (see Chapter 8, Bradshaw for more details on this pedigree) (Woo 1996). This study was conducted to measure ozone sensitivity in 16 different clones with the ultimate objective of attempting to explain the various underlying physiological mechanisms and their genetic control. Six replicates of each clone were studied; three replicates were exposed to 100 ppb ozone for 6 h each day. The other 3 replicates were exposed to charcoal filtered air. The response in daily net assimilation shown in this figure was observed by measuring net assimilation every 2 h for 24 h approximately 40 d after initiation of exposure. Net assimilation in the ozone exposed ramets is expressed as a percentage of the mean of 3 ramets exposed to charcoal-filtered air. The *P. deltoides* parent appeared more sensitive than the *P. trichocarpa* parent, and the F_1 parents somewhat intermediate, whereas the F_2 clones range from extremely sensitive (e.g., 331–1130) to almost as insensitive as the

Fig. 2. The total daily net carbon exchange of ozone treated clones is expressed as a percentage of that observed in the control (i.e., charcoal filtered air). Carbon exchange was measured on a recent mature leaf of each of 3 replicates of each clone. Clones were placed in open top chambers and given either ozone (90–110 ppb) or charcoal filtered air for a 9-wk period (figure redrawn from Woo 1996).

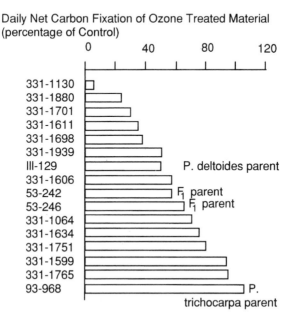

P. trichocarpa parent (e.g., 331–1765). Knowing that this much variation exists may elicit three reactions: *excitement*, as this variation provides the foundation from which real mechanistic understanding might be derived and linked, via marker maps and other molecular tools, to other morphogenic and physiological processes operating in *Populus*; *caution*, as this variation must be accounted for in sampling and scaling efforts; and finally *skepticism,* as simple axioms of function (e.g., maximum stomatal conductance and ozone sensitivity are highly related) may not exist, as variation is explored at a level of detail not heretofore possible, or in genetically reshuffled genomes. The various chapters in this section discuss the nature and degree of such genetic variation in terms of process or structure. *Populus* offers a platform for the next generation of studies on which genetic and physiological information can be combined in models and in scaling exercises.

References

Woo, S.-Y. 1996. Physiological differences between resistant and sensitive clones of hybrid poplar (*Populus trichocarpa* × *Populus deltoides*) to ozone exposure. Unpublished Ph.D. dissertation, University of Washington, Seattle, WA.

CHAPTER 12
Leaf growth physiology

Elizabeth Van Volkenburgh and Gail Taylor

Introduction

The growth and productivity of trees, whether in agricultural or natural settings, depends in large part on the expansion and display of photosynthetic surface area, the leaves. The timing of bud break in the spring, the pattern of cell formation and expansion in the growing leaves, the rate of leaf expansion, and the duration and efficiency of photosynthetic activity through the growing season all affect the amount of carbon that can be fixed, and ultimately the size of the tree. All of these traits vary to lesser or greater degree among species and genotypes of *Populus*. The genetic variation in leaf growth characteristics gives rise to possibilities of novel combinations of traits that may optimize leaf growth and function for particular habitats or silvicultural practices. Each characteristic of leaf growth is complex, involving many biochemical processes which are largely undescribed, and the regulation of leaf growth at the cellular level is a problem of considerable interest and potential for discovery. The following presents a framework for considering the development and growth of dicotyledonous leaves in general, with specific examples where they are known from the genus *Populus*.

Leaf development in *Populus*

Leaf formation and bud break

Leaf growth is initiated at the apical meristem of stems or branches, where leaf primordia are formed by production of new cell layers that generate a planar form. The newly formed leaves may directly develop and expand to full size, as occurs for most of the leaves produced by poplar trees during a growing season.

E. Van Volkenburgh. Botany Department, University of Washington, Seattle, WA 98195-1330, USA.
G. Taylor. School of Biological Sciences, University of Sussex, Falmer, Brighton BN1 9QG, UK.
Correct citation: Van Volkenburgh, E., and Taylor, G. 1996. Leaf growth physiology. *In* Biology of *Populus* and its implications for management and conservation. Part II, Chapter 12. *Edited by* R.F. Stettler, H.D. Bradshaw, Jr., P.E. Heilman, and T.M. Hinckley. NRC Research Press, National Research Council of Canada, Ottawa, ON. pp. 283–299.

Alternatively, development of the primordia may be arrested, giving rise to dormant lateral or terminal buds. In the early spring, "bud break" reinitiates growth of preformed leaves which are primarily responsible for providing photosynthate for stem growth during the beginning of the growing season. The timing of bud break is crucial for optimizing photosynthetic production while minimizing the hazard of experiencing a late frost. Although it is known that the timing of bud break is genetically regulated, that it relies on the perception of day length by the pigment phytochrome, and varies among genotypes, the mechanisms regulating the release of bud dormancy are not yet understood (Mohr and Schopfer 1995).

Leaf shape and anatomy

During the formation of the leaf primordium and expansion of the lamina, the orientation of planes of cell division as well as the directionality of cell expansion produce an anatomical structure unique to each species or genotype (Green 1980). Leaves develop basipetally, with the meristematically active cells leaving the tip of the leaf to mature while the base of the leaf is formed (Avery 1933; Dale 1988). Variations among parental species and hybrids of poplar with respect to leaf development and photosynthetic characteristics have been described by Isebrands and Larson (1973) and Ceulemans et al. (1987). Leaves of *Populus* range in shape from lanceolate to deltoid (Hitchcock and Cronquist 1973). The width-to-length ratio reported in one study of poplar genotypes ranged from 0.4 to 0.7 for sylleptic and current terminal leaves of *P. trichocarpa*, compared to 0.8–1.2 in *P. deltoides*, and the ratio was highly variable among individuals in the F_2 generation created from their interspecific cross (Wu et al. 1996). These differences in leaf shape most likely reflect differences in the duration of cell division in the meristematic region at the basal end of the leaf.

Developmental differences between leaves of *P. trichocarpa* and *P. deltoides* are clear from estimates of the number of epidermal cells on the adaxial surface of leaves monitored over the course of leaf expansion (Fig. 1). These data indicate that cell division is completed very early in the development of a *P. trichocarpa* leaf (cell number reached a maximum when the leaves were less than 5% full size), whereas in *P. deltoides* cell number per leaf continued to increase until the leaf reached full size. The size of *P. trichocarpa* leaves, therefore, must be obtained primarily by prolonged expansion of cells after they are formed, whereas leaves of *P. deltoides* expand by continually producing and enlarging new cells. Consistent with this is the observation that epidermal cells of *P. trichocarpa* are generally much larger than those of *P. deltoides* (Ridge et al. 1986). During maturation of the leaf, continued meristematic activity at the base of the *P. deltoides* leaf contributes to widening the leaf base.

Leaves of *P. trichocarpa* are displayed nearly horizontally, compared to the vertical display of *P. deltoides*. This difference in orientation with respect to sunlight is reflected by the arrangement of photosynthetic cell layers within the

Fig. 1. Epidermal cell number per adaxial leaf surface for *P. trichocarpa* (∗, ▲) and *P. deltoides* (⊠, ■) grown from cuttings in pots in a growth chamber at low $(150–300 \ \mu mol \cdot m^{-2} \cdot s^{-1}$; ▲, ■) or high fluence rate $(700 \ \mu mol \cdot m^{-2} \cdot s^{-1}$; ∗, ⊠). Leaf impressions were made nondestructively with dental paste so that individual leaves could be followed through time; cells were counted at 440× on transferred acrylic impressions. (Data from Ms. Pamela Clum, Botany Department, University of Washington.)

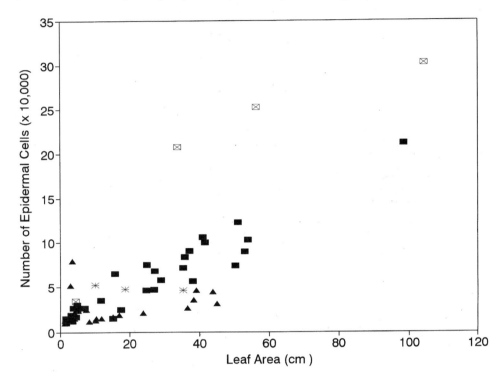

leaves (Fig. 2). In *P. trichocarpa* (Fig. 2A), two layers of palisade mesophyll are formed above an amorphous layer of spongy mesophyll that is penetrated by veins. Within the spongy layer, it is possible to discern a horizontal contiguity among cells, which may be related to the paraveinal mesophyll described in legumes (Fisher 1967; Kevekordes et al. 1988) and thought to be involved in assimilate transport (Franceschi and Giaquinta 1983). The photosynthetic cell layers are situated above a layer of achlorophyllous, extremely elongate cells that define large intercellular airspaces. The presence of airspaces is correlated with the reflectivity of the underside of these leaves, which look white and spongy. The functional advantage of such large airspaces has not been quantified, although similar anatomical structure has been noted for species growing in flooded habitats (McDougall 1931) and has been discussed with regard to gas exchange (Givnish 1979) and reflectivity (DeLucia et al. 1996). Stomatal distribution is highly asymmetric, with most stomates developing on the lower epidermal surface (E. Van Volkenburgh and R.F. Stettler, unpublished data).

Fig. 2A. Scanning electron microscope images of cross sections of leaves from *P. trichocarpa*. Bar indicates 100 μm. (Tissue preparation and photographs by Ms. Dale Blum, Botany Department, University of Washington.)

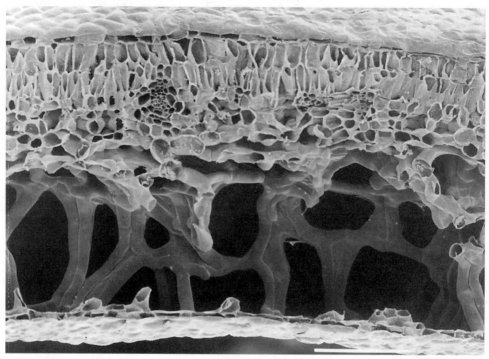

Leaves of *P. deltoides* (Fig. 2B) contain a "double palisade" with two layers of palisade mesophyll under both epidermal surfaces, connected by an amorphous cell layer that may function as paraveinal mesophyll (as described above). All of these cell layers contain chloroplasts, and there is no layer comparable to that in *P. trichocarpa* containing large intercellular air spaces. As a result, the underside of these leaves is green, and stomatal distribution is more symmetrical with about one-third as many stomates on the upper surface as on the lower (E. Van Volkenburgh and R.F. Stettler, unpublished data). Sidedness of the leaves can be easily determined by both stomatal distribution and vein architecture; the veins protrude from the lower surface.

The leaf anatomy of hybrids between *P. trichocarpa* and *P. deltoides* is variable. The structure of F_1 leaves is intermediate between those of the parents, whereas the F_2 leaves exhibit a wide range of traits spanning from one parent to the other. Generally the relative amount of air space in the lower mesophyll is correlated with the "greenness" of the underside of the leaf, and the stomatal distribution is correlated with the degree of development of the lower mesophyll into palisade tissue.

Fig. 2B. Scanning electron microscope images of cross sections of leaves from *P. deltoides*. Bar indicates 100 µm. (Tissue preparation and photographs by Ms. Dale Blum, Botany Department, University of Washington.)

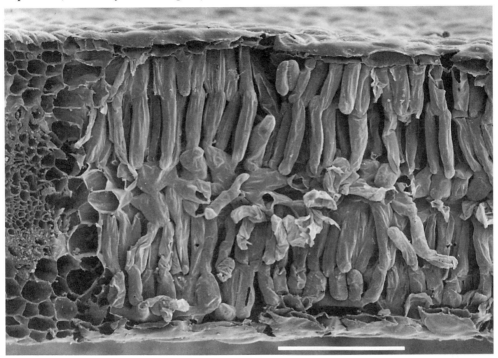

Contribution of cell division and cell expansion

The final size and shape of leaves will depend both on the total number of cells produced, and their volume. Although it can be argued that plant morphology is dictated by a higher organizational principle than the production of cells alone (Haber and Foard 1963; Kaplan and Hagemann 1991), it has frequently been observed within a species that leaf size differs from one environment to the next because of changes in either cell number or cell size (Humphries and Wheeler 1963; Verbelen and De Greef 1979; Smit et al. 1989). In a comparative study of leaf growth characteristics of *P. trichocarpa* and *P. deltoides* and their hybrids, Ridge et al. (1986) suggested that leaves of *P. trichocarpa* exert a genetic restraint on cell division, with leaf area being obtained largely from cell expansion, whereas leaves of *P. deltoides* exert a constraint on cell expansion, with leaf area being obtained from cell division. The expression of hybrid vigor in their progeny, it was suggested, includes removal of these restraints allowing hybrid leaves to produce many, large cells. From such observations, the underlying limitations to cell division or enlargement become important to understand. In a review of this subject, Dale (1988) cites several factors as controlling cell division, including supply of photosynthate, nutrition, photomorphogenic factors, and the underlying genetic program of the plant. But he as well as others also point out that prior to division, cells must expand, and that "growth cannot occur without cell expansion" (Dale 1988). With this view, it is necessary to understand the constraints and regulation of leaf cell enlargement.

Physiology of leaf cell growth

For some plantations of poplar trees, stem volume produced in a season is highly correlated both with the total leaf area, and with the area of individual leaves, which in turn is related to leaf growth rate (Larson and Isebrands 1972; Ridge et al. 1986). Factors that limit leaf growth rate, including both genetic and environmental constraints, will ultimately reduce leaf area and photosynthesis. In poplars, plant productivity is more closely linked to total leaf area than is the case for plants harvested for grain, fruit, or tuber production (Gifford and Evans 1981). As described above, the rate of leaf expansion is ultimately determined by the rate at which cells increase in volume.

Biophysical principles

Cell growth in plants may be analyzed using the conceptual framework put forward by Lockhart (1965) in which he stated that the relative growth rate of plant cells (dV/Vdt) is a function of the amount that the osmotic potential gradient from outside to inside of the cell ($d\pi$) exceeds the yield threshold of the cell walls (Y) or ($d\pi - Y$), wall extensibility (m), and hydraulic conductance (L_p):

[1] $$\frac{dV}{Vdt} = \frac{mL_p (d\pi - Y)}{m + L_p}$$

In tissues for which hydraulic conductance is large, and therefore not limiting, the function reduces to:

[2] $$\frac{dV}{Vdt} = m (P - Y)$$

which shows that growth rate is dependent on cell wall extensibility, and the amount of turgor pressure exerted on the walls (P) in excess of the yield threshold; turgor pressure is equal to the osmotic potential gradient ($d\pi$) from Eq. 1. For tissues limited by hydraulic conductance, growth rate is not dependent on wall extensibility, but rather on the water potential gradient from outside to inside the cell, and the conductance of the tissue. In the latter case, turgor will be less than the amount that the osmotic potential gradient could generate, and the limitation for growth will be in generating sufficiently low cellular water potential to cause inward water flux. Most analyses of the rate of leaf expansion have focused on the parameters in Eq. 2.

When leaves expand, newly formed cells increase in volume by as much as 50 times their initial size (Maksymowych 1973; Dale 1988). This increase in volume is driven by turgor pressure generated by water flux into the growing cells; water flux occurs down a water potential gradient determined in part by the osmotic concentration of the cell contents, and the water potential of the leaf apoplast. As long as turgor exceeds the yield threshold for cell walls, the pressure will cause extension of the viscoelastic material that makes up the wall. The cell walls, on the other hand, may be either very extensible, or stiff, depending on the regulations imposed by the cells on biochemical reactions that loosen, tighten, and synthesize new walls. At a given turgor pressure, growth rate will be determined by the extensibility of the walls, and by the yield threshold. Conversely, with constant wall properties, growth rate will be a function of changes in turgor. To determine the limiting factors for the growth rate of any leaves, each of these variables must be measured.

Several different methods have been used to assess cell wall extensibility, each being indirect and subject to difficulties of interpretation (Cosgrove 1986). Likewise, yield threshold can easily be determined in single-celled algae (Green 1968), but is more difficult to assess in tissues (Van Volkenburgh and Cleland 1986; Okamoto and Okamoto 1994). Turgor pressure can be indirectly estimated by measurement of osmotic concentration of expressed sap (Roden et al. 1990), by psychrometric measurement of frozen-thawed tissue (Matthews et al. 1984), or by direct measurement with the pressure probe (Cosgrove et al. 1984).

Regulation of turgor and cell wall properties

Efforts to determine which parameter is most limiting to leaf growth have been undertaken on leaves of various species. Perhaps the most surprising, and most consistent result, of these investigations has been the lack of correlation between growth rate and turgor pressure, or effective turgor (P − Y). For primary bean leaves growing over a period of 8 days to mature size, most rapid growth rate is associated with the lowest effective turgor (Van Volkenburgh and Cleland 1986). On the other hand, for leaves of both bean and poplar, the slowing of growth and final cessation at mature size has been correlated with a gradual loss of effective turgor, and stiffening of cell walls (Van Volkenburgh and Cleland 1986; Frost et al. 1991; Taylor et al. 1995).

More often correlated with growth rate is wall extensibility. For beans (Van Volkenburgh and Cleland 1980; Van Volkenburgh and Cleland 1986) and poplar (Frost et al. 1991; Taylor et al. 1995), most rapid growth rates are correlated with higher extensibility and lower yield thresholds. The cellular mechanisms controlling cell wall extensibility in leaves are only partly known. In primary bean leaves stimulated to grow by light, enhanced growth rate and wall extensibility depend on acidification of the cell wall (Van Volkenburgh and Cleland 1980). When bean leaves reach mature size, the reduced growth rate is related to a loss of the cell walls' capacity to extend when acidified (Van Volkenburgh et al. 1985). Similar results were obtained with bean leaves when growth was inhibited by abscisic acid (Van Volkenburgh and Davies 1983) and with corn when growth was inhibited by water deficit (Van Volkenburgh and Boyer 1985). Light-stimulated growth of the sun-loving species *Betula* is correlated with increased wall extensibility and acidification (Taylor and Davies 1985). On the other hand, when bean leaf growth is stimulated by application of gibberellin or cytokinin, wall extensibility is increased without acidification (Brock and Cleland 1989). For the shade-tolerant species *Acer*, changes in growth rate were not correlated with changes in wall properties nor wall acidification (Taylor and Davies 1985).

In recent years, the biochemical basis of cell wall loosening and extensibility has been unraveled a little further. The complexity of this problem is great, given that the cell wall is composed of co-extensive networks of polysaccharides and structural proteins and that identification of the load-bearing bonds in such a structure has proved to be extremely difficult (Taiz 1984). In dicotyledonous plants, such as poplar, McCann and Roberts (1994) propose that there are two interpenetrating networks, one of cellulose and xyloglucan and one of pectin, and that xyloglucan cross-links are the principal load-bearing molecules in the longitudinal axis of an elongating cell. Fry and co-workers (Fry et al. 1992) suggest that xyloglucan acts to "tether" cellulose microfibrils together and that the enzyme xyloglucan endotransglycosylase (XET) acts to cut out and reform these tethering molecules, thereby allowing cell wall loosening and extension. There are only limited data on the role of this enzyme in controlling the growth of leaves, and as yet, no data for poplars. However, data on the model bean leaf system suggest that rapid leaf growth occurs when XET activity is maximal (Ranasinghe and Taylor 1996).

Environmental effects on leaf expansion

Plants rarely experience optimal conditions for growth. Understanding how the environment influences plant growth is of key interest to physiologists and plant breeders alike. For poplar, Hensen (1992) has illustrated the importance of selection and breeding, utilizing clonal responses to constrain environmental variables. For leaves, light, nutrients, and water all play an important role in determining the rate of leaf expansion (Van Volkenburgh et al. 1985; Taylor et al. 1993; McDonald and Stadenberg 1993; Roden et al. 1990; Smit et al. 1989). Of these the effects of flooding and water deficit will be discussed with specific reference to how leaf expansion is affected. In addition to these classical environmental factors, two emerging and important factors will be considered: tropospheric ozone and carbon dioxide.

Water and leaf growth: hypoxia and soil moisture deficit

Perhaps the most pervasive environmental variable limiting plant productivity is water, and although water deficit is a most significant stress for plants (Boyer 1982), excess water or flooding also reduces plant performance (Wadman-van Schravendijk and van Andel 1985). In a study of *P. trichocarpa* × *P. deltoides* hybrids, Smit et al. (1989) showed that the rate of leaf expansion was inhibited within hours of imposition of hypoxia at the roots. Final leaf size was reduced by 35–60%, primarily due to reductions in cell area. Cell number per leaf was reduced only in those leaves that experienced hypoxic stress very early in development. No differences in leaf water potential were detected between hypoxic and aerated plants, but the slower-growing leaves on hypoxic plants accumulated more solutes and as a consequence maintained even higher turgor pressure in the leaves than did controls. Wall extensibility, measured by extensiometer (Cleland 1967), was considerably reduced in the stressed leaves. However, the mechanism by which cell wall loosening was inhibited (or wall stiffening promoted) was not determined.

Similar results were obtained in a study of *P. trichocarpa*, *P. deltoides*, and their interspecific F_1 hybrids, grown in irrigated and unirrigated plots (Roden et al. 1990). Stem volume, total leaf area, individual leaf area, and leaf growth rate were reduced by water deficit. Stomatal conductance was also reduced in the stressed plants, and osmotic adjustment occurred such that turgor was fully maintained in the growing leaves of all 3 genotypes on unirrigated plots. As in the hypoxia treatment, reduction of leaf growth rate was correlated with lower values of wall extensibility. It remains to be seen how these inhibitory environmental conditions are translated by cells into down-regulation of wall loosening processes.

Elevated CO_2 and leaf growth

The concentration of carbon dioxide in the atmosphere has risen from a preindustrial value of approximately 270 ppm ($\mu mol \cdot mol^{-1}$) to a present day concentration of 355 ppm. A continued increase into the next century is likely, given current predictions for population growth, fossil fuel consumption and the destruction of tropical rain forests (Houghton et al. 1992). This aspect of environmental change is extremely important for all C_3 plants since CO_2, fixed by the enzyme Rubisco, acts as the source of carbohydrate for growth and metabolism. Additional plant carbon may have a number of direct and indirect effects on plant growth and development, influencing both cell production and cell expansion. A number of studies on woody species (see reviews by Jarvis 1989; Ceulemans and Mousseau 1994) have shown that exposure to CO_2, at the concentrations likely to exist in the next century, may result in increased tree biomass, including stimulation in leaf growth. For example, total tree leaf areas for several *P. trichocarpa* × *P. deltoides* clones were increased following exposure to elevated CO_2 in glasshouse chambers (Radoglou and Jarvis 1990), while Gaudilliere and Mousseau (1989) showed that leaf size was stimulated when a hybrid was exposed to elevated CO_2. Until recently, the mechanisms underlying such effects on leaf growth were poorly understood although, given the link between leaf area development and stem wood production (Ridge et al. 1986), their importance was clear.

Research in Taylor's laboratory has illustrated the positive effect that elevated CO_2 may have on the expansion rate of individual leaves, and has also identified the cellular mechanism likely to explain such a response. Figure 3A illustrates the effect of elevated CO_2 on the rate of leaf extension for 'Unal' (*Populus trichocarpa* × *P. deltoides*). Hardwood cuttings were grown in a controlled environment (CE) experiment and exposed to either ambient, present day CO_2 (355 $\mu mol \cdot mol^{-1}$) or elevated CO_2 (700 $\mu mol \cdot mol^{-1}$). For this hybrid, exposure to elevated CO_2 resulted in an increase in the rate of leaf extension. The next stage in this study was to establish whether cell production or cell expansion

Fig. 3A. Effect of elevated CO_2 on 'Unal' (*P. trichocarpa* × *P. deltoides*) on leaf extension rate ($mm \cdot d^{-1}$). Trees were exposed to either ambient (open bars) or elevated (hatched bars) CO_2. Asterisks indicate the results of a one-way ANOVA where *$P < 0.05$, **$P < 0.01$, ***$P < 0.001$.

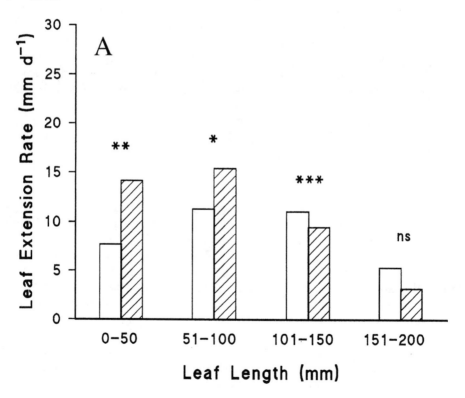

was responsible for increased growth. In a complementary study, using primary bean leaves as a model system (Van Volkenburgh and Cleland, 1980), it has been shown that large stimulations in leaf area development in elevated CO_2 are usually associated with cell expansion, rather than cell division (Ranasinghe and Taylor 1996). For poplars, this also appears to be true, as shown in Fig. 3B, for 'Unal.' Epidermal cell size was enhanced for CO_2-treated leaves. Using the biophysical concepts formalized by Lockhart and described in the previous section, we have also shown that this effect on cell expansion is usually associated with altered cell wall extensibility (m), rather than through any change in leaf cell water relations. Data for tensiometric plasticity are also given in Fig. 3B, again for 'Unal,' and confirm this finding, since plasticity (the irreversible extension of the wall, an estimate of m) was stimulated following exposure to elevated CO_2. Similar effects on wall plasticity were also documented for other clones. Spatial maps of extensibility across the expanding lamina have revealed

Fig. 3B. Effect of elevated CO_2 on 'Unal' (*P. trichocarpa* × *P. deltoides*) on cell wall properties and epidermal cell size. Trees were exposed to either ambient (open bars) or elevated (hatched bars) CO_2. Asterisks indicate the results of a one-way ANOVA where *P < 0.05, **P < 0.01, ***P < 0.001.

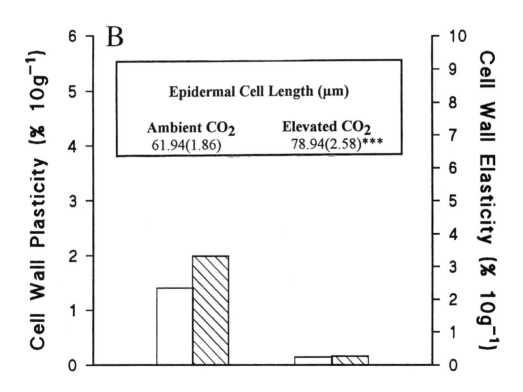

that exposure to elevated CO_2 may accelerate the normal developmental sequence of leaf growth and maturation. Extensibility was not only higher when leaves grew in elevated CO_2, but also followed a more basipetal pattern of increase, compared with the response in ambient CO_2 (Gardner et al. 1995). Further research is needed to investigate this effect of increased carbon supply on leaf cell expansion in poplars, but in the model bean leaves, the activity of the putative wall loosening enzyme, xyloglucan endotransglycosylase (XET), was enhanced in elevated CO_2 (Taylor et al. 1994).

As indicated earlier, additional carbon may also stimulate cell production in expanding leaves (Dale 1976). Research on roots has confirmed that sucrose may have a central role in the regulation of the plant cell cycle (the passage of cells through mitosis and the intervening phase of DNA synthesis), since addition of sucrose to the bathing medium in the studies of Van't Hof and Kovacs (1972) resulted in a stimulation of the mitotic index. Work is now in progress to identify specific genes or gene products that may control different points in this cycle (Francis 1992) and which may be sensitive to carbon supply. For poplars, the study by Gaudilliere and Moussseau (1989) suggested that enhanced leaf size in elevated CO_2 was the result of increased cell production, rather than cell expansion, while Radoglou and Jarvis (1990) suggested that effects on expansion were of overriding importance.

Our own data show that cell production may be stimulated in elevated CO_2, since at full expansion, epidermal cell size was unaltered for both 'Primo' (*P. deltoides* × *P. nigra*) and 'Beaupré' (*P. trichocarpa* × *P. deltoides*) clones, while leaf size was still enhanced. This was confirmed in the model bean leaf system where primary leaves were exposed to elevated CO_2 during the division phase, which resulted in a large stimulation in leaf cell production (Ranasinghe and Taylor 1996).

Are other aspects of leaf growth physiology influenced when poplars are exposed to elevated CO_2? Studies on herbaceous plants have shown that leaf production and tillering are often stimulated in elevated CO_2, perhaps reflecting an effect of carbon on meristematic activity. Leaf size and total tree leaf area are both known to influence light interception and biomass production in poplar (Chen et al. 1994) and our data suggest that leaf production may also be stimulated by elevated CO_2, depending on clone. All *P. trichocarpa* × *P. deltoides* clones studied ('Unal,' 'Beaupré,' 'Boelare') showed no effect of elevated CO_2 on leaf production; in contrast, the *P. deltoides* × *P. nigra* 'Primo' showed a consistent stimulation of leaf production in elevated CO_2 (Bosac et al. 1995).

These data on leaf growth and elevated CO_2 suggest that in the future, the rate of leaf expansion, leaf production, and total tree leaf area may be stimulated for a wide range of hybrid poplar clones. Increased leaf cell expansion is an important mechanism contributing to this effect, although cell production and meristematic activity, leading to the formation of new leaves, are also enhanced. The effects of such increases in leaf growth for stem wood biomass may be large. Taylor et al. (1995) have recorded a 30% increase in stem biomass in elevated CO_2, for 'Boelare,' while Ceulemans et al. (1996) have reported increases in stem volume of 38 and 55% for the hybrids 'Beaupré' and 'Robusta,' respectively.

Sensitivity to tropospheric ozone

Poplars are among the most sensitive of woody plants to the atmospheric pollutant, tropospheric ozone. This has been documented as premature leaf senescence and abcission, and leaf flecking and necrosis (Keller 1988; Reich and Lassoie 1985). The effects of premature leaf loss on seasonal stem productivity may be great. Gardner (1996) reported a 42% decline in stem biomass following exposure to 100 ppb O_3 during a single growing season. This study also revealed significant effects of ozone on leaf growth physiology, confirming an earlier report by Taylor and Frost (1992) that ambient concentrations of ozone were sufficient to inhibit leaf extension in *P. deltoides* × *P. nigra* hybrids. This inhibition in leaf extension was observed in both studies at the beginning of the growing season. As the season progressed, however, ozone exposure resulted in the premature senescence and loss of older leaves and a simultaneous stimulation in leaf extension of young leaves, suggesting that compensatory growth of young leaves occurred following redistribution of nutrients from older leaves. Nevertheless, the overall effect of exposure to ozone was negative, when considering the stem wood production by hybrids in these studies. A biophysical analysis of leaf growth conducted by Gardner (1996) revealed a number of significant effects of ozone. Firstly, leaf cell water potential was lowered significantly and this was associated with a lower cell solute potential, perhaps reflecting a reduced capacity to regulate ion and water fluxes following damage to the transport systems of the plasmamembrane. Ozone may affect both protein and lipid components of this membrane. The second effect, interestingly, was on cell walls. Cell wall extensibility was enhanced by exposure to ozone, particularly in young, expanding leaves, although the explanation for this effect, given the reductions in the rate of leaf extension, has yet to be elucidated. Recent data from SuYoung Woo (unpublished) suggest considerable variation in the response of hybrids from a single *P. trichocarpa* × *P. deltoides* pedigree (families 53 and 331) to ozone. *Populus* may prove extraordinarily useful as study material by which the genetic control of ozone damage, recovery and resistance may be better understood. This subject is also covered in Chapter 17 by Neuman and others.

Solving questions using *Populus*

The display of leaves by plants most often is matched to the availability of environmental resources such as water, nutrients, and light. Plants adapted to harsh environments, especially xerophytes, develop small leaf surface areas, and leaves are constructed so as to maximize water retention by the plant. In nutrient poor conditions, leaf development is restricted which in turn reduces the potential for photosynthetic production. It is clear that integrated plant behavior is accomplished by environmental inputs to all organs, and that leaf expansion is regulated by a complexity of environmental and internal signals.

Perhaps the best way to discover both the genetic basis, and the biochemical control of flexibility in leaf development, is to assess variability among trees that grow in contrasting habitats. This can be accomplished by studying natural variability among wild species: for example, by measuring differences in leaf anatomy, expansion rate, sensitivity to environmental stresses in *P. trichocarpa* growing in moist habitats of western Washington compared with dry habitats of eastern Washington (e.g., Dunlap et al. 1995). More precisely, genetic information may be gained and correlated with specific traits using controlled crosses and mapped quantitative trait loci (as described by Bradshaw in Chapter 8).

Because the developmental patterns of leaves of *P. trichocarpa* and *P. deltoides* are distinctly different, it is likely that genetic investigation of these species will shed light on the genetic control of cell division, and perhaps also on the development of intercellular spaces. Furthermore, by monitoring performance of trees in a variety of environments particularly contrasting in water availability, it may be possible to find out the role of intercellular spaces and especially the adaptive nature of the caverns in leaves of *P. trichocarpa*.

The response of *Populus* trees to environmental stresses will in the end determine their longevity and potential for photosynthetic production. As long as we are able to maintain genetic diversity in this genus, the plants will continue to display their ability to adapt and grow in a variety of cultural and natural settings. Leaf area production may become a subject of manipulation in our effort to design ideotypes (see Chapter 19 by Dickmann and Keithley); if we intend to do this we will need to understand much more thoroughly the biochemical regulations employed by plants to control leaf expansion. In part these regulations are organ-autonomous, with the perception of, and response to, the environment occurring within individual leaves. On the other hand, it is quite clear that soil conditions affecting development and function of roots are parlayed into effects on expanding leaves. Communication between roots and shoots, besides being ultimately substrate-driven as in supply of water and nutrients, is accomplished by "signals" of as yet unknown nature. The defined genetic system being developed in *Populus*, and the fact that in this species we are working with long-lived plants of basically wild "temperament" (i.e., not yet domesticated beyond natural adaptability), will allow us to determine the relative importance of such signaling processes, and perhaps their nature.

References

Avery, G.S. 1933. Structure and development of the tobacco leaf. Amer. J. Bot. **20**: 565–592.

Bosac, C., Gardner, S.D.L., Taylor, G., and Wilkins, D. 1995. Elevated CO_2 and hybrid poplar: a detailed investigation of root and shoot growth and physiology of *Populus* euramericana, "Primo." For. Ecol. Manage. **74**: 103–116.

Boyer, J.S. 1982. Plant productivity and environment. Science (Washington, DC), **218**: 443–448.

Brock, T.G., and Cleland, R.E. 1989. Role of acid efflux during growth promotion of primary leaves of *Phaseolus vulgaris* L. by hormones and light. Planta, **177**: 476–482.

Ceulemans, R., and Mousseau, M. 1994. Effects of elevated atmospheric CO_2 on woody plants. New Phytol. **127**: 425–446.

Ceulemans, R., Impens, I., and Steenackers, V. 1987. Variations in photosynthetic, anatomical, and enzymatic leaf traits and correlations with growth in recently selected *Populus* hybrids. Can. J. For. Res. **17**: 273–283.

Ceulemans, R., Shao, B.Y., Jiang, X.N., and Kalina, J. 1996. First and second year aboveground growth and procuctivity of two *Populus* hybrids grown at ambient and elevated CO_2. Tree Physiol. **16**: 61–68.

Chen, S.G., Ceulemans, R., and Impens, I. 1994. A fractal based *Populus* canopy structure model for calculation of light interception. For. Ecol. Manage. **69**: 97–110.

Cleland, R.E. 1967. Extensibility of isolated cell walls: measurement and changes during cell elongation. Planta, **74**: 197–209.

Cosgrove, D.J. 1986. Biophysical control of plant cell growth. Annu. Rev. Plant Physiol. **37**: 377–405.

Cosgrove, D.J., Van Volkenburgh, E., and Cleland, R.E. 1984. Stress relaxation of cell walls and the yield threshold for growth. Demonstration and measurement by micro-pressure probe and psychrometer techniques. Planta, **162**: 46–54.

Dale, J.E. 1976. Cell division in leaves. *In* Cell division in higher plants. *Edited by* M.M. Yeoman. Academic Press, London and New York. pp. 315–345.

Dale, J.E. 1988. The control of leaf expansion. Annu. Rev. Plant Physiol. Plant Mol. Biol. **39**: 267–295.

DeLucia, E.H., Nelson, K., Vogelmann, T.C., and Smith, W.K. 1996. Contribution of intercellular reflectance to photosynthesis in shade leaves. Plant Cell Environ. **19**: 159–170.

Dunlap, J.M., Heilman, P.E., and Stettler, R.F. 1995. Genetic variation and productivity of *Populus trichocarpa* and its hybrids. VIII. Leaf and crown morphology of native *P. trichocarpa* clones from four river valleys in Washington. Can. J. For. Res. **25**: 1710–1724.

Fisher, D.B. 1967. An unusual layer of cells in the mesophyll of the soybean leaf. Bot. Gaz. **128**: 215–218.

Franceschi, V.R., and Giaquinta, R.T. 1983. Specialized cellular arrangements in legume leaves in relation to assimilate transport and compartmentation. Comparison of the paraveinal mesophyll. Planta, **159**: 415–422.

Francis, D. 1992. The cell cycle in plant development. New Phytol. **121**: 1–22.

Frost, D.L., Taylor, G., and Davies, W.J. 1991. Biophysics of leaf growth of hybrid poplar: impact of ozone. New Phytol. **118**: 407–415.

Fry, S.C., Smith, R.C., Renwick, K.F., Martin, D.J., Hodge, S.K., and Matthews, K.J. 1992. Xyloglucan endo-transglycosylase, a new wall loosening enzyme activity from plants. Biochem. J. **282**: 821–828.

Gardner, S.D.L. 1996. The effects of elevated CO_2 and ozone on hybrid poplar. Ph.D Dissertation, University of Sussex, UK.

Gardner, S.D.L., Taylor, G., and Bosac, C. 1995. Leaf growth of hybrid poplar following exposure to elevated CO_2. New Phytol. **131**: 81–90.

Gaudilliere, J.P., and Mousseau, M. 1989. Short-term effect of CO_2 enrichment on leaf development and gas exchange of young poplars (*Populus euramericana*). Acta Oecol. **10**: 95–105.

Gifford, R.M., and Evans, L.T. 1981. Photosynthesis, carbon partitioning, and yield. Annu. Rev. Plant Physiol. **32**: 485–509.

Givnish, T.J. 1979. On the adaptive significance of leaf form. *In* Topics in plant population biology. *Edited by* O.T. Solbrig, S. Jain, B. Johnson, and P.H. Raven. Columbia University Press, New York, NY. pp. 375–407.

Green, P.B. 1968. Growth physics in *Nitella*: A method for continuous *in vivo* analysis of extensibility based on a micro-manometer technique for turgor pressure. Plant Physiol. **43**: 1169–1184.

Green, P.B. 1980. Organogenesis — a biophysical viewpoint. Annu. Rev. Plant Physiol. **31**: 51–82.

Haber, A.H., and Foard, D.E. 1963. Nonessentiality of concurrent cell divisions for degree of polarisation of leaf growth. II. Evidence from untreated plants and from chemically induced changes of the degree of polarisation. Amer. J. Bot. **50**: 937–944.

Hensen, E.A. 1992. Poplar woody biomass yields: A look to the future. Biomass Bioenergy, **1**: 1–7.

Hitchcock, C.L., and Cronquist, A. 1973. Flora of the Pacific Northwest. University of Washington Press, Seattle, WA. 730 p.

Houghton, J.T., Callander, B.A., and Varney, S.K. 1992. Intergovernmental panel on climate change. Climate Change 1992: the supplementary report to the scientific assessment. Cambridge University Press, Cambridge, UK.

Humphries, E.C., and Wheeler, A.W. 1963. The physiology of leaf growth. Annu. Rev. Plant Physiol. **14**: 385–410.

Isebrands, J.G., and Larson, P.R. 1973. Anatomical changes during leaf ontogeny in *Populus deltoides*. Amer. J. Bot. **60**: 199–208.

Jarvis, P.G. 1989. Atmospheric carbon dioxide and forests. Phil. Trans. R. Soc. Lond. Ser. B **324**: 369–392.

Kaplan, D.R., and Hagemann, W. 1991. The relationship of cell and organism in vascular plants. Bioscience, **41**: 693–703.

Keller, T. 1988. Growth and premature leaf fall in American aspen as bioindication for ozone. Environ. Pollut. **52**: 183–192.

Kevekordes, K.G., McCully, M.E., and Canny, M.J. 1988. The occurrence of an extended bundle sheath system (paraveinal mesophyll) in the legumes. Can. J. Bot. **66**: 94–100.

Larson, P.R., and Isebrands, J.G. 1972. The relation between leaf production and wood weight in first-year root sprouts of two *Populus* clones. Can. J. For. Res. **2**: 98–104.

Lockhart, J.A. 1965. An analysis of irreversible plant cell elongation. J. Theor. Biol. **8**: 264–275.

Matthews, M.A., Van Volkenburgh, E., and Boyer, J.S. 1984. Acclimation of leaf growth to low water potentials in sunflower. Plant Cell Environ. **7**: 199–206.

Maksymowych, R. 1973. Analysis of leaf development. Cambridge University Press, Cambridge, UK.

McCann, M.C., and Roberts, K. 1994. Changes in cell wall architecture during elongation. *In* Growth in Planta. J. Exp. Bot. **45**: 1683–1691.

McDonald, A.J.S., and Stadenberg, I. 1993. Diurnal pattern of leaf extension in *Salix viminalis* relates to the difference in leaf turgor before and after stress relaxation. Tree Physiol. **13**: 311–318.

McDougall, W.B. 1931. Plant ecology. Lea and Febiger, Philadephia, PA. pp. 53–58.

Mohr, H., and Schopfer, P. 1995. Plant physiology. Springer-Verlag, Berlin. pp. 419–422.

Okamoto, H., and Okamoto, A. 1994. The pH-dependent yield threshold of the cell wall in a glycerinated hollow cylinder (in vitro system) of cowpea hypocotyl. Plant Cell Environ. **17**: 979–983.

Radoglou, K.M., and Jarvis, P.G. 1990. Effects of CO_2 enrichment on four poplar clones. I. Growth and anatomy. Ann. Bot. **65**: 617–626.

Ranasinghe, C.S., and Taylor, G. 1996. Mechanism for increased leaf growth in elevated CO_2. J. Exp. Bot. **47**: 349–358.

Reich, P.B., and Lassoie, J.P. 1985. Influence of low concentrations of ozone on growth, biomass partitioning and leaf senescence in young hybrid poplar plants. Environ. Pollut. **39**: 39–51.

Ridge, C.R., Hinckley, T.M., Stettler, R.F., and Van Volkenburgh, E. 1986. Leaf growth characteristics of fast growing poplar hybrids *Populus trichocarpa* × *P. deltoides*. Tree Physiol. **1**: 209–216.

Roden, J.E., Van Volkenburgh, E., and Hinckley, T.M. 1990. Cellular basis of limitation of poplar leaf growth by water deficit. Tree Physiol. **6**: 211–220.

Smit, B., Stachowiak, M., and Van Volkenburgh, E. 1989. Cellular processes limiting leaf growth in plants under hypoxic root stress. J. Exp. Bot. **40**: 89–94.

Taiz, L. 1984. Plant cell expansion: regulation of cell wall mechanical properties. Annu. Rev. Plant Physiol. **35**: 585–657.

Taylor, G., and Davies, W.J. 1985. The control of leaf growth of *Betula* and *Acer* by photoenvironment. New Phytol. **101**: 259–268.

Taylor, G., and Frost, D.L. 1992. Impact of gaseous air pollution on leaf growth of hybrid poplar. For. Ecol. Manage. **51**: 151–162.

Taylor, G., McDonald, A.J.S., Stadenberg, I., and Freer-Smith, P.H. 1993. Nitrate supply and the control of leaf growth in *Salix viminalis*. J. Exp. Bot. **44**: 155–164.

Taylor, G., Ranasinghe, C.S., Bosac, C., and Gardner, S.D.L. 1994. Elevated CO_2 and plant growth: Cellular mechanisms and responses of whole plants. *In* Growth in Planta, J. Exp. Bot. **45**: 1761–1774.

Taylor, G., Gardner, S.D.L., Bosac, C., Flowers, T.J., Crookshanks, M., and Dolan, L. 1995. Effects of elevated CO_2 on cellular mechanisms, growth and development of trees with particular reference to hybrid poplar. Forestry, **68**: 379–390.

Van't Hof, J., and Kovacs, C.J. 1972. Mitotic cycle regulation in the meristem of cultured roots: The principal control point hypothesis. *In* The dynamics of meristem cell populations. *Edited by* M.W. Miller and C.C. Kuehnert. Plenum Press, New York, NY. pp. 15–33.

Van Volkenburgh, E., and Boyer, J.S. 1985. Inhibitory effects of water deficit on maize leaf elongation. Plant Physiol. **77**: 190–194.

Van Volkenburgh, E., and Cleland, R.E. 1980. Proton excretion and cell expansion in bean leaves. Planta, **167**: 37–43.

Van Volkenburgh, E., and Cleland, R.E. 1986. Wall yield threshold and effective turgor in growing bean leaves. Planta, **167**: 37–43.

Van Volkenburgh, E., and Davies, W.J. 1983. Inhibition of light-stimulated leaf expansion by abscisic acid. J. Exp. Bot. **34**: 835–845.

Van Volkenburgh, E., Schmidt, M.G., and Cleland, R.E. 1985. Loss of capacity for acid-induced wall loosening as the principal cause of the cessation of cell enlargement in light-grown bean leaves. Planta, **163**: 500–505.

Verbelen, J.-P., and De Greef, J.A. 1979. Leaf development of *Phaseolus vulgaris* L. in light and darkness. Amer. J. Bot. **66**: 970–976.

Wadman-van Schravendijk, H., and van Andel, O.M. 1985. Interdependence of growth, water relations and abscisic acid level in *Phaseolus vulgaris* during waterlogging. Physiol. Plant. **63**: 215–220.

Wu, R., Bradshaw, H.D., Jr., and Stettler, R.F. 1996. Molecular genetics of growth and development in *Populus*. V. Mapping quantitative trait loci affecting leaf variation. Amer. J. Bot. In press.

CHAPTER 13
Physiology of secondary tissues of *Populus*

F.W. Telewski, R. Aloni, and J.J. Sauter

Introduction

This chapter provides an overview of the complex physiology of secondary tissues in *Populus*. The general field of knowledge with regard to any of the topics covered within is fairly large and it is not the intention of the authors to conduct an exhaustive review of any particular physiological function. Rather, we focus specifically on studies conducted within the genus *Populus* with select references to link these studies to work in the broader field.

The topic of secondary tissues covers broad areas in terms of both structures and functions. Tissues that will be covered include the vascular cambium from which the secondary xylem and secondary phloem develop, as well as the phellogen and its derivatives, the phelloderm, and phellem. It should be noted that these tissues are not limited to the stem but function in the branches of the canopy and in the roots. Functions include metabolism, differentiation, translocation, structural support, anchorage, and storage. The secondary xylem of poplar, as for most forest trees, is the most economically important secondary tissue of the tree, supplying wood and fiber as well as a feed stock for a variety of chemical processes. Therefore, a large portion of literature on the physiology of secondary tissues has focused on xylem structure and function, and on cambial

Abbreviations: ABA, Abscisic Acid; GA3, Gibberellin; IAA, Indole-3-acetic acid; IAAsp, Indole-3-acetylaspartic acid; NAA, Naphthalene-acetic acid; NPA, N-1-Naphtylphalamic acid; OxIAA, Oxindole-3-acetic acid; OxIAAsp, Oxindole-3-acetylaspartic acid; TIBA, Tri-iodobenzoic acid; ZR, Zeatin riboside.
F.W. Telewski. W.J. Beal Botanical Garden, Department of Botany and Plant Pathology, Michigan State University, East Lansing, MI 48824, USA.
R. Aloni. Department of Botany, The George S. Wise Faculty of Life Sciences, Tel Aviv University, Tel Aviv 69978, Israel.
J.J. Sauter. Botanisches Institut und Botanischer Garten, Cristian-Albrechts-Universität Kiel, Olshausenstrasse 40, D-24098 Kiel, Germany.
Correct citation: Telewski, F.W., Aloni, R., and Sauter, J.J. 1996. Physiology of secondary tissues of *Populus*. *In* Biology of *Populus* and its implications for management and conservation. Part II, Chapter 13. *Edited by* R.F. Stettler, H.D. Bradshaw, Jr., P.E. Heilman, and T.M. Hinckley. NRC Research Press, National Research Council of Canada, Ottawa, ON. pp. 301–329.

differentiation leading to the formation of wood (xylogenesis). The characterization of the xylem and its physiology will be covered in this chapter (with the exception of water transport to be covered in Chapter 18 and part of Chapter 19). The functions to be covered include differentiation, translocation, mechanical support, and the storage function of the xylem of poplar. In addition, aspects of the physiology of other secondary tissues including the secondary phloem and the products of the cork cambium (phellogen), phelloderm, and phellem will be addressed.

The vascular cambium and control of vascular differentiation

Any student of the vascular cambium will recognize the complex nature of this secondary meristem and its association with the apical meristem, primary vascular architecture of the leaves and stems, and the procambium as parts of a continuum from primary to secondary meristems, as demonstrated in *Populus deltoides* Bartr. ex Marsh. (Larson 1979), and presented in a more general context (Larson 1982). For a detailed description of the development and structure of the vascular cambium, the reader is encouraged to review the monograph by Larson (1994).

The foundation for our current knowledge regarding the development of secondary tissues from primary tissues in woody plants is based upon the detailed analysis of foliar and vascular development within the stem of *P. deltoides* (Larson and Isebrands 1971, 1974; Larson 1975, 1976b, 1977, 1979, 1980; Goffinet and Larson 1981, 1982; Meicenheimer and Larson 1983; DeGroote and Larson 1984). The transition zone between the primary and secondary tissues within the stem of *Populus* occurs in the internode just beneath the most recently matured leaf and is identified anatomically as having "... occurred when fibers with birefringent walls were first detected both within and between adjacent traces forming the vascular cylinder, with the exception of those traces last to enter the stem and all traces situated between them." (Larson and Isebrands 1974). The transition is not uniform within the internode, but proceeds according to the phyllotaxy of the bundles and progresses in a counterclockwise direction around the stem. The position of the transition zone advances acropetally in the stem in close association with leaf maturation and is further influenced by plant vigour (Larson 1980). A woody cylinder, composed of the anastomosing vascular system, forms within the secondary transition zone as a result of the coalescence of the primary vascular bundles (Larson and Isebrands 1974).

In dormant *Populus grandidentata* Michx, the transition zone was observed to occur at or near the base of the terminal bud (Larson 1976a). During the induction of dormancy in *P. deltoides*, the size and number of the xylem elements and overall development of associated leaf traces are reduced at the level of the vascular transition. The reduced size of the vascular tissue is directly related to the reduction in mature leaf size and vigour associated with the onset of dormancy (Goffinet and Larson 1981). As the transition approaches the base of the apical bud, the acropetal development of the vascular cambium becomes confined to a few active leaf traces, i.e., the metacambium (an intermediate stage between procambium and cambium) persisting within the remaining traces (Larson 1976*b*; Goffinet and Larson 1981). The transition advanced to the second or third internode below the bud, never advancing to the bud base.

Reactivation of metaxylem in the spring occurred above the transition zone located in the second or third internode below the expanding apical bud (Goffinet and Larson 1982). Traces leading to abscised leaf positions or to bud-scale leaf positions were reactivated with the bidirectional differentiation of secondary xylem vessels which were functionally continuous with metaxylem vessels differentiating acropetally. Xylem fiber differentiation followed the position and sequence exhibited by vessels and lagged behind vessel differentiation. Vessel differentiation was associated with leaf expansion and fiber differentiation was associated with leaf maturation (Goffinet and Larson 1982). With subsequent leaf maturation in the newly expanding shoot, the primary–secondary transition zone advanced acropetally to the position of the bud base (now the expanding shoot) and then into the new shoot. Development of the vascular tissues continued in the new shoot in association with phyllotaxy as previously described (Larson and Isebrands 1971, 1974; Larson 1975, 1976*b*, 1977, 1979, 1980).

Cambial activity and vascular differentiation are induced and controlled by hormonal signals (see Aloni 1987, 1988, 1995; Fukuda 1996; Savidge 1996 for a review). The highest level of endogenous auxin within the elongating stems of *P. deltoides* was recorded to coincide with the primary–secondary vascular transition zone. Dormancy inducing short day treatments also resulted in declining levels of endogenous auxin as the transition zone developed acropetally. Long day treatments resulted in a resumption of high auxin activity in the transition zone (DeGroote and Larson 1984). The changes observed in the differentiation of vessels during the transition from primary to secondary tissues (Meicenheimer and Larson 1983) and from juvenile to mature wood in *Populus* appear to be consistent with the "six-point" hypothesis for the control of vascular conduit size and density along the tree axis as proposed by Aloni and Zimmermann (1983). The hypothesis is centered on the role of basipetal polar flow of auxin, which is suggested to establish a gradient of decreasing auxin concentration from leaves to roots. The flow of auxin can be altered by local structural or physiological obstructions resulting in a local increase in auxin concentration. The amount of auxin flowing through the differentiating cells of the vascular cambium is determined by the distance from the source of auxin. The rate of vessel differentiation is positively correlated with the amount of auxin that the differentiating cells receive, creating a gradient from rapid differentiation within the canopy to relatively slow differentiation near the roots. The ultimate size of the vessel is determined by the rate of differentiation, therefore resulting in an increase in vessel size along the axial length of the stem and also observed as an increase in vessel size in the radial direction. Finally, conduit density is also controlled by and positively correlated with auxin concentration. An auxin concentration gradient from a high level near the apical meristem to a correspondingly low level near the roots of *Populus* is suggested as regulating the transition from juvenile to mature wood (Aloni 1991).

The influence and requirement for translocated regulatory substances from shoots and the associated expanding leaves on cambial activity in *Populus tremuloides* Michx. was studied by Evert and Kozlowski (1967). Patches of bark, which were isolated from phloem transport but remained attached to the tree, exhibited continued but altered cambial activity. If isolation was established during the dormant season, xylem differentiation was inhibited. Fusiform cambial initials were subdivided into strands of parenchymatous elements. Newly formed phloem was affected by premature death of sieve elements. Isolation of bark and cambial tissues after xylogenesis was initiated resulted in the curtailment of secondary wall formation. Cells of the cambial region were shortened, resulting in the production of shorter vessel members. Although the authors did not consider the influence of wounding and ethylene effects at the time this study was conducted, the study does suggest a role for basipitally transported growth factors, most likely auxin, influencing cambial development.

The regulation of vessel size in *Populus* was modelled to be regulated by leaf development and maturation, and the distance to the differentiating xylem mother cells (Meicenheimer and Larson 1983). This model was confirmed in a study on *P. deltoides* xylogenesis and the role of auxins (Meicenheimer and Larson 1985). Removal of leaf laminae resulted in the differentiation of fewer vessels with reduced mean vessel area (transverse lumen area of individual vessels) which resulted in a significant reduction in total transverse vessel area per transverse section. Application of exogenous indoleacetic acid (IAA) or naphthalene-acetic acid (NAA) to replace excised laminae, when compared to excised treatments, increased the number of vessels differentiating in the associated central leaf traces of the stem, but had no influence on the cross-sectional area of vessels. Application of auxin, regardless of concentration, only generated half the mean vessel area observed in intact controls. Treatment of intact leaf petioles with the auxin transport inhibitor N-1-naphthylphthalamic acid (NPA) reduced mean vessel area and the number of differentiating vessels. These interesting results, specifically a reduction in vessel number and size in response to removal of an auxin source, are not consistent with the six-point hypothesis.

Transgenic hybrid aspen (*Populus tremula* L. × *P. tremuloides* Michx.) expressing IAA-biosynthetic genes produced xylem at the base of the stem characterized by narrower vessels in higher density when compared to nontransgenic hybrid controls (Tuominen et al. 1995). These results support the six-point hypothesis (Aloni and Zimmermann 1983). Auxin production in the transgenic hybrids was determined to be significantly greater in root tissue, but not significantly different in stems when compared to controls (based on whole stem tissue fresh weight). The role of IAA in division, differentiation, and maturation within transgenic hybrids at site specific tissue concentrations needs to be studied further. The application of transgenic hybrids to future research is discussed by Sundberg et al. (1996). The methods and results of both Tuominen et al. (1995) and Meicenheimer and Larson (1985) will be useful in further testing the six-point hypothesis of the control of vascular element size and density.

Recently, advances have been made in the characterization of the metabolic pathway for IAA in *Populus*. Indole-3-acetylaspartic acid (IAAsp) and oxindole-3-acetylaspartic acid (OxIAAsp) were identified as new metabolites of exogenously applied IAA during the induction of adventitious root formation in greenwood cuttings of *P. tremula* (Plüss et al. 1989). The metabolic pathway for IAA in apical shoots of *P. tremula* L. × *P. tremuloides* Michx. was subsequently elucidated (Tuominen et al. 1994). IAA is converted to IAAsp, oxidized to OxIAAsp, and then hydrolysed to oxindole-3-acetic acid (OxIAA).

The differentiation of xylem fibers in *Populus,* as well as other species, appears to require the presence of both auxin and gibberellin (Digby and Wareing 1966; Aloni 1985, 1987, 1988). There is very little information regarding the role of cytokinins in cambial development, and currently there are no studies in the *Populus* system. Cytokinin is a controlling factor in vessel regeneration around wounds and appears to be required during the early stages of xylem fiber differentiation in herbaceous plants (see Aloni 1993, 1995 for a review). Zeatin riboside (ZR) is synthesised in the roots of poplars and its translocation to shoots is retarded by the flooding induced hypoxia (Smit et al. 1990).

For vessels and fibers, the amount of time a cell takes to mature within the cambial zone will ultimately determine cell size and cell wall thickness (see Aloni 1987, 1988, 1991 for a review). During the early phase of primary radial cell wall maturation within newly-laid cell walls in the cambial zone, identification of a middle lamella and primary wall is impossible (Catesson and Roland 1981; Catesson et al. 1994). As maturation and development continues, the intrawall reorganization results in the formation of a true middle lamella. Subsequent changes in primary radial cell wall chemical composition after division from the cambial initial can be used to study the early steps in cell differentiation to produce either a xylem or phloem element (Catesson et al. 1994). Xylans (Simson and Timell 1978), xylan synthetase activity (Dalessandro and Northcote 1981), and xyloglucans (Baba et al. 1992) all increased during primary wall development in xylem derivatives of *Populus*. The different chemical compositions observed within the primary radial walls were used to interpret repartitioning of enzymatic activity within xylem and phloem mother cells. Pectin methylesterase and xylan synthase were interpreted as being highly active in differentiating xylem cells, with only moderate xylan synthase activity in cambial initials. Cellulose synthase was suggested to be highly active in developing phloem cells, but not in the cambium initial or developing xylem initials. Moderated pectin methylesterase activity and no xylan synthase activity appears to be present in phloem initials (Catesson et al. 1994).

The timing of the process of secondary cell wall development, including the deposition of cellulose and lignin, was the focus of a study by Bobák and Nečesaný (1967). They reported an alternating pattern over a 24-h period for the formation of cell walls in fibers of *Populus nigra* var. *italica* (Moench.) Koehne which included the following diurnal cycle: cellulose deposition from 12 noon to 6 pm, lignin deposition from midnight to 4 am, no deposition activity from 6 to 10 am.

Seasonal changes in carbohydrate composition also occur during the transition between an active vascular cambium and the cambium at rest. In the spring, the cell walls of cambial cells and their phloem derivatives from *Populus* ×*euramericana* (Dode) Guinier (*P. deltoides* × *P. nigra*) contain 9% uronic acids by dry weight. This value drops to 7% during the summer and winter. Methylated galacturonic acids were only 2% dry weight in spring. The acidic pectins accumulate over the course of the active season to reach a high of 35% dry weight in the winter. During the winter, arabins and xylans were the main cell wall carbohydrates, whereas in spring and summer xylans were scarce and arabinans were common (Baïer et al. 1994). Calcium ions were also found in the cambium, located mainly in cell junctions. Calcium ions were distributed throughout the entire cell wall of the phloem derivatives (Baïer et al. 1994).

Lignification within secondary phloem tissues by isoperoxidases appears to be mostly restricted to fiber cells as only fiber cells, within the secondary phloem are able to lignify. In the spring and summer, the fiber cell walls of *Populus* ×*euramericana* are able to oxidize syringaldazine, ferulic acid salts and *p*-phenylenediamine-pyrocatechol (Baïer et al. 1993), three substrates which closely resemble natural lignin precursors and can be only oxidized *in situ* by cell wall lignifying isoperoxidases (Goldberg et al. 1983; Pang et al. 1989). Nonfiber tissue cell walls are only capable of oxidizing *p*-phenylenediamine-pyrocatechol, with the greatest activity prior to the onset of dormancy (Baïer et al. 1993).

Trockenbrodt (1991) described the early development of secondary growth tissues from primary tissues in the bark of *Populus tremula* L. Later stages of development were reported in a subsequent paper (Trockenbrodt 1994); the percentage of sclereid present in the bark of *P. tremula* increased for the first 6 years of development and then remained more or less constant. The development of conduction and storage tissues remained constant for the first 6 years and then decreased. Phloem rays were 1–28 cells high, with the average between 5 and 15 cells in height. Ray size did not vary with age. Secondary phloem fibers were reported to vary between the north and south side of trees, with a range in length from 590 to 1460 µm (average length 1000 µm), with an increase in fiber length correlated with an increase in tree age. Sieve tube members ranged between 160 and 310 µm in length when generated by a young vascular cambium (1–5 years) and reached a maximum of 660 µm in length when produced by an older vascular cambium, whereas the diameter increased from 34 µm in the first year's growth to 48 µm in the fifth year's growth. After 5 years, the diameter of the sieve tubes varied around the higher value (Trockenbrodt 1994).

The xylem

Xylem structure

As is the case for the wood of most dicotyledonous angiosperms, the xylem of poplar is a complex combination of cell types including vessel elements, libriform wood fibers, and axial and ray parenchyma. The wood of the genus *Populus* is characterized as being diffuse-porous, where the earlywood vessels are not conspicuously large and the vessels of the latewood are indistinct to the naked eye (Panshin and deZeeuw 1980). Based on gross anatomical features, the wood is very similar to that of the willows (*Salix* spp.), who share the same family (Salicaceae). However, the two different genera can be positively separated based on the structure of the wood rays (*Salix* spp. rays being essentially heterocellular, *Populus* spp. rays being essentially homocellular). Within the genus, it is almost impossible to differentiate between the various species based on anatomical features. Panshin and DeZeeuw (1980) separate the *Populus* spp. of the section *Aigeiros* (cottonwoods or black poplars) from *Populus* spp. in the section *Populus* (formerly the section *Leuce*), subsection *Trepidae* (aspens) on the density of vessels per square millimetre of wood cross section and on the range of vessel diameter (75–150 µm for cottonwoods, 50–100 µm for aspens). Vessel element length varies between species from 67 µm in *P. tremuloides* to 58 µm in *Populus trichocarpa* Torr. & A. Gray. Fibers vary in length from 1.38 mm for *P. trichocarpa* to 1.32 mm in *P. tremuloides*.

The parenchyma of poplar wood is composed of a variety of cells which differ in anatomy and in physiological function (Sauter and van Cleve 1989*b*). The parenchyma cells are contained within both the rays of the radial system of tissues and as axial parenchyma, arranged vertically within the woody tissue. Both radial and axial systems are physically and functionally connected. Functions include the storage and distribution of food materials (Kramer and Kozlowski 1979), exchange of solutes with the water conducting elements (Sauter 1966, 1972, 1980, 1981*a*, 1981*b*), the excretion and secretion of products, or the compartmentalization and sealing of wounds (Shigo 1984).

Within the uniseriate, 'homogeneous' rays of *Populus*, parenchyma cells can be classified as contact cells, isolation cells, and cells of the contact cell rows

(Sauter and van Cleve 1989*b*). The differentiation and identification of the different cell types is based on position within the ray and the physiological function as related to storage product mobilization and deposition. Contact cells facilitate exchange of solutes with vessels by means of numerous extraordinary large pits called 'contact pits' and exhibit higher levels of respiratory and enzymatic activity (Sauter 1972). Contact cells, like other ray parenchyma cells, are capable of storing starch in amyloplasts, protein in the form of protein bodies, and fat within oleosomes, although premature disappearance of starch from these cells has been found (Sauter and van Cleve 1989*b*, 1990). Isolation cells are centrally located within the ray and are adapted for radial translocation, as is evidenced by the increased frequency of plasmodesmata along tangential walls (Sauter and Kloth 1986). Both isolation cells and cells of the contact row lack contact with vessels, and contain especially large plastids filled with starch grains (Sauter and van Cleve 1989*b*).

Juvenile wood

Juvenile wood is formed during the rapid early growth of a tree. However, it is not restricted to the early seedling and sapling stages of tree growth. Juvenile wood development is associated with the relatively high level of auxin produced in the tree crown by apical meristems and foliage (Larson 1962). Therefore, a central core of juvenile wood is maintained within a tree by the growth of the apical meristem. Wood developing directly below the apical meristematic region will possess characteristics of juvenile wood regardless of the tree age (see Zobel and van Buijtenen 1989 for a review of the literature on juvenile wood development). Juvenile wood is generally weak, lower density, and composed of cells which are shorter in length and vessels narrower in diameter when compared to those found in mature wood.

Comparative studies on juvenile and mature anatomical features within the genus *Populus* are extensive and in some cases contradictory. Ray volume remained uniform from pith to cambium in the *P. deltoides* clones studied by Cheng and Bensend (1979) and Onilude (1982), but was observed to decrease from juvenile to mature wood in the *Populus* ×*euramericana* clones studied by Peszlen (1994). Fiber and vessel lengths increased in the transition from juvenile to mature wood in all *Populus* species studied (Boyce and Kaiser 1964; Kaiser and Boyce 1965; Marton et al. 1968; Bendtsen 1978; Holt and Murphey 1978; Cheng and Bensend 1979; Murphey et al. 1979; Bendtsen et al. 1981; Yanchuck et al. 1984; Phelps et al. 1985; Bendtsen and Senft 1986; Peszlen 1994).

In a comparative anatomical analysis between 18-year-old *P. deltoides* and 12-year-old hybrid NE-237 (*P. deltoides* × *P. nigra* cv. 'Volga'), Bendtsen et al. (1981) reported variations between and within species for mechanical and anatomical features. Juvenile fibers were shorter in the hybrid than in the pure species; however, the mature fibers were about the same length in both taxa.

Bendtsen et al. (1981) reported an increase for both vessel diameter and vessel-lumen diameter in the transition from juvenile to mature wood. Similar results were reported for vessels by Peszlen (1994). Fibril angle from the vertical also decreased markedly from the pith to the cambium in both species, representing the juvenile to mature wood transition. However, the mature wood of the hybrid appeared to have the lowest fibril angles. Within the juvenile wood of 6-year-old *Populus* 'Tristis #1' hybrid (*P. tristis* Fisch. × *P. balsamifera* cv. 'Tristis #1'), fibers averaged 0.75 mm in length, vessel elements 0.45 mm and specific gravity averaged 0.35 g/mL (Phelps et al. 1985).

Mechanically, juvenile wood was reported to be between 62 to 79% of the strength recorded for mature wood in *P. deltoides* (Bendtsen and Senft 1986). Specific gravity, maximum crushing strength, modulus of elasticity, and modulus of rupture all increased from juvenile wood to mature wood in *P. deltoides* (Bendtsen et al. 1981; Bendtsen and Senft 1986).

In general, the differences between juvenile and mature wood in *Populus* does not appear to have as significant an impact on pulp and paper production as it does in coniferous species; however, the difference in density and fibril angle between juvenile and mature wood results in a low stability of solid wood products derived from juvenile wood (Zobel and van Buijtenen 1989).

Tension wood

As in other woody dicotyledonous angiosperms, *Populus* spp. also produce tension wood. Tension wood is a type of reaction wood which develops on the upper side of branches or of displaced stems in response to gravity or possibly tension induced by bending. Tension wood functions by magnifying an internal tensional force, resulting in the bending of the displaced stem back towards a vertical position. This type of wood is characterized by the presence of gelatinous fibers, which have lower lignin and higher cellulose content compared to normal fibers. Detailed anatomical (Onaka 1949; Perem 1964; Hughes 1965; Ohta 1979), chemical (Timell 1969) and ultrastructural (Cote and Day 1962; Cote et al. 1969; Mia 1968) analyses have been previously presented. Rays in the tension wood of *Populus monilifera* Henry tend to be fewer in number, but are composed of more cells with a greater average height (Kučera and Nečesaný 1970). The tension wood of *P. tremula* was characterized as having fewer rays per mm^2 than normal wood (Ollinmaa 1959).

Tension wood has been reported in straight trees of *Populus* (Kaeiser 1955). Kaeiser and Boyce (1965) reported that the amount of gelatinous fibers present in a stem can be minimized by selecting for straight trees with good form. However, it has been suggested that tension wood can be induced under the conditions of rapid growth without displacement relative to the gravitational vector (Berlyn 1961; White and Robards 1965). The presence of gelatinous fibers was found to be high and extremely variable in the stems of rapidly grown plantation *P. deltoides* which were selected for their straight, vertically oriented stems, free of any bends or sweep (Isebrands and Bensend 1972) verifying earlier reports. Similar results were reported for *Populus balsamifera* L. by Kroll et al. (1992). Isebrands and Bensend (1972) suggest that the common occurrence of gelatinous fibers in nonleaning trees may be the result of a vascular cambium which is very sensitive to the stimulus involved in the initiation of tension wood formation. They also go on to suggest that there may be a genetic component to the disposition for tension wood formation which may be selected against in *Populus* breeding programs.

Reaction wood formation is generally hypothesised to develop in response to an internal gradient of auxin, with tension wood developing in the region of low auxin concentration. In the branches of woody angiosperms, the formation of tension wood is inhibited by the application of auxin on the upper side of the branch, and induced in vertical stems opposite the side of application. Auxin antagonists also induce tension wood formation at the site of application. These results suggest tension wood formation is regulated by an auxin deficiency (see Timell 1986 for a review). In *Populus*, the application of 0.25 and 1.0% of the auxin transport inhibitor tri-iodobenzoic acid (TIBA) resulted in the formation of tension wood at the point of application (Blum 1970).

Wood of coppiced stems

Populus taxa develop coppice shoots (stump sprouts) from the bases of felled trees. The juvenile anatomical characteristics of coppiced stems is significantly different from the juvenile anatomical characteristics from the same first rotation trees (Phelps et al. 1987). Wood specific gravity and wood fiber length decrease in coppice shoots, whereas bark specific gravity and bast fiber lengths increase in coppice shoots. Despite the significant differences reported in these properties, the authors state that the differences were of little to no importance to industrial applications compared to the value of increased woody biomass production after coppicing.

Environmental influence on xylem development

Environmental conditions and silvicultural practices can have a significant influence on the development of secondary tissues, especially xylem. The environmental influence was clearly demonstrated in the study on *Populus euphratica* Oliv. growing on dry and wet sites in Israel. Although the wood of all *Populus* spp., including *P. euphratica*, is classified as diffuse porous, ring porous wood can be produced in *P. euphratica* when grown on dry sites where water is limiting for growth (Liphschitz and Waisel 1970). Evolutionarily, the ring-porous wood pattern originated from diffuse-porous wood (Aloni 1991). Aloni (1991) suggested that continuous selective pressures in limiting environments finally resulted in the development of the specialized ring-porous wood that maximized the efficiency of water conduction. Evidence which supports the 'limiting-growth hypothesis' (Aloni 1991) shows that the selection for ring-porous wood has led to a decrease in the intensity of vegetative growth, accompanied by reduced levels of growth regulators. The latter was followed by an increase in the sensitivity of the cambium to relatively low levels of auxin stimulation. These physiological changes created the special internal conditions that enable the differentiation of wide and long earlywood vessels during spring in ring-porous trees. It has been found that in both a diffuse-porous poplar (*P. euphratica)* (Liphschitz and Waisel 1970; Liphschitz 1995) and a ring-porous oak (*Quercus ithaburensis*) (Liphschitz 1995) when extension growth is intensive under favorable environmental conditions, wide rings with diffuse-porous wood are produced. However, when extension growth is suppressed under stress conditions, narrow rings characterized by ring-porous wood are formed. These results support the 'limiting-growth hypothesis' (Aloni 1991).

Several studies have reported an influence of cultural practices or management strategies on *Populus* wood properties, especially specific gravity and fiber length; rapid growth usually results in wood of lower specific gravity and shorter fibers (Kennedy 1957; Kennedy and Smith 1959; Marton et al. 1968; Dickson et al. 1974). More recently, Blankenhorn et al. (1988) supported these earlier observations by observing that management strategy and site influence wood properties of a first rotation *Populus* hybrid (NE-388, *P. maximowiczii* Henry × *P. trichocarpa* Torr. & Gray). The effect is more pronounced and significant in second rotation growth (Blankenhorn et al. 1992). The highest specific gravity values were observed in control and irrigated trees, whereas supplemental fertilization and a combination of fertilization and irrigation resulted in lower specific gravities. Site differences in wood specific gravity were also observed. Trees growing on a sandy loam soil, considered an "unfavorable" site by the authors (Blankenhorn et al. 1988), yielded consistently higher specific gravities for all management strategies than for trees grown on the "favorable" silt loam soil site (Blankenhorn et al. 1992).

Peszlen (1994) studied the influence of site on anatomical properties of *Populus* ×*euramericana*. The better of the two sites was characterized as having a wet, histosol "bog soil" compared to a well-drained, vertisol "meadow forest soil" of the poorer site. Maturation of anatomical properties, and the transition from juvenile to mature wood, were observed to be accelerated on the "better" site, but larger diameter and longer fibers and vessels were observed in the mature wood from trees growing on the "poor" site.

Wind also influences the secondary growth of some *Populus* clones, e.g., stem taper and root development in *P. deltoides* × *P. trichocarpa* (Harrington and DeBell 1996). Wind may also influence the development of tension wood in vertical stems of *P. balsamifera* since the occurrence of tension wood along a specific radius of a stem was strongly correlated with prevailing winds during the growing season (Kroll et al. 1992).

Flooding reduces gas exchange between the root, soil, and atmosphere, creating hypoxic conditions. In *P. trichocarpa* (Harrington 1987) and hybrid 11–11 of *P. trichocarpa* × *P. deltoides* (Smit and Stachowiak 1988), this condition results in increased resistance to root water uptake and water flux through roots. Flooding also influences development by stimulating the formation of hyper-hydric tissues in the bark of *P. tremuloides* (Angeles 1990) and *P. trichocarpa* and adventitious roots with aerenchyma in *P. trichocarpa* (Harrington 1987) and *P. trichocarpa* × *P. deltoides* hybrids (Smit and Stachowiak 1988). After 5 d of flood treatment, phellogen activity was increased in the region of lenticels resulting in an increased production of phelloderm, pushing the lenticel filling tissue outwards (Angeles 1990). After 9 d of flood treatment, continued phellogen activity extended beyond the lenticels, creating patches of hyper-hydric tissues covering most of the stem. The new cells produced by the phellogen were thin walled, unsuberized, and enucleate. The cells were firmly attached at the radial cell wall to other cells in the radial file; however, the connections to neighboring cells in adjacent radial files were via a few points of knob-like projections. After 22 d of treatment, cortical parenchyma and rays of the secondary phloem were stimulated to produce large cells rich in starch with aerenchyma spaces. These changes resulted in an increase in bark thickness and porosity. The type of tissue development observed in response to flooding is consistent with observations on the role of ethylene in the flood response in *Ulmus americana* L. (Yamamoto et al. 1987).

Populus trichocarpa and its hybrids appear well adapted to flooding stress, although not as well adapted as red alder (*Alnus rubra* Bong.) (Harrington 1987). Both species are adapted to riparian habitats which are exposed to periodic flooding. Flooding induced root dieback (Harrington 1987) and reduced leaf growth (Smit et al. 1990), but plants were able to survive the stress of flooding.

Heartwood formation

Heartwood begins to form in *P. tremuloides* after 5 years of growth. Tree age is an important factor in determining the width and basal area of both sapwood and heartwood. The basal area of both types of wood increase linearly with age. The width of the sapwood initially increases with age, but tapers off after about 50 years of age and then begins to decline (Yang and Hazenberg 1991).

The boundary between heartwood and sapwood is determined by the absence or presence of living parenchyma cells in the inner sapwood. The general pattern of ray parenchyma cell mortality has been observed in numerous coniferous species; however, only a limited number of angiosperm species have been studied. These patterns were classified by Nobuchi et al. (1979) into 3 types. Type I, with all ray parenchyma cells remaining alive up to the sapwood/heartwood boundary; these species have very distinct colored boundaries in cross section. Type II, where there are some dead ray parenchyma cells in the middle of the sapwood, with the percent of dead cells rapidly increasing from the middle sapwood toward the sapwood/heartwood boundary. These species have been reported to possess a rather diffuse or indistinctly colored sapwood/heartwood boundary in cross section. Type III is characterized as having some dead ray parenchyma cells in the outer sapwood near the vascular cambium, with the number of dead cells increasing towards the sapwood/heartwood boundary. Type III species show no color transition from the sapwood to the heartwood.

Yang (1992) reported an exponential decline in the percent of living parenchyma cells from about the third growth ring from the cambium to the sapwood/heartwood boundary in 26-year-old *P. tremuloides*. The rate of ray parenchyma cell death was greater in cells located centrally (isolation cells) in the rays as compared to cells located on the margins of the rays (contact cells and cells of the contact row). Using the classification system based on the general pattern of ray parenchyma cell death developed by Nobuchi et al. (1979), Yang (1992) classified *P. tremuloides* as a type II species. Yang (1992) reported this finding consistent with the colored appearance of the transition from sapwood to heartwood in *P. tremuloides*.

Yang (1992) also documented the percent of ray parenchyma cells with irregularly shaped nuclei (a condition which exists just prior to disintegration of the nucleus and cell death) across the sapwood. The highest percent of cells with irregularly shaped nuclei was in the sapwood growth ring closest to the sapwood/heartwood boundary, with an exponential decline in the number of irregularly shaped nuclei extending into the middle of the sapwood. There were no irregularly shaped nuclei in ray parenchyma cells in the first few growth rings closest to the vascular cambium. Ethylene and peroxidase activity have been correlated with heartwood formation in *Pinus radiata* D. Don. (Shain and Hillis 1973), *Juglans nigra* L. and *Prunus serotina* Ehrh. (Nelson 1978); however, there have been no studies on the role of ethylene in heartwood formation in *Populus* taxa.

Wounding responses and compartmentalization

The response of secondary tissues to wounding, characterized by the release of ethylene and the resulting compartmentalization to prevent desiccation and further damage or infection due to pathogens, is an important physiological and developmental response. Depending on the location of the wound, different tissues may respond including secondary xylem, vascular cambium, and bark. Wounding of the secondary vascular tissue increases substantially the ray system (Lev-Yadun and Aloni 1995) resulting in compartmentalization of the wound region from the rest of the vascular tissues. *Populus* responds to wounding in a fashion similar to other hardwood species (Kuroda and Shimaji 1985; Schmitt and Liese 1993).

Using the pinning method of marking the vascular cambium (Wolter 1968; Yoshimura et al. 1981*a*, *b*), Kuroda and Shimaji (1985) observed cytodifferentiation in response to wounding in the xylem and cambial zone of *P.* ×*euramericana*. They observed a mitotic reactivation of ray parenchyma cells. Within the region of secondary wall thickening and maturation of the cambial zone, the new ray parenchyma derivatives invaded the wound opening and adjacent cells and had the appearance of septate fibers. Within the xylem mother cell region of the cambial zone, wounding stimulated transverse divisions of fusiform cells, creating abnormal septate cells, but prevented the differentiation into vessel elements. Wounding induced the formation of septate fusiform cells within the region of the cambial initials, the result of transverse division within the cambial initials themselves. Ray cells in the region of the wound contained druse-type crystals and appeared yellow-brown, probably because of the presence of phenolics (Kuroda and Shimaji 1985).

The suberization of ray and axial parenchyma cells within the xylem of *Populus* forms a distinct boundary around a wound (Schmitt and Liese 1993). Vessels and fibers within and directly behind the affected area have an occluded appearance with the vessels containing suberized tyloses (Kuroda and Shimaji 1985; Schmitt and Liese 1993). Schmitt and Liese (1993) go on to describe the mechanism for suberin deposition. The process involves the cisternae of the endoplasmic reticulum, where suberin compounds are extruded by cytoplasmic vesicles which fuse with the plasma membrane. The structure of the suberin layer was described as lamellate. Continuity between the suberized cells was maintained by the plasmodesmata through the suberin layer. Wounding induced during the winter did not stimulate suberization of the injured tissues (Schmitt and Liese 1993).

Barrier zone formation or compartmentalization was defined as a mechanism by which trees can defend against vascular pathogens (Tippett and Shigo 1981; Shigo 1984). Clones within *P. deltoides* × *P. trichocarpa* vary in their response to wounding by producing different volumes of compartmentalized discolored wood, suggesting that compartmentalization was under moderate to strong genetic control (Garrett et al. 1976; Shigo et al. 1977). Hybrids of *P. deltoides* × *P. trichocarpa* with strong-, moderate-, and weak-compartmentalizing ability were analysed to determine the relationship of the degree of compartmentalization with differences in anatomical structure (Eckstein et al. 1979). The wood of clones with the strongest ability to compartmentalize were characterized by fewer and smaller diameter vessels with fewer connections to other vessels than weak-compartmentalizing clones. The strong-compartmentalizing clones also had the highest percent mean xylem composition of parenchyma than moderate- or weak-compartmentalizing clones. Eckstein et al. (1979) suggested these characteristics could be used to select for strong-compartmentalizing clones resistant to defects as part of an early genetic evaluation program. Noh et al. (1986) observed that the rate of discoloration due to artificial wound induction and heartwood development was higher in *P. nigra* × *P. maximowiczii* clones than in *Populus koreana* Rehder × *P. nigra* var. *italica* clones. They calculated broad sense heritability in the *P. nigra* × *P. maximowiczii* clones of 0.41 for discoloration and 0.32 for rate of heartwood formation suggesting both are under moderate genetic control.

Xylem translocation and phloem transport

Although water transport is covered in Chapter 16 and part of Chapter 17 of this volume, the xylem also plays an important role in the translocation of organic compounds from the roots to developing leaves. The growth regulators abscisic acid (ABA) and zeatin riboside (ZR) are conducted in the xylem sap of *Populus*. When the roots of *P. trichocarpa* × *P. deltoides* (Hybrid 11–11) were deprived of oxygen, inducing root hypoxia, there was a marked decrease in the flux of both ABA and ZR in xylem sap. Phloem transport of assimilates to the roots continued under the hypoxic conditions, and appeared necessary to maintain root health during this period of stress (Smit et al. 1990).

In *Populus*, glutamine is the major organic nitrogen compound translocated from the roots to shoots in the xylem sap. Several other amino compounds are also present at a lesser concentration (Dickson 1979; Sauter and van Cleve 1992). Xylem to phloem transfer via rays occurs at all levels of the stem and is most pronounced at nodal junctions (Fisher et al. 1983). In the upper shoot, the glutamine is translocated to the phloem via metaxylem parenchyma, secondary xylem parenchyma and rays. These tissues uptake the glutamine conducted in the xylem vessels in the stem. The accumulation of the glutamine in the phloem occurred in internodes subtending recently mature leaves, but does not enter the mature leaf xylem or phloem. Instead it moves to developing tissues of the upper shoot, indicating that translocation of nitrogen compounds in *Populus* does not totally follow the transpirational stream. Instead, there is an efficient uptake and transfer system for organic nitrogen stored in roots that supports developing tissues (Dickson et al. 1985). A 32-kDa storage protein which accumulates in the parenchyma cells is also translocated in the form of glutamine in the xylem sap after it is mobilized in spring (Sauter and van Cleve 1992).

The leaf gaps, important in the translocation of photosynthate from the leaves to the stem, consist mainly of heavily pitted, thick-walled parenchyma cells that retain their protoplast. These cells are birefringent in polarized light (Larson and Richards 1981; Larson and Fisher 1983). Fisher and Larson (1983) suggest that these cells function similarly to transfer cells (Pate et al. 1971; Gunning 1977); however, no evidence was found to support this hypothesis (Fisher et al. 1983). In *P. deltoides*, very little photosynthate is translocated to mature cells in the gap region of the central leaf trace due to the lack of differentiated rays. The central leaf trace does contribute photosynthate to the cambium-like region which adds cells to the trace when the axillary branch elongates. Photosynthate is also translocated from the central leaf trace to the branch trace in the nodal region (Fisher et al. 1983). Photosynthate is translocated from the branch trace acropetally in the branch, basipetally in the stem or laterally to adjacent stem traces. Photosynthate can also be translocated into the gap region via rays, where it accumulates as stored starch during predormancy. Photosynthate is not transferred between unrelated leaf traces of the stem (Fisher et al. 1983).

Calcium plays an important role in the translocation and partitioning in phloem of *P. trichocarpa* (Schulte-Baukloh and Fromm 1993). Photosynthate in calcium-starved plants was maintained at higher levels in foliar tissues and at decreased levels in stems and roots. Photosynthate in the stem was restricted to the phloem, whereas in noncalcium starved plants, photosynthate was translocated from leaves to the stem and unloaded from the phloem to the middle part of the stem. In calcium starved *Populus*, the concentration of calcium in the phloem was greatly reduced, whereas the concentrations of both magnesium and phosphorus increased in sieve element walls and cytoplasm. Deficiencies in potassium and magnesium had no influence on phloem translocation or unloading (Schulte-Baukloh and Fromm 1993).

Bark structure and rhytidome development

The tissues located to the outside of the vascular cambium are referred to as bark. The bark can be separated into two categories: the living 'inner bark' consisting of secondary phloem up to the last (inner) formed periderm, and the dead 'outer bark' or rhytidome, consisting of old secondary phloem, old periderms, cortex, and primary phloem (Trockenbrodt 1990). The phloem is a complex tissue containing sieve tube elements and their associated companion cells, parenchyma cells, fibers, and sclereids. The periderm is characterized by the formation of the phellogen or cork cambium (a bifacial cambial tissue) and the subsequent secondary tissues, phelloderm to the inside and phellem (cork) to the outside. In *Populus*, the suberized dead cork cells arranged in radial rows and lacking intercellular spaces are characterized as the thin walled type. As in other woody species, the phelloderm cells resemble cortical parenchyma cells and remain alive, capable of photosynthesis and storage. The total bark photosynthesis in *P. deltoides* amounts to only 5% of the tree total (Schaedle 1975). The first periderm develops within the cortex, a primary stem tissue. The development of a rhytidome is characterized by the transition from a smooth barked stem with a continuous periderm to a rough or furrowed barked stem with the formation of additional arcs of periderm developing in the phloem below the first periderm. Initiation of the rhytidome occurs after 8 years in *P. tremula* (Trockenbrodt 1994). The bark of *Populus* also exhibits juvenile and mature characteristics in addition to the formation of the rhytidome (Trockenbrodt 1994).

Lev-Yadun and Aloni (1990) proposed that auxin and ethylene are the major factors controlling periderm formation. They suggested that moderate auxin flow retards periderm formation, while a high auxin level promotes ethylene production and therefore indirectly results in periderm formation. Thus, leaves and buds determine the patterns of periderm ontogeny by producing auxin, which polarly inhibits periderm formation below them. On the other hand, ethylene stimulates periderm formation in wounds, in waterlogged plants and under different environmental stresses.

Lev-Yadun and Aloni (1990) also propose a positive feedback control mechanism which promotes phellogen activity and rhytidome formation. This mechanism is based on the fact that the first-formed periderm creates a barrier for further ethylene release from the inner tissues of the plant to its outside surrounding atmosphere. Thus, in the bark tissues inside the first-formed cork relatively high ethylene levels accumulate. This enhanced hormonal stimulation increases phellogen activity, which results in the formation of rhytidome.

Storage in secondary tissues

Starch, sugars, protein, and fat accumulate seasonally and are stored in the parenchymatous cells of the wood and bark in both the stem and root. These cells serve as essential vegetative storage tissues maintaining a reserve of accumulated photosynthate during winter months, functioning in cold acclimation, winter respiration and, after mobilization, supplying the tree with photosynthate for the initiation of new growth and reproduction (Ziegler 1964; Kramer and Kozlowski 1979; Kozlowski 1992). The amount of stored reserves within the wood of *Populus* is also an important determinant of field performance and establishment of hardwood cuttings (Fege and Brown 1984), the major method of propagation of *Populus* clones. An understanding of the storage and metabolism of storage products will be useful in optimizing harvesting and storage of hardwood cuttings (Fege and Brown 1994). Within the xylem of *Populus*, this specialized vegetative storage parenchyma is located almost exclusively within rays (Sauter 1966, 1982; Sauter and Kloth 1987; Sauter and van Cleve 1989*b*). Micromorphometric studies report that during the winter, between 20–30% of the cell area is occupied by amyloplasts, 12% by protein storing vacuoles, and 1.1–1.4% by oleosomes (Sauter and van Cleve 1989*a*). The seasonal levels of the different storage products varies and can be correlated to environmental stimuli and physiological demands. The different storage products appear to accumulate and be metabolized independently of each other in the wood storage parenchyma (Sauter and van Cleve 1994).

Translocation, specifically of accumulated carbohydrates, is very rapid between storage ray parenchyma cells. Sauter and Kloth (1986) determined the minimum radial flow rate of sugars across tangential walls was 81 $pmol \cdot cm^{-2} \cdot s^{-1}$,

$400-800$ pmol·cm^{-2}·s^{-1} for pit fields and $1.0-1.7 \times 10^{-7}$ pmol plasmodesma^{-1}·s^{-1} per plasmodesma. From these data, they concluded that radial translocation in the ray tissue must proceed via plasmodesmata rather than by a transmembrane flux mechanism. Ray cell parenchyma pit fields contain 39 plasmodesmata/μm^2 and tangential walls 8.0 plasmodesmata/μm^2 to make up a total of 1.98% of the wall area occupied by plasmodesmata. Cells along the margins of rays have a lower frequency (1.16% area) of plasmodesmata on their tangential walls (Sauter and Kloth 1986).

Storage, mobilization, and translocation of stored reserves in the roots of *Populus* are addressed by Pregitzer and Friend (Chapter 14, Part II), whereas carbon allocation is covered by Isebrands and Ceulemans (Chapter 15, Part II).

Carbohydrates (starch and sugars)

The majority of carbohydrates are stored within the plastids of the isolation cells and cells of the contact rows in rays (Sauter and van Cleve 1989*b*). The accumulation and metabolism of carbohydrates within secondary tissues is seasonal, with a buildup of sugars towards the end of the growing season, followed by a subsequent decline from winter into spring (Fege and Brown 1984). Seasonally, as carbohydrate pools are supplemented by photosynthesis, shifts in concentration among monosaccharides, polysaccharides and starch occur within the ray cells. Sucrose was observed to be at the highest level, 10% of dry weight, during March. The levels of fructose, glucose, myo-inositol and melibiose remained constant throughout the entire year at a level of less than 1% of dry weight. The wood of *Populus* contains a pool of maltose (Sauter and Kloth 1987), which builds up from late summer (August) until mid-autumn (October) with a maximum concentration of 0.8% total xylem dry weight (Sauter and van Cleve 1993). After the abscission of leaves, the maltose pool vanishes, with a conversion to sucrose and its galactosides, raffinose and stachyose (Sauter and van Cleve 1991, 1993). The process of starch–sugar conversion appears to be associated with a decrease in temperature rather than the actual abscission of foliar tissue, since lower temperatures facilitate a more complete starch hydrolysis with conversion to sucrose, raffinose, and stachyose (Sauter and van Cleve 1991). The synthesis of sucrose, raffinose, and stachyose from the hydrolysis of starch in late autumn is consistent with the measured changes in starch content. During the month of December, the concentrations of raffinose and stachyose within 1-year-old coppice stems (bark, wood, and pith), increased to 6 and 7% stem dry weight, respectively (Fege and Brown 1984) and within the xylem, total sucrose, raffinose, and stachyose concentration increased to 1.5–2.8% wood dry weight (Sauter and van Cleve 1993) during the month of December. The levels of both galactosides was less than 0.2% total stem dry weight in September and May (Fege and Brown 1984).

Starch content and the time of peak accumulation in the autumn can vary among clones (Fege and Brown 1984). Starch accumulation is stimulated by an 8-h photoperiod with a day:night temperature regime of 20°C:14°C (Nelson and Dickson 1981); starch concentrations increased in the lower stem during the third week of exposure and continued for an 8-wk treatment period. Accumulation of total nonstructural carbohydrates (starch and sugars) was linear after the first week of short day exposure (Nelson and Dickson 1981). The concentration of starch begins to build up from late spring (Sauter and Neumann 1994) and reaches a maximum in the autumn, then begins to decline after leaf abscission. The decline is the result of amylase hydrolysis of the starch which is stimulated by lower temperatures (Sauter and van Cleve 1991). The starch levels remain low through the winter (Fege and Brown 1984; Sauter and van Cleve 1993). The intensity and level of starch deposition could be increased under conditions (induced by ringing) of increased photosynthate availability (Sauter and Neumann 1994).

Within the ray parenchyma cells, the interconversion of starch–fat and sugars–starch appears to be regulated by temperature (Jeremias 1968; Sauter 1988; Sauter and van Cleve 1991). With a decrease in temperatures in the autumn, there is at first a transient increase in maltose and later in maltase which is suggested to deliver the hexose moieties for the increasing synthesis of sucrose and its galactosides (Sauter and van Cleve 1993). The maltose pool remains prevalent at temperatures around 10°C. However, a subsequent increase in sucrose and maltose galactosides at temperatures below 5°C (the result of maltase activity) suggests a regulating role of temperature on maltase synthesis (Sauter 1988; Sauter and van Cleve 1993). However, no significant starch–fat conversion was observed during this stage of carbohydrate metabolism (Sauter and van Cleve 1991). Instead, there was a decrease in fat and an increase in the concentration of glycerol within the cells (see section on fat storage below).

Proteins

The storage of protein occurs as protein bodies within all ray cells of the wood (Sauter and van Cleve 1989*a*, 1990) and within the vascular cambium, cortical parenchyma, and phloem parenchyma (Wetzel et al. 1989). Within wood ray cells, the protein bodies are the exclusive storage site of a 32-kDa polypeptide (Sauter and van Cleve 1990). Within the cortical and phloem parenchyma, there are 3 storage proteins, a 32-, 36-, and 38-kDa polypeptide (Coleman et al. 1991, 1992, 1993; Stepien et al. 1992; Wetzel et al. 1989). The 32-kDa wood storage protein is structurally very similar to the 32-kDa bark storage protein (Stepien et al. 1992). Preceding the accumulation of wood storage protein is a large, transient appearance of corresponding mRNA (Clausen and Apel 1991). Accumulation of all proteins in secondary tissues occurs primarily during the late summer/fall and is associated with the period of leaf yellowing prior to abscission (Clausen and Apel 1991; Sauter and Neumann 1994). During fall and winter, transient changes in protein content coincided with increased populations of vesicles and/or tubular membrane cisternae within ray contact cells, suggesting transfer of protein from the storage pool to the structural pool of membrane associated proteins (Sauter and van Cleve 1990). The level of protein in both tissues remains high throughout the winter months and declines in the spring and summer as protein bodies are replaced by large central vacuoles within the storage cells (Sauter and van Cleve 1990; Wetzel et al. 1989). The process of proteolysis that occurs in the spring has been reported to be similar to that observed in seeds (Sauter and van Cleve 1990).

The mechanism for storage protein deposition in xylem or bark has been suggested to be under photoperiodic control (Nelson and Dickson 1981; Coleman et al. 1991, 1992, 1993; Langheinrich and Tischner 1991), or influenced by nitrogen level (van Cleve and Apel 1993; Sauter and Neumann 1994; Coleman et al. 1994; Stepien et al. 1994), temperature (van Cleve and Apel 1993; Sauter and Neumann 1994; Stepien et al. 1994), or available photosynthate (Sauter and Neumann 1994). After 2 weeks' treatment with an 8-hour photoperiod, free amino acids increased in concentration (Nelson and Dickson 1981). The increase in free amino acids can in part be accounted for by a redistribution of proteins from senescent foliar tissue, translocated to bark tissues as amino acids and resynthesized into storage proteins (Côté and Dawson 1986). Coleman et al. (1991, 1992, 1993) reported increased synthesis and accumulation of 32-kDa bark storage protein in the inner bark parenchyma when trees were exposed to short-day (8 hrs of light) conditions. Langheinrich and Tischner (1991) assumed the photoperiodic control for synthesis of 32-and 36-kDa proteins was regulated by the phytochrome system. However, recent studies report storage protein synthesis can be stimulated under long day conditions by the application of nitrogen (van Cleve and Apel 1993; Coleman et al. 1994; Sauter and Neumann 1994; Stepien et al. 1994), by low-temperature treatment (van Cleve and Apel 1993; Stepien et al. 1994), or by ringing (Sauter and Neumann 1994; Stepien and Sauter 1994). These data indicate increased storage protein synthesis is not under the control of a single mechanism, such as photoperiod, temperature, or available nitrogen, and that the accumulation of storage proteins is ultimately related to the actual levels of available photosynthate (Sauter and Neumann 1994). Bark storage protein appears to function both as a seasonal store for nitrogen and as a short-term nitrogen storage during periods of excess nitrogen availability (Coleman et al. 1994).

Bark storage protein degradation does not appear to be under photoperiodic control in dormant *Populus*, indicating that increased day length associated with late winter and early spring is not sufficient to mobilize these resources for spring growth. Dormant trees need to be exposed to low temperatures in order to break dormancy and degrade the stored bark proteins (Coleman et al. 1993).

Fats

Initially, *Populus* was classified as a "fat storing" species (Sinnott 1918; Mia 1972). As first suggested by Nelson and Dickson (1981) and further supported by the recent wealth of information on storage products, *Populus* is not strictly a lipid-storing species, and in fact relies on carbohydrates as the predominant form of winter food reserve. Lipids do function as a winter storage product, primarily in the cortical tissues of the bark (Nelson and Dickson 1981).

Peak deposition of fat does occur in secondary ray tissues mainly in late summer for woody tissues (Sauter and Neumann 1994) and was reported to occur during winter in bark tissues (Jeremias 1968; Nelson and Dickson 1981). The deposition of fats appears to be ultimately regulated by the actual levels of available photosynthate, in a manner similar to carbohydrates and proteins (Sauter and Neumann 1994). Fat formation in the wood rays of *Populus* proceeds when high concentrations of soluble carbohydrates within the stem are available, primarily during a short period in the spring and then in mid- to late-summer (Sauter and Neumann 1994). As mentioned earlier in this chapter, no starch–fat conversion has been observed in *Populus* woody tissues during the fall and winter months.

Concluding remarks

We hope this chapter will stimulate future studies for finding treatments and techniques which will increase wood production in poplar trees. As summarized in this chapter, the physiology of secondary tissues in the genus *Populus* is a complex topic covering many different structures and functions. The xylem and cambial zone are the most studied tissues for reason of their obvious economic benefits. The genus has been used extensively as a model system for the study of vascularization and the development of meristematic tissues in woody plants (Larson 1994). Although our knowledge about the regulation of cambial development by plant growth regulators is limited for the genus, the increased availability of different clones and transgenic hybrids should present *Populus* as a model system for elucidation of the role of growth regulators in xylogenesis, phloem differentiation, and rhytidome development.

There are still many questions related to xylogenesis and cambial activity that can be answered, building on the solid foundation of existing knowledge. For example, what factors are involved in determining the allocation of new photosynthate to twigs, stems, and roots? What is the influence of the leaf biomass on the xylogenesis and actual differentiation of vessels? Could the amount of vessels produced be altered? Could vessel morphology be altered in such a way that the conducting system would become less vulnerable at a given tension to cavitation? As discussed in the Water relations chapter, considerable genetic based variation in xylem vulnerability has been observed. Various hormonal combinations of both hormones increased cambial activity, fiber differentiation and wood production in *Populus* (Aloni 1985; R. Aloni unpublished). Aloni (1985) has shown that treating young *Populus alba* with both auxin and gibberellin substantially increased fiber production. In this study, a low level (10 mg·L^{-1} NAA) of auxin was required in conjunction with gibberellin (100 mg·L^{-1} GA3), which is the major regulator for fiber differentiation (Aloni 1987). Gibberellin is a key regulator in cambium activity and wood formation (Aloni 1987, 1988, 1991).

The use of transgenic *Populus* as a research tool has already been demonstrated (Sundberg et al. 1996). Therefore, we suggest the production of transgenic trees expressing other growth regulator genes, including the gibberellin genes, would be useful in further defining their role in xylogenesis. Resulting alterations in wood structure could also be used to address some of the functional questions with regard to hydraulic conductivity and mechanical support, as well as economic issues such as wood yield and quality. It is possible that these transgenic plants will possess high rates of cambial activity and will result in increased production of wood and fibers.

The xylem of *Populus* has also served as a model system for studies of the storage and metabolism of different compounds, including carbohydrates, proteins, and lipids. These studies are creating a new understanding of the phenology of growth, development, and propagation in woody plants. Additional questions to address include determining the influence of the level of stored material in twigs, especially protein levels, on new growth in spring. Does the protein level influence the biomass of new leaves and new shoots, thereby determining the capacity of the photosynthetic apparatus in the following year? What triggers, and what determines the maximum accumulation of storage products in the twigs? Could fertilization at certain periods be used for optimalization of these events? Could the heartwood–sapwood boundary be altered in order to increase the area for storage and translocation? What specific factors determine the rate of cold acclimation and deacclimaiton and the degree of frost hardiness? These factors are essential to restricting the areas where poplars can be cultivated.

The development of cell walls and the role of carbohydrate and phenolic biochemistry has been studied extensively in the poplar system. How does the chemistry of cell walls alter wood and fiber quality? Can the chemistry of cell walls be altered to improve the economic quality of xylem without negatively influencing the biological and mechanical function of this critical tissue in the living tree?

As is the case for so many plants, there is the need for more studies on the physiology of root tissues. It is evident that, in addition to absorption of water and nutrients and storage of photosynthate, roots respond to environmental stresses, e.g., flooding and wind sway. The mechanism and potential inheritance of these responses need to be further elucidated. Could the root system and the branching intensity of the roots be increased and how would such a modification influence above-ground growth? In addition, it will be important to know if similar patterns in ray parenchyma tissue storage, so well detailed for stem tissues, exists in the ray parenchyma of roots. Due to the increased economic importance of the genus, studies to better understand the patterns of growth and development of poplars will continue to produce exciting results. Poplar is rapidly developing as a model system for studying woody plant physiology.

References

Aloni, R. 1985. Plant growth method and composition. United States patent no. 4 507 144 issued March 26, 1985.

Aloni, R. 1987. Differentiation of vascular tissues. Annu. Rev. Plant Physiol. **38**: 179–204.

Aloni, R. 1988. Vascular differentiation within the plant. *In* Vascular differentiation and plant growth regulators. *Edited by* L.W. Roberts, P.B. Gahan, and R. Aloni. Springer-Verlag, Berlin. pp. 39–59.

Aloni, R. 1991. Wood formation in deciduous hardwood trees. *In* Physiology of trees. *Edited by* A.S. Raghavendra. John Wiley & Sons, Inc., New York, NY. pp. 175–198.

Aloni, R. 1993. The role of cytokinin in organized differentiation of vascular tissues. Aust. J. Plant Physiol. **20**: 601–608.

Aloni, R. 1995. The induction of vascular tissues by auxin and cytokinin. *In* Plant Hormones: Physiology, Biochemistry and Molecular Biology, 2nd ed. *Edited by* P.J. Davis. Kluwer Academic, Dordrecht, The Netherlands. pp. 531–546.

Aloni, R., and Zimmermann, M.H. 1983. The control of vessel size and density along the plant axis — a new hypothesis. Differentiation, **24**: 203–208.

Angeles, G. 1990. Hyperhydric tissue formation in flooded *Populus tremuloides* seedlings. IAWA (Int. Assoc. Wood Anat.) Bull. n.s. **11**: 85–96.

Baba, K., Sone, Y., Misaki, A., Shibuya, N., Hyashi, T., and Itoh, T. 1992. Immunocytochemistry on the cell wall polysaccharides in the woody plants. *In* Plant cell walls as biopolymers with physiological functions. *Edited by* Y. Masuda. Yamada Science Foundation, Osaka, Japan. pp. 327–331.

Baïer, M., Goldberg, R., Catesson, A.M., Francesch, C., and Rolando, C. 1993. Seasonal changes of isoperoxidases from poplar bark tissues. Phytochemistry, **32**: 789–793.

Baïer, M., Goldberg, R., Catesson, A.M., Liberman, M., Bouchemal, N., Michon, V., and du Penhoat, C.H. 1994. Pectin changes in samples containing poplar cambium and inner bark in relation to the seasonal cycle. Planta, **193**: 446–454.

Bendtsen, B.A. 1978. Properties of wood from improved and intensively managed trees. For. Prod. J. **28**: 61–72.

Bendtsen, B.A., Maeglin, R.R., and Deneke, F. 1981. Comparison of mechanical and anatomical properties of eastern cottonwood and *Populus* hybrid NE-237. Wood Sci. **14**: 1–14.

Bendtsen, B.A., and Senft, J. 1986. Mechanical and anatomical properties in individual growth rings of plantation-grown eastern cottonwood and loblolly pine. Wood Fiber Sci. **18**: 23–38.

Berlyn, G.P. 1961. Factors affecting the incidence of reaction tissue in *Populus deltoides*, Bartr. Iowa State J. Sci. **35**: 367–424.

Blankenhorn, P.R., Bowersox, T.W., Strauss, C.H., Stimely, G.L., Stover, L.R., and Di-Cola, M.L. 1988. Effects of management strategy and site on selected properties of first rotation *Populus* hybrid NE-388. Wood Fiber Sci. **20**: 74–81.

Blankenhorn, P.R., Bowersox, T.W., Strauss, C.H., Kessler, K.R., Stover, L.R., Kilmer, W.R., and DiCola, M.L. 1992. Effects of management strategy and site on specific gravity of a *Populus* hybrid clone. Wood Fiber Sci. **24**: 274–279.

Blum, W. 1970. Über die experimentelle Beeinflussung der Reaktionsholzbildung bei Fichten und Pappeln. Ber. Schweiz. Bot. Ges. **80**: 225–252.

Bobák, M., and Nečesaný, V. 1967. Changes in the formation of the lignified cell wall within a twenty-four hour period. Biol. Plant. (Praha), **9**: 195–201.

Boyce, S.G., and Kaeiser, M. 1964. Improve wood quality in eastern cottonwood by breeding and selecting for straight, vertical stems. South. Lumberman, **209**: 115–118.

Catesson, A.M., and Roland, J.C. 1981. Sequential changes associated with cell wall formation and fusion in the vascular cambium. IAWA (Int. Assoc. Wood Anat.) Bull. n.s. **2**: 151–162.

Catesson, A.M., Funada, R., Robert-Baby, D., Quinet-Szély, M., Chu-Bâ, J., and Goldberg, R. 1994. Biochemical and cytochemical cell wall changes across the cambial zone. IAWA (Int. Assoc. Wood Anat.) J. **15**: 91–101.

Cheng, W.W., and Bensend, D.W. 1979. Anatomical properties of selected *Populus* clones grown under intensive culture. Wood Sci. **11**: 182–187.

Clausen, S., and Apel, K. 1991. Seasonal changes in the concentration of the major storage protein and its mRNA in xylem ray cells of poplar trees. Plant Mol. Biol. **17**: 69–678.

Coleman, G.D., Chen, T.H.H., Ernst, S.G., and Fuchigami, L.H. 1991. Photoperiod control of poplar bark storage protein accumulation. Plant Physiol. **96**: 686–692.

Coleman, G.D., Chen, T.H.H., and Fuchigami, L.H. 1992. Complementary DNA cloning of poplar bark storage protein accumulation. Plant Physiol. **98**: 687–693.

Coleman, G.D., Englert, J.M., Chen, T.H.H., and Fuchigami, L.H. 1993. Physiological and environmental requirements for poplar (*Populus deltoides*) bark storage protein degradation. Plant Physiol. **102**: 53–59.

Coleman, G.D., Bañados, M.P, and Chen, T.H.H. 1994. Poplar bark storage protein and a related wound-induced gene are differentially induced by nitrogen. Plant Phyiol. **106**: 211–215.

Côté, B., and Dawson, J.O. 1986. Autumnal changes in total nitrogen, salt-extractable proteins and amino acids in leaves and adjacent bark of black alder, eastern cottonwood and white basswood. Physiol. Plant. **67**: 102–108.

Cote, W.A., Jr., and Day, A.C. 1962. The G-layer in gelatinous fibers: Electron microscopic studies. For. Prod. J. **12**: 333–338.

Cote, W.A., Jr., Day, A.C., and Timell, T.E. 1969. A contribution to the ultrastructure of tension wood fibers. Wood Sci. Technol. **3**: 257–271.

Dalessandro, G., and Northcote, D.H. 1981. Increase of xylan synthetase activity during xylem differentiation of the vascular cambium of sycamore and poplar trees. Planta, **151**: 61–67.

DeGroote, D.K., and Larson, P.R. 1984. Correlations between net auxin and secondary xylem development in young *Populus deltoides*. Physiol. Plant. **60**: 459–466.

Dickson, R.E. 1979. Xylem translocation of amino acids from roots to shoots in cottonwood plants. Can. J. For. Res. **9**: 374–378.

Dickson, R.E., Larson, P.R., and Isebrands, J.G. 1974. Differences in cell wall chemical composition among three-year-old *Populus* hybrid clones. *In* Proc. 9th Central States For. Tree Improv. Conf. 1974, Ames, IA. pp. 21–34.

Dickson, R.E., Vogelmann, T.C., and Larson, P.R. 1985. Glutamine transfer from xylem to phloem and translocation to developing leaves of *Populus deltoidies*. Plant Physiol. **77**: 412–417.

Digby, J., and Wareing, P.F. 1966. The effect of applied growth hormones on cambial division and the differentiation of the cambial derivatives. Ann. Bot. **30**: 539–548.

Eckstein, D., Liese, W., and Shigo, A.L. 1979. Relationship of wood structure to compartmentalization of discolored wood in hybrid poplar. Can. J. For. Res. **9**: 205–210.

Evert, R.F., and Kozlowski, T.T. 1967. Effect of isolation of bark on cambial activity and development of xylem and phloem in trembling aspen. Amer. J. Bot. **54**: 1045–1054.

Fege, A.S., and Brown, G.N. 1984. Carbohydrate distribution in dormant *Populus* shoots and hardwood cuttings. For. Sci. **30**: 999–1010.

Fisher, D.G., and Larson, P.R. 1983. Structure of the leaf/branch gap parenchyma and associated tissues in *Populus deltoides*. Bot. Gaz. **144**: 73–85.

Fisher, D.G., Larson, P.R., and Dickson, R.E. 1983. Phloem translocation from a leaf to its nodal region and axillary branch in *Populus deltoides*. Bot. Gaz. **144**: 481–490.

Fukuda, H. 1996. Xylogenesis: initiation, progression and cell death. Annu. Rev. Plant Physiol. Plant Mol. Biol. **47**: 299–328.

Garrett, P.W., Shigo, A.L., and Carter, J. 1976. Variation in diameter of central columns of discoloration in six poplar clones. Can. J. For. Res. **6**: 475–477.

Goffinet, M.C., and Larson, P.R. 1981. Structural changes in *Populus deltoides* terminal buds and in the vasular transition zone of the stems during dormancy induction. Amer. J. Bot. **68**: 118–129.

Goffinet, M.C., and Larson, P.R. 1982. Xylary union between the new shoot and old stem during terminal bud break in *Populus deltoides*. Amer. J. Bot. **69**: 432–446.

Goldberg, R., Catesson, A.M., and Czaninski, Y. 1983. Some properties of syringaldazine oxidase, a peroxidase specifically involved in the lignification process. Z. Pflanzenphysiol. **110**: 267–279.

Gunning, B.E.S. 1977. Transfer cells and their roles in transport of solutes in plants. Sci. Progr. **64**: 539–568.

Harrington, C.A. 1987. Response of red alder and black cottonwood seedlings to flooding. Physiol. Plant. **69**: 35–48.

Harrington, C.A., and DeBell, D.S. 1996. Above- and below-ground characteristics associated with wind toppling in young *Populus*. Trees, **11**(2): 109–118.

Holt, D.H., and Murphey, W.K. 1978. Properties of hybrid poplar juvenile wood affected by silvicultural treatments. Wood Sci. **10**: 198–203.

Hughes, F.E. 1965. Tension wood: A review of literature. For. Abstr. **26**: 2–9, 179–186.

Isebrands, J.G., and Bensend, D.W. 1972. Incidence and structure of gelatinous fibers within rapid-growing eastern cottonwood. Wood Fiber Sci. **4**: 61–71.

Jeremias, K. 1968. Die Veränderungen des Fettgehaltes in den Rinden der Pappelsorten Oxford, Rochester und Androscoggin im Verlauf eines Jahres. Mitt. Ver. Forstl. Standortskde. Forstpflanzensüchtg. **18**: 95–97.

Kaeiser, M. 1955. Frequency and distribution of gelatinous fibers in eastern cottonwood. Amer. J. Bot. **42**: 331–336.

Kaeiser, M., and Boyce, S.G. 1965. The relationship of gelatinous fibers to wood structure in eastern cottonwood. Amer. J. Bot. **52**: 711–715.

Kennedy, R.W. 1957. Fibre length of fast- and slow-grown black cottonwood. For. Chron. **33**: 46–50.

Kennedy, R.W., and Smith, J.H.G. 1959. The effects of some genetic and environmental factors on wood quality in poplar. Pulp Pap. Mag. Can. **60**: 35–36.

Kozlowski, T.T. 1992. Carbohydrate sources and sinks in woody plants. Bot. Rev. **58**: 107–222.

Kramer, P.J., and Kozlowski, T.T. 1979. Physiology of woody plants. Academic Press, New York, NY.

Kroll, R.E., Ritter, D.C., Gertjejansen, R.O., and Au, K.C. 1992. Anatomical and physical properties of balsam poplar (*Populus balsamifera* L.) in Minnesota. Wood Fiber Sci. **24**: 13–24.

Kučera, L., and Nečesaný, V. 1970. The effect of dorsiventrality on the amount of wood rays in the branch of fir (*Abies alba* Mill.) and poplar (*Populus monilifera* Henry). Part I. Some wood ray characteristics. Drev. Vysk. **15**: 1–6.

Kuroda, K., and Shimaji, K. 1985. Wound effects on cytodifferentiation in hardwood xylem. IAWA (Int. Assoc. Wood Anat.) Bull. **6**: 107–118.

Langheinrich, U., and Tischner, R. 1991. Vegetative storage proteins in poplar. Induction and characterization of a 32- and 36- kilodalton polypeptide. Plant Physiol. **97**: 1017–1025.

Larson, P.R. 1962. A biological approach to wood quality. TAPPI (Tech. Assoc. Pulp Pap. Ind.), **45**: 443–448.

Larson, P.R. 1975. Development and organization of the primary vascular system in *Populus deltoides* according to phyllotaxy. Amer. J. Bot. **62**: 1084–1099.

Larson, P.R. 1976a. Development and organization of the secondary vessel system in *Populus grandidentata*. Amer. J. Bot. **63**: 369–391.

Larson, P.R. 1976b. Procambium vs. cambium and protoxylem vs. metaxylem in *Populus deltoides* seedlings. Amer. J. Bot. **63**: 1332–1348.

Larson, P.R. 1977. Phyllotactic transitions in the vascular system of *Populus deltoides* Bartr. as determined by 14C labeling. Planta, **134**: 241–249.

Larson, P.R. 1979. Establishment of the vascular system in seedlings of *Populus deltoides* Bartr. Amer. J. Bot. **66**: 452–462.

Larson, P.R. 1980. Interrelations between phyllotaxis, leaf development and the primary-secondary vascular transition in *Populus deltoides*. Ann. Bot. **46**: 757–769.

Larson, P.R. 1982. The concept of cambium. *In* New Perspectives in Wood Anatomy. *Edited by* P. Bass. Nijhoff/Junk, The Hague. pp. 85–92.

Larson, P.R. 1994. The Vascular Cambium. Springer-Verlag, Berlin.

Larson, P.R., and Fisher, D.G. 1983. Xylary union between elongating lateral branches and the main stem in *Populus deltoides*. Can. J. Bot. **61**: 1040–1051.

Larson, P.R., and Isebrands, J.G. 1971. The plastochron index as applied to developmental studies on cottonwood. Can. J. For. Res. **1**: 1–11.

Larson, P.R., and Isebrands, J.G. 1974. Anatomy of the primary-secondary transition zone in stems of *Populus deltoides*. Wood Sci. Tech. **8**: 11–26.

Larson, P.R., and Richards, J.H. 1981. Lateral branch vascularizaion: its circularity and its relation to anisophylly. Can. J. Bot. **59**: 2577–2591.

Lev-Yadun, S., and Aloni, R. 1990. Polar patterns of periderm ontogeny, their relationship to leaves and buds, and the control of cork formation. IAWA (Int. Assoc. Wood Anat.) Bull. n.s. **11**: 289–300.

Lev-Yadun, S., and Aloni, R. 1995. Differentiation of the ray system in woody plants. Bot. Rev. **61**: 49–88.

Liphschitz, N. 1995. Ecological wood anatomy: changes in xylem structure in Israeli trees. *In* Proceedings Inter. Symp. Tree. Anatomy and Wood Formation, Tianjin, China. Wood anatomy research 1995. *Edited by* W. Shuming. International Academic Publishers, Beijing. pp. 12–15.

Liphschitz, N., and Waisel, Y. 1970. Effect of environment on relations between extension and cambial growth of *Populus euphratica* Oliv. New Phytol. **69**: 1059–1064.

Marton, R., Stairs, G.R., and Schreiner, E.J. 1968. Influence of growth rate and clonal effects on wood anatomy and pulping properties of hybrid poplars. TAPPI (Tech. Assoc. Pulp Pap. Ind.), **51**: 230–235.

Meicenheimer, R.D., and Larson, P.R. 1983. Empirical models for xylogenesis in *Populus deltoides*. Ann. Bot. **51**: 491–502.

Meicenheimer, R.D., and Larson, P.R. 1985. Exogenous auxin and N-1-naphthylphthalamic acid effects on *Populus deltoides* xylogenesis. J. Exp. Bot. **163**: 320–329.

Mia, A.J. 1968. Organization of tension wood fibers with special reference to the gelatinous layer in *Populus tremuloides* Michx. Wood Sci. **1**: 105–115.

Mia, A.J. 1972. Fine structure of the ray parenchyma cells in *Populus tremuloides* in relation to senescence and seasonal changes. Tex. J. Sci. **24**: 245–260.

Murphey, W.K., Browersox, T.W., and Blankenhorn, P.R. 1979. Selected wood properties of young *Populus* hybrids. Wood Sci. **11**: 263–267.

Nelson, E.A., and Dickson, R.E. 1981. Accumulation of food reserves in cottonwood stems during dormancy induction. Can. J. For. Res. **11**: 145–154.

Nelson, N.D. 1978. Xylem ethylene, phenol-oxidizing enzymes, and nitorgen and heartwood formation in black walnut and black cherry. Can. J. Bot. **56**: 626–634.

Nobuchi, T., Takahara, S., and Harada, H. 1979. Studies on the survival rate of ray parenchyma cells with aging process in coniferous secondary xylem. Bull. Kyoto Univ. For. **51**: 239–246.

Noh, E.R., Lee, S.K, and Koo, Y.B. 1986. Compartmentalizing ability of poplar clones for discoloration and decay associated with artificial wounds. Res. Rep. Inst. For. Gen. Korea. **22**: 21–25. (English abstract, in Korean.)

Ohta, S. 1979. Tension wood from the stems of Poplars (*Populus ×euramericana* C.V.) with various degrees of leaning. I. The macroscopic identification and distribution of tension wood within stems. J. Jap. Wood Res. Soc. **25**: 610–614. (Translation from the USDA Forest Products Laboratory.)

Ollinmaa, P.J. 1959. Reaktion puututkimuksa (Study on reaction wood). Acta For. Fenn. **72**: 1–54.

Onilude, M.A. 1982. Quantitative anatomical characteristics of plantation grown loblolly pine (*Pinus taeda* L.) and cottonwood (*Populus deltoides* Bart. ex Marsch.) and their relationship to mechancial properties. Ph.D. Dissertation. *In* Forestry and forest products. VPI & SU, Blacksburg, VA. 175 p.

Onaka, F. 1949. Studies on compression wood and tension wood. Wood Res. Bull. **1**, Wood Res. Inst., Kyoto Univ., Kyoto, Japan. 99 p.

Pang, A., Catesson, A.M., Francesch, C., Rolando, C., and Goldberg, R. 1989. On substrate specificity of peroxidases involved in the lignification process. J. Plant Physiol. **135**: 325–329.

Panshin, A.J., and deZeeuw, C. 1980. Textbook of wood technology, 4th ed. McGraw-Hill, New York, NY.

Pate, J.S., Gunning, B.E.S., and Milliken, F.F. 1971. Function of transfer cells in the nodal regions of stems, particularly in relation to the nutrition of young seedlings. Protoplasma, **71**: 313–334.

Perem, E. 1964. Tension wood in Canadian hardwoods. Can. Dep. For. Publ. 1057. 38 p.

Peszlen, I. 1994. Influence of age on selected anatomical properties of *Populus* clones. IAWA (Int. Assoc. Wood Anat.) J. **15**: 311–321.

Phelps, J.E., Isebrands, J.G., Einspahr, D.W., Crist, J.B., and Sturos, J.A. 1985. Wood and paper properties of vacuum airlift segregated juvenile poplar whole-tree chips. Wood and Fiber Sci. **17**: 529–539.

Phelps, J.E., Isebrands, J.G., and Teclaw, R.M. 1987. Raw material quality of short-rotation, intensively cultured *Populus* clones. II. Wood and bark from first-rotation stems and stems grown from coppice. IAWA (Int. Assoc. Wood Anat.) Bull. n.s. **8**: 182–186.

Plüss, R., Jenny, T., and Meier, H. 1989. IAA-induced adventitious root formation in greenwood cuttings of *Populus tremula* and formation of 2-indolone-acetylaspartic acid, a new metabolite of exogeneously applied indole-3-acetic acid. Physiol. Plant. **75**: 89–96.

Sauter, J.J. 1966. Untersuchungen zur Physiologie der Pappelholzstrahlen. I. Jahresperiodischer Verlauf der Stärkespeicherung im Holzstrahlparenchym. Z. Pflanzenphysiol. **55**: 246–258.

Sauter, J.J. 1972. Respiratory and phosphatase activities in contact cells of wood rays and their possible role in sugar secretion. Z. Pflanzenphysiol. **67**: 135–145.

Sauter, J.J. 1980. Seasonal variation of sucrose content in the xylem sap of *Salix* ×*smithiana*. Z. Pflanzenphysiol. **98**: 377–391.

Sauter, J.J. 1981*a*. Seasonal variation of amino acids and amids in the xylem sap of *Salix*. Z. Pflanzenphysiol. **101**: 399–411.

Sauter, J.J. 1981*b*. Sucrose uptake in the xylem of *Populus* ×*canadensis*. Z. Pflanzenphysiol. **103**: 165–168.

Sauter, J.J. 1982. Transport in Markstrahlen. Ber. Dtsch. Bot. Ges. **95**: 593–618.

Sauter, J.J. 1988. Temperature-induced changes in starch and sugars in the stem of *Populus* ×*canadensis* 'robusta'. J. Plant Physiol. **132**: 608–612.

Sauter, J.J., and Kloth, S. 1986. Plasmodesmatal frequency and radial translocation rates in ray cells of poplar (*Populus* ×*canadensis* Monench 'robusta') Planta, **168**: 377–380.

Sauter, J.J., and Kloth, S. 1987. Changes in carbohydrates and ultrastructure in xylem ray cells of *Populus* in response to chilling. Protoplasma, **137**: 45–55.

Sauter, J.J., and Neumann, U. 1994. The accumulation of storage materials in ray cells of poplar wood (*Populus* ×*canadensis* 'robusta'): Effect of ringing and defoliation. J. Plant Physiol. **143**: 21–26.

Sauter, J.J., and van Cleve, B. 1989*a*. Immunochemical localization of a willow storage protein with a poplar storage protein antibody. Protoplasma, **149**: 175–177.

Sauter, J.J., and van Cleve, B. 1989*b*. Micromorphometric determination of organelles and of storage material in wood ray cells-a useful method for detecting differentiation within a tissue. IAWA (Int. Assoc. Wood Anat.) Bull. **10**: 395–403.

Sauter, J.J., and van Cleve, B. 1990. Biochemical, immunochemical, and ultrastructural studies of protein storage in poplar (*Populus* ×*canadensis* 'robusta') wood. Planta, **183**: 92–100.

Sauter, J.J., and van Cleve, B. 1991. Biochemical and ultrastructural results during starch-sugar-conversion in ray parenchyma cells of *Populus* during cold adaptation. J. Plant Physiol. **139**: 19–26.

Sauter, J.J., and van Cleve, B. 1992. Seasonal variation of amino acids in the tracheal sap of *Populus* ×*canadensis* and its relation to protein body mobilization. Trees, **7**: 26–32.

Sauter, J.J., and van Cleve, B. 1993. Occurrence of a maltose pool and of maltase in poplar wood (*Populus* ×*canadensis* 'robusta') during fall. J. Plant Physiol. **141**: 248–250.

Sauter, J.J., and van Cleve, B. 1994. Storage, mobilization and interrelatoins of strach, sugars, protein and fat in the ray storage tissues of poplar trees. Trees, **8**: 297–304.

Savidge, R.A. 1996. Xylogenesis, genetic and environmental regulation — a review. IAWA (Int. Assoc. Wood Anat.) J. **17**: 269–310.

Schaedle, M. 1975. Tree photosynthesis. Annu. Rev. Plant Physiol. **26**: 101–115.

Schmitt, U., and Liese, W. 1993. Response of xylem parenchyma by suberization in some hardwoods after mechanical injury. Trees, **8**: 23–30.

Schulte-Baukloh, C., and Fromm, J. 1993. The effect of calcium starvation on assimilate partitioning and mineral distribution of the phloem. J. Exp. Bot. **44**: 1703–1707.

Shain, K., and Hillis, N.E. 1973. Ethylene production in xylem of *Pinus radiata* in relation to heartwood formation. Can. J. Bot. **51**: 1331–1335.

Shigo, A.L. 1984. Compartmentalization: a conceptual framework for understanding how trees grow and defend themselves. Annu. Rev. Phytopathol. **22**: 189–214.

Shigo, A.L., Shortle, W.C., and Garrett, P.W. 1977. Genetic control suggested in compartmentalization of discolored wood associated with tree wounds. For. Sci. **23**: 179.

Simson, B.W., and Timell, T.E. 1978. Polysaccharides in cambial tissues of *Populus tremuloides* and *Tilia americana*. I. Isolation, fractionation, and chemical composition of the cambial tissues. Cellul. Chem. Technol. **12**: 39–50.

Sinnott, E.W. 1918. Factors determining character and distribution of food reserves in woody plants. Bot. Gaz. **66**: 162–175.

Smit, B.A., and Stachowiak, M.L. 1988. Effects of hypoxia and elevated carbon dioxide concentration on water flux through *Populus* roots. Tree Physiol. **4**: 153–165.

Smit, B.A., Neuman, D.S., and Stachowiak, M.L. 1990. Root hypoxia reduces leaf growth. Role of factors in the transpiration stream. Plant Physiol. **92**: 1021–1028.

Stepien, V., and Sauter, J.J. 1994. Ringing induces the accumulation of vegetative storage proteins in poplar bark. Trees, **9**: 88–92.

Stepien, V., Sauter, J.J., and Martin, F. 1992. Structural and immunological homologies between storage proteins in the wood and the bark of poplar. J. Plant Physiol. **140**: 247–250.

Stepien, V., Sauter, J.J., and Martin, F. 1994. Vegetative storage proteins in woody plants. Plant Physiol. Biochem. **32**: 185–192.

Sundberg, B., Touminen, H., Nilsson, O., Moritz, T., Little, C.H.A., Sandberg, G., and Olsson, O. 1996. Alteration of growth and development in transgenic *Populus*: current status and potential applications. *In* Micropropagatoin, genetic engineering, and molecular biology of *Populus. Edited by* N.B. Klopfenstein, Y.W. Chun, and M.R. Ahuja. U.S. Dep. Agric., For. Serv., Rocky Mt. For. Range Exp. Stn. Fort Collins, CO.

Timell, T.E. 1969. The chemical composition of tension wood. Sven. Papperstidn. **72**: 173–181.

Timell, T.E. 1986. Compression Wood in Gymnosperms. Vol 2. Springer Verlag, Berlin.

Tippett, J.T., and Shigo, A.L. 1981. Barrier zone formation: A mechanism of tree defense against vascular pathogens. IAWA (Int. Assoc. Wood Anat.) Bull. **2**: 163–168.

Trockenbrodt, M. 1990. Survey and discussion of the terminology used in bark anatomy. IAWA (Int. Assoc. Wood Anat.) Bull. n.s. **11**: 141–166.

Trockenbrodt, M. 1991. Qualitative structural changes during bark development in *Quercus robur, Ulmus glabra, Populus tremula* and *Betula pendula*. IAWA (Int. Assoc. Wood Anat.) Bull. **12**: 5–22.

Trockenbrodt, M. 1994. Quantitative changes of some anatomical characters during bark development in *Quercus robur, Ulmus glabra, Populus tremula* and *Betula pendula*. IAWA (Int. Assoc. Wood Anat.) J. **15**: 387–398.

Tuominen, H., Östin, A., Sandberg, G., and Sundberg, B. 1994. A novel metabolic pathway for Indole-3-acetic acid in apical shoots of *Populus tremula* (L.) × *Populus tremuloides* (Michx.) Plant Physiol. **106**: 1511–1520.

Tuominen, H., Sitbon, F., Jacobsson, C., Sandberg, G., Olsson, O., and Sundberg, B. 1995. Altered growth and wood characteristics in transgenic hybrid aspen (*Populus tremula* L. × *P. Tremuloides* Michx.) expressing agrobacterium tumefaciens T-DNA indolacetic acid-biosynthetic genes. Physiol. Plant. **109**: 1179–1189.

Van Cleve, B., and Apel, K. 1993. Induction by nitrogen and low temperature of storage-protein synthesis in poplar trees exposed to long days. Planta, **189**: 157–160.

Wetzel, S., Demmers, C., and Greenwood, J.S. 1989. Seasonally fluctuating bark proteins are a potential form of nitrogen storage in three temperate hardwoods. Planta, **178**: 275–281.

White, D.J.B., and Robards, A.W. 1965. Gelatinous fibers in ash. Nature (London), **205**: 818.

Wolter, K.E. 1968. A new method for marking xylem growth. For. Sci. **14**: 102–105.

Yamamoto, F., Angeles, G., and Kozlowski, T.T. 1987. Effects of ethrel on stem anatomy of *Ulmus americana* seedlings. IAWA (Int. Assoc. Wood Anat.) Bull. n.s. **8**: 3–9.

Yanchuk, A.D., Dancik, B.P., and Micko, M.M. 1984. Variation and heritability of wood density and fiber length of trembling aspen in Alberta, Canada. Silvae Genet. **10**: 65–70.

Yang, K.C. 1992. Survival rate and nuclear irregularity index of sapwood ray parenchyma cells in four tree species. Can. J. For. Res. **23**: 673–679.

Yang, K.C., and Hazenberg, G. 1991. Relationship between tree age and sapwood/heartwood width in *Populus tremuloides* Michx. Wood Fiber Sci. **23**: 247–252.

Yoshimura, K., Hayashi, S., Itoh, T., and Shimaji, K. 1981*a*. Studies on the improvement of the pinning method for marking xylem growth. I. Minute examination of pin marks in Taeda pine and other species. Wood Res. **67**: 1–16.

Yoshimura, K., Itoh, T., and Shimaji, K. 1981*b*. Studies on the improvement of the pinning method for marking xylem growth. II. Pursuit of the time sequence of abnormal tissue formation in loblolly pine. Mokuzai Gakkaishi, **27**: 755–760.

Ziegler, H. 1964. Storage, mobilization, and distribution of reserve material in trees. *In* The formation of wood in forest trees. *Edited by* M.H. Zimmermann. Academic Press, New York, NY. pp. 303–320.

Zobel, B.J., and van Buijtenen, J.P. 1989. Wood variation. Its causes and control. Springer-Verlag, Berlin, Heidelberg.

CHAPTER 14

The structure and function of *Populus* root systems

Kurt S. Pregitzer and Alexander L. Friend

Introduction

Populus is a diverse genus with species living in flood plain, temperate, boreal and montane forests throughout the northern hemisphere. Genetic variation is high in the genus, but very little is known about how the genome influences the growth and physiology of roots. Many species are aggressive colonists following disturbance and the root systems play an interesting and critical role in the establishment of trees in nature. In some habitats, the aspens (section *Populus*) persist for millennia following repeated wildfires and the new shoots arise from the persistent root system.

We know that roots consume a tremendous portion of the tree's carbon budget and that root processes can limit whole plant growth. The small-diameter "feeder" roots of *Populus* are very dynamic in terms of growth and life span and the soil environment can play an important role in the allocation of carbon to roots and shoots. Unfortunately, the root systems of *Populus* are without question the most poorly understood portion of the plant. The interesting life-histories of species within the genus, and the important role that root systems play in whole-tree physiology are potent motivators for a general understanding of the structure and function of *Populus* root systems. Thus, biologists and growers are beginning to place more emphasis on research in this area. The objectives of this review are to synthesize current information on root system structure, root growth and mortality, and root physiology. The significance of current findings to research and management are discussed, and key unanswered questions are identified.

K.S. Pregitzer. School of Forestry and Wood Products, Michigan Technological University, 1400 Townsend Drive, Houghton, MI 49931-1295, USA.
A.L. Friend. Department of Forestry, Box 9681, Mississippi State University, Mississippi State, MS 39762-9681, USA.
Correct citation: Pregitzer, K.S., and Friend, A.L. 1996. The structure and function of *Populus* root systems. *In* Biology of *Populus* and its implications for management and conservation. Part II, Chapter 14. *Edited by* R.F. Stettler, H.D. Bradshaw, Jr., P.E. Heilman, and T.M. Hinckley. NRC Research Press, National Research Council of Canada, Ottawa, ON. pp. 331–354.

Root system structure

Root system habit

Within 12 h of seed germination, a delicate brush of hairs develops around the base of the *Populus* hypocotyl and these hairs become attached to the soil. During this period and until the primary root has grown into the soil, seedlings are very susceptible to desiccation. About 5 d after germination, the primary root begins to grow slowly; after 12 d it may only be 1.5 mm long (Schreiner 1974). Slow growth of the primary root continues for a few weeks, presumably because there is little photosynthetic capacity to fuel root development and the seeds of *Populus* are so small. Although early root growth is slow relative to larger-seeded trees, it can be remarkably fast and extensive for germinants with so little carbon available for growth. After 46 d, *P. deltoides* ×*balsamifera* seedlings accumulated only 1 cm² of leaf area, but had roots more than 17 cm in length in a simulated riparian environment (Mahoney and Rood 1991). Downward root elongation, in pursuit of a deep or declining water table, is of great adaptive significance for *Populus* species dependent upon disturbed, coarse-textured, riparian substrates for regeneration.

As the tree grows and develops, multiple orders of lateral roots arise from the primary axis. Up to 7 orders of lateral roots have been reported in trees (Sutton and Tinus 1983). However, as is the case for most taxa of temperate deciduous trees, there is little known about the branch architecture of *Populus* root systems. We are unaware of any published reports of the number of orders of lateral roots in mature trees of *Populus*, and we do not understand much about the architecture of whole root systems (number of root orders, branching angle, average internodal distance, average root length and width, variability in these parameters). Formal description of the ontogeny and morphology of the entire root system of field-grown trees is an area sorely in need of further investigation.

Ignoring relationships between morphology and physiology can result in confusion. Most physiological studies are conducted on "new roots," "fine roots," "structural roots," or "suberized roots." Yet these physiological state or size classes are arbitrary and often quite ambiguous. For example, many studies are conducted on "fine roots," e.g., all roots less than 2 mm in diameter. In *Populus*, roots 0–2 mm in diameter can include new roots, brown (older) roots and woody roots (Pregitzer et al. 1990). Roots 0–2 mm in diameter could range from a second order root to the highest order of root on the tree. An analogy from the shoot system would be attempting to understand differences in the photosynthetic capacity of individual leaves by lumping together all sun and shade leaves that were similar in size. Most physiologists are very careful to describe leaf structure and the position of a leaf on the shoot system when reporting rates of photosynthesis. It seems rather fruitless to try to understand the physiology of tree root systems without the proper morphological context, but this is our current state of affairs.

Fortunately, there are a few general descriptions of *Populus* root systems. This taxon is characterized by strongly developed horizontal roots that grow radially away from the taproot (Gilman 1988). Friend et al. (1991) found that 80% of the roots of two different *Populus trichocarpa* × *P. deltoides* hybrids exhibited root angles between 0 (horizontal) and 30°. Horizontal roots are generally found between 5 and 20 cm from the soil surface (Fig. 1). As trees grow in height, roots also extend horizontally. Both Faulkner (1976) and Hansen (1981) found significant linear relationships between tree height and length of horizontal roots. Horizontal roots can be found several tree lengths away from the base of the stem. Graham et al. (1963, p. 12) illustrate an aspen sucker 25.3 m from the nearest parent tree; Buell and Buell (1959) excavated one horizontal root that was 31.7 m long. The "effective soil foraging area" for a tree must have a diameter greater than total tree height, a factor often ignored in experiments (Hansen 1981).

Vertical "sinker" roots branch from the horizontal roots and explore the soil to depths of 1 to >3 m (Graham et al. 1963; Barnes 1966; Faulkner 1976; Strong and La Roi 1983, Heilman et al. 1994*a*). Sinker roots seem to be a common attribute of all taxa within the genus *Populus* (Fig. 1). Their functional significance remains largely unexplored, but they may be important in water uptake during periods when the soil surface contains little available water. Strong and La Roi (1983) report that the overall form of an aspen root system is plastic, being strongly influenced by soil type.

Aspen root systems

The clonal nature of aspen species (section *Populus*) and their root systems deserve special mention. Following disturbances such as clearcutting or wildfire, North American aspens (*Populus tremuloides* and *P. grandidentata*) regenerate largely by vegetative suckering from the residual root systems of the previous stand (Strothmann and Zasada 1957; Graham et al. 1963; Mitton and Grant 1996). These adventitious shoots ("suckers") grow very rapidly, and depend heavily upon the parent root system for growth for at least the first 25 years following establishment (Zahner and DeByle 1965).

As young suckers develop, the cambium and secondary vascular system aligns with that portion of the parent root immediately on the distal side of the sucker from the original location of the parent tree (Brown 1935). Diameter growth of the parent root in this region is nearly equal to that of the stem just above the root, with the result that the root on the distal side of the sucker becomes greatly enlarged, while that on the proximal side grows slowly (Zahner and Debyle 1965). At the base of each sucker, varying numbers of new roots develop within a few years, and these gradually become the major root system for the individual

Fig. 1. Sketches of *Populus* root systems depicting the structural (woody) roots. **A.** The general appearance of the root system of an Euramerican *Populus* clone grown under short-rotation, intensive culture (from Faulkner 1976). **B.** The vertical and partial horizontal root system of *Populus tremuloides* (modified from Strong and La Roi 1983).

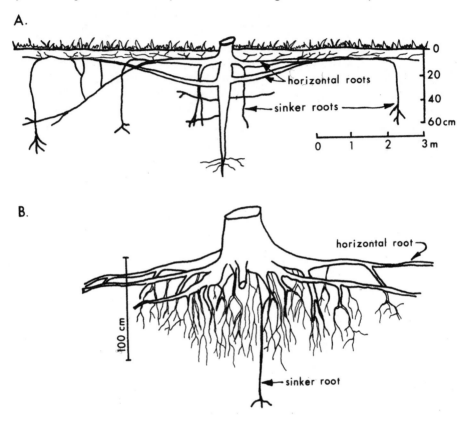

stems (Fig. 2). However, the old parent root system remains alive and functioning for at least 40–50 years (Debyle 1964). Schier (1982) reports that unhealthy clones of *Populus tremuloides* in Utah expand the parental root system in priority over the establishment of new roots. Carbon translocation to the distal portion of the parental root appears to have the highest priority in aspen root systems (Schier 1982) and this results in the characteristic form of aspen roots as depicted in Fig. 2.

Because the root systems of *Populus* are not visible it is often difficult to appreciate just how much carbon is invested in them and how wide-ranging and extensive they can be. Figure 3 shows the extensive and interconnected nature of the root system of a single *Populus tremuloides* clone. One-hundred-six ramets of this clone were hydraulically excavated, staked in place, and mapped (Barnes 1966). There was no predictable, systematic arrangement of the ramets, and the

Fig. 2. *Populus grandidentata* root system from age 6 to 24 years (from Zahner and DeByle 1965).

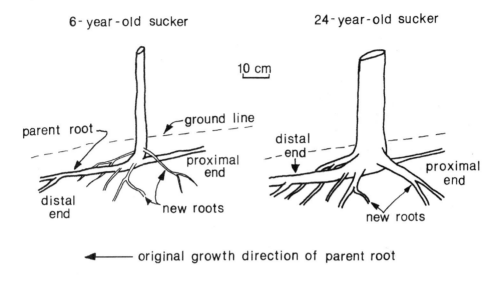

horizontal roots "were a tangled mass" (Graham et al. 1963). It is important to reiterate here that the research done by Graham, Spurr, Zahner, Barnes, and DeByle at the University of Michigan Biological Station in the 1950s and early 1960s clearly demonstrated the importance of the root system to the coppice vigor of aspen sucker sprouts. The parental (residual) root system can remain important to the growth of coppice regeneration for at least 20 years following removal of the shoots.

Fine roots and mycorrhizal fungi

The small-diameter "short" or "fine" roots of *Populus* are, in comparison to many other trees (especially conifers), long, relatively unbranched, and very thin (Brundrett et al. 1990). Most ephemeral fine roots are only 0.1–0.3 mm in diameter (Brundrett et al. 1990; Pregitzer et al. 1990), and specific root length can be very high (42–51 $m \cdot g^{-1}$, Heilman et al. 1994a). Figure 4 is a scaled drawing of hybrid poplar fine roots made from a minirhizotron image. We believe this drawing is characteristic of the fine roots of several different taxa of *Populus*. Brundrett et al. (1990) reported that species within the genus *Populus* have very similar fine root anatomy and morphology. Minirhizotron investigations of several different taxa (Pregitzer et al. 1993; Zak et al. 1993; Pregitzer et al. 1995) all support the idea that the fine roots of *Populus* are long, thin, and relatively unbranched. Brundrett et al. (1990) also notes that the structure of *Populus* fine roots is similar to that of *Salix nigra*. All these taxa are in the family Salicaceae. Heilman et al. (1994a) report surface soil root length densities ranging from 2.4 to 6.3 $cm \cdot cm^3$ for *Populus trichocarpa*, *P. deltoides*, and their hybrids. These

Fig. 3. The network of roots of a clone of *Populus tremuloides*. The soil has been washed away and the roots are held in place by stakes. Note the mass of horizontal roots in an individual clone. The distal ends of the parental roots (see Fig. 2) are also sometimes clearly evident. The root system was excavated by Burton V. Barnes and the late Stephen H. Spurr (shown in the photograph) at The University of Michigan Biological Station in northern lower Michigan. (Photograph July 1956, courtesy of Burton V. Barnes.)

root length densities are high compared to those reported for other tree species (Kalisz et al. 1987).

The fine roots of *Populus* may be associated with both ectomycorrhizal and vesicular-arbuscular mycorrhizal fungi. Fine roots are most often ectomycorrhizal, particularly those of the aspens (Vozzo and Hacskaylo 1974; Brundrett et al. 1990; Cripps and Miller 1995). The ectomycorrhizal mantel may be either thick or thin. *Populus deltoides* roots can be dominated by either ectomycorrhizal or vesicular-arbuscular mycorrhizal fungi (Vozzo 1969; Vozzo and Hacskaylo 1974). However, the species and its hybrids are usually dominated by vesicular-arbuscular mycorrhizal fungi (Godbout and Fortin 1985; Schultz et al. 1983). The functional significance of root-mycorrhizal associations in *Populus* is an area that desperately needs further investigation. Vozzo and Hacskaylo (1974) report higher rates of root respiration and protein synthesis in mycorrhizal vs. nonmycorrhizal roots.

Fig. 4. Tracing of the fine roots of *Populus tristis* × *P. balsamifera* 'Tristis' made from a minirhizotron image. Arrows indicate ectomycorrhizal roots. (Original minirhizotron image courtsey of Dr. Mark Coleman.)

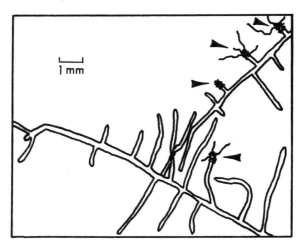

Root growth and mortality

Root growth

The fine roots of *Populus* have the potential to grow very rapidly. Average rates of root length entension can exceed 10 mm/d when soil moisture and fertility are high (Pregitzer et al. 1995). There are at least three aspects of fine root growth in *Populus* that deserve emphasis. First, it appears that rapid rates of root length extension are correlated with rapid above ground growth, at least in young trees. In 1968, Eliasson demonstrated that root elongation of rooted cuttings of *Populus tremula* was directly dependent upon current photosynthate (Eliasson 1968). Excision of the leaves or steam-killing of the stem tissue below the leaves was followed by cessation of root growth within 24 h. Several independent investigators have subsequently reported that early leaf and root growth are strongly correlated (Pallardy and Kozlowski 1979; Tschaplinski and Blake 1989*a*; Rhodenbaugh and Pallardy 1993; Heilman et al. 1994*a*). Rhodenbaugh and Pallardy (1993) explain that rapid early leaf growth is more strongly related to rates of root length extension than is photosynthetic capacity per unit leaf area. Pregitzer et al. (1995) also present data that suggest that rapid rates of root length extension in juvenile trees are related to rapid rates of leaf area accumulation. The practical implications for short rotation intensive culture of *Populus* seem clear. Fast juvenile growth may depend on rapid early root establishment (Rhodenbaugh and Pallardy 1993), especially at the base of the cutting (Heilman et al. 1994*b*). Just how rates of root and shoot growth are correlated over longer periods of time remains to be determined.

It is interesting to note that deteriorating clones of *Populus tremuloides* in Utah had little capacity to produce new roots (Schier 1982). An interesting report from Armenia describes the relationship between crown morphology and root morphology in healthy and dying *Populus deltoides* (Kazaryan and Shakhazizyan 1980). As branches die back and the crown becomes "stagheaded" and decrepit, fine roots furthest from the base of the tree perish, and the root system also becomes decrepit and concentrated around the base of the tree. Taken together, these results strongly suggest a close coupling between the growth and development of the above and below ground portions of the tree. However, as trees mature, there may be a tendency for above and below ground growth to become somewhat uncoupled. Older trees should have greater capacity to store nonstructural carbohydrates and nutrients, and root growth in older trees may depend less upon current photosynthate. Just how coupled above and below ground growth and development are in mature trees is not well documented. However, the literature suggests a strong correspondence between fine root and leaf area growth regardless of tree age.

A second important aspect of root growth in *Populus* is the fact that carbon allocation to root growth appears to be under strong genetic control. Reports of clonal and species-level differences in root growth and root physiology are common (Pallardy and Kozlowski 1979; Schier 1982; Tschaplinski and Blake 1989*a*; Nguyen et al. 1990; Pregitzer et al. 1990; Reighard and Hanover 1990; Liu and Dickmann 1992; Rhodenbaugh and Pallardy 1993; Heilman et al. 1994*a*). It is well-known that shoot growth and disease resistance are under strong genetic control in *Populus*, so it is not surprising that root growth and physiology also exhibit genetic variability. The heritablity of root growth needs to be quantified and eventually related to the long-term performance of field-grown genotypes. Most experimental studies have only utilized a few genotypes. These studies have demonstrated that patterns of root growth are strongly related to genotype under widely ranging environmental conditions (e.g., Pregitzer et al. 1990; Rhodenbaugh and Pallardy 1993). However, it would be useful to conduct experiments that actually enable the calculation of heritability of root growth. These experiments would require the use of a much larger number of genotypes. Building upon the results of Heilman et al. (1994*b*), it might be useful to measure initial rates of root length extension from the base of cuttings of a uniform size and weight for a large number of genotypes. Concurrrent measurements of early leaf area accumulation could also prove quite interesting, since it appears that early root and leaf production are related to rapid tree growth.

The third important aspect of fine root growth centers on understanding plasticity in response to different environments. In a study of the effects of elevated atmospheric CO_2 and nitrogen availability on *Populus ×euramericana* cv. Eugenei, Pregitzer et al. (1995) found root production and mortality to be highly responsive to both atmospheric CO_2 and soil N (Fig. 5). Nitrogen played the most important role in regulating growth and mortality of roots. Figure 5

Fig. 5. Cumulative fine root production (a) (length of roots <0.5 mm diameter) and cummulative fine root mortality (b) of *Populus ×euramericana* cv. Eugenei trees grown in ambient (squares) and twice ambient (circles) atmospheric CO_2 at low (open symbols) and high (closed symbols) soil nitrogen availability (from Pregitzer et al. 1995).

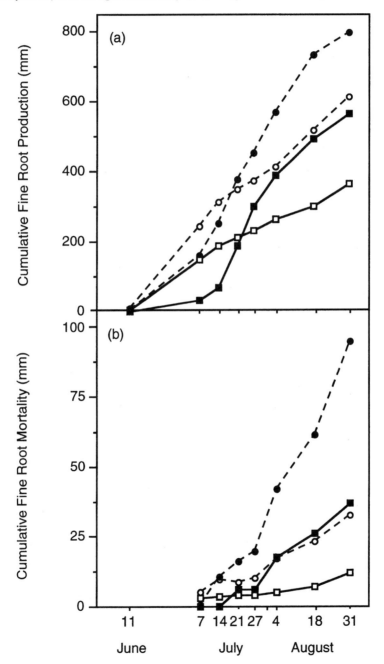

demonstrates why estimates of carbon allocated to fine roots made by destructively harvesting at a single point in time can be misleading. In some cases, *Populus* root production appears to be very plastic and responsive to changing environmental conditions.

How do we resolve the fact that allocation of carbon to *Populus* roots seems to be both highly heritable and plastic in response to changing environments? While this question begs definitive research, it appears that certain genotypes (e.g., *Populus ×euramericana* cv. Eugenei) are particularily plastic in terms of root growth (Nguyen et al. 1990; Pregitzer et al. 1990; Pregitzer et al. 1995), while others (e.g., *Populus tristis* × *P. balsamifera* cv. Tristis #1) seem to be much less responsive to altered soil conditions (Nguyen et al. 1990; Pregitzer et al. 1990). We speculate that inherent root growth plasticity may be an important attribute of fast-growing genotypes, i.e., phenotypic amplitude is under genetic control and root growth plasticity is correlated with rapid shoot growth.

Root:shoot ratios

No discussion of root growth would be complete without mention of root: shoot ratios and root biomass. Trends in root:shoot ratios and root biomass for fast-growing woody plants, including the genus *Populus*, were reviewed by Dickmann and Pregitzer (1992). Values can be highly variable when different studies are compared. Root:shoot ratios in *Populus* decline with age (Shepperd and Smith 1993). They are also likely influenced by genome and environmental variation. But most of all, the wide range reported in the literature probably reflects experimental conditions (Thomas and Strain 1991) and sampling methods. Figure 3 and the literature (Vogt et al. 1986) leave no doubt that root systems represent a major carbon sink. But the utility of describing root:shoot ratios is rather limited in terms of actually understanding carbon allocation to roots. An approach that seems better than simply reporting root:shoot ratios would be to develop and refine mechanistic models of carbon allocation to roots and then validate these models with harvest data. Fine root length per unit leaf area is sometimes much more revealing than simple mass-based root:shoot ratio (Pregitzer et al. 1990).

Root mortality

One of the surprises about root systems that has emerged in the past several years is the highly dynamic nature of fine root mortality. It has become evident that the time-scale over which senescence, mortality, and turnover occurs may be very short. Although this new perspective on roots is firmly established, the plant and environmental controls over the processes of senescence and mortality are not. Mortality rates of *Populus* roots during the growing season vary between about 0.5 and 1.0% loss of length per day, when calculated over periods of greater than 30 d (after Pregitzer et al. 1993; and Coleman et al. 1996). If these rates represented a steady-state, the range in lifetime of *Populus* roots would be between 100 and 200 d. However, mortality is not constant over time, there is often a high mortality rate during the first 30 d in the life of a root cohort during the growing season, followed by a decrease in mortality rate, followed by a period of little or no mortality during the winter months. Rapid early mortality may be attributable to a greater vulnerability of young roots to pathogenic fungi or herbivory when compared to older roots. As the roots age, they become protected by greater concentrations of secondary metabolites, e.g., phenolics and tannins.

Very limited *Populus* data suggest that mortality rates may be greater during late summer and early fall months than during spring or winter months (Coleman et al. 1996). It is interesting to note that mortality may also depend upon differences in nutrient availability in the soil matrix. When patches of soil were enriched with nitrogen, *Populus* roots in the patches lived longer than those in the relatively infertile soil matrix (Pregitzer et al. 1993). When the entire soil matrix was manipulated to provide greater available nitrogen, however, root longevity decreased (Fig. 5). These contrasting results demonstrate the plastic nature of root longevity. The genetic and environmental controls over production and mortality of *Populus* roots appear to be quite complicated. To date, there are no simple, universal responses to changing environmental conditions. Modeling of root senescence is also made difficult by the common observation of episodic rather than gradual root disappearance (Hendrick and Pregitzer 1992). But one thing seems quite clear: *Populus* fine root production and mortality can be very dynamic in time, much more so than the production and mortality of leaves.

Root physiology

Water uptake

Although leaf processes have generally been the focus of investigations into water stress adaptation in *Populus* (cf., Braatne et al. 1992), roots may also play a large role in whole plant resistance to water stress. The most obvious variable in this regard is root proliferation and spatial distribution. In order to supply water to the shoot, roots must be present in moist zones of soil. The ability to produce wide-spreading and prolific root systems has been associated with water stress resistance in *Populus*. For example, Tschaplinski and Blake (1989b) provide evidence to support the concept that early investment into root system mass, even at the expense of foliar mass, improves the juvenile growth of *Populus deltoides* × *P. nigra* hybrids by reducing water stress. The potential importance of large root systems to water stress resistance is also illustrated by comparison of two contrasting hybrids in which the root-intensive hybrid, "Tristis #1," resisted water stress and recovered from water stress to a greater extent than the leaf-intensive hybrid, "Eugenii" (Mazzoleni and Dickmann 1988). At a microenvironmental scale, preferential growth of roots in wet zones of soil may also be important to water stress resistance by compensating for roots in dry zones of soil (Pregitzer et al. 1993). The role of deep sinker roots in the water relations of *Populus* has not been carefully studied.

The mere presence of roots in moist soil, however, does not ensure water influx into the plant. Suberization and permeability are two factors that may limit water movement into roots.[1] Partially-suberized roots are generally believed to be the most permeable roots of woody plants (Kramer 1983, p. 241); however suberized roots may actually contribute more to whole plant water uptake due to their abundance (cf., VanRees and Comerford 1990), although this question has not been specifically addressed for *Populus*.

The permeability of flooded roots has been described. Root hydraulic conductance of one *Populus trichocarpa* × *P. deltoides* hybrid was strongly reduced by both low O_2 and by high CO_2 concentrations, such as would be associated with flooded soil conditions (Smit and Stachowiak 1988). The authors suggest that morphological plasticity (adventitious rooting and aerenchyma formation) is necessary to sustain water uptake by *Populus* under flooded conditions (see also Krasny et al. 1988). Low temperatures may also negatively influence root hydraulic conductance (cf., Kramer 1983, p. 256).

Ion uptake

Populus species are generally reported to be rapid accumulators of nutrients compared with other tree species (see also Chapter 20). Bargali and Singh (1991) report twice the rates of nutrient uptake in *Populus* compared with *Eucalyptus* of similar age and net primary productivity. This trait may not be desirable from the perspective of management costs as *Populus* is often nutrient demanding, but, this is desirable from the perspective of maintaining groundwater quality. For instance, *Populus* ×*canadensis* roots are reported to substantially reduce the concentration of NO_3 in simulated groundwater depletion studies, with root rather than shoot accumulation of N credited for this ability during 2 mo of continuous NO_3 additions (O'Neill and Gordon 1994). Related research reports greater improvements to groundwater quality from cuttings planted 1.5 m deep compared with those planted 0.3 m deep; presumably, deep planting resulted in more rapid establishment of deep roots and subsequently more effective nutrient uptake than with shallow planting (Licht 1992).

Rapid nutrient accumulation in *Populus* is accounted for by a combination of high growth-induced demand (c.f., Hui-jun and Ingestad 1984), and either high root length densities, and/or high rates of ion uptake. Ion uptake kinetics have

---

[1]The term suberization is used here to denote a series of morphological and physiological transformations that can occur as roots age (Richards and Considine 1981). Be aware that various specific developmental changes in roots could influence root permeability in different ways.

received little attention in *Populus*, with most published research directed to the question of ammonium versus nitrate utilization in the arctic (Chapin et al. 1986; Table 1). Although Chapin et al. (1986) reported greater rates of ammonium than nitrate uptake in both *P. balsamifera* and *P. tremuloides*, this may not be a safe generalization for temperate *Populus* species. Hui-jun and Ingestad (1984) found nitrate used in approximately equal proportion to ammonium in *Populus simonii*. Nutrient balance in *Populus* was similar to that of other hardwoods, i.e., optimum nutrient proportions (by weight) were: N = 100, K = 70, P = 14, Ca = 7, Mg = 7 (Hui-jun and Ingestad 1984). Ion uptake kinetics are likely to be much less important than proliferation of root length density (cf., Smethurst and Comerford 1993) in effecting the rapid nutrient accumulation observed in many *Populus* species. However, in accounting for rapid nutrient uptake, the issue of increased ion uptake kinetics vs. rates of root proliferation is an area that deserves additional study (Caldwell et al. 1992).

From the perspective of basic plant nutrition, *Populus trichocarpa* has been used successfully as a model system to study the interactions among calcium, magnesium, and potassium. Schulte-Baukloh and Fromm (1993) reported substantial decreases in carbon allocation to roots and stems of *P. trichocarpa* rooted cuttings deficient in Ca^{2+} but not in those deficient in Mg^{2+} or K^+ and, linked Ca^{2+} directly with the loading process of sugars into sieve elements of the phloem. A separate study used rooted cuttings of *P. trichocarpa* to investigate an antagonism between K^+ in the soil solution and Ca^{2+} and Mg^{2+} nutrition (Diem and Godbold 1993). It was found that K^+ interfered with the export of Ca^{2+} and Mg^{2+} from the roots rather than the uptake of these elements into roots.

In addition to nutrient ions, the dynamics of heavy metals have also been investigated in *Populus* roots. Using rooted cuttings of a horticultural *Populus* clone, Gingas et al. (1988) found generally similar amounts of Cd accumulation in *Populus* and *Helianthus*. Greater Cd accumulation was found in foliage compared to roots. This result contrasted with a previous *Populus* study that found more Cd accumulation in roots rather than foliage of seedlings. Whether these results reflect genetic differences in ion uptake and distribution, or differences in experimental technique (e.g., seedlings vs. rooted cuttings), remains unknown. Another study of heavy metals found evidence for translocation of Pb as electron-dense grains through the plasmodesmata, although many details of how Pb is translocated remain unresolved (Idzikowska 1988).

Root carbon and nitrogen dynamics

There is significant temporal variation in the amount of carbon translocated from leaves to root systems in *Populus*, with bud dormancy having the largest influence on this process. First-year rooted cuttings translocated from 2 to 63% of leaf exported carbon to structural root systems (Isebrands and Nelson 1983). Variation was associated with clone and with sampling date, but there was a

Table 1. Ion uptake properties of *Populus* roots in comparison with those of other deciduous tree species at approximately 20°C.

Species	Ion	C_1 (μM)	I_{net} (μmol·gdw^{-1}·h^{-1})	Source
P. balsamifera	PO_4^{2-}	20	4–7	Chapin (1986)
	NH_4^+	4000	30–40	Chapin (1986)
	NO^{3-}	600	1–3	Chapin (1986)
	K^+(Rb)	70	2–3	Chapin (1986)
P. tremuloides	PO_4^{2-}	20	2–3	Chapin (1986)
	NH_4^+	4000	40–60	Chapin (1986)
	NO^{3-}	600	3	Chapin (1986)
	K^+(Rb)	70	2	Chapin (1986)
Various from eastern USA	PO_4^{2-}	10	0.1–0.6	Lajtha (1994)
	NH_4^+	1000	10–60	Lajtha (1994)
	NO^{3-}	1000	10–40	Lajtha (1994)

Note: Parameters listed are based on the following equation: $I_{net} = I_{max}(C_1 - C_{min})/(K_m + C_1 - C_{min})$ where I_{net} is the net ion influx rate, I_{max} is the maximum ion influx rate, C_1 is the solution concentration, C_{min} is the minimum solution concentration for positive I_{net}, and K_m is the solution concentration at which net ion influx is one–half of I_{max}. Representative values for the parameters not presented in the table are as follows: NH_4^+ C_{min} 3–4 μM, K_m 50–60 μM; NO^{3-} C_{min} 4 μM, K_m 10 μM; PO_4^{2-} C_{min} 0.1–1.0 μM, K_m 2–16 μM; K^+ C_{min} 1–5 μM, K_m 7–29 μM (Barber 1995, p. 66).

7-fold variation in carbon allocation to roots within a given clone depending upon sampling date. The previous authors documented the key role of budset in shifting carbon allocation from aboveground to belowground sinks, but recent work has shown this shift to begin even before bud formation in certain clones (Coleman et al. 1995). Environment undoubtedly also has an influence on the proportion of carbon allocated to roots (c.f., Friend et al. 1994); however this is not well quantified for *Populus*. The proportion of post-budset carbon allocation to roots (including the associated cutting) decreases dramatically with plant development. More than 75% of leaf exported carbon was allocated to roots and cuttings for 1st-year plants (*Populus tristis* × *P. balsamifera*, Isebrands and Nelson 1983). Less than 25% of leaf-exported carbon was allocated to roots and cuttings of 2nd-year plants (*Populus trichocarpa* × *P. deltoides*, Friend et al. 1991). The reason for this change is an increase in the capacity of stem carbohydrate storage with increasing plant age (see also Chapters 15, 17).

Populus root systems can have a tremendous capacity to store and release carbohydrates. Nguyen et al. (1990) found total nonstructural carbohydrate (TNC) concentrations above 42% (dry-weight basis) in structural roots of a 1st-year *P. nigra* × *P. deltoides* hybrid ("Eugenii"), with 80% of whole plant TNC contained in the root system by November. This clone had a tremendous ability to load carbohydrates into roots, with a 20-fold increase in TNC content of root systems from August to November, and a 75-fold increase in starch concentration of fine roots (<0.5 mm) over the same period. Another clone (*Populus tristis* × *P. balsamifera*, "Tristis #1") studied by the same authors had much less variability in carbohydrate storage, only a 2-fold increase in TNC content of root systems from August to November of the same year, despite an earlier date of budset. Although the mobility and fate of carbon allocated to roots for storage has not been extensively investigated in *Populus*, information from studies of cuttings is comparable with the dynamics of storage and release in coarse roots. Fege (1983) labeled *Populus* "Tristis 1" hybrids with ^{14}C in the fall (after budset) and ^{14}C was traced during the following growing season. Calculations from Fege's data indicate that 75% of the ^{14}C accumulated in the cutting is lost during the 1st mo of spring growth. These data reflect the tremendous ability of roots and associated tissues to store carbon and use it later to fuel growth and tissue maintenance.

Data from the Fege (1983) study also indicated that stored carbon was used more in support of new stem and leaf growth than for new root growth. Ten days after planting, the proportion of total ^{14}C accumulated by stems and leaves was roughly equivalent to the proportion of total biomass accumulated by stems and leaves. In roots, however, the proportion of total ^{14}C accumulated was only one-third of the proportion of biomass accumulated. These data are consistent with the concept that new root growth is largely supported by current rather than stored assimilates (van den Driessche 1991; Horwath et al. 1994), but they also illustrate that stored reserves are used to support new root growth of *Populus*, even if their contribution is less than that of new shoot growth.

Nitrogen storage is also substantial in *Populus* roots and has been investigated for the same two contrasting clones described above. Clonal differences were not as evident for N as for TNC. Eugenii accumulated more total N than Tristis, but the timing of N storage in root systems was similar between the two clones (Pregitzer et al. 1990). Pregitzer et al. (1990) found N concentrations as high as 1.6% (dwt basis) in structural roots and up to a 2-fold seasonal increase in the concentration of root N. The percent of whole-plant N stored in roots was similar to that of TNC (70–80%); however the mechanism of movement was different: carbohydrates primarily accumulated from allocation of current photosynthate into root storage rather than shoot growth, while nitrogen accumulated as a result of remobilization of leaf N and translocation to the roots. *Populus* may be remarkably conservative of N, with up to 70% of prelitterfall N retained in the plant through the dormant season (Pregitzer et al. 1990).

The propensity of roots to effectively store carbon and nitrogen is of obvious benefit to spring growth in the following year. The controls of storage in roots are known to be generally related to the onset of bud dormancy and to the process of cold hardening (Nguyen et al. 1990); however, the causes for observed clonal differences and the significance of these differences to plant performance remain largely unexplored. One obvious management implication is that dormant-season harvest should result in the most vigorous coppice regrowth.

Root carbon loss

Carbon allocated to root systems may be lost along several different pathways: remobilization and export to the shoot, various types of root respiration, herbivory, root mortality, exudation, and sloughage (Fig. 6). Of all the aspects of root physiology this may be the least understood because these fluxes are very difficult to isolate and measure in realistic environments (Horwath et al. 1994). If a steady state is assumed, however, the net efflux of total respiration (Rt, Fig. 6) can be estimated by measuring CO_2 efflux from the soil surface. This was done by Jurik et al. (1991) who found little variation in Rt for *Populus grandidentata* (ages 11–70 years) among diverse site conditions using soda lime to trap CO_2. During the growing season, Rt was between 1600–2400 $nmol \cdot m^{-2} \cdot s^{-1}$. On an annual basis, the magnitude of Rt fluxes from the soil surface was very similar to that reported for aboveground net primary production; both Rt and aboveground net primary production were approximately 30 mol $C \cdot m^{-2} \cdot yr^{-1}$ (excluding the least productive stand, assuming biomass is 50% carbon, correcting Rt for CO_2 from litterfall decomposition, using data from Jurik et al. 1991). These results emphasize the extent to which root systems consume whole plant carbon.

Measurements of root respiration under realistic soil conditions are clouded by methodological uncertainty. For instance, a recent study by Cheng et al. (1993) with *Triticum* has shed considerable doubt on the assumption that root respiration dominates Rt. These authors estimated that up to 59% of Rt is accounted for by $r_{microb.}$ due to rapid microbial metabolism of root exudates. Special precautions may be needed to avoid overestimation of root respiration due to this potential artifact. An additional concern is that root systems are a dynamic assemblage of different size classes and functional categories of roots. Thus, specific root respiration values ($nmol \cdot g^{-1} \cdot s^{-1}$) will be influenced by accumulation of secondary growth by roots and by the relative abundance of fine and coarse roots. A rapidly growing root system may appear to have a lower specific respiration rate than a slow growing root system due to the dilution of the rate by secondary growth in roots (Coleman et al. 1996). Roots of woody plants provide a greater challenge than herbaceous or crop plants in this regard. The few reports of specific root respiration for *Populus* appear to conflict to a greater extent than can be explained by plant and environmental factors. One study reports peak root respiration for entire root systems of 5-mo-old *P. balsamifera* and *P. tremuloides* between 4–5 nmol $CO_2 \cdot g^{-1} \cdot s^{-1}$ @ 25°C (Lawrence and Oechel 1983), which compares favorably with similar measurements recorded for *Pinus taeda* (5–7 $nmol \cdot g^{-1} \cdot s^{-1}$ @ 26°C; Edwards 1991). But a more recent study reports a range for *Populus tremuloides* of 10–70 $nmol \cdot g^{-1} \cdot s^{-1}$ @ 14–22°C (Coleman et al.

Fig. 6. Simplified conceptual model of pools and fluxes of carbon allocated to the root system. Boxes indicate pools and arrows indicate fluxes. The system can be viewed from two levels: net import of carbon to the root system and its loss to remobilization of stored carbon and soil respiration (Rt); or, as the complex pools and fluxes within the root-soil system; including three types of root respiration: r_{maint} (maintenance respiration), r_{growth} (growth respiration), and r_{uptake} (respiration associated with ion uptake); two types of soil respiration: $r_{microb.}$ (microbial respiration), and r_{fauna} (respiration of soil fauna); and movement into and out of the three pools shown: root and mycorrhizal biomass, soil carbon, and soil faunal biomass.

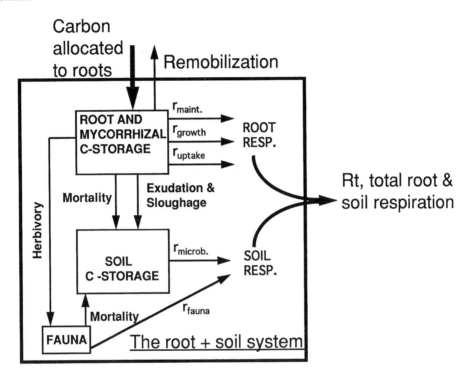

1996). Variation in specific root respiration rates may be explained by variations in the magnitude of r_{maint}, r_{growth}, and r_{uptake}; however, order-of-magnitude discrepancies may also reflect methodological inaccuracies, differences in root size classes, and the influence of root age.

To date, separating root respiration into maintenance, growth, and ion uptake components (cf., Bloom et al. 1992) has been limited to controlled environments. If the magnitude and control of these respiration components can be determined under controlled conditions, it will require a modeling approach, together with new field techniques, to estimate the components of root respiration in natural soil environments. Little information of this sort exists for *Populus* (but see Horwath et al. 1994).

Root × *shoot interactions*

Root systems cannot be viewed as autonomous organs functioning only as the environment permits. More and more evidence has been collected to demonstrate the influence of shoots over root processes and of roots over shoot processes. Three general categories of root × shoot interaction may be identified. First is the obvious coordination required between the growth of roots — with their nutrient supply to leaves and shoot — and the growth of leaves and shoots — with their carbon supply to the roots. Second is the influence that numerous stresses in the root environment have on shoot physiology, which appears to be independent of direct influences on water or nutrient supply to the shoot (e.g., Davies and Zhang 1991). Third is the control over movement of carbon and nutrient reserves between root and shoot associated with dormancy and budbreak.

Growth coordination between leaves and roots appears to be influenced by the relative availability of above and belowground resources. In general, the percentage of carbon allocated to root growth increases when the supply of water or nutrients is low, and carbon allocation to shoot growth increases when the supply of carbon or nutrients is high; however, the mechanism by which this is accomplished varies with the scale of resource availability and with species' phylogenetic constraints (Grime 1993). It is very important to note that absolute trends in carbon allocation do not necessarily follow relative trends. Root:shoot ratios can be very deceptive. They show relative trends at one point in time, but absolute carbon flux to roots can be quite another story (Pregitzer et al. 1995).

The balance between root and shoot carbon allocation in *Populus* is nicely illustrated by defoliation experiments. Bassman and Dickmann (1985) found carbon allocated to roots from the defoliated zone of foliage was 0% of ^{14}C recovered 3 d after defoliation (compared with 3% in the control), and 2% of ^{14}C recovered 23 d after defoliation (compared with 15% in the control plants). Trends in biomass partitioning following defoliation are consistent with the use of ^{14}C (Bassman and Dickmann 1982). Presumably defoliated plants compensate for the loss in foliage by increasing carbon allocated to foliage, while decreasing that allocated to roots (Bassman and Zwier 1993). Environmental controls over root–shoot allocation of carbon are reviewed in detail elsewhere (see Neuman et al., Chapter 17).

Communication of stress in the root environment to the shoot, independent of resource availability, has been extensively studied using *Populus trichocarpa* × *Populus deltoides* hybrids and oxygen stress. Lack of oxygen, or hypoxia, is known to decrease root growth, stimulate the production of adventitious roots (a potential adaptation to this stress), and decrease whole-plant growth. Recent work with *Populus* has demonstrated remarkably rapid shoot responses to hypoxia: leaf growth decreases within a few hours after decreasing the oxygen concentration of the root environment (Smit et al. 1989). Although the mechanism of this root–shoot communication has not been clearly elucidated, it has been shown that a substance is produced in the roots and transported to the shoot in the transpiration stream, and that this substance is the cause of reduced leaf growth (Smit et al. 1990; Neuman and Smit 1993). Root–shoot communication is currently an active area of research for numerous plant species and soil stresses (Davies and Zhang 1991; see also Dickmann and Keathley, Chapter 19). The significance of this phenomenon to root physiology is that regulation of plant growth may be independent of resource availability. Root–shoot communication offers the potential, therefore, to manipulate plant responses to accommodate unique environmental situations.

The control of shoot–root communication over the storage and release of root system carbon in *Populus* is evident from studies of shoot phenology and carbon allocation to cuttings and roots (e.g., Fege 1983; Isebrands and Nelson 1983). At the simplest level, there appear to be switching mechanisms related to shoot phenology that trigger roots to invest carbon into storage rather than growth in the fall and trigger roots to release carbon from storage in the spring. At a more complex level, there appear to be other messages communicated from the shoot, together with environmental cues, which trigger interconversions of storage carbohydrates (cf., Fege 1983). The aforementioned mechanisms have not yet been studied in detail; yet, they illustrate the ways in which shoots may control root physiology independently of resource availability. Clearly, the study of root physiology must acknowledge the potential regulatory role of the shoot.

Conclusions

The root systems of *Populus* represent a large portion of the tree's carbon economy. The genus is characterized by wide-ranging horizontal roots that can eventually extend more than 30 m from the stem. Deep "sinker" roots branch from horizontal roots and their functional significance remains unknown. Fine roots are thin and relatively unbranched compared to many other taxa of temperate trees. The production and mortality of fine roots is very dynamic in time, more so than leaves. Genetic and environmental controls over production and mortality are poorly understood, but it is clear that both genome and environment play important roles in regulating the allocation of carbon to fine roots, and in controlling the longevity of individual roots. The functional significance of mycorrhizal fungi in the genus *Populus* is very poorly understood. Roots may be associated with both ecto- and VA mycorrhizal fungi, but the importance of these mycorrhizal associations remains mostly unexplored.

The root system has been shown to be important in water stress resistance, but it is not yet clear how important root proliferation or deep rooting are in conferring resistance. Accompanying rapid rates of growth, for which the genus *Populus* is renowned, are very high nutrient demands. Unfortunately, little is known about ion uptake kinetics. Rates of ion uptake may be very high based on the observation that the fine roots of *Populus* are thin and relatively un-branched, and the knowledge that the canopy is often supplied with large quantities of nutrients from the soil. This hypothesis needs to be tested. Alternatively, high root length density may be the key to supplying nutrients to the canopy. The role of mycorrhizae in ion uptake has not been studied in *Populus*. More information is needed before we can begin to understand the mechanisms of rapid nutrient supply to the demanding canopy.

The root system of *Populus* serves an important storage function for both non-structural carbohydrates and nutrients. The seasonal rhythm of loading roots in the fall and utilizing stored reserves in the spring is well documented and has important management implications. In the aspens (section *Populus*), the parental root system can remain functional for at least 20 years following removal of the shoots and plays a pivotal role in coppice regrowth. The role of the root system in regulating productivity and harvestable yield is still largely a mystery. The literature strongly suggests close coordination between shoot and root growth. Fast-growing genotypes appear to be those that exhibit both rapid leaf area accumulation and root length extension, but this is more speculation than fact. The heritability of important attributes of the root system is unknown in a genus where genetic variation is the hallmark of understanding aboveground productivity. There is much to learn about *Populus* root systems. Those portions of the tree that are out of sight should not be out of mind. Understanding belowground processes is one of the cornerstones of the house known as forest productivity.

Acknowledgements

Results reviewed in this manscript and its preparation were partially supported by NSF grants BSR 890565 and DEB 92-21003, the DOE (NIGEC and 93PER61666), USDA Competitive Grant 90-37290-5668, and USDA-CSREES McIntire-Stennis projects MISZ-0601 and MISZ-0606. We gratefully acknowledge the support for our research. We would also like to thank Burton Barnes, Andrew Burton, Mark Coleman, Margaret Gale, Mark Kubiske, Jennifer Maziasz, Trish Steman and two anonymous reviewers for improving early drafts of the manuscript.

References

Bargali, S.S., and Singh, S.P. 1991. Aspects of productivity and nutrient cycling in an 8-year-old *Eucalyptus* plantation in a moist plain area adjacent to central Himalaya, India. Can. J. For. Res. **21**: 1365–1372.

Barnes, B.V. 1966. The clonal growth habit of American aspens. Ecology, **47**: 439–447.

Bassman, J.H., and Dickmann, D.I. 1982. Effects of defoliation in the developing leaf zone on young *Populus* ×*euramericana* plants. I. Photosynthetic physiology, growth, and dry weight partitioning. For. Sci. **28**: 599–612.

Bassman, J.H., and Dickmann, D.I. 1985. Effects of defoliation in the developing leaf zone on young *Populus* ×*euramericana* plants. II. Distribution of ^{14}C-photosynthate after defoliation. For. Sci. **31**: 358–366.

Bassman, J.H., and Zwier, J.C. 1993. Effect of partial defoliation on growth and carbon exchange of two clones of young *Populus trichocarpa* Torr. & Gray. For. Sci. **39**: 419–431.

Bloom, A.J., Sukrapanna, S.S., and Warner, R.L. 1992. Root respiration associated with ammonium and nitrate absorption and assimilation by barley. Plant Physiol. **99**: 1294–1301.

Braatne, J.H., Hinckley, T.M., and Stettler, R.F. 1992. Influence of soil water on the physiological and morphological components of plant water balance in *Populus trichocarpa*, *Populus deltoides*, and their F1 hybrids. Tree Physiol. **11**: 325–339.

Brown, A.B. 1935. Cambial activity, root habit and sucker shoot development in two species of poplar. New Phytol. **34**: 163–179.

Brundrett, M., Murase, G., and Kendrick, B. 1990. Comparative anatomy of roots and mycorrhizae of common Ontario trees. Can. J. Bot. **68**: 551–578.

Buell, M.F., and Buell, H.F. 1959. Aspen invasion of prarie. Bull. Torrey Bot. Club, **86**: 264–265.

Caldwell, M.M., Dudley, L.M., and Lilieholm, B. 1992. Soil solution phosphate, root uptake kinetics and nutrient acquisition: implications for a patchy soil environment. Oecologia, **89**: 305–309.

Chapin, F.S., III, Van Cleve, K., and Tryon, P.R. 1986. Relationship of ion absorption to growth rate in tiaga trees. Oecologia, **69**: 238–242.

Cheng, W., Coleman, D.C., Carroll, C.R., and Hoffman, C.A. 1993. *In situ* measurement of root respiration and soluble C concentrations in the rhizosphere. Soil. Biol. Biochem. **25**: 1189–1196.

Coleman, M.D., Dickson, R.E., Isebrands, J.G., and Karnosky, D.F. 1995. Carbon allocation and partitioning in aspen clones varying in sensitivity to tropospheric ozone. Tree Physiol. **15**: 593–604.

Coleman, M.D., Dickson, R.E., Isebrands, J.G., and Karnosky, D.F. 1996. Root growth and physiology of potted and field-grown trembling aspen exposed to tropospheric ozone. Tree Physiol. **16**: 145–152.

Cripps, C.L., and Miller, J. 1995. Ectomycorrhizae formed *in vitro* by quaking aspen: including *Inocybe lacera* and *Amanita pantherina*. Mycorrhiza, **5**: 357–370.

Davies, W.J., and Zhang, J. 1991. Root signals and the regulation of growth and development of plants in drying soil. Annu. Rev. Plant Physiol. Mol. Biol. **42**: 55–76.

DeByle, N.V. 1964. Detection of functional intra-aspen root connections by tracers and excavation. For. Sci. **10**: 386–396.

Dickmann, D.I., and Pregitzer, K.S. 1992. The structure and dynamics of woody plant root systems. *In* Ecophysiology of short rotation forest crops. *Edited by* C.P. Mitchell, J.B. Ford-Robertson, T. Hinkley, and L. Sennery-Forsse. Elsevier Applied Science, New York, NY. pp. 95–123.

Diem, B., and Godbold, D.L. 1993. Potassium, calcium and magnesium antagonism in clones of *Populus trichocarpa*. Plant Soil, **156**: 411–414.

Edwards, N.T. 1991. Root and soil respiration responses to ozone in *Pinus taeda* L seedlings. New Phytol. **118**: 315–321.

Eliasson, L. 1968. Dependence of root growth on photosynthesis in *Populus tremula*. Physiol. Plant. **21**: 806–810.

Faulkner, H.G. 1976. Root distribution, amount, and development from 5-year-old *Populus* ×*euramericana* (Dode) Guinier. M.S.F. thesis, University of Toronto, Toronto, ON. 130 p.

Fege, A.S. 1983. Changes in *Populus* carbohydrate reserves during induction of dormancy, cold storage of cuttings, and development of young plants. Ph.D. Dissertation, University of Minnesota, St. Paul, MN.

Friend, A.L., Scarascia-Mugnozza, G., Isebrands, J.G., and Heilman, P.E. 1991. Quantification of two-year-old hybrid poplar root systems: morphology, biomass, and ^{14}C distribution. Tree Physiol. **8**: 109–119.

Friend, A.L., Coleman, M.D., and Isebrands, J.G. 1994. Carbon allocation to root and shoot systems of woody plants. *In* Biology of adventitious root formation. *Edited by* T.D. Davis and B.E. Haissig. Plenum Press, New York, NY. pp. 245–273.

Gilman, E.F. 1988. Tree root spread in relation to branch dripline and harvestable root ball. Hortscience, **23**: 351–353.

Gingas, V.M., Sydnor, T.D., and Weidensaul, T.C. 1988. Effects of simulated acid rain on cadmium mobilization in soils and subsequent uptake and accumulation in poplar and sunflower. J. Amer. Soc. Hort. Sci. **113**: 258–261.

Godbout, C., and Fortin, J.A. 1985. Synthesized ectomycorrhizae of aspen: fungal genus level of structural characterization. Can. J. Bot. **63**: 252–262.

Graham, S.A., Harrison, R.P., Jr., and Westell, C.E., Jr. 1963. Aspens: Phoenix Trees of the Great Lakes Region. The University of Michigan Press, Ann Arbor, MI. 272 p.

Grime, J.P. 1993. Stress, competition, resource dynamics and vegetation processes. *In* Plant adaptation to environmental stress. *Edited by* L. Fowden, T. Mansfield, and J. Stoddart. Chapman & Hall, London. pp. 45–63.

Hansen, E.A. 1981. Root length in young hybrid *Populus* plantations: Its implication for border width of research plots. For. Sci. **27**: 808–814.

Heilman, P.E., Ekuan, G., and Fogle, D. 1994*a*. Above-and below-ground biomass and fine roots of 4-year-old hybrids of *Populus trichocarpa* × *Populus deltoides* and parental species in short-rotation culture. Can. J. For. Res. **24**: 1186–1192.

Heilman, P.E., Gorden, E., and Fogle, D.B. 1994*b*. First-order root development from cuttings of *Populus trichocarpa* × *P. deltoides* hybrids. Tree Physiol. **14**: 911–920.

Hendrick, R.L., and Pregitzer, K.S. 1992. The demography of fine roots in a northern hardwood forest. Ecollogy, **73**: 1094–1104.

Horwath, W.R., Pregitzer, K.S., and Paul, E.A. 1994. [14]C Allocation in tree-soil systems. Tree Physiol. **14**: 1163–1176.

Hui-jun, J., and Ingestad, T. 1984. Nutrient requirements and stress response of *Populus simonii* and *Paulownia tomentosa*. Physiol. Plant. **62**: 117–124.

Idzikowska, K. 1988. Preliminary research on lead absorption and translocation in root tip cells of *Populus nigra* "Italica" Moench. Acta Soc. Bot. Pol. **57**: 217–222.

Isebrands, J.G., and Nelson, N.D. 1983. Distribution of ^{14}C-labeled photosynthates within intensively cultured *Populus* clones during the establishment year. Physiol. Plant. **59**: 9–18.

Jurik, T.W., Briggs, G.M., and Gates, D.M. 1991. Soil respiration of five aspen stands in northern lower Michigan. Amer. Midl. Nat. **126**: 68–75.

Kalisz, P.J., Zimmerman, R.W., and Muller, R.N. 1987. Root density, abundance, and distribution in the mixed mesophytic forest of eastern Kentucky. Soil Sci. Soc. Amer. J. **51**: 220–225.

Kazaryan, V.O., and Shakhazizyan, R.S. 1980. Basipetal displacement of the active zone of the roots in woody plants during ontogenesis. Fiziol. Rast. **27**(3): 607–611.

Kramer, P.J. 1983. Water relations of plants. Academic Press, New York, NY. 489 p.

Krasny, M.E., Zasada, J.C., and Vogt, K.A. 1988. Adventitious rooting of four Salicaceae species in response to a flooding event. Can. J. Bot. **66**: 2597–2598.

Lajtha, K. 1994. Nutrient uptake in eastern deciduous tree seedlings. Plant Soil, **160**: 193–199.

Lawrence, W.T., and Oechel, W.C. 1983. Effects of soil temperature on the carbon exchange of taiga seedlings. I. Root respiration [*Alnus crispa, Populus balsamifera, Populus tremuloides, Betula papyrifera*, Alaska]. Can. J. For. Res. **13**: 840–849.

Licht, L.A. 1992. Salicaceae family trees in sustainable agroecosystems. For. Chron. **68**: 214–217.

Liu, Z., and Dickmann, D.I. 1992. Responses of two hybrid *Populus* clones to flooding, drought, and nitrogen availability. I. Morphology and growth. Can. J. Bot. **70**: 2265–2270.

Mahoney, J.M., and Rood, S.B. 1991. A device for studying the influence of declining water table on poplar growth and survival. Tree Physiol. **8**: 305–314.

Mazzoleni, S., and Dickmann, D.I. 1988. Differential physiological and morphological responses of two hybrid *Populus* clones to water stress. Tree Physiol. **4**: 61–70.

Mitton, J.B., and M.C. Grant. 1996. Genetic variation and the natural history of quacking aspen. BioScience, **46**: 25–31.

Neuman, D.S., and Smit, B.A. 1993. Root hypoxia-induced changes in the pattern of translatable mRNAs in popular leaves. J. Exp. Bot. **44**: 1781–1786.

Nguyen, P.V., Dickmann, D.I., Pregitzer, K.S., and Hendrick, R. 1990. Late-season changes in allocation of starch and sugar to shoots, coarse roots, and fine roots in two hybrid poplar clones. Tree Physiol. **7**: 95–105.

O'Neill, G.J., and Gordon, A.M. 1994. The nitrogen filtering capability of Carolina Poplar in an artificial riparian zone. J. Environ. Qual. **23**: 1218–1223.

Pallardy, S.G., and Kozlowski, T.T. 1979. Early root and shoot growth of *Populus* clones. Silvae Genet. **28**: 153–156.

Pregitzer, K.S., Dickmann, D.I., Hendrick, R., and Nguyen, P.V. 1990. Whole-tree carbon and nitrogen partitioning in young hybrid poplars. Tree Physiol. **7**: 79–93.

Pregitzer, K.S., Hendrick, R.L., and Fogel, R. 1993. The demography of fine roots in response to patches of water and nitrogen. New Phytol. **125**: 575–580.

Pregitzer, K.S., Zak, D.R., Curtis, P.S., Kubiske, M.E., Teeri, J.A., and Vogel, C.S. 1995. Atmospheric CO_2, soil nitrogen and turnover of fine roots. New Phytol. **129**: 579–585.

Reighhard, G.L., and Hanover, J.W. 1990. Shoot and root development and dry matter partitioning in *Populus grandidentata*, *P. tremuloides*, and *P.* ×*smithii*. Can. J. For. Res. **20**: 849–852.

Rhodenbaugh, E.J., and Pallardy, S.G. 1993. Water stress, photosynthesis and early growth patterns of cuttings of three *Populus* clones. Tree Physiol. **13**: 213–226.

Richards, D., and Considine, J.A. 1981. Suberization and browning of grapevine roots. *In* Structure and function of plant roots. *Edited by* R. Brouwer. Martinus Nijhoff/Dr Junk Publishers, The Hague, The Netherlands. pp. 111–115.

Schier, G.A. 1982. Sucker regeneration in some deteriorating Utah aspen stands: development of independent root systems. Can. J. For. Res. **12**: 1032–1035.

Schreiner, E.J. 1974. *Populus* L. *In* Seeds of woody plants in the United States. *Edited by* C.S. Schopmeyer. U.S. Dep. Agric., Agric. Handb. No. 450. U.S. Dep. Agric. For. Serv.,Washington, DC. pp. 645–655.

Schulte-Baukloh, C., and Fromm, J. 1993. The effect of calcium starvation on assimilate partitioning and mineral distribution of the phloem. J. Exp. Bot. **44**: 1703–1707.

Schultz, R.C., Isebrands, J.G., and Kormanik, P.P. 1983. Mycorrhizae of poplars. *In* Intensive plantation culture: 12 years of research. *Edited by* E.A. Hansen. Gen. Tech. Rep. NC-91. U.S. Dep. Agric. For. Serv., St. Paul, MN. pp. 17–28.

Shepperd, W.D., and Smith, F.W. 1993. The role of near-surface lateral roots in the life cycle of aspen in the central Rocky Mountains. For. Ecol. Manage. **61**: 157–170.

Smethurst, P.J., and Comerford, N.B. 1993. Simulating nutrient uptake by single or competing and contrasting root systems. Soil Sci. Soc. Amer. J. **57**: 1361–1367.

Smit, B., and Stachowiak, M. 1988. Effects of hypoxia and elevated carbon dioxide concentration on water flux through *Populus* roots. Tree Physiol. **4**: 153–165.

Smit, B.A., Stachowiak, M., and Van Volkenburgh, E. 1989. Cellular processes limiting leaf growth in plants under hypoxic root stress. J. Exp. Bot. **40**: 89–94.

Smit, B.A., Neuman, D.S., and Stachowiak, M.L. 1990. Root hypoxia reduces leaf growth. Role of factors in the transpiration stream. Plant Physiol. **92**: 1021–1028.

Strong, W.L., and La Roi, G.H. 1983. Root-system morphology of common boreal forest trees in Alberta, Canada. Can. J. For. Res. **13**: 1164–1173.

Strothmann, R.O., and Zasada, Z.A. 1957. Silvical characteristics of quaking aspen (*Populus tremuloides*). U.S. Dep. Agric. For. Ser. Lake States For. Expt. Sta. Pap. 49. 26 p.

Sutton, R.F., and Tinus, R.W. 1983. Root and root system terminology. For. Sci. Monogr. **24**. 135 p.

Thomas, R.B., and Strain, B.R. 1991. Root restriction as a factor in photosynthetic acclimation of cotton seedlings grown in elevated carbon dioxide. Plant Physiol. **96**: 627–634.

353

Tschaplinski, T.J., and Blake, T.J. 1989*a*. Correlation between early root production, carbohydrate metabolism, and subsequent biomass production in hybrid poplar. Can. J. Bot. **67**: 2168–2174.

Tschaplinski, T.J., and Blake, T.J. 1989*b*. Water relations, photosynthetic capacity, and root/shoot partitioning of photosynthate as determinants of productivity in hybrid poplar. Can. J. Bot. **67**: 1689–1697.

van den Driessche, R. 1991. New root growth of Douglas-fir seedlings at low carbon dioxide concentration. Tree Physiol. **8**: 289–295.

Van Rees, K.C.J., and Comerford, N.B. 1990. The role of woody roots of slash pine seedlings in water and potassium absorption. Can. J. For. Res. **20**: 1183–1191.

Vogt, K.A., Grier, C.C., and Vogt, D.J. 1986. Production, turnover, and nutritional dynamics of above- and below-ground detritus of world forests. Adv. Ecol. Res. **15**: 303–317.

Vozzo, J.A. 1969. Endotrophic mycorrhizae found on *Populus deltoides*. For. Sci. **15**: 158.

Vozzo, J.A., and Hacskaylo, E. 1974. Endo- and ectomycorrhizal associations in five *Populus* species. Bull. Torrey Bot. Club, **101**: 182–186.

Zahner, R., and N.V. DeByle. 1965. Effect of pruning the parent root on growth of aspen suckers. Ecology, **46**: 373–375.

Zak, D.R., Pregitzer, K.S., Curtis, P.S., Teeri, J.A., Fogel, R.F., and Randlett, D.L. 1993. Elevated atmospheric CO_2 and feedback between carbon and nitrogen cycles. Plant Soil, **151**: 105–117.

CHAPTER 15
Carbon acquisition and allocation

R. Ceulemans and J.G. Isebrands

Introduction

Because the harvest yield in tree crops is vegetative rather than reproductive and most of the biomass consists of carbon compounds, clonal selection for increased harvest yield is intimately linked to changes in the photosynthetic fixation of CO_2 per unit of land area. The photosynthetic basis for increasing harvest yield involves maximizing the amount of light intercepted by the canopy, as well as maximizing the conversion efficiency of intercepted light to photosynthetic products. Larson and Gordon (1969) stated that "it is necessary to increase photosynthesis and at the same time increase that proportion of available photosynthate directed to the stem. This task may be accomplished by manipulating crown shape and structure, leaf display, or photosynthetic efficiency." Ecophysiological research and continuous breeding efforts provide numerous possibilities for improving plant material and for selecting genotypes that can utilize the environment more efficiently (Hüttermann 1993). These possibilities can be broadly classified into two groups: (*i*) more efficient coverage of the land area to optimize conversion of solar energy into chemical energy (by photosynthetic energy conversion), and (*ii*) more efficient use of the growing season.

To maximize yields of short rotation forestry (SRF), close spacing is used for complete utilization of the site by the plants at the earliest possible age. Within this context photosynthetic CO_2 uptake and the allocation of the photosynthates into the harvestable product play a key role. The photosynthetic process is essentially a double conversion process, i.e., a transformation of solar energy into chemical energy, and a conversion of simple inorganic molecules (CO_2 and

R. Ceulemans. Department of Biology, University of Antwerpen (UIA), Universiteitsplein 1, B-2610 Wilrijk, Belgium. Fax: 32-3-820-2271; e-mail address: rceulem@uia.ua.ac.be
J.G. Isebrands. USDA Forest Service, North Central Forest Experiment Station, Rhinelander, WI 54501, USA. Fax: 715-362-1166; e-mail address: jisebran@newnorth.net

Corrrect citation: Ceulemans, R., and Isebrands, J.G. 1996. Carbon acquisition and allocation. *In* Biology of *Populus* and its implications for management and conservation. Part II, Chapter 15. *Edited by* R.F. Stettler, H.D. Bradshaw, Jr., P.E. Heilman, and T.M. Hinckley. NRC Research Press, National Research Council of Canada, Ottawa, ON. pp. 355–399.

H_2O) into more complex organic substances. Selections can be made to achieve a balance among such factors as plant form, leaf display, leaf retention, and photosynthetic efficiency to maximize light interception and photosynthetic CO_2 uptake. Furthermore, allocation of the resulting photosynthate products into the final harvestable yield components should be optimized. In the SRF practice, initial selections should be for those trees (or genotypes) that can produce maximum yields at relatively close spacings.

In this chapter we begin by analyzing the photosynthetic performance of a wide range of *Populus* species and finally narrow down to a comparison of the performance of a number of superior hybrids for SRF with that of their parental species. When discussing the various factors that directly or indirectly affect carbon acquisition, our main focus will be on factors such as leaf orientation and display, branch type, and effect of elevated atmospheric CO_2 that are of direct relevance to one or several of the detailed models that are available for poplar. Specific models for poplar are also described and illustrated. Although factors such as leaf anatomy, leaf nitrogen content, water status, and stress conditions are very important for carbon acquisition, they are not dealt with in total in the context of the present chapter. Besides the carbon uptake performance, the distribution of photosynthetic production over the plant organs (allocation) is also a crucial factor in yield enhancement. An array of key functional and structural components affecting growth and productivity of superior hybrids are elucidated with a focus on light interception, photosynthetic processes, and allocation strategies.

We believe it is important to clearly define our terminology pertaining to photosynthesis advised by the Crop Science Society of America (CSSA) Committee on Crop Terminology (Shibles 1976), and used throughout this chapter. Photosynthetically active radiation (PAR) is the radiation in the 400–700 nm waveband. Photosynthetic photon flux density (PPFD) is the photon flux density of PAR, and gives the number of photons (400–700 nm) incident per unit time on a unit surface. The units of PPFD are µmols of photons area^{-1}·time^{-1} (or µmol·m^{-2}·s^{-1}). The accepted and advised terms relating to the process of photosynthesis, and to photosynthetic (and respiratory) rate are the following: apparent photosynthesis is the photosynthesis estimated indirectly and uncorrected for respiratory activity (Shibles 1976), for example, photosynthesis measured by ^{14}C analysis (Michael et al. 1985). CO_2 exchange rate (CER) is the net rate of carbon dioxide diffusion out from (–) or in to (+) an entity such as a plant tissue, organ or canopy, or a soil surface. Photosynthetic capacity is the maximum CO_2 exchange rate at saturating light, 20°C and 350 µmol·mol^{-1} CO_2. Integrated whole tree photosynthesis is defined as whole leaf integrated over the entire tree for a given time period (Isebrands et al. 1988). To maintain similarity for time and linear dimensions among fluxes of energy, water, CO_2, and with resistances to flux of water and CO_2, we use the specific units for leaf and canopy CO_2 (and water) flux densities to be $\mu mol \cdot m^{-2} \cdot s^{-1}$.

Carbon assimilation

Genotypic variation in photosynthetic and respiration rates: an overview

Although many factors can affect the magnitude of photosynthetic productivity attained in the field, CER is fundamental to plant productivity. However, the hypothesis that increased photosynthesis results in increased plant productivity and crop yield can only be validated after the quantitative relationship between photosynthesis and plant productivity has been considered (Gifford et al. 1984; Coombs et al. 1985). Genetic differences in CER and related photosynthetic traits have been studied in woody plants and trees for over 50 years (Kozlowski and Keller 1966; Schaedle 1975). There have been perhaps more photosynthetic studies on *Populus* than on any other tree genus, with the possible exception of apple (genus *Malus*; Avery 1977). Evidence for the large clonal variation in photosynthetic traits in *Populus* is shown in Table 1 (see also Larcher 1969; Isebrands et al. 1988; Ceulemans and Saugier 1991). This table gives a detailed review of the numerous photosynthetic and gas exchange studies published on poplar and summarizes the photosynthetic capacity and dark respiration rates in controlled environments, glasshouses, and in the field for a wide range of *Populus* species and hybrids.

Not only are there a large number of studies on poplar, but also there is a significantly large variation in photosynthetic rates and capacities and dark respiration rates among various *Populus* species, hybrids, and genoytpes within species (Ceulemans and Impens 1980; Nelson and Michael 1982; Ceulemans and Impens 1983). The range of maximum CER (and whole leaf photosynthesis) reported values for poplars is well within the range determined for most agricultural and herbaceous C_3 plants (Gifford et al. 1984; Coombs et al. 1985). Because poplars exhibit some of the highest CER's and photosynthetic capacities among woody plants, the widely held idea that woody plants are inherently low in photosynthetic capacity apparently does not apply to this genus (Nelson 1984). Pedigreed material would be ideal for studying genetic variation in photosynthetic rates in poplars, but to our knowledge has not yet been done.

Integrated whole tree photosynthesis

Because of the intimate relation between foliar growth and wood formation in poplar (Ridge et al. 1986), the most promising starting point for a yield improvement is to increase photosynthetic capacity by cultural and genetic improvement. This approach has been successful in agronomic crops, and has been confirmed in the genus *Populus*. In general, the more productive selected poplar hybrids

Table 1. Average photosynthetic capacity (A_{max}) and dark respiration rates (R_d) of different *Populus* species and hybrids.

Species or hybrid	Factors studied	A_{max}	R_d	Experimental conditions	Method	References
Populus Species						
P. deltoides	leaf development	9.5	2.2	G/G	I	Dickmann 1971
P. deltoides	leaf development	8	1.1	G/L	R	Larson et al. 1972
P. deltoides	flooding	14.8	0.8	N/L	I	Regehr et al. 1975
P. deltoides	leaf orientation & whole-tree photosynthesis	15.9	—	N/N and G/G	R and C	Isebrands and Michael 1986
P. deltoides	water stress	13.4	—	N/L	I	Scarascia-Mugnozza et al. 1986
P. deltoides	clonal variation & leaf orientation	13–16	0.9	N/N and G/G	I	Isebrands et al. 1988; Wiard 1987
P. grandidentata	light	8.8	—	G/L	I	Pollard 1970
P. grandidentata	development	—	—	G/?	?	Okafo and Hanover 1978
P. grandidentata	CO₂ concentration	14–20	—	N/N	I	Jurik et al. 1984
P. grandidentata	site quality	17.3	—	N/N	I	Briggs et al. 1986
P. maximowiczii	clonal variation, light & temperature	9	0.1	G/L	I	Luukkanen and Kozlowski 1972
P. maximowiczii	nutrition	13.6	—	G/G	I	Fei et al. 1990
P. nigra	clonal variation, light & temperature	5–10	0.9–1	G/L	I	Luukkanen and Kozlowski 1972

Table 1. (*continued*).

Species or hybrid	Factors studied	A_{max}	R_d	Experimental conditions	Method	References
P. nigra	clonal variation & selection criteria	10	1.2	G/L and N/G	I	Ceulemans and Impens 1980
P. nigra	air pollutants rooting	11.4	—	OTC	I	Ceulemans et al. 1987; Ballach et al. 1992
P. tremula	cuttings shade	1.3	0.6	G/G	I	Okoro and Grace 1976
P. tremula	tolerance	9	—	N/N	I	Tsel'niker 1979
P. tremula	ozone	12	—	OTC	I	Matyssek et al. 1993*a* and 1993*b*
P. tremuloides	light	6.8	—	G/L	I	Pollard 1970
P. tremuloides	seasonal trends	20–22	0.7	N/N	I	Foote and Schaedle 1974
P. tremuloides	plant development	21	—	G/?	?	Okafo and Hanover 1978
P. trichocarpa	clonal variation, light & temperature	10.7	1.5	G/L	I	Luukkanen and Kozlowski 1972; Ceulemans and Impens 1980
P. trichocarpa	clonal variation & selection criteria	10–12	1.7	G/L and N/G	I	Ceulemans et al. 1987
P. trichocarpa	water stress	10	—	G/L	I	Pezeshki and Hinckley 1982
P. trichocarpa	water stress	15	—	N/L	I	Scarascia-Mugnozza et al. 1986
P. trichocarpa	whole-tree photosynthesis	14–15	—	N/N and G/G	I and C	Isebrands and Michael 1986; Isebrands et al. 1988

Table 1. (*continued*).

Species or hybrid	Factors studied	A_{max}	R_d	Experimental conditions	Method	References
P. trichocarpa	clonal variation & leaf orientation	20–22	—	N/N and G/G	I	Wiard 1987
Populus hybrids						
P. deltoides × *P. trichocarpa*	light & leaf age	23	—	G/G	I	Reich 1984
P. deltoides × *P. trichocarpa*	ozone	18	—	L/L	I	Reich and Amundson 1985
P. ×euramericana	light, temperature & humidity	9.5–17	2.5	G/L	I	Furukawa 1973 & 1975
P. ×euramericana	clonal variation	12.6	0.9	G/L	I	Gjerstad 1975
P. ×euramericana	leaf development	6.2	1.5	G/G	R	Dickmann and Gordon 1975
P. ×euramericana	development	8.2	1.2	G/G	I	Dickmann et al. 1975
P. ×euramericana	clonal variation & selection criteria	9.5	—	G/G	I	Gordon and Promnitz 1976
P. ×euramericana	rooting cuttings	1.3	0.4	G/G	I	Okoro and Grace 1976
P. ×euramericana	light	9	0.5	G/L	I	Fasehun 1978

Table 1. (*continued*).

Species or hybrid	Factors studied	A_{max}	R_d	Experimental conditions	Method	References
P. ×euramericana	clonal variation & correlation with growth	10–16	0.6–1.5	G/L	I	Ceulemans and Impens 1980; Ceulemans et al. 1987
P. ×euramericana	defoliation effects	9.5–15.8	—	G/G	R	Bassman and Dickmann 1982
P. ×euramericana	herbicide	6.1	—	G/G	R	Akinyemiju and Dickmann 1982
P. ×euramericana	infection by fungal foliar parasite	10–15	—	G/G	I	Maurer et al. 1987
P. ×euramericana	clonal variation & development	20.4	—	N/N	R	Michael et al. 1985
P. ×euramericana	CO_2 concentration	20	0.6	G/L	I	Gaudillère and Mousseau 1989
P. grandidentata × P. alba	clonal variation & light	8.8	—	N/G	I	Gatherum et al. 1967
P. maximowiczii × P. nigra	clonal variation, light & temperature	12	1.6	G/L	I	Luukkanen and Kozlowski 1972
P. maximowiczii × P. trichocarpa	herbicide	8.4	—	G/G	R	Akinyemiju and Dickmann 1982
P. nigra × P. maximowiczii	light, temperature & water	5.7–15.8	0.8	G/L	I	Furukawa 1972
P. nigra × P. laurifolia	late season effects	18	—	N/N	R	Nelson and Isebrands 1983

Table 1. (*continued*).

Species or hybrid	Factors studied	A_{max}	R_d	Experimental conditions	Method	References
P. tristis × P. balsamifera	clonal variation & selection criteria	6	—	G/G	I	Gordon and Promnitz 1976
P. tristis × P. balsamifera	growth room versus field conditions	11 (G) 20 (N)	—	G/G and N/N	I	Nelson and Ehlers 1984
P. tristis × P. balsamifera	late season & shoot type	25	—	N/N	R	Nelson and Michael 1982; Nelson 1984
P. tristis × P. balsamifera	senescence	13.6	—	G/L	I	Nelson 1985
P. tristis × P. balsamifera	clonal variation & seasonal development	23.8	—	N/N	R	Nelson and Michael 1982; Michael et al. 1985; Michael et al. 1988
P. trichocarpa × P. balsamifera	somaclonal variation and callus culture	8.8	2.3	G/G	I	Saieed et al. 1994a
P. trichocarpa × P. deltoides	light & leaf age	23–25	1.2–1.9	G/G	I	Reich 1984
P. trichocarpa × P. deltoides	clonal variation & correlation with growth	10.5	1.2	G/L	I	Ceulemans and Impens 1980; Isebrands and Michael 1986
P. trichocarpa × P. deltoides	whole-tree photosynthesis	17	—	N/N and G/G	I and C	Isebrands et al. 1988
P. trichocarpa × P. deltoides	water stress	13.9	—	N/L	I	Scarascia-Mugnozza et al. 1986

Table 1. (*concluded*).

Species or hybrid	Factors studied	A_{max}	R_d	Experimental conditions	Method	References
P. trichocarpa × P. deltoides	clonal variation & leaf orientation	14–17	1.0	N/N and G/G	I	Wiard 1987

Note: Photosynthetic capacity values refer to maximum CO_2 exchange rates at saturating light, 20–25°C and 350 $\mu mol \cdot mol^{-1}$ CO_2. Dark respiration rates were measured at (or corrected to) atmospheric temperatures of 20–25°C. All values of A_{max} and R_d are expressed in $\mu mol \cdot m^{-2} \cdot s^{-1}$ of single leaf area. Subject matter, experimental conditions and methodology are also indicated. Experimental growth/measurement conditions: the first letter indicates growth conditions of plant material and the second letter indicates experimental measurement conditions. N = natural conditions (field, nursery, or *in situ*), G = controlled environment (growth chamber or glasshouse), L = laboratory, OTC = open top chambers. Method used: I = infrared gas analysis, R = radioactive $^{14}CO_2$ technique, C = whole-plant enclosure technique.

often have enhanced CERs, photosynthetic capacities, and integrated whole tree photosynthesis (Table 1). Much research on increasing photosynthetic production and directing it into usable plant parts and products is currently underway (Saieed et al. 1994*a*, *b*; Stettler and Bradshaw 1994).

When integrated over the whole plant and over time, instantaneous CER measurements give a good estimation of whole-tree photosynthesis (Wolf et al. 1995 *a*, *b*), (also see Heilman et al., Chapter 18, Production Physiology). Small differences in CER (expressed per unit leaf area) of different poplar clones when integrated over the entire leaf and over the whole plant become more significant as an indicator of a leaf's (and tree's) overall photosynthetic performance (Isebrands et al. 1988). In general, the fast growing and high yielding intersectional hybrids of *P. trichocarpa* and *P. deltoides* combine high maximum photosynthesis (per leaf) with large leaves and long leaf area durations (Ceulemans and Impens 1980; Isebrands et al. 1988; Table 1). Net photosynthesis at the whole tree and stand level are discussed in Chapter 18 by Heilman et al.

Several poplar genotypes exhibit significant photosynthetic activity and photosynthate distribution late in the growing season (Nelson and Isebrands 1983). A superior poplar genotype might be one in which the period of cambial activity extends well beyond that of height growth or the production of new leaves. Because mature leaves are the highest producers of photosynthates for wood formation (Nelson 1985; Ceulemans et al. 1987), selections for long leaf retention combined with extended cambial activity is advisable (Larson and Gordon 1969). Therefore, increasing biomass yield in selected species and/or genotypes can be accomplished by extending the effective growing season and increasing the carbon acquisition rate (see Zsuffa et al., Chapter 20).

Maximum wood production does not always necessarily follow maximum photosynthetic production (Griese et al. 1993). However, maximum photosynthetic production does not imply that the wood will be most efficiently distributed within the tree (e.g., above-ground versus below-ground). Numerous possibilities for influencing the distribution of photosynthates have been suggested. Although there are few examples of correlations of photosynthetic traits with yield in trees, there are some good examples in *Populus* (Huber and Polster 1955; Gatherum et al. 1967; Fasehun 1978; Ceulemans and Impens 1983); in particular, there is good correlation where photosynthesis has been integrated over the whole tree and/or entire season (Isebrands and Michael 1986; Wiard 1987; Isebrands et al. 1988). While instantaneous measurements of photosynthesis may sometimes be misleading (Zelitch 1982), yield is closely related to net photosynthetic assimilation throughout an entire season and over the entire canopy area. Our own experimental observations in Antwerp (Belgium), Rhinelander (Wisconsin) and Seattle (Washington) as well as those of other studies reported in the literature suggest that the prospects for genetic improvement of *Populus* (and perhaps other tree crops) based on photosynthetic characteristics has more potential for improving yields than in some agronomic crops because their harvest index is vegetative rather than reproductive.

Leaf morphology, orientation, and display

Leaf orientation has a significant impact on the CER of the leaf because of the photosynthetic photon flux density (PPFD) interception capacity and the effect on the leaf's energy balance. Leaf orientation in poplar can be quantified by measures (and calculations) of basically three different leaf angles, i.e., (i) the leaf lamina angle, an index of directional leaf lamina tilt (vertical tilt of the leaf with the top of the leaf facing left or right); (ii) the midrib angle which is an index of the leaf "hang" (pointing up or hanging downwards), and (iii) the azimuth angle, which represents the azimuth directional orientation of the leaf toward north. A fourth angle, the zenith angle, is the angle between the leaf normal and the zenith, and is calculated from the three aforementioned angles. Zenith angle gives an overall index of the position of the leaf relative to the sun (Isebrands and Michael 1986). With the exception of the azimuth angle, all angles show significant species and clonal differences that might partly explain clonal differences in CERs, among other factors. *P. deltoides* generally has a significant leaf lamina tilt from horizontal, that is reflected in the more even distribution of angle frequency diagrams (Fig. 1) and may be related to adaptations to the climatic conditions in the region of origin. *P. trichocarpa* genotypes have leaves that are predominantly horizontal and show very little deviation from horizontal, while interspecific hybrid clones generally show orientation of leaves somewhere in between (Fig. 1). Different poplar species and hybrids have varying leaf orientation (Fig. 2). Zenith angle seems to be related to the orientation of the branches on which the leaves are positioned (Fig. 3). In Fig. 3 we refer to proleptic and sylleptic branches (see Sylleptic vs. proleptic branches).

Some examples of the effects of a contrasting leaf display on the light interception and CER rates of different poplar genotypes have been given (Wiard 1987; Dickmann et al. 1990; Hinckley et al. 1992). No significant effects of differences of midrib angle were observed on CER on clear days for various poplar clones. On cloudy days, however, *P. trichocarpa* leaves had higher CER rates than leaves of *P. deltoides*. These differences were due to both the greater interception of PPFD and the higher quantum yields at PPFD levels between 100 and 500 $\mu mol \cdot m^{-2} \cdot s^{-1}$ for *P. trichocarpa* as compared to *P. deltoides*. It should be also noted that *P. deltoides* has a green leaf undersurface (Figliola 1986) and vertical orientation (Fig. 1) that collects much PPFD on the abaxial leaf side.

Sylleptic versus proleptic branches

Syllepsis is a phenomenon that is recognized as a common form of development in crowns of many tropical trees, but of only a few north-temperate trees, including *Populus* (Hallé et al. 1978; Powell and Vescio 1986). Sylleptic shoots result from the neoformed development of a newly initiated lateral axis without

Fig. 1. Relative frequency distributions of leaf lamina angle classes for *P. deltoides* clone Ill–005 (from Illinois), *P. trichocarpa* clone 1–12 (from British Columbia) and two *P. trichocarpa* × *P. deltoides* hybrids (clones 11–11 and 44–136). The leaf lamina angle gives an indication of the absolute tilt of the lamina from horizontal, rather than directional tilt.

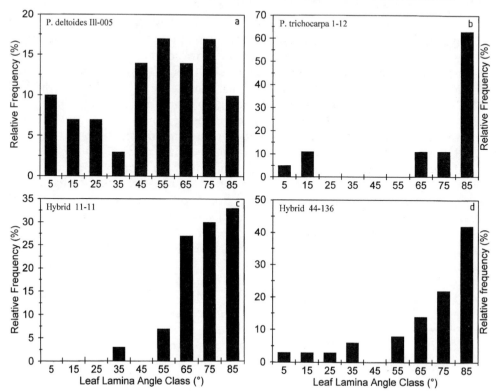

the apical meristem of that axis having had an intervening rest period (Remphrey and Powell 1985). In contrast, proleptic shoots (branches) are the result of the discontinuous development of a new lateral axis, with its apical meristem having experienced some intervening period of rest (Hallé et al. 1978). Certain poplar genotypes (e.g., *P. trichocarpa* clone 12–106 from Oregon, *P. deltoides* clone Ill–005 from Illinois, and interspecific hybrid clone 11–11) produce sylleptic branches early in the growing season (Fig. 4), while other clones (e.g., interspecific hybrid clone 44–136 and *P. trichocarpa* clone 1–12 from British Columbia) show the syllepsis phenomenon much later (Ceulemans et al. 1990).

An optimal crown structural arrangement is essential for a maximization of the PAR interception by the tree or canopy. Leaf area and leaf biomass distribution by vertical strata, by branch type and by current terminal shoots versus branches all show significant differences among clones, and these differences are important with respect to the carbon allocation patterns in the tree (Ceulemans et al. 1990; Isebrands and Nelson 1983). The relative proportion of leaf area on sylleptic

Fig. 2. Lamina angles of an entire-leaf series for 1-yr-old *Populus* trees (cv. Tristis and Eugenei) plotted by position (Tristis = *P. tristis* × *P. balsamifera*, Eugenei = *P.* ×*euramericana* cv. Eugenei). LPI denotes leaf plastochron index. LPI 0 is the first leaf below apex ≥ 3 cm in lamina length (from Isebrands et al. 1983).

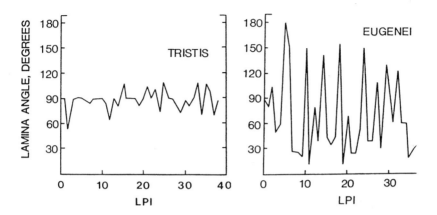

branches in the interspecific hybrid clones is generally lower than that of the two parental species, *P. trichocarpa* and *P. deltoides*, at both the beginning and end of the growing season, but often exceed them through July and August (Fig. 5). In the early part of the growing season (May and June) the majority of the total leaf area of the two *P. trichocarpa* clones is carried on the first- and second-year sylleptic branches (e.g., 67% in clone 1–12 and 55% in clone 12–106) (Fig. 5). For the interspecific hybrid clones 11–11 and 44–136 these values are 39 and 42%, respectively (Scarascia-Mugnozza 1991). These patterns have implications for carbon allocation patterns (see Branches — proleptic and sylleptic).

Leaf area distribution on the main stem versus sylleptic and proleptic branches shows pronounced seasonal variation and clonal differences. Leaf area on sylleptic branches may range from 67% early in the growing season to 31% near the end of the growing season. Because sylleptic branches export more carbon than proleptic branches, seasonal patterns in leaf area on branches have important implications for whole tree carbon allocation patterns (see Whole Plant). The leaf area on the current terminal early in the season is generally low and ranges from 2 to 7%. In the middle of the season, the relative proportion of leaf area on the current terminal increases slightly (up to 22% in *P. trichocarpa* clone 12–106), but remains a minor part of the total leaf area of the tree until the end of the growing season. After August, as budset and senescence has occured, leaves on 1st-yr sylleptic branches gradually abscise, while leaf area of the sylleptic branches of the third year increase. For example, in hybrid clone 11–11 and *P. trichocarpa* clone 1–12 almost 50% of the total leaf area consisted of

Fig. 3. Mean normal leaf zenith angle as a function of branch angle for one clone of *Populus deltoides* (clone Ill–005 from Illinois), one clone of *P. trichocarpa* (clone 12–106 from Orgeon) and one *P. trichocarpa* × *P. deltoides* hybrid (clone 11–11). Open symbols are for proleptic branches and closed symbols refer to sylleptic branches. Fine vertical bars represent standard error of the mean. Sylleptic branches arise in axil of parent leaf without undergoing dormant period.

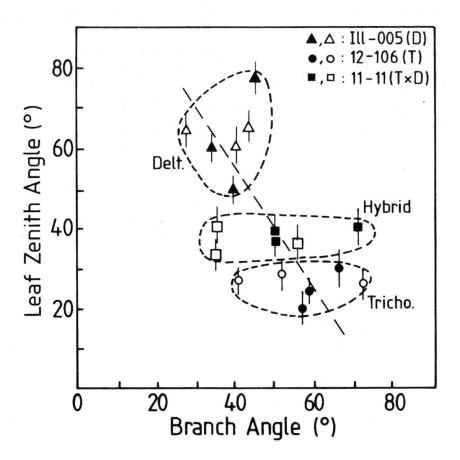

leaves on branches whereas in hybrid clone 44–136 and *P. deltoides* clone Ill–005 this proportion was only 15%.

Significant differences also exist in the leaf area distribution and structure between different interspecific hybrids, illustrated by Fig. 5 for two *P. trichocarpa* × *P. deltoides* hybrids from different families. Leaf area on the current terminal near the end of the growing season is particularly prominent in hybrid clone 44–136 (43%), *P. trichocarpa* clones 12–106 (41%) and 1–12 (35%). In hybrid clone 11–11, however, two-thirds of the total leaf area at that time are

Fig. 4. Sylleptic branches on current terminal shoot of *Populus trichocarpa* × *deltoides* hybrid.

still present on the major, second-year proleptic branches, and only 9% on the current terminal (Fig. 5). Late in the season the upper current terminal leaves contribute significant amounts of photosynthate to the tree (Wiard 1987). These photosynthates are important for growth and reserves storage in the stem and roots (Nelson and Isebrands 1983). In addition to the differences between leaves on sylleptic and proleptic branches, differences between sun and shade leaves play an important role in the overall photosynthetic performance of the tree (Fig. 6).

Branch orientation and branchiness also play an important role to the overall PAR interception by the tree and by the forest canopy. Branch structures with acute branch angles, low biomass, and high leaf area density per unit of branch length occupy minimum radial space and attain maximum leaf area under SRF conditions. In each yearly height growth increment for most of the clones examined, the number of sylleptic branches is larger than the number of proleptic branches. Although sylleptic branches are usually smaller in size than proleptic branches, they usually carry a considerable part of the total leaf area of a tree, and may play a critical role in the superior productivity of the hybrid poplar genotypes through the large amount of PAR they intercept (Zavitkovski 1982; Ceulemans et al. 1990).

Fig. 5. Leaf area distribution at different times during the third growing season for five intensively cultured poplar clones, i.e., *P. deltoides* clone Ill–005 (from Illinois), *P. trichocarpa* clones 1–12 (from British Columbia) and 12–106 (from Oregon) and *P. trichocarpa* × *P. deltoides* clones 11–11 and 44–136. All leaf areas are expressed as percentage of the whole-tree leaf area. Early growing season: May–June; Mid: July–August; Late: September–October.

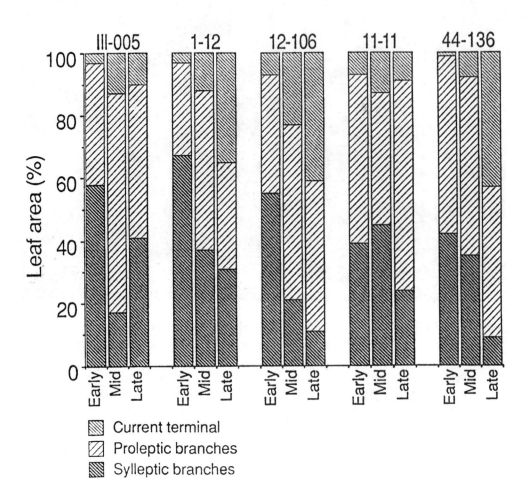

Based upon these and other results, the desirable branch characteristics that emerge as important traits that should be included in the list of productivity determinants for poplar ideotypes (see Dickmann and Keathley, Chapter 19), are (*i*) narrow crowns with few and small branches combined with a higher proportion of sylleptic branches, (*ii*) acute branch angles, and (*iii*) high leaf area density on the branches. Given the diversity of branch characteristics found in natural populations of *Populus* and their high degree of genotypic control, clonal

Fig. 6. Representative response curve of CO_2 exchange rate (expressed per unit single leaf area) to incoming photosynthetic photon flux density for a sun-type leaf and a shade-type leaf of an interspecific hybrid poplar clone.

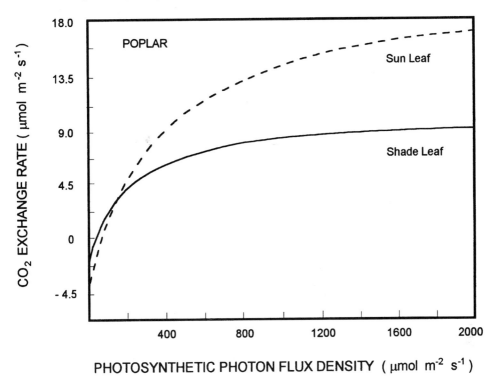

selection for desirable types is quite effective, especially if carried out under typical plantation conditions. Branch traits can, therefore, be incorporated into the ensemble of physiological traits as ideotypes are formulated for SRF (Dickmann 1985; see also Dickmann and Keathley, Chapter 19). As better defined and better studied genetic materials become available it will be possible to more carefully assess the tradeoffs among various traits such as branch characteristics, optimal PAR interception and other physiological components of productivity — a necessary prerequisite for any approach to breeding for high productivity.

Effects of elevated atmospheric CO_2

When poplar plants are treated with elevated concentrations of CO_2 CER, as well as growth, in general, is significantly enhanced (Ceulemans and Mousseau 1994). This observation has been demonstrated for different treatment conditions (e.g., open top chambers and glasshouse cabinets) as well as for different poplar genotypes (Fig. 7). In an experiment were the responses of growth and gas exchange rates of different clones were compared for plants grown and treated

Fig. 7. Net CO_2 exchange rates of *P. deltoides* × *P. nigra* (euramerican; top figure) and *P. trichocarpa* × *P. deltoides* (interamerican; bottom figure) poplar hybrids grown under ambient and elevated atmospheric CO_2 concentrations in open top chambers or in controlled glasshouse cabinets. Open bars represent photosynthesis measurements made in low CO_2 concentration (ca. 350 $\mu mol \cdot mol^{-1}$) and hatched bars represent measurements made under high CO_2 concentrations (ca. 700 $\mu mol \cdot mol^{-1}$). All data are mean values of at least ten replications, and fine vertical lines represent single standard error of the mean (from R. Ceulemans, Antwerp and G. Taylor, Sussex, UK). "Euramerican" and "interamerican" are traditional terms used on the continent to denote hybrids.

a) Euramerican

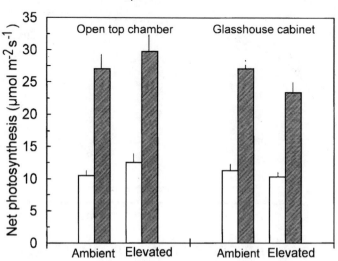

b) Interamerican

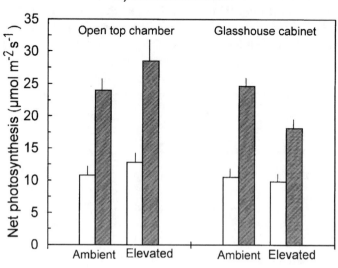

in open top chambers (Fig. 8) with those from glasshouse cabinets, PPFD saturated CERs and their responses to elevated CO_2 were shown to be rather similar (Ceulemans et al. 1994). This result is evident from Fig. 7 for the euramerican (*P. deltoides* × *P. nigra*) clones Primo and Robusta as well as for the *P. trichocarpa* × *P. deltoides* clone Beaupré. For the ambient CO_2 treatment no differences in absolute values of PPFD-saturated CER were found between open top chambers and glasshouse cabinets, but CER of plants in the elevated CO_2 treatment was always significantly higher in the open-top chambers than in the glasshouse cabinets.

In the same study it was also shown that the short-term (i.e., measurement) effect of high CO_2 on CER was different from the long-term (i.e., growth) effect, both in the open top chambers and the glasshouse cabinets (Fig. 7). When considering CER measured at high (700 $\mu mol \cdot mol^{-1}$) CO_2, values were higher for the elevated CO_2 treatment than for the ambient treatment in the open top chamber experiment. However, the result was the reverse in the experiment in the glasshouse cabinets: slightly lower net photosynthesis in the elevated treatment than in the ambient treatment when measurements were made in high CO_2 (Fig. 7), indicating some acclimation of photosynthesis. However, CER measured under high CO_2 (hatched bars on Fig. 7) was found to be at least twice that measured under low atmospheric CO_2 concentration (open bars). This result was true for both growth treatments (ambient and elevated) and for euramerican as well as for interamerican hybrid clones. Genotypic differences in CER are also illustrated on Fig. 7 and were observed both in the open top chambers and the glasshouse cabinets. Thus, we can say that the measurement conditions are of more importance for CER than the experimental growth (or treatment) conditions, and in this regard all clones in the study had a similar response (Ceulemans et al. 1994). Clonal differences in CER are also shown in Fig. 7. These differences were observed in both open top and glasshouse experiments.

Carbon allocation and partitioning

Introduction and terminology

Maintaining a positive carbon balance is key to the growth and survival of all plants. Thus, knowledge of carbon allocation processes is essential to understanding plant growth in the environment where they experience interacting environmental stresses (Pearcy et al. 1987; Dickson 1989; Dickson and Isebrands 1991). Unfortunately, knowledge of the mechanisms and controls of the dynamic processes of carbon allocation in trees and forests is largely lacking (Landsberg et al. 1991; Gower et al. 1995). However, there is perhaps more known about the overall whole tree carbon allocation patterns and processes in *Populus* than any other tree genera. Much research has been done on carbon allocation within the poplar plant at all scales including at the shoot apex, leaf,

Fig. 8. Experimental study on the effects of elevated atmospheric CO_2 on the gas exchange and growth of poplar. Second year of growth of hybrid poplar in 6 m tall open top chambers on the campus of the University of Antwerpen, Wilrijk (Belgium).

branch, whole tree, and stand level (see reviews by Dickson 1986; Hinckley et al. 1989; Ceulemans 1990). In fact, carbon allocation coefficients within *Populus* trees have been incorporated in ecophysiological process models of poplar trees (Rauscher et al. 1990). It is noteworthy that many of the key studies of carbon allocation in *Populus* have been conducted as part of research programs on growing poplars under SRF.

In this paper we use the terms "carbon allocation," "carbon partitioning," and "component biomass accumulation" (i.e., leaf, stem and root biomass) according to the process-based definitions of Dickson and Isebrands (1993). Our terminology originated from basic radiotracer studies of carbon fixation and allocation in *Populus* and has been adopted by other researchers of carbon allocation and partitioning in trees (Friend et al. 1994; Gower et al. 1995). Carbon allocation is the process of distribution of C within the plant to different parts (i.e., source to "sink"). Carbon partitioning is the process of C flow into and among different chemical fractions (i.e., different molecules, different storage and transport pools). Biomass component accumulation is the end product of the process of C accumulation at a specific sink. Allocation, partitioning, and distribution are relative terms (e.g., percent of total), whereas growth and accumulation reflect absolute size (e.g., dry weight, moles of C, etc.).

Radiotracer experiments

Much of the progress in understanding carbon allocation processes in *Populus* has been made through radiotracer experimentation. Structural-functional development in poplar has been studied thoroughly with [14]C (Isebrands et al. 1983; Dickson 1989). In fact, poplar experimentation formed the bases of a comprehensive general review paper of radiotracer techniques and applications in forest tree ecophysiology aimed at understanding carbon allocation and partitioning (Isebrands and Dickson 1991). Those tracer techniques have been used extensively by poplar physiologists to expand our understanding of carbon allocation processes in seedlings to field situations as outlined below.

Whole plant

One of the most significant research developments arising from poplar research that has resulted in a better understanding of the overall structural–functional processes of whole plants (and poplars) was the concept of the plastochron index (Larson and Isebrands 1971). This concept developed for poplar allowed investigators to (*i*) study plants at different morphological stages by adjusting the plant's developmental stages to a standardized morphological time scale, and to (*ii*) predict developmental processes (such as carbon fixation and allocation) and events from simple nondestructive measurements. As a result, the plastochron index has been used extensively in poplar research and has become the standard for analyses and interpretation of poplar carbon allocation and partitioning studies at all scales (Fig. 9).

Fig. 9. Diagram of a 16-leaf cottonwood plant showing the leaf plastochron index (LPI) numbering system, developing and mature leaf zones, areas of primary and secondary vascularization, and photosynthate transport. Arrows at left indicate direction of transport; arrow size indicates relative quantities (from Dickson, 1986).

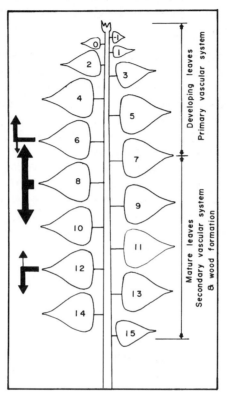

The published literature on carbon allocation and partitioning in *Populus* is exhaustive and too voluminous to be reviewed in total here. The reader is referred to Dickson (1986, 1989, 1991), Dickson and Isebrands (1991), and Friend et al. (1994) for thorough review chapters on the subject. In poplar, whole plant carbon allocation and partitioning patterns have been worked out from the shoot apex (Larson 1977) to the root (Pregitzer et al. 1990; Friend et al. 1991) and from the growth chamber, greenhouse, to the field (Isebrands and Nelson 1983; Isebrands et al. 1983; Nelson and Isebrands 1983; Nelson and Ehlers 1984; Hinckley et al. 1992).

We highlight the most significant contributions on carbon allocation and partitioning from poplar experimentation. Probably the most unique contributions in this regard were the studies of Larson and co-workers on the carbon allocation patterns associated with the phyllotacic development of the leaf-node-stem continuum in poplar (Larson and Dickson 1973; Isebrands et al. 1976; Isebrands and Larson 1977; Larson 1977, 1980; Isebrands and Larson 1980; Larson et al. 1980). They showed that leaves control the development of the structural-functional relationships in poplars and that phyllotacic vascular traces connect every organ and tissue in the developing plant (including roots), thereby controlling carbon allocation patterns and other essential products throughout the plant (Dickson and Isebrands 1991).

Research on poplar also was important for understanding the carbon allocation and partitioning patterns within the development of leaves of indeterminately growing plants in general (Larson et al. 1972; Dickson and Larson 1981). Carbon transport occurs initially from the tips of developing leaves and proceeds basipetally within the leaf over the course of leaf development. Radiotracer experiments in poplar are also used to illustrate classical carbon transport patterns for indeterminate shoots. Developing leaves transport carbon primarily upward to other developing leaves and the apex; recently mature leaves transport both upward and downward to the lower stem and roots; while mature leaves transport primarily downward (Dickson 1986). These general patterns hold for poplars grown in controlled environment and in the field (Isebrands and Nelson 1983; Nelson and Isebrands 1983), or under stress conditions (Dickson and Isebrands 1991; Coleman et al. 1995a, b, 1996).

Other unique carbon allocation experiments include the work of Friend et al. (1991) who conducted ^{14}C labelling of lateral branches to quantify the transport of carbon to root systems in 2-year-old poplar trees; Horwath et al. (1994) who labeled 3 m hybrid poplar trees in a large chamber to study tree-soil systems, and the work of Fege (1983) who used whole tree ^{14}C-labelling of coppiced poplar shoots to follow the transport of ^{14}C-labelled reserves from the roots during regrowth in the spring. Moreover, Nelson and Isebrands (1983) used ^{14}C to show the importance of autumn retention of leaves in poplars for radial stem and root growth and for storage reserves in hybrid poplars.

Branches — proleptic and sylleptic

Little information on carbon allocation and partitioning is available in field-grown trees because of problems with tree size and the experimental difficulties associated with it (Dickson 1986). But, there is perhaps more known about carbon allocation and partitioning within the crown of *Populus* than most trees, again due to extensive radiotracer studies (Isebrands and Dickson 1991).

Carbon allocation and partitioning patterns become more complicated as trees age because of the addition of proleptic and sylleptic branches, thereby increasing the complexity of crown morphology, whole tree photosynthetic and other physiological gradients, and the interactions of environmental variables and cultural practices. Studies of 2-year-old poplars illustrate the contribution of leaves at various crown positions to whole-tree carbon partitioning (Isebrands 1982; Isebrands et al. 1983). Subsequent studies of older trees give similar conceptual results. Mature leaves on the current terminal (CT) have the highest photosynthetic capacity in the crown and maintain that capacity throughout the growing season. Photosynthetic efficiency of leaves from mid and lower crown are always lower than those of the CT and declined rapidly during the season. During active shoot growth, photosynthate from CT mature leaves is translocated primarily to developing leaves and stem within the CT. After bud set, progressively more ^{14}C-photosynthate is translocated to main-stem and roots and less ^{14}C was retained in CT stem and leaves. Very little ^{14}C is translocated to lateral branches. By September nearly 50% of the ^{14}C from CT leaves is translocated to roots (Dickson 1986).

Carbon allocation from mature leaves on lateral branches is similar for each position in the crown, although important quantitative differences were present (Fig. 10). During active growth, ^{14}C is translocated to developing leaves, stem of the treated lateral, and to the main-stem below the lateral. After bud set, more ^{14}C is translocated to main-stem and roots. Little ^{14}C is translocated from a lateral branch to the CT or to other lateral branches. From upper to lower crown, the percentage of ^{14}C remaining in the treated branch increased and that recovered in main-stem and roots decreased with branch position. Thus, the CT and upper laterals are the major contributors of photosynthate for main-stem and root growth. The magnitude of this relationship would change depending on the light environment of laterals in upper and lower crown (Dickson 1986).

Although lateral branches in poplar are important sources of photosynthate for wood and root production, little photosynthate from lateral branches is exported to the current terminal shoot. This finding suggests that lateral branches do not directly contribute to height growth in poplar. Moreover, lateral branches do not export appreciable photosynthate to other lateral branches. Thus, the contributions of lateral branches in poplar are somewhat independent in terms of lateral branch growth. However, the main stem and roots receive localized photosynthate contributions from each of the numerous lateral branches above (Isebrands et al. 1983).

Research on carbon allocation and partitioning from sylleptic branches in tree crowns is unique to *Populus*. The radiotracer studies of Scarascia-Mugnozza (1991) that have been reviewed by Hinckley et al. (1989) and Ceulemans (1990) show the importance of sylleptic branches to early growth and productivity of *Populus trichocarpa* and *P. deltoides* and their interspecific hybrids (Fig. 11). Large clonal differences exist in the carbon allocation patterns from sylleptic branches. Clones with large root/shoot biomass ratios have greater below ground contributions from sylleptic branches. Carbon transport from sylleptic branches increases over the growing season and leaves on those branches contribute more to the lower stem and roots than other leaves. In subsequent years, clones having sylleptic branches outperform clones without sylleptic branches. Again, there is little transport of carbon among branches. Studies of carbon allocation from

Fig. 10. Cumulative percent of exported ^{14}C from mature leaves at 4 positions within the crown of 2-year-old intensively cultured *P. tristis* trees treated during the course of the season. Recovery positions include current terminal (CT) leaves, CT stem, treated lateral (TL) leaves, TL stem, lateral branches other than TL, stem internodes of HGI 1, and roots. A. Mature leaves of CT shoot treated; B. Mature leaves of 5th first-order lateral branch from top (UL) of HGI 1 treated; C. Mature leaves of middle first-order branch (ML) on HGI treated; D. Mature leaves of 5th first-order branch from base (LL) of HGI 1 treated. Area under the curves estimates the proportion of ^{14}C that would be recovered in a season from a given treatment position. Arrows denote date of budset of treatment position (A through D) and leaf abscission date.

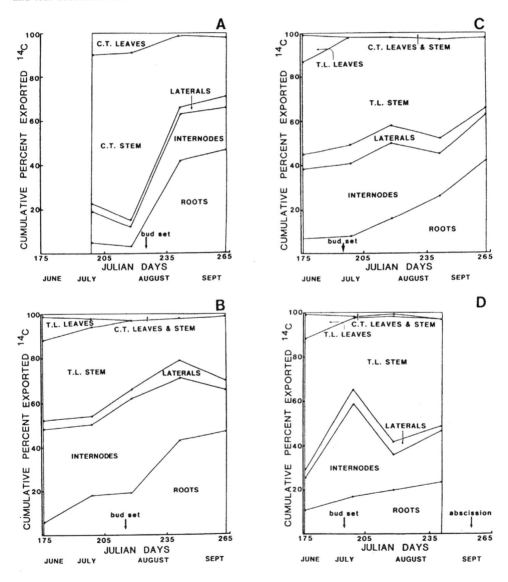

Fig. 11. Patterns of exported ^{14}C-assimilates in trees of *P. trichocarpa* × *P. deltoides* hybrid clones 11–11 and 44–136 treated early in the growing season (A) and after budset (B). In both cases, recently mature leaves were treated. Arrows indicate direction and approximate quantity of export (Scarascia-Mugnozza 1991).

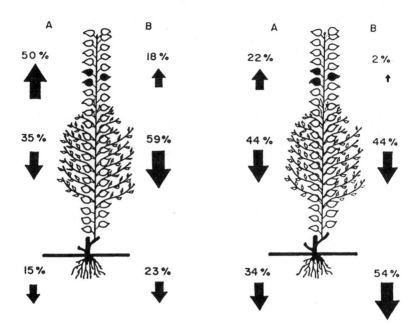

sylleptic branches in poplars may lead to their eventual use as traits in poplar selection and breeding programs with pedigrees.

Interacting effects of tropospheric ozone and CO₂

Tropospheric ozone (O_3), a potent phytotoxin, and the greenhouse gas, CO_2, are increasing yearly in the atmosphere. The result of this increase is a growing air pollution problem in the forests of the US, especially near metropolitan areas. Numerous studies have been conducted on the effects of O_3 on *Populus* growth and physiological processes (Table 2). In general, O_3 has a negative effect on productivity of poplars that is correlated with decreased chlorophyll content, photosynthesis, premature leaf abscission, and decreases in total carbon acquisition and accumulation (Reich 1983; Matyssek et al. 1993*b*; Coleman et al. 1995*a, b*).

Moreover, clonal differences have been identified in the growth and physiological responses of poplars to ozone stress and its interaction with CO_2 (Kull et al. 1996). Poplar clones also show differences in the effects of O_3 alone on carbon allocation and partitioning within trees with a pronounced negative effect on

Table 2. List of studies on effects of tropospheric ozone on carbon assimilation, allocation, and partitioning in *Populus*.

Parentage	Subject	Reference
P. nigra; P. maximowiczii ×		
P. nigra	Carbohydrates	Bücker and Ballach 1992
P. deltoides	Foliar chemistry; insect feeding	Jones and Coleman 1989
P. tremuloides	Photosynthesis	Coleman et al. 1995*a*
P. tremuloides	Carbon allocation and partitioning	Coleman et al. 1995*b*
P. tremuloides	Carbon allocation; roots	Coleman et al. 1996
P. ×*euramericana*	Leaf area and growth	Frost et al. 1991
P. tremuloides	Carbon gain; leaf area; photosynthesis	Greitner et al. 1994
P. tremuloides	Carbon allocation, clonal variation	Karnosky et al. 1992
P. tremuloides	Carbon allocation; leaf biomass	Karnosky et al. 1996
P. tristis × *P. balsamifera*	Foliar chemistry; insect feeding	Lindroth et al. 1993
P. tremuloides	Photosynthesis; interaction with CO_2	Kull et al. 1996
P. ×*euramericana*	Chemical fractions; leaf chemistry	Landolt et al. 1994
P. tremula	Photosynthesis	Matyssek et al. 1993*a*
P. ×*euramericana*	Carbon allocation and partitioning	Matyssek et al. 1993*b, c*
P. deltoides × *P. trichocarpa*	Leaf area and biomass	Noble and Jensen 1980
P. tremuloides	Carbon allocation	Pell et al. 1995
P. deltoides × *P. trichocarpa*	Chlorophyll; photosynthesis	Reich 1983
P. tristis × *P. balsamifera*	Photosynthesis; respiration	Tjoelker et al. (1993)
P. tristis × *P. balsamifera*	Chlorophyll; photosynthesis	Volin et al. 1993

allocation and partitioning to roots in the autumn season (Coleman et al. 1995*b*). Although much of the poplar ozone research has been with the aspen (Table 2), recent research has shown that hybrid poplar clones are also differentially affected by O_3 and its interaction with CO_2 just as with the aspens (D. Karnosky et al., unpublished). Noteably, some of the most productive hybrid poplar clones (namely *P.* ×*euramericana* and *P. nigra* × *P. maximowiczii*) are most affected. Results to date suggest that (*i*) CO_2 does not compensate for the deleterious effects of O_3 on poplars, (*ii*) CO_2 increases the sensitivity of some clones to O_3, and (*iii*) insect larval feeding habits on poplars are affected by interacting O_3 and CO_2 concentrations (Jones and Coleman 1989; Lindroth et al. 1993). These ozone-based results have led researchers to two major conclusions concerning the importance of the poplar to fundamental and applied global change research programs. They are: (*i*) the differential research findings of poplars to stresses may help researchers better understand the mechanisms of O_3 toxicity in plants (Pell et al. 1995) and its interaction with elevated CO_2, and (*ii*) poplars may be useful as bioindicator for O_3 pollution in the environment. In fact, poplars are currently being tested as bioindicator plants as part of a biomonitoring program in the Lake States, USA (Jepsen 1994).

Carbon storage

Woody plants store reserves when excess carbon production occurs, and they store them as carbohydrates, lipids, and other chemical compounds. Reserves are used for respiration and maintenance as well as new growth. Carbohydrates are the major storage compounds in deciduous trees such as poplar. In branches and stems of poplars, the concentration of total nonstructural carbohydrates increases in late summer and autumn, decreases slightly during the winter, then decreases rapidly in early spring during the spring growth flush. Starch concentration usually has two maxima, one in autumn and one in spring. Starch is stored in late summer and autumn as growth demands of the tree decrease. Associated with cool temperatures in late autumn, the stored starch is hydrolyzed to soluble sugar (e.g., sucrose, raffinose, stachyose), which may increase cold-hardiness. Starch is again formed in early spring with warm temperatures before bud break and is then depleted during the spring growth flush. In poplar the major form of lipid storage is in the form of triglycerides that can reach 1–3% of the residue dry weight during dormancy (Dickson 1991).

Root–shoot feedback

Root–shoot feedbacks are involved in the dynamic processes of root–shoot carbon allocation in plants under varying environmental conditions. A number of mechanisms have been proposed to account for the root–shoot communication in plants. They include functional equilibrium, metabolic control, and hormonal signals. These plausible mechanisms have been reviewed by Friend et al. (1994) and are beyond the scope of this review. No specific mechanism has been found for poplar.

^{13}C stable isotope

During the process of photosynthesis the ^{13}C stable isotope that occurs naturally in the environment is discriminated against relative to ^{12}C (O'Leary 1988). This discrimination allows the plant physiologist to investigate many carbon processes in the environment (Brugnoli et al. 1988). The basis for the use of the isotope ratio ($^{13}C/C^{12}$) lies in its role as an integrated index of intercellular/ambient CO_2 ratio (Ehleringer et al. 1993). It is useful for studying photosynthesis, light interception, water use efficiency, salinity, and air pollution effects (Farquhar et al. 1989; Svejcar et al. 1990). Moreover, there are also potential uses of ^{13}C as a tool for selection for genetic improvement in some plant genera (Ehleringer et al. 1993). Such approaches have not been widely applied in tree genetics and selection programs, except for *Eucalyptus*, because of the lack of available information on sources of variation in ^{13}C within and between trees growing in the field. However, it is known that ^{13}C patterns can vary dramatically within a tree crown (Schleser 1992) and that ^{13}C can be affected by such variables as species, location, climate, elevation, salinity, age, plant origin, and sex (Farquhar et al. 1989). Little information is available in the literature on

Table 3. Carbon isotope discrimination of (δ^{13}) of *Populus* clones growing at the Italian Centro Sperim. Agric. Forestale, SAF, Populetum Mediterraneum, Cesuri, Italy, and natural stands throughout Italy in 1989.

Parentage	Sex	Latitude/Longitude	Altitude (m)	$\delta^{13}C(‰)$
Populus alba				
42–57	M			−25.1
GA–1	M	44°04′ 10°20′	200	−25.3
Casa Oia	—			−25.2
Chianciano	F	43°03′ 11°50′	350	−25.5
Casole 56B	F			−24.9
DI–1	M	43°58′ 10°24′	70	−25.3
Shirazi	—			−26.3
PE–179	F			−26.5
PE–92	F			−26.3
58–57	F			−25.0
Algeria	F			−26.3
44–64	—			−26.0
Mean				−25.6
Range				−24.9 to −26.5
Populus alba				
(natural stands)				
Policoro, Italy			0	−27.0
Ponte Calare (Eboli.)			150	−27.4
Francavilla			400	−27.0
Mean				−27.2
Range				−26.4 to −28.3
Populus nigra				
N–005	M	46°15′ 13°03′	184	−24.9
N–022	F	46°09′ 11°14′	482	−24.9
N–058	M	45°33′ 11°09′	470	−25.5
N–094	M	44°59′ 9°41′	89	−26.0
N–117	F	44°49′ 9°56′	157	−26.5
N–124	F	44°45′ 8°08′	271	−25.0
N–134	F	44°40′ 8°11′	244	−25.5
N–160	M	44°19′ 7°40′	575	−25.8
N–187	F	43°15′ 10°57′	245	−26.0
N–270	M	40°35′ 15°05′	48	−26.6
76368				−24.9
77153				−25.1
Mean				−25.6
Range				−24.9 to −26.6

Populus, but given the large body of genetic knowledge on poplars and the availability of pedigreed material (see Stettler et al., Chapter 4), the potential of using ^{13}C as a selection criteria in poplars should be explored further. We conducted a survey of ^{13}C variation in exotic and natural forests of poplars in 1989 in Italy. Carbon isotope discrimination (δ^{13}) in *Populus alba* and *Populus nigra* clones growing in a populetum near Rome ranged from –24.9 to –26.5 averaging –25.6 for *P. alba*, to –24.9 to –26.6 for *P. nigra*, averaging –25.6, respectively (Table 3). Although the range of variation was significant among clones within the populetum, there was no evidence of differences due to geographic origin, latitude/longitude, altitude, or sex. Moreover, samples from natural *P. alba* stands ranged from –26.4 to –28.3 and averaged –27.2 across a rather diverse set of latitudinal and elevation gradients.

T. Hinckley and co-workers (unpublished) from the University of Washington, Seattle conducted a study of ^{13}C variation in a common garden study of *Populus trichocarpa* and *P. deltoides* and some of their hybrids growing in western Washington on wet and dry plots (Fig. 12). They found a narrow range of variability from –27 to –30 $\delta^{13}C$ ($^o/_{oo}$) and their data suggested that there was less discrimination on the dry plot than wet plot due to an improvement in water use efficiency. However, they found no correlation of ^{13}C values with productivity, but their sample size was limited. The aforementioned studies represent a rather limited, although unique data set suggesting that there is variation in carbon isotope discrimination in poplars that may be useful if explored in more extensive and controlled experiments.

Models

Why whole-tree growth models specifically for poplar?

Through the years researchers have attempted to understand, define, and quantify plant processes and structures in order to simulate, predict, and improve yield. As a result of these efforts different crop growth simulation models have been developed to consolidate and integrate large bodies of information and the current knowledge of a specific species or crop into a predictive management, research, and breeding tool (e.g., Evans 1975; De Wit 1978). In contrast to agronomic crops, relatively few physiologically based growth models have been developed for trees, largely due to a lack of fundamental physiological information and the difficulty of working with complex perennial plants (Isebrands et al. 1982; Ford 1985; Rauscher et al. 1988).

Because of the large silvicultural (Hansen et al. 1983), anatomical (Larson 1980), and physiological (Isebrands and Nelson 1983; Isebrands and Michael 1986; Ceulemans et al. 1987) databases that are available for selected *Populus* hybrids and clones, a number of physiologically based growth models have been developed specifically for poplar. These models include basic information on

Fig. 12. Carbon isotope discrimination (δ^{13}) of 1-year-old *Populus* clones growing at Washington State University experimental farm in Puyallup, WA under irrigated and unirrigated conditions. Values based upon mature leaf tissue. Range = –24.5 to –29.0^{13}C (‰). Pd = *Populus deltoides*, Pt = *Populus trichocarpa*, Hybrid = *P. trichocarpa* × *P. deltoides* hybrid (T. Hinckley et al., University of Washington, Seattle, unpublished).

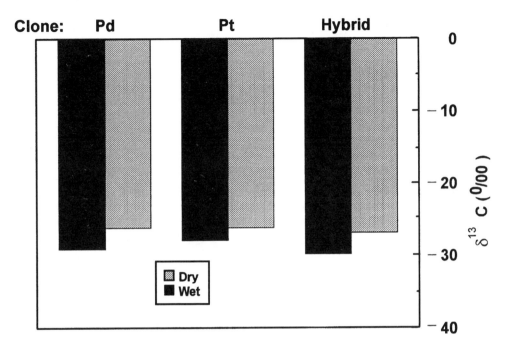

carbon uptake and allocation in poplar, as well as components and parameters of leaf display and crown structure outlined above. Out of these models, three of the more important ones are briefly described and illustrated in this chapter. All three models simulate carbon uptake, carbon allocation, growth and/or light interception in poplar and incorporate some specific parameters of leaf display, position in the tree, and branch structure described above.

FRACPO

The FRACPO model, a fractal-based poplar model developed at the Department of Biology of the University of Antwerpen, Belgium, is a three-dimensional crown model that has specifically been developed to generate individual poplar tree structures within a stand (Chen et al. 1994a). The model has been based on experimentally observed and simplified growth rules of young poplar trees grown under SRF. Stems and branches are considered as the structural carriers of leaves, and the spatial distribution of leaves is mainly determined by the structure and architecture of stems and branches. Several branch characteristics have been measured in detail on young poplar trees grown under SRF, and a computer graphical model was developed to simulate the dynamics of crown architecture. The structural crown development is then coupled with a leaf sub-model to generate and simulate the three-dimensional leaf distribution in the canopy. The leaf submodel primarily integrates information on leaf initiation, leaf longevity, expansion, and distribution of leaf sizes within the canopy (Chen et al. 1994*b*; 1994*c*). These parameters determine the instantaneous leaf area index (LAI) and the temporal variation of LAI. The spatial distribution of leaf area depends on the positioning of leaves within the canopy and on the variations in leaf size with their positions on stem and branches. The resulting three-dimensional leaf distribution within the canopy can then be used for calculations of canopy radiative transfer, light interception, and photosynthesis (e.g., the CANLIP model described below).

As described above and because of its fractal nature, the FRACPO model requires a large number of detailed input parameters such as (*i*) length and number of internodes for each height growth increment (HGI), (*ii*) length, number, and angles of sylleptic and proleptic branches, (*iii*) successive branch growth increments, (*iv*) insertion height of branches and relationship between height of insertion and branch length, (*v*) leaf area index and leaf longevity as a function of the time of leaf initiation, and so on. The output that is generated by the FRACPO model includes (*i*) the dynamics of the three-dimensional distribution of leaf area density, (*ii*) both profile and top-down projections of the high-density poplar canopy, and (*iii*) the distribution of light irradiance on leaf surfaces (combined with the three-dimensional radiative model, CANLIP described below). The time scale of the output is optional and can range from 3 to 4 d, to an output over an entire growing season.

To simplify the descriptive rules and processes of branching, internode formation, etc., a number of assumptions are made in the model. With regard to sylleptic branches, for example, it is assumed in the model that all sylleptic branches start to develop only in the middle of the growing season. Furthermore, they are produced acropetally, but not continuously. And finally, several sylleptic branches arise from the preformed buds in one surge of growth. Other main assumptions in the model include: (*i*) the time period required for the formation of one internode is identical for all internodes, (*ii*) the average length of a branch growth increment is a fraction of that of the previous one, (*iii*) the average internode length within a branch growth increment of order k is a fraction of that of the previous order $k - 1$. This assumption means that the sequential pattern of internode length in a branch growth increment is symmetrical, (*iv*) branch curvature is constant, thus the branch angles do not vary with height or height growth increment, and (*v*) the timing of leaf initiation is the same as its associated internode formation.

The FRACPO model can simulate tree growth and crown architecture on the individual tree level as well as on the stand scale (Fig. 13). Canopy architecture

Fig. 13. A simplified poplar tree stand as generated by the FRACPO model. Each of the parameters was given a random deviation from its mean value. As an approximation the interferences between branches were not considered.

is basically determined by crown architecture and the arrangement of the individual tree frames in the canopy (this is true in SRF as well as in widely spaced plantations). A stand of trees is simulated by the FRACPO model with individual trees being regularly distributed over the plantation area. Each of the above described input parameters is then given a random deviation from its mean (default) value, and interferences between branches are not considered. This simplification is reasonable because all parameters used were originally obtained on trees that were actually grown in a SRF stand. Thus, several of these parameters already incorporate the information on branch angles and branch curvature that reflect to a large extent these branch interferences.

One of the novel applications of the FRACPO model is that the model can be used to study the distribution of canopy gap fractions by means of a digitalized image analysis system (Chen et al. 1993*a*). Top-down projections of different layers of a simulated canopy can then be used to estimate the downward cumulated LAI. Furthermore, relationships between branch angle and leaf orientation (compare with Fig. 3) or between branch orientation and spacing density can also be examined.

CANLIP

The model CANLIP is a three-dimensional radiative transfer model of canopy light interception developed by S.G. Chen at the University of Antwerpen,

Belgium. It calculates canopy light interception and gross photosynthesis of poplar at various periods during the growing season. The canopy of the CANLIP model is divided into a large number of spatially defined rectangular cells. Given the leaf area density for each of the cells and assuming a distribution function of the leaf normal orientation, the radiative transfer in a finite number of directions is computed. The penetration of direct and diffuse light within the canopy and the irradiance on the exposed leaf surfaces in each cell are calculated (Chen et al. 1994*b*, 1994*c*). The results are predictions of PPFD on all leaf surfaces in the canopy. Consequently this information can be used to calculate photosynthetic rates in each cell because the process of photosynthesis is primarily driven by the interception of PPFD by the foliage. Because the single leaf photosynthetic rate is a nonlinear function of PPFD irradiance (Ceulemans et al. 1980), the spatial and temporal variation of the PPFD irradiance on all leaf surfaces are simulated in the photosynthesis submodel to accurately estimate canopy photosynthetic rate as a spatial integral of the photosynthetic rates of all individual cells. So, the CANLIP model estimates the instantaneous and daily canopy gross photosynthetic rates by coupling the intercepted PAR with a leaf photosynthetic response function. When incorporated with submodels of respiration and allocation of photosynthates, this model can be used to simulate overall poplar growth and biomass production. Final validation of the model is in progress.

Dimensions (such as height, length, width) and azimuth orientation of the stand are the basic input parameters for the CANLIP model, together with three-dimensional leaf area density and leaf orientation distribution functions. For these three-dimensional distribution functions of leaf area density and leaf orientation the output of the above described FRACPO model can be used as the main input source of the CANLIP model. However, the CANLIP model is entirely independent of FRACPO. Chen et al. (1993*b*), for example, have compared different vertical leaf area density functions in combination with horizontally, random, and clumping patterns to quantify the effects of canopy structure on light interception and photosynthesis. For each canopy layer, a random matrix was generated by the CANLIP model and various leaf normal distribution functions were used. Other input parameters that are required in the CANLIP model include (*i*) latitudinal position of the specific location and date, (*ii*) ratio of diffuse radiation to total incoming radiation, and (*iii*) some basic parameters of the single leaf photosynthesis-PAR response curve (such as PAR-saturated photosynthesis, quantum yield, and convexity coefficient for sun and shade leaves).

The output that is generated by the CANLIP model refers to (*i*) the vertical and horizontal distributions of penetration and interception of direct and diffuse light within a SRF poplar stand, and (*ii*) the instantaneous and daily canopy light interception and photosynthesis of the canopy. The main assumptions that are made in the CANLIP model are (*i*) PAR represents 50% of the global incoming radiation and the leaves do not scatter PAR, (*ii*) diffuse sky light is isotropic, and (*iii*) the leaf area density within a given cell is randomly distributed. The spatial scale of the CANLIP model ranges from individual leaf surfaces within the crown to light interception of the entire stand. A canopy space is divided into a number of rectangular cells. The time scale of the output of the model is one day (Chen et al. 1993*b*, 1994*b*).

The three-dimensional radiative transfer model CANLIP is well-suited to simulate the canopy light regime of a SRF poplar stand at a very detailed level. The seasonal three-dimensional light interception can be easily extrapolated to SRF *Populus* plantations. Another useful application of the CANLIP model is that it allows the simulation of the photosynthetic strategy of a poplar tree; gradual changes in sun/shade behaviour of leaves and of accompanying photosynthetic adaptations in the canopy can be simulated in the model by using different photosynthesis-light response functions at different positions in the canopy. Overall canopy photosynthesis is estimated in the model from basic leaf photosynthetic functions. Furthermore, the effect of leaf orientation (quantified by leaf angle) on radiation penetration can be examined by using CANILP as in the FRACPO model. The model CANLIP in combination with FRACPO has been applied to indirectly simulate the effects of elevated atmospheric CO_2 on the light interception and canopy growth of densely planted poplar clones in open top chambers (Fig. 8) through its effects on crown architectural components.

ECOPHYS

ECOPHYS is a physiologically based poplar growth model developed by researchers of the USDA Forest Service in Rhinelander, Wisconsin (USA) in cooperation with the University of Minnesota, Duluth. In the ECOPHYS model the individual leaf is maintained as the principal biological unit, and the integrated contribution of individual leaves within a tree are used to determine the quantity of photosynthate available for tree growth (Michael et al. 1988, 1990; Rauscher et al. 1988, 1990). The model evaluates the influence of leaf orientation, light interception, temperature, and sink demand on the photosynthetic capacity of individual leaves, and then integrates the performance of individual leaves to obtain an estimate of the overall performance of the tree. The resultant tree can be graphically displayed and viewed in three dimensions when plotted (Michael et al. 1988; Rauscher et al. 1988, 1990).

The ECOPHYS model is a mechanistic whole-tree model that simulates growth of poplar in its establishment year (Host et al. 1990*a*). The main objectives of ECOPHYS are (*i*) to provide a means to evaluate and rapidly screen the growth potential of newly acquired poplar clones for SRF, (*ii*) to develop morphological and physiological criteria for early selection, (*iii*) to assess the impacts of multiple environmental stresses on tree growth, and (*iv*) to provide a framework for integrating observations into ecophysiological theory. The model has been specifically parameterized for various poplar genotypes including hybrids of poplar and aspen.

The two principal environmental driving variables for ECOPHYS are hourly PPFD and hourly temperature. Further necessary input includes assorted morphological, physiological, and phenological parameters which characterize poplar and aspen clones. In the model, individual leaf geometry is uniquely maintained in a three-dimensional coordinate system to allow accurate calculation of mutual shading. Once leaf and shade area for each leaf has been calculated, the PPFD is calculated for the shaded and unshaded regions of each leaf as a function of leaf orientation, leaf position in the crown, atmospheric attenuation of radiation, and solar position, thus accounting for both direct and diffuse light. Consequently apparent photosynthesis, a genetically dependent function, is estimated (Michael et al. 1985, 1990). Apparent photosynthesis is different for each of four leaf age classes (i.e., expanding, recently mature, mature, and over-mature) and is also corrected for the influence of temperature. Then to calculate the quantity of photosynthate available for growth it is necessary to correct apparent photosynthesis for maintenance and growth respiration. The temperature influence on respiration rates is also simulated. In ECOPHYS growth of the poplar tree is directly related to the resultant available photosynthates.

The output of the model provides overall carbon assimilation rates, height, and diameter growth of the individual plant and biomass production by component (including leaves, shoot, roots). No empirical relationships are used to force growth to occur at a certain rate or to set lower or upper bounds to growth. Biomass accretion is simulated for the following components: immature leaves, mature leaves, immature internodes, mature internodes, cutting, large roots, and small roots. In addition, dimensional growth is computed so that leaf area, tree height, and tree diameter are estimated. Optionally, hourly atmospheric CO_2 concentrations and ozone levels can be used as input parameters for the model.

The basic time step of the ECOPHYS model is one hour. Some of the main assumptions of the model are that moisture and nutrients are maintained at optimal levels as is the case in many SRF plantations. However, future versions of the model will also be able to simulate moisture and nutrient stress. That model is specifically suited (*i*) for screening poplar clones for enhanced growth potential, (*ii*) for a sensitivity analysis of factors related to growth, (*iii*) for assessing climate change effects (such as elevated atmospheric CO_2 and/or increased air temperature), and (*iv*) for providing input to models at larger spatial or temporal scales. ECOPHYS has also been used for a conceptual framework for modeling coppice willow plantations (Host et al. 1996; Isebrands et al. 1996).

Field data from SRF poplar plantations in Michigan, Washington, and Wisconsin (USA) have been used to validate the photosynthesis submodel of ECOPHYS (Host et al. 1990*b*). For this purpose five clones, representing a range of morphological, phenological, and physiological characteristics were planted on the same date at three different sites (Host and Isebrands 1994). The model generally predicted height growth within a standard deviation of the field plantation means. Only three simulations of the 15 clone × site interactions were significantly different from the plantation means, and the median error between predicted and observed values was 5%.

Growth process models such as the ones highlighted above, provide important tools for assessing the relative influence of different genetically selectable traits. Because clonal responses differ among different sites, final selection depends on particular site conditions. In addition, theoretically based process models can provide a means of predicting stress response, and studies are ongoing to demonstrate how the ECOPHYS model, for example, can effectively account for the interactions between clones and environment under unstressed as well as stressed conditions. Further applications of these models include their educational value, that allow us to improve our understanding of the basic structures, functions, and processes that govern growth and productivity of poplars scaled to forest plantations.

Summary: future needs and direction

More is perhaps known about carbon acquisition and allocation in *Populus* than in any other tree genus. Thus, research on *Populus* carbon physiology has been significant in setting the stage for and shaping parallel research in other plant genera. In this chapter the sources of variation in carbon assimilation and allocation in poplar have been reviewed with emphasis on major *Populus* research contributions and on physiologically based genetics and breeding work in the genus. It has been demonstrated and illustrated that the genus *Populus* has some unique features in terms of carbon acquisition and allocation, i.e.,

- there is a large database of information available from various studies

- the genus shows some exceptionally high carbon uptake rates, comparable to those of agricultural crops

- large genetic (clonal) variation exists among different species, hybrids, and/or clones

- autumnal photosynthesis activity has significant implications on stem diameter and root growth

- sylleptic branches are important in relation to the overall photosynthetic performance of an individual tree and might partly explain large differences in photosynthesis among various species and/or hybrid clones

- integration of whole tree photosynthesis allows to scale-up from the leaf level to the entire tree level.

Special attention has been given to a number of inherent genetic factors affecting photosynthesis and carbon allocation, such as leaf morphology, orientation and display, position in the tree crown, including sylleptic branches, and overall structure. The effect of important external factors such as elevated CO_2 and their interaction on leaf photosynthesis and carbon allocation has also been discussed. Because there is much information on the carbon physiology available for poplar, some useful physiological process models for poplar have been developed and evaluated that are comparable to some agricultural crop models. Much of our basic understanding of carbon assimilation and allocation (from ^{14}C studies) on poplar has been integrated in the ECOPHYS model to simulate growth and development in the early stages of poplar growth. Consequently, the basic information on leaf display, branch type, crown structure, and plant architecture in later growth stages (up to the 4th year), is incorporated into the three-dimensional FRACPO model using an approach based on fractal geometry and allowing to simulate canopy structure in SRF plantations. When coupled with the three-dimensional radiative transfer model CANLIP, the canopy light regime and growth of SRF poplar plantations can be studied and simulated in detail.

The overall focus of the chapter has been on the potential of using physiologically based selection and breeding approaches to improving *Populus* for biomass plantations. The examples of this chapter on photosynthetic processes and structural traits illustrate how physiologically based selection and breeding programs can be applied in *Populus*. Further integration of molecular genetics, breeding, physiology, production culture, and modelling shows much promise for improving poplar and should be encouraged. Pedigreed materials that are now available with poplars (see Bradshaw, Chapter 8) represent an ideal opportunity for physiologists and geneticists to work together on fundamental studies. These materials provide linkage to plant molecular biology that no doubt will facilitate larger future gains in tree health and productivity. For example, research underway by the Poplar Molecular Genetic Cooperative at University of Washington, Seattle in close collaboration with other research laboratories is providing the link among leaf, branch, and whole plant traits to genetic loci and/or molecular markers. These advances will enable researchers to make large gains in their understanding of carbon acquisition and allocation in poplars. This understanding will lead to increased biomass yields of SRF poplar plantations.

References

Akinyemiju, O.A., and Dickmann, D.I. 1982. Contrasting effects of simazine on the photosynthetic physiology and leaf morphology of two *Populus* clones. Physiol. Plant. **55**: 402–406.

Avery, D.J. 1977. Maximum photosynthetic rate — a case study in apple. New Phytol. **78**: 55–63.

Ballach, H.J., Mooi, J., and Wittig, R. 1992. Premature aging in *Populus nigra* L. after exposure to air pollutants. Angew. Bot. **66**: 14–20.

Bassman, J.H., and Dickmann, D.I. 1982. Effects of defoliation in the developing leaf zone on young *Populus* ×*euramericana* plants. I. Photosynthetic physiology, growth and dry weight partitioning. For. Sci. **28**: 599–612.

Briggs, G.M., Jurik, T.W., and Gates, D.M. 1986. A comparison of rates of aboveground growth and carbon dioxide assimilation by aspen on sites of high and low quality. Tree Physiol. **2**: 29–34.

Brugnoli, E., Hubick, K.T., Von Caemmerer, S., Wong, S.C., and Farquhar, G.D. 1988. Correlation between the carbon isotope discrimination in leaf starch and sugars of C_3 plants and the ratio of intercellular and atmospheric partial pressures of carbon dioxide. Plant Physiol. **88**: 1418–1424.

Bücker, J., and Ballach, H.J. 1992. Alterations in carbohydrate levels in leaves of *Populus* due to ambient air pollution. Physiol. Plant. **86**: 512–517.

Ceulemans, R. 1990. Genetic variation in functional and structural productivity determinants in poplar. Thesis Publishers, Amsterdam, The Netherlands. 100 p.

Ceulemans, R., and Impens, I. 1980. Leaf gas exchange processes and related characteristics of several poplar clones under laboratory conditions. Can. J. For. Res. **10**: 429–435.

Ceulemans, R., and Impens, I. 1983. Net CO_2 exchange rate and shoot growth of young poplar (*Populus*) clones. J. Exp. Bot. **34**: 866–870.

Ceulemans, R., and Mousseau, M. 1994. Tansley Review : Effects of elevated atmospheric CO_2 on woody plants. New Phytol. **127**: 425–446.

Ceulemans, R., and Saugier, B. 1991. Photosynthesis. Chapter 2. *In* Physiology of trees. *Edited by* A.S. Raghavendra. J. Wiley & Sons Inc., New York, NY. pp. 21–50.

Ceulemans, R., Impens, I., and Moermans, R. 1980. Describing the response of CO_2 exchange rate to photosynthetic photon flux density for several *Populus* clones under laboratory conditions. Photosynth. Res. **1**: 137–142.

Ceulemans, R., Impens, I., and Steenackers, V. 1987. Variations in photosynthetic, anatomical and enzymatic leaf traits and correlations with growth in recently selected *Populus* hybrids. Can. J. For. Res. **17**: 273–283.

Ceulemans, R., Stettler, R.F., Hinckley, T.M., Isebrands, J.G., and Heilman, P.E. 1990. Crown architecture of *Populus* clones as determined by branch orientation and branch characteristics. Tree Physiol. **7**: 157–167.

Ceulemans, R., Jiang, X.N., and Shao, B.Y. 1994. Growth and physiology of one-year old poplar *(Populus)* under elevated atmospheric CO_2. Ann. Bot. **75**: 609–617.

Chen, S.G., Impens, I., Ceulemans, R., and Kockelbergh, F. 1993*a*. Measurement of gap fraction of fractal generated canopies using digitalized image analysis. Agric. For. Meteorol. **65**: 245–259.

Chen, S.G., Shao, B.Y., and Impens, I. 1993*b*. A computerised numerical experimental study of average solar radiation penetration in plant stands. J. Quant. Spectrosc. Radiat. Transfer, **49**: 651–658.

Chen, S.G., Ceulemans, R., and Impens, I. 1994*a*. A fractal-based *Populus* canopy structure model for the calculation of light interception. For. Ecol. Manage. **69**: 97–110.

Chen, S.G., Ceulemans, R., and Impens, I. 1994*b*. Is there a light regime determined tree ideotype? J. Theor. Biol. **169**: 153–161.

Chen, S.G., Shao, B.Y., Impens, I., and Ceulemans, R. 1994*c*. Effects of plant canopy structure on light interception and photosynthesis. J. Quant. Spectrosc. Radiat. Transfer, **52**: 115–123.

Coleman, M.D., Isebrands, J.G., Dickson, R.E., and Karnosky, D.F. 1995*a*. Photosynthetic productivity of aspen clones varying in sensitivity to tropospheric ozone. Tree Physiol. **15**: 585–592.

Coleman, M.D., Dickson, R.E., Isebrands, J.G., and Karnosky, D.F. 1995*b*. Carbon allocation and partitioning in aspen clones varying in sensitivity to tropospheric ozone. Tree Physiol. **15**: 593–604.

Coleman, M.D., Dickson, R.E., Isebrands, J.G., and Karnosky, D.F. 1996. Root growth and physiology of potted and field-grown trembling aspen exposed to tropospheric ozone. Tree Physiol. **16**: 145–152.

Coombs, J., Hall, D.O., Long, S.P., and Scurlock, J.M.O. (*Editors*). 1985. Techniques in bioproductivity and photosynthesis, 2nd ed. Pergamon Press, Oxford, England. 298 p.

Dawson, D.H., Zavitkovski, J., and Isebrands, J.G. 1980. Managing forests for maximum biomass production. *Edited by* G.R. Lightsey. American Institute of Chemical Engineers. AIChE Symp. Ser. 195, **76**: 36–42.

De Wit, C.T. 1978. Simulation of assimilation, respiration and transpiration of crops. Halsted Press, J. Wiley & Sons, New York, NY. 140 p.

Dickmann, D.I. 1971. Photosynthesis and respiration by developing leaves of cottonwood (*Populus deltoides* Bartr.). Bot. Gaz. **132**: 253–259.

Dickmann, D.I. 1985. The ideotype concept applied to forest trees. *In* Trees as crop plants. *Edited by* M.G.R. Cannell and J. Jackson. Institute of Terrestrial Ecology, Huntington, U.K. pp. 89–101.

Dickmann, D.I., and Gordon, J.C. 1975. Incorporation of ^{14}C-photosynthate into protein during leaf development in young *Populus* plants. Plant Physiol. **56**: 23–27.

Dickmann, D.I., Gjerstad, D.H., and Gordon, J.C. 1975. Developmental patterns of CO_2 exchange, diffusion resistance and protein synthesis in leaves of *Populus* ×*euramericana*. *In* Environmental and biological control of photosynthesis. *Edited by* R. Marcelle. Dr. W. Junk Publishers, The Hague, The Netherlands. pp. 171–181.

Dickmann, D.I., Michael, D.A., Isebrands, J.G., and Westin, S. 1990. Effects of leaf display on light interception and apparent photosynthesis in two contrasting *Populus* cultivars during their second growing season. Tree Physiol. **7**: 7–20.

Dickson, R.E. 1986. Carbon fixation and distribution in young *Populus* trees. *In* Proc. Crown and Canopy Structure in Relation to Productivity. *Edited by* T. Fujimori and D. Whitehead. Forestry and Forest Products Research Institute, Ibaraki, Japan. pp. 409–426.

Dickson, R.E. 1989. Carbon and nitrogen allocation in trees. Ann. Sci. For. **46**(suppl.): 631s–647s.

Dickson, R.E. 1991. Assimilate distribution and storage. *In* Physiology of trees. *Edited by* A.S. Raghavendra. John Wiley & Sons, Inc., New York, NY. pp. 51–85.

Dickson, R.E., and Isebrands, J.G. 1991. Leaves as regulators of stress response. *In* Response of plants to multiple stresses. *Edited by* H.A. Mooney, W.E. Winner, and E.J. Pell. Academic Press, Inc., San Diego, CA. pp. 3–34.

Dickson, R.E., and Isebrands, J.G. 1993. Carbon allocation terminology: should it be more rational? Bull. Ecol. Soc. Amer. **74**: 175–177.

Dickson, R.E., and Larson, P.R. 1981. ^{14}C fixation, metabolic labeling patterns, and translocation profiles during leaf development in *Populus deltoides*. Planta, **152**: 461–470.

Ehleringer, J.R., Hall, A.E., and Farquhar, G.D. 1993. Stable isotopes and plant carbon-water relations. Academic Press, Inc., New York, NY. 555 p.

Evans, L.T. 1975. Crop physiology in some case studies. Cambridge University Press, Cambridge, UK. 374 p.

Farquhar, G.D., Ehleringer, J.R., and Hubick, K.T. 1989. Carbon isotope discrimination and photosynthesis. Annu. Rev. Plant. Physiol. Plant Mol. Biol. **40**: 503–537.

Fasehun, F.E. 1978. Effect of irradiance on growth and photosynthesis of *Populus* ×*euramericana* clones. Can. J. For. Res. **8**: 94–99.

Fege, A.S. 1983. Changes in *Populus* carbohydrate reserves during induction of dormancy, cold storage of cuttings, and development of young plants. Ph.D. thesis, University of Minnesota, St. Paul, MN. 128 p.

Fei, H., Godbold, D.L., Wang, S.S., and Hüttermann, A. 1990. Gas exchange in *Populus maximowiczii* in relation to potassium and phosphorous nutrition. J. Plant Physiol. **135**: 675–679.

Figliola, A.L. 1986. Studies in the physiology, morphology, and anatomy of *Populus trichocarpa, Populus deltoides*, and their hybrids. M.Sc. thesis, University of Washington, Seattle, WA. 119 p.

Foote, K., and Schaedle, M. 1974. Seasonal field rates of photosynthesis and respiration in stems of *Populus tremuloides*. Plant Physiol. **53**(Suppl.): 352.

Ford, E.D. 1985. Branching, crown structure and the control of timber production. *In* Attributes of trees as crop plants. *Edited by* M.G.R. Cannell and J.E. Jackson. Titus Wilson & Son Ltd., Cumbria, UK. pp. 228–252.

Friend, A.L., Scarascia-Mugnozza, G., Isebrands, J.G., and Heilman, P.E. 1991. Quantification of two-year-old hybrid poplar root systems: morphology, biomass, and ^{14}C distribution. Tree Physiol. **8**: 109–119.

Friend, A.L., Coleman, M.D., and Isebrands, J.G. 1994. Carbon allocation to root and shoot systems of woody plants. *In* Biology of adventitious root formation. *Edited by* T.D. Davis and B.E. Haissig. Plenum Press, New York, NY. pp. 245–273.

Frost, D.L., Taylor, G., and Davies, W.J. 1991. Biophysics of leaf growth of hybrid poplar: impact of ozone. New Phytol. **118**: 407–415.

Furukawa, A. 1972. Photosynthesis and respiration in poplar plant. J. Jap. For. Soc. **54**: 80–84.

Furukawa, A. 1973. Carbon dioxide compensation points in poplar plants. J. Jap. For. Soc. **55**: 95–99.

Furukawa, A. 1975. Comparison of photosynthesis, postillumination CO_2 outburst, and CO_2 compensation in poplar varieties, sunflower and bean. J. Jap. For. Soc. **57**: 268–274.

Gatherum, G.E., Gordon, J.C., and Broerman, B.F.S. 1967. Effects of clone and light intensity on photosynthesis, respiration and growth of aspen-poplar hybrids. Silvae Genet. **16**: 128–132.

Gaudillère, J.P., and Mousseau, M. 1989. Short term effect of CO_2 enrichment on leaf development and gas exchange of young poplars (*Populus euramericana* cv. I-214). Oecol. Plant. **10**: 95–105.

Gifford, R.M., Thorne, J.H., Hitz, W.D., and Giaquinta, R.T. 1984. Crop productivity and photoassimilate partitioning. Science (Washington, DC), **225**: 801–808.

Gjerstad, D. 1975. Photosynthesis, photorespiration and dark respiration in *Populus* ×*euramericana*: Effects of genotype and leaf age. Ph.D. thesis, Iowa State University, Ames, IA. 70 p.

Gordon, J.C., and Promnitz, L.C 1976. Photosynthetic and enzymatic criteria for the early selection of fast-growing *Populus* clones. *In* Tree physiology and yield improvement. *Edited by* M.G.R. Cannell and F.T. Last. Academic Press, London, UK. pp. 79–97.

Gower, S.T., Isebrands, J.G., and Sheriff, D.W. 1995. Carbon allocation and accumulation in conifers. *In* Resource physiology of conifers, acquisition, allocation, and utilization. *Edited by* W.K. Smith and T.M. Hinckley. Academic Press, New York, NY. pp. 217–254.

Greitner, C.S., Pell, E.J., and Winner, W.E. 1994. Analysis of aspen foliage exposed to multiple stresses: ozone, nitrogen deficiency, and drought. New Phytol. **127**: 579–589.

Griese, C., Leonhardt, A., and Larsen, B. 1993. Okophysiologische Untersuchungen an verschiedenen Pappelklonen. *In* Anbau von Pappel bei mittlerer Umtriebszeit. *Edited by* A. Hüttermann, Band 110. J.D. Sauerlander's Verlag, Frankfurt am Main, Germany. pp. 20–37.

Hallé, F.R., Oldeman, A.A., and Tomlinson, P.B. 1978. Tropical trees and forests. An architectural analysis. Springer-Verlag, Berlin & New York.

Hansen, E., Moore, L., Netzer, D., Ostry, M., Phipps, H., and Zavitkovski, J. 1983. Establishing intensively cultured hybrid poplar plantations for fuel and fiber. U.S. Dep. Agric., For. Serv. Gen. Tech. Rep. NC-78. 24 p.

Hinckley, T.M., Ceulemans, R., Dunlap, J.M., Figliola, A., Heilman, P.E., Isebrands, J.G., Scarascia-Mugnozza, G., Schulte, P.J., Smit, B., Stettler, R.F., Van Volkenburgh, E., and Wiard, B.M. 1989. Physiological, morphological and anatomical components of hybrid vigor in *Populus*. *In* Structural and functional responses to environmental stresses. *Edited by* K.H. Kreeb, H. Richter, and T.M. Hinckley. SPB Academic Publishing, The Hague, The Netherlands. pp. 199–217.

Hinckley, T.M., Braatne, J., Ceulemans, R., Clum, P., Dunlap, J.M., Newman, D., Smit, B., Scarascia-Mugnozza, G., and Van Volkenburgh, E. 1992. Growth dynamics and canopy structure. *In* Ecophysiology of short rotation forest crops. *Edited by* C.P. Mitchell, J.B. Ford-Robertson, T.M. Hinckley, and L. Sennerby-Forsse. Elsevier Science Publish. Ltd., London & New York. pp. 1–34.

Horwath, W.R., Pregitzer, K.S., and Paul, E.A. 1994. ^{14}C allocation in tree-soil systems. Tree Physiol. **14**: 1163–1176.

Host, G.E., and Isebrands, J.G. 1994. An interregional validation of ECOPHYS, a growth process model of juvenile poplar clones. Tree Physiol. **14**: 933–945.

Host, G.E, Rauscher, H.M., Isebrands, J.G., Dickmann, D.I., Dickson, R.E., Crow, T.R., and Michael, D.A. 1990*a*. The Microcomputer Scientific Software Series #6 : The ECOPHYS User's Manual. U.S. Dep. Agric., For. Serv. Gen. Tech. Rep. NC-131.

Host, G.E., Rauscher, H.M., Isebrands, J.G., and Michael, D.A. 1990*b*. Validation of the photosynthate production in ECOPHYS, an ecophysiological growth process model of *Populus*. Tree Physiol. **7**: 268–296.

Host, G.E., Isebrands, J.G., Theseira, G.W., Kiniry, J.R., and Graham, R.L. 1996. Temporal and spatial scaling from individual trees to plantations: a modeling strategy. Biomass Bioenergy, **11**: 233–243.

Huber, B., and Polster, H. 1955. Zur Frage der physiologische Ursachen der unterschiedlichen Stofferzeugung von Pappelklonen. Biol. Zentralbl. **74**: 370–420.

Hüttermann, A. (*Editor*). 1993. Anbau von Pappel bei mittlerer Umtriebszeit. Band 110. J.D. Sauerlander's Verlag, Frankfurt am Main, Germany. 199 p.

Isebrands, J.G. 1982. Toward a physiological basis of intensive culture of poplar. *In* Proceedings of the Technical Association of the Pulp and Paper Industry. August 29 – September 1, 1982, TAPPI Press, Asheville, NC. pp. 81–90.

Isebrands, J.G., and Dickson, R.E. 1991. Measuring carbohydrate production and distribution: radiotracer techniques and applications. *In* Techniques and approaches in forest tree ecophysiology. *Edited by* J.P. Lassoie and T.M. Hinckley. CRC Press, Boca Raton, FL. pp. 358–392.

Isebrands, J.G., and Larson, P.R. 1977. Vascular anatomy of the nodal region in *Populus deltoides* Bartr. Amer. J. Bot. **64**: 1066–1077.

Isebrands, J.G., and Larson, P.R. 1980. Ontogeny of major veins in the lamina of *Populus deltoides* Bartr. Amer. J. Bot. **67**: 23–33.

Isebrands, J.G., and Michael, D.A. 1986. Effects of leaf morphology and orientation on solar radiation interception and photosynthesis in *Populus*. *In* Crown and canopy structure in relation to productivity. *Edited by* T. Fujimori and D. Whitehead. Forestry and Forest Products Research Institute, Ibaraki, Japan. pp. 359–381.

Isebrands, J.G., and Nelson, N.D. 1983. Distribution of ^{14}C-labelled photosynthates within intensively cultured *Populus* clones during their establishment year. Physiol. Plant. **59**: 9–18.

Isebrands, J.G., Dickson, R.E., and Larson, P.R. 1976. Translocation and incorporation of ^{14}C into the petiole from different regions within developing cottonwood leaves. Planta, **128**: 185–193.

Isebrands, J.G., Ek, A.R., and Meldahl, R.S. 1982. Comparison of growth model and harvest yields of short rotation intensively cultured *Populus*: a case study. Can. J. For. Res. **12**: 58–63.

Isebrands, J.G., Nelson, N.D., Dickmann, D.I., and Michael, D. 1983. Yield physiology of short rotation intensively cultured poplars. *In* Intensive plantation culture: 12 years research. U.S. Dep. Agric. For. Serv., NCFES, Gen. Tech. Rep. NC-91. pp. 77–93.

Isebrands, J.G., Ceulemans, R., and Wiard, B.M. 1988. Genetic variation in photosynthetic traits among *Populus* clones in relation to yield. Plant Physiol. Biochem. **26**: 427–437.

Isebrands, J.G., Host, G.E., Bollmark, L., Porter, J., Philippot, S., Stevens, E., and Rushton, K. 1996. A strategy for process modelling of short rotation *Salix* coppice plantations. Biomass Bioenergy, **11**: 245–252.

Jepsen, E. 1994. Ozone and acid deposition gradients and biomonitoring site selection in Wisconsin. *In* Proc. 16th Int. Meeting "Air Pollution Effects on Forest Ecosystems," September 7–9, 1994, Fredericton, NB. pp. 23.

Jones, C.G., and Coleman, J.S. 1989. Biochemical indicators of air pollution effects in trees: unambiguous signals based on fast-growing species? *In* Biological markers of air-pollution stress and damage in forests. National Academy Press, Washington, DC. pp. 261–273.

Jurik, T.W., Weber, J.A., and Gates, D.M. 1984. Short-term effects of CO_2 on gas exchange of leaves of bigtooth aspen (*Populus grandidentata*) in the field. Plant Physiol. **75**: 1022–1026.

Karnosky, D.F., Gagnon, Z.E., Reed, D.D., and Witter, J.A. 1992. Growth and biomass allocation of symptomatic and asymptomatic *Populus tremuloides* clones in response to seasonal ozone exposures. Can. J. For. Res. **22**: 1785–1788.

Karnosky, D.F., Gagnon, Z.E., Dickson, R.E., Coleman, M.D., Lee, E.H., and Isebrands, J.G. 1996. Changes in growth, leaf abscission, and biomass associated with seasonal tropospheric ozone exposures of *Populus tremuloides* clones and seedlings. Can. J. For. Res. **25**: 23–37.

Kozlowski, T.T., and Keller, T. 1966. Food relations in woody plants. Bot. Rev. **32**: 293–382.

Kull, O., Sober, A., Coleman, M.D., Dickson, R.E., Isebrands, J.G., Gagnon, Z.E., and Karnosky, D.F. 1996. Photosynthetic responses of aspen clones to simultaneous exposures of ozone and CO_2. Can. J. For. Res. **26**: 639–648.

Landolt, W., Gunhardt-Goerg, M., and Pfenninger, I. 1994. Ozone-induced microscopical changes and quantitative carbohydrate contents of hybrid poplar (*Populus ×euramericana*). Trees, **8**: 183–190.

Landsberg, J.J., Kaufmann, M.R., Binkley, D., Isebrands, J.G., and Jarvis, P.G. 1991. Evaluating progress toward closed forest models based on fluxes of carbon, water and nutrients. Tree Physiol. **9**: 1–15.

Larcher, W. 1969. The effect of environmental and physiological variables on the carbon dioxide gas exchange of trees. Photosynthetica, **3**: 167–198.

Larson, P.R. 1977. Phyllotacic transitions in the vascular system of *Populus deltoides* Bartr. as defined by ^{14}C labeling. Planta, **134**: 241–249.

Larson, P.R. 1980. Interrelations between phyllotaxis, leaf development and the primary-secondary vascular transition in *Populus deltoides*. Ann. Bot. **46**: 757–769.

Larson, P.R., and Dickson, R.E. 1973. Distribution of imported ^{14}C in developing leaves of eastern cottonwood according to phyllotaxy. Planta, **111**: 95–112.

Larson, P.R., and Gordon, J.C. 1969. Photosynthesis and wood yield. Agricult. Sci. Rev. **7**: 7–14.

Larson, P.R., and Isebrands, J.G. 1971. The plastochron index as applied to developmental studies of cottonwood. Can. J. For. Res. **1**: 1–11.

Larson, P.R., Isebrands, J.G., and Dickson, R.E. 1972. Fixation patterns of ^{14}C within developing leaves of eastern cottonwood. Planta, **107**: 301–314.

Larson, P.R., Isebrands, J.G., and Dickson, R.E. 1980. Sink to source transition of *Populus* leaves. Ber. Deutsch. Bot. Ges. **93**: 79–90.

Lindroth, R.L., Reich, P.B., Tjoelker, M.G., Volin, J.C., and Oleksyn, J. 1993. Light environment alters response to ozone stress in seedlings of *Acer saccharum* and hybrid *Populus* L. III. Consequences for performance of gypsy moth. New Phytol. **124**: 647–651.

Luukkanen, O., and Kozlowski, T.T. 1972. Gas exchange in six *Populus* clones. Silvae Genet. **21**: 220–229.

Matyssek, R., Keller, T., and Koike, T. 1993*a*. Branch growth and leaf gas exchange of *Populus tremula* exposed to low ozone concentrations throughout two growing seasons. Environ. Pollut. **79**: 1–7.

Matyssek, R., Günthardt-Goerg, M.S., and Keller, T. 1993*b*. Physiological effects of air pollutants in low concentration on trees. *In* Air pollution and interactions between organisms in forest ecosystems. *Edited by* M. Tesche and S. Feiler. Proceedings IUFRO Centennial, Tharandt/Dresden, Germany. pp. 76–86.

Matyssek, R., Günthardt-Goerg, M.S., Landolt, W., and Keller, T. 1993*c*. Whole-plant growth and leaf formation in ozonated hybrid poplar (*Populus ×euramericana*). Environ. Pollut. **81**: 207–212.

Maurer, P., Dreyer, E., and Pinon, J. 1987. Evolution de la photosynthèse du peuplier au cours d'un cycle d'infection par *Marssonina brunnea*: comparaison de 3 clones. Ann. Sci. For. **44**: 135–152.

Michael, D.A., Dickmann, D.I., Gottschalk, K.W., Nelson, N.D., and Isebrands, J.G. 1985. Determining photosynthesis of tree leaves in the field using a portable [14]CO_2 apparatus: procedures and problems. Photosynthetica, **19**: 98–108.

Michael, D.A., Isebrands, J.G., Dickmann, D.I., and Nelson, N.D. 1988. Growth and development during the establishment year of two *Populus* clones with contrasting morphology and phenology. Tree Physiol. **4**: 139–152.

Michael, D.A., Dickmann, D.I., Isebrands, J.G., and Nelson, N.D. 1990. Photosynthesis patterns during the establishment year within two *Populus* clones with contrasting morphology and phenology. Tree Physiol. **6**: 11–27.

Nelson, N.D. 1984. Woody plants are not inherently low in photosynthetic capacity. Photosynthetica, **18**: 600–605.

Nelson, N.D. 1985. Photosynthetic life span of attached poplar leaves under favorable controlled environment conditions. For. Sci. **31**: 700–705.

Nelson, N.D., and Michael, D. 1982. Photosynthesis, leaf conductance, and specific leaf weight in long and short shoots of *Populus Tristis* 1 grown under intensive culture. For. Sci. **28**: 734–744.

Nelson, N.D., and Isebrands, J.G. 1983. Late-season photosynthesis and photosynthate distribution in an intensively cultured *Populus nigra* ×*laurifolia* clone. Photosynthetica, **17**: 537–549.

Nelson, N.D., and Ehlers, P. 1984. Comparative carbon dioxide exchange for two *Populus* clones grown in growth room, greenhouse, and field environments. Can. J. For. Res. **14**: 924–932.

Noble, R.D., and Jensen, K.F. 1980. Effects of sulfur dioxide and ozone on growth of hybrid poplar leaves. Amer. J. Bot. **67**: 1005–1009.

Okafo, O.A., and Hanover, J.W. 1978. Comparative photosynthesis and respiration of trembling and bigtooth aspens in relation to growth and development. For. Sci. **24**: 103–109.

Okoro, O.O., and Grace, J. 1976. The physiology of rooting of *Populus* cuttings. I. Carbohydrates and photosynthesis. Physiol. Plant. **36**: 133–138.

O'Leary, M.H. 1988. Carbon isotopes in photosynthesis. BioScience, **38**: 328–336.

Pearcy, R.W., Björkman, O., Caldwell, M.M., Keeley, J.E., Monson, R.K., and Strain, B.R. 1987. Carbon gain by plants in natural environments. BioScience, **37**: 21–29.

Pell, E.J., Sinn, J.P., and Johansen, C.V. 1995. Nitrogen supply as a limiting factor determining the sensitivity of *Populus tremuloides* to ozone stress. New Phytol. **130**: 437–446.

Pezeshki, S.R., and Hinckley, T.M. 1982. The stomatal response of red alder and black cottonwood to changing water status. Can. J. For. Res. **12**: 761–771.

Pollard, D.F.W. 1970. The effect of rapidly changing light on the rate of photosynthesis in largetooth aspen (*Populus grandidentata*). Can. J. Bot. **48**: 823–829.

Powell, G.R., and Vescio, S.A. 1986. Syllepsis in *Larix laricina*: occurrence and distribution of sylleptic long shoots and their relationships with age and vigour in young plantation-grown trees. Can. J. For. Res. **16**: 597–607.

Pregitzer, K.S., Dickmann, D.I., Hendrick, R., and Nguyen, P.U. 1990. Whole tree carbon and nitrogen partitioning in young hybrid poplars. Tree Physiol. **7**: 79–93.

Rauscher, H.M., Isebrands, J.G., Crow, T.R., Dickson, R.E., Dickmann, D.I., and Michael, D.A. 1988. Simulating the influence of temperature and light on the growth of juvenile poplars in their establishment year. *In* Proceedings IUFRO Forest Growth Modelling and Prediction Conference. U.S. Dep. Agric., For. Serv. Gen. Tech. Rept. NC-120, Minneapolis, MN.

Rauscher, H.M., Isebrands, J.G., Host, G.E., Dickson, R.E., Dickmann, D.I., Crow, T.R., and Michael, D.A. 1990. ECOPHYS: an ecophysiological growth process model for juvenile poplar. Tree Physiol. **7**: 255–781.

Regehr, D.L., Bazzaz, F.A., and Boggess, W.R. 1975. Photosynthesis, transpiration and leaf conductance of *Populus deltoides* in relation to flooding and drought. Photosynthetica, **9**: 52–61.

Reich, P.B. 1983. Effects of low concentrations of O_3 on net photosynthesis, dark respiration, and chlorophyll contents in aging hybrid poplar leaves. Plant Physiol. **73**: 291–296.

Reich, P.B. 1984. Relationships between leaf age, irradiance, leaf conductance, CO_2 exchange and water use efficiency in hybrid poplar. Photosynthetica, **18**: 445–453.

Reich, P.B., and Amundson, R.G. 1985. Ambient levels of ozone reduce net photosynthesis in tree and crop species. Science (Washington, DC), **230**: 566–570.

Remphrey, W.R., and Powell, G.R. 1985. Crown architecture of *Larix laricina* saplings: sylleptic branching on the main stem. Can. J. Bot. **63**: 1296–1302.

Ridge, C.R., Hinckley, T.M., Stettler, R.F., and Van Volkenburgh, E. 1986. Leaf growth characteristics of fast-growing poplar hybrids *Populus trichocarpa* × *P. deltoides*. Tree Physiol. **1**: 209–216.

Saieed, N.T., Douglas, G.C., and Fry, D.J. 1994*a*. Induction and stability of somaclonal variation in growth, leaf phenotype and gas exchange characteristics of poplar regenerated from callus culture. Tree Physiol. **14**: 1–16.

Saieed, N.T., Douglas, G.C., and Fry, D.J. 1994*b*. Somaclonal variation in growth, leaf phenotype and gas exchange characteristics of poplar: utilization of leaf morphotype analysis as a basis for selection. Tree Physiol. **14**: 17–26.

Scarascia-Mugnozza, G. 1991. Physiological and morphological determinants of yield in intensively cultured poplars (*Populus* spp.). Ph.D. thesis, University of Washington, Seattle, WA. 164 p.

Scarascia-Mugnozza, G., Hinckley, T.M., and Stettler, R.F. 1986. Evidence for nonstomatal inhibition of net photosynthesis in rapidly dehydrated shoots of *Populus*. Can. J. For. Res. **16**: 1371–1375.

Schaedle, M. 1975. Tree photosynthesis. Annu. Rev. Plant Physiol. **26**: 101–115.

Schleser, G.H. 1992. $\delta^{13}C$ pattern in a forest tree as an indicator of carbon transfer in trees. Ecology, **73**: 1922–1925.

Shibles, R. 1976. Terminology pertaining to photosynthesis. Crop Sci. **16**: 437–439.

Stettler, R.F., and Bradshaw, H.D., Jr. 1994. The choice of genetic material for mechanistic studies of adaptation in forest trees. Tree Physiol. **14**: 781–796.

Svejcar, T.J., Boutton, T.W., and Trent, J.D. 1990. Assessment of carbon allocation with stable carbon isotope labeling. Agron. J. **82**: 18–21.

Tjoelker, M.G., Volin, J.C., Oleksyn, J., and Reich, P.B. 1993. Light environment alters response to ozone stress in seedlings of *Acer saccharum* Marsh and hybrid *Populus* L. I. In situ net photosynthesis, dark respiration and growth. New Phytol. **124**: 627–636.

Tsel'niker, Y.L. 1979. Resistances to CO_2 uptake at light saturation in forest tree seedlings raised under various artificial shade. Photosynthetica, **13**: 124–129.

Volin, J.C., Tjoelker, M.G., Oleksyn, J., and Reich, P.B. 1993. Light environment alters response to ozone stress in seedlings of *Acer saccharum* M. and hybrid *Populus* L. II. Diagnostic gas exchange and leaf chemistry. New Phytol. **124**: 637–646.

Wiard, B.M. 1987. Growth of selected *Populus* clones as affected by leaf orientation, light interception, and photosynthesis. M.Sc. thesis, University of Washington, Seattle, WA. 131 p.

Wolf, A.T., Burk, T.E., and Isebrands, J.G. 1995*a*. Estimation of daily and seasonal whole-tree photosynthesis using Monte Carlo integration techniques. Can. J. For. Res. **25**: 253–260.

Wolf, A.T., Burk, T.E., and Isebrands, J.G. 1995*b*. Evaluation of sampling schemes for estimating instantaneous whole-tree photosynthesis in *Populus* clones: a modeling approach. Tree Physiol. **15**: 237–244.

Zavitkovski, J. 1982. Characterization of light climate under conditions of intensively cultured hybrid poplar plantations. Agric. Meteorol. **25**: 245–255.

Zelitch, I. 1982. The close relationship between net photosynthesis and crop yield. BioScience, **32**: 796–802.

CHAPTER 16
Water relations

T.J. Blake, J.S. Sperry, T.J. Tschaplinski, and S.S. Wang

Introduction

Despite reviews of water relations from the level of the cell to the stand (Hinckley et al. 1991; Blake and Tschaplinski 1992; Hinckley et al. 1994), it is perhaps timely to review water relations of *Populus* since many long-held water relations concepts have recently been questioned, including the mechanism of water transport in xylem transport, cavitation and the significance of water relations adjustments, water use efficiency, and root–shoot signals. In this chapter, these topics are considered insofar as they influence growth and productivity of poplars.

The tendency for water to move rapidly through the soil–plant–air continuum is usually quantified in terms of water potential (ψ), measured as the sum of its major components ($\psi = P + \pi$), where P is the turgor or pressure potential and π, the osmotic potential, the contribution of dissolved solutes. Matric potential and gravity are usually ignored since they are assumed to be relatively small in magnitude. Changes in ψ provide a sensitive indicator of daily and seasonal changes in plant water status. Midday depression of ψ reflects the loss of water in transpiration which varies with the ambient environment. Stomatal conductance (g_s) and transpiration rates (E) decline overnight, causing ψ to increase to a maximum, which is usually measured in terms of predawn xylem pressure

Abbreviations: A, net photosynthesis; E, transpiration rate; g_s, stomatal conductance; H, height; RCD, root collar diameter; MII, membrane injury index; π, osmotic potential; π_0, osmotic potential at saturation; ψ, water potential; ψ_x, xylem pressure potential; ψ_{pd}, predawn xylem pressure potential; WUE, water use efficiency.
T.J. Blake. Faculty of Forestry, University of Toronto, Toronto, ON M5S 3B3, Canada.
J. Sperry. Department of Biology, University of Utah, Salt Lake City, UT 84112, USA.
T.J. Tschaplinski. Environmental Sciences Division, Oak Ridge National Laboratory, Oak Ridge, TN 37831, USA.
S.S. Wang. Experimental Center of Forest Biology, Beijing Forestry University, 100083 Beijing, People's Republic of China.
Correct citation: Blake, T.J., Sperry, J., Tschaplinski, T.J., and Wang, S.S. 1996. Water relations. *In* Biology of *Populus* and its implications for management and conservation. Part II, Chapter 16. *Edited by* R.F. Stettler, H.D. Bradshaw, Jr., P.E. Heilman, and T.M. Hinckley. NRC Research Press, National Research Council of Canada, Ottawa, ON. pp. 401–422.

potential (ψ_{pd}). Although changes in this parameters reflects changes in soil water status, ψ_{pd} itself also influences maximum g_s and P_n in plants. In contrast to the suggestion of Davies et al. (1989), that g_s is controlled solely by soil water availability, it is also regulated, in part, by shoot water relations (Kramer 1988).

Leaf ψ is a sensitive indicator of plant water status of poplars, and ψ_{pd} changes varied depending on the species, level of irrigation and duration of drought. Data from irrigation experiments (S.S. Wang, unpublished) revealed a similar ψ_{pd} (−0.10 MPa) prior to drought in young leaves of four poplar species. However, when soil water content was lowered to 30% of full field capacity, ψ_{pd} declined to −0.6 MPa in *Populus* ×*euramericana* cv. '*I–214*' and *P.* ×*euramericana* Guinier cv. '*Robusta*' during the first day of drought. Loss of lower leaves was correlated with increase in ψ_p after 4–7 d of drought, but levels returned to prestress values after 21 d. Since loss of leaves reduced total leaf area, water balance would be restored by the resulting reduction in water loss. By contrast, ψ_{pd} of *P. popularis* and *P. berolinensis* declined marginally (to 0.2–0.3 MPa) after one day's drought, but quickly recovered without loss of leaves. Although the reason for the smaller decline in the latter species was not determined, osmotic or elastic adjustment may have aided recovery.

Recently, challenges have been made to several long-held viewpoints: (*a*) ψ and other hydraulic parameters have long been thought to influence g_s and growth. However, metabolic signals are thought by some to be more important. ABA transmitted to leaves from roots in drying soil appears more highly correlated with adjustments to g_s, P_n and growth rates under drought than changes in ψ (see section on 'Root to shoot signals'). (*b*) The type of adjustments made by trees as they respond to drought vary, depending on the species and its ecological requirements. A number of studies have suggested the importance of osmotic adjustment, although the importance of this adaptation has been questioned by Munns (1988) and others have found that elastic adjustment may be an order of magnitude greater, depending on the species (Fan et al. 1994). The relative importance of different types of adjustments (e.g., the possibility that cell wall elasticity and cell shrinkage may be more importance for turgor maintenance) is considered in the section on 'Water use efficiency (WUE).' (*c*) According to the cohesion–tension theory, water is pulled through trees by the evaporative power of air and cohesive properties of water. Scholander et al. (1965) provided the first clear quantitative evidence of negative pressures sufficient to move water from soil to leaves. Xylem tensions were determined using the pressure chamber, which determined the balance pressure required to force water from the more elastic living cells of leaves back into more rigid xylem of cut shoots. Experiments of Zimmermann et al. (1994) and Smith (1994), using the pressure probe, suggested that negative pressures (−0.5 MPa) were far too small to explain water movement through trees. However, Pockman et al. (1995) have suggested that insertion of glass capillary tubes through the walls of xylem conduits induce cavitation, which would keep xylem pressures above −0.5 MPa. Indirect confirmation of the cohesion–tension theory has also been provided by Holbrook et al. (1995) and Pockman et al. (1995) who observed sustained, negative pressures, comparable to those observed with the pressure chamber. Tensions were generated using rotating arms which imposed a centrifugal force similar in magnitude to the balance pressures measured with the pressure chamber technique. Depending on the species, negative pressures of between −0.5 and −3.5 MPa were observed in intact stems and xylem conduits remained water filled and conductive at tensions ranging from −1.2 to below −3.5 MPa, depending on the species, without interruption to water conduction by xylem cavitation (Pockman et al. 1995).

Xylem transport

Xylem structure

Populus species (and other members of the *Salicaceae*) are diffuse-porous or semi-ring porous in some *P. deltoides genotypes* (Panshin and de Zeeuw 1980; Wang et al. 1992) and xylem conduits consist solely of vessels with simple perforation plates. More than 90% of the vessels are shorter than 15 cm in length with some extending beyond 30 cm in the two species examined, *P. tremuloides* and *P. balsamifera (*Zimmermann and Jeje 1981; Sperry and Sullivan 1992). The secondary xylem fibers, at least in *P. tremuloides*, become air-filled with age (Sperry et al. 1991). This feature is rarely reported because it can only be observed in fresh-cut material sectioned when fully hydrated (to avoid air-filling during sectioning). It may be important in maximizing the capillary water-storage of the xylem (Zimmermann 1983).

Most diffuse–porous species conduct water in several annual rings. However *Populus* trees often conduct only in the most recent 1 or 2 years' radial growth (at least in vigorous branches, Sperry et al. 1991, 1994). Despite having less conducting area, the hydraulic conductance of the stem xylem per leaf area (leaf–specific conductance) in *Populus* species (*P. grandidentatam*, Zimmermann 1978; *P. tremuloides, P. balsamifera*, Sperry et al. 1994) was similar to or even higher than that of other diffuse–porous genera. This observation is consistent with the high transpiration rates yet similar to leaf xylem pressures in *Populus* species relative to other mesic deciduous trees (Pallardy and Kozlowski 1981; Table 1). Part of the reason for high leaf-specific conductances may be a tendency for *Populus* xylem to have a slightly higher hydraulic conductance per functional xylem vessel compared to that in other diffuse–porous genera (Pallardy et al. 1995). In addition, *Populus* species may have a somewhat lower leaf area per unit stem basal area than other diffuse–porous trees (Kaufmann and Troendle 1981; Waring et al. 1977; Sperry et al. 1994), although data for comparison are scanty.

Xylem cavitation

During transpiration-driven water uptake, xylem pressures become subatmospheric and drop substantially below the vapour pressure of water. The water remains in a metastable liquid state owing to the absence of nucleating sites for vaporization within the xylem (Pickard 1981). There is a limit, however, to how negative xylem pressures can become before vaporization ("cavitation") is initiated. Cavitation inhibits water transport because it results in a gas-filled (embolized) conduit. If extensive, cavitation will reduce hydraulic conductance of xylem and result in increased stomatal closure and/or declining ψ. If all transport becomes blocked in this manner, the shoot will desiccate and die.

Most evidence supports the view that xylem cavitation occurs when interconduit pits allow air to enter functional xylem conduits from neighbouring air-filled ones (Crombie et al. 1985; Sperry et al. 1995; Tyree et al. 1994; Jarbeau et al. 1995). This phenomenon is the air-seeding mechanism proposed by Zimmermann (1983). Air-filled conduits are always present in the vascular system from the openings in the xylem caused by leaf abscission, death of fine roots, and other causes. Normally, this air is prevented from passing between conduits because of the self-sealing properties of interconduit pit membranes. In the interconduit pits of most angiosperms, the pit membrane is a uniformly thin and porous structure that traps a gas–water meniscus by capillary forces. However, this capillary seal can fail when xylem pressures drop low enough to pull the gas–water meniscus through the pit membrane pores. The entry of gas then "seeds" cavitation. Alternatively, any degradation of the pit membrane structure that enlarges its pores can result in a weakened capillary seal allowing water vapour and air to propagate from one conduit to the next. As mentioned above, this can lead to loss of transport in aging sapwood of *P. tremuloides*.

Cavitation and loss of water transport places obvious theoretical limits on plant survival from both static (soil drought-induced) and dynamic (transpiration-induced) water stress. Even if stomata completely close during a drought, static water stress will increase as xylem pressure drops in response to decreased soil water potential. If pressures drop low enough to cavitate all xylem conduits, the plant will not be able to conduct water when the drought is over. When stomata are open and transpiration is occurring, the pressure drop associated with xylem transport cannot be great enough to trigger "runaway cavitation" because unstable positive feedback occurs between decreasing hydraulic conductance from cavitation and decreasing xylem pressure from dynamic water stress (Tyree and Sperry 1988).

The susceptibility of the stem xylem (stem diameters ca. 0.25–1.0 cm) to cavitation was investigated in six *Populus* species (Fig. 1). In these studies, cavitation was determined from the loss of hydraulic conductance caused either by progressively lowering xylem pressures in dehydrating stems or progressively

Fig. 1. The relationship between the loss of hydraulic conductance in stem xylem vs. xylem pressure ("vulnerability curves") in six species of *Populus*. Data was obtained either by dehydration or air-injection methods (see Sperry et al. 1995). Open circles: *Populus deltoides*, data pooled from Tyree et al. 1992 (northern Vermont), Cochard et al. 1992 (France), Alder and Sperry, unpublished data (Utah). Open diamonds: *Populus deltoides* from Tyree et al. 1994 (western Canada). Open triangles: *Populus augustifolia* from Tyree et al. 1994 (western Canada). Open squares: *Populus balsamifera*, data pooled from Tyree et al. 1994 (western Canada) and Hacke and Sauter 1995 (northern Germany). Open inverted triangles: *Populus fremontii* from Pockman and Sperry, in preparation (southern Arizona). Solid circles: *Populus tremuloides* from Sperry et al. 1991 (Utah). Solid triangles: *Populus trichocarpa* from Alder and Sperry, unpublished data.

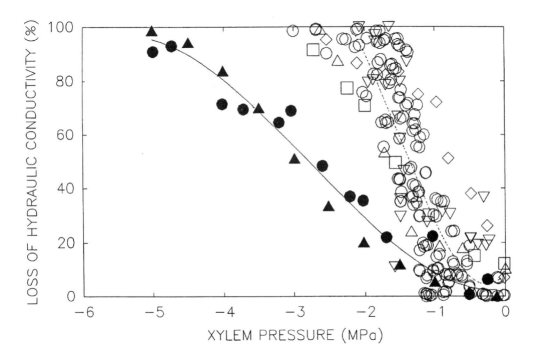

raising air pressures in hydrated stems. The two techniques give the same result because whether the air is pushed or pulled into the conduits, the pressure difference required is the same (Sperry et al. 1995). The results were plotted as "vulnerability curves." The six species can be represented by two basic curves. One group (Fig. 1, dashed line; *P. deltoides, P. angustifolia, P. balsamifera, P. fremontii*) showed most loss of xylem conductance occurring between –1 and –2 MPa (Tyree et al. 1994, 1992; Hacke and Sauter 1995; Pockman et al. in review; N.N. Alder and J.S. Sperry, unpublished). The second group (Fig. 1, solid line; *P. tremuloides, P. trichocarpa*) had more resistant xylem that lost conductance primarily between –2 and –5 MPa (Sperry et al. 1991; N.N. Alder and J.S. Sperry, unpublished).

These curves can be used to determine the minimum stable xylem pressure in the stem induced by transpiration at any given soil water potential, assuming steady state conditions (Tyree and Sperry 1988; Jones and Sutherland 1991; Alder et al. 1995). Xylem pressure cannot decline below this value because runaway cavitation is induced. A computer program for estimating this critical pressure is described in Alder et al. (1995). Results of this program for the six *Populus* species in Fig. 1 are shown in Table 1. The critical stem pressures reported are for soil water potentials of 0 MPa. The four most vulnerable species have critical values between –1.05 and –1.45 MPa; the two more resistant species have lower critical values of –2.27 and –2.38 MPa. As soil water potential decreases, the critical pressure decreases until the soil water potential approaches the xylem pressure, causing 100% loss of xylem conductance.

The type of curve and its critical pressure reflects the moisture availability of the species' habitat. The four most vulnerable species are typically found in riparian zones (*P. fremontii, P. deltoides, P. angustifolia*), and in other perennially-moist habitats (*P. balsamifera;* Elias 1980). Their vulnerability curves are very similar to other strict riparian species (*Salix* spp., *Betula occidentalis, Alnus incana*; Pockman et al. in review; Sperry et al. 1994). Their critical stem pressures are actually a few tenths of a megapascal above the minimum leaf xylem pressures reported for nondroughted material (Table 1). However, stem xylem pressure will be less negative than leaf pressure, and leaf xylem may be more resistant to cavitation than stem xylem (as inferred for *Betula occidentalis*, Sperry and Saliendra 1994). Therefore, leaf pressures may drop below the critical value for stems without inducing runaway cavitation. Furthermore, although leaf pressures cited in Table 1 were measured under nondrought conditions, the soil water potential was probably not zero in all cases. A lower soil water potential would correspond to a lower critical stem pressure. Nevertheless, the data in Table 1 suggest these four most vulnerable species maintain a very slight margin of safety from runaway cavitation in their native habitat.

The two cavitation-resistant species (*P. tremuloides, P. trichocarpa*) are found in riparian zones but also in seasonally drier habitats (Elias 1980). In general, xylem pressures under well-watered conditions are similar, in these species, to those of the more vulnerable group (Table 1), suggesting they maintain a much larger margin of safety from runaway cavitation under well-watered conditions.

There are parallels between the safety margin from cavitation and the species' stomatal response to water stress. *P. deltoides* has a relatively sensitive stomatal response to water stress, as would be expected for a species with a narrow safety margin (Schulte et al. 1987; Hinckley et al. 1989). *P. balsamifera*, which has a vulnerability curve similar to that of *P. deltoides*, is also very effective in limiting xylem pressures to above ca. –1.0 MPa by adjustment of stomatal conductance, despite widely varying climatic conditions (Hacke and Sauter 1995). On the other hand, *P. trichocarpa,* which maintains a much larger safety margin,

Table 1. Critical xylem pressures (in MPa) for soil water potential of 0 MPa in six *Populus* species from Fig. 1.

Species	Critical stem pressure (MPa)	Minimum leaf pressure (Mpa)
Populus deltoides	−1.05	−1.45[a]
Populus balsamifera	−1.32	−0.9, 1.5[a, e]
Populus angustifolia	−1.45	−1.5 (−2.0)[a]
Populus fremontii	−1.20	−1.5[b]
Populus trichocarpa	−2.27	−1.4[c]
Populus tremuloides	−2.38	−1.8[d]

Note: This pressure is the minimum (most negative) allowable in stem xylem without the induction of runaway cavitation (Tyree and Sperry 1988; Jones and Sutherland 1991). Values were calculated from vulnerability curves in Fig. 1 (pooled by species) using the computer program described in Alder et al. (1995). Also shown are minimum leaf xylem pressures under well-watered conditions.

[a]Tyree et al. (1994); [b]W.T. Pockman and J.S. Sperry (unpublished data); [c]Pezeshki and Hinckley (1982); [d]Sperry and Sullivan (1992); [e]Hacke and Sauter (1995).

exhibits relatively poor stomatal control over transpiration and xylem pressure in response to decreased soil moisture or humidity, with stomata remaining open at leaf water potentials below −3.5 MPa (Schulte and Hinckley 1987a; Schulte et al. 1987).

These observations indicate that the vulnerability of poplar xylem to cavitation provides an understanding of species' ecological tolerance for water stress and the regulation of water use via stomatal control. For example, the conspicuous dieback seen in riparian *Populus* species following dam-mediated reductions in stream flow (Tyree et al. 1994) may ultimately be linked to their relatively small margins of safety from critical cavitation levels. In a controlled drought experiment, Braatne et al. (1992) found that survival of *P. trichocarpa, P. deltoides,* and two F_1 hybrids was related more to the loss of hydraulic conductance from xylem cavitation than to dehydration tolerance of leaf tissue. Dieback resulting from cavitation could result either directly from runaway cavitation, or indirectly by chronic stomatal closure required to avoid critical cavitation (Tyree et al. 1994), as discussed in the section on 'Stomatal adjustment.'

While most studies of cavitation have been done on stem xylem, there is some evidence that cavitation in root xylem is even more important for determining how a species responds to water stress. Critical pressures in root xylem can be approached well before those in stem xylem (Alder et al. 1995). Unfortunately there are no data on root cavitation in any *Populus* species.

There is little or no information on genetic vs. environmental influence on a species' vulnerability to cavitation. Studies of three of the species in Fig.1 have been replicated at least twice in widely varying habitats (*P. deltoides, P. balsamifera, P. tremuloides*). The data for *P. balsamifera* at sites in northern

Germany (Hacke and Sauter 1995) vs. western Canada (Tyree et al. 1994) were indistinguishable (Fig. 1, open squares). *P. tremuloides* was also similar between sites in interior Alaska and northern Utah (Sperry et al. 1994; Sperry et al. 1991; Fig. 1, closed circles). Data for *P. deltoides* were consistent between sites in Vermont (Tyree et al. 1992), Utah (greenhouse stock, N.N. Alder and J.S. Sperry, unpublished) and France (Cochard et al. 1992), although material from western Canada was somewhat more vulnerable (Fig. 1, open diamonds vs. open circles; Tyree et al. 1994). These results suggest relatively little intra-specific variation in stem cavitation. A study of hybrids between *P. deltoides* and *P. trichocarpa* (which have widely divergent vulnerability curves, Fig. 1) is underway and may provide insight into the genetic control of cavitation vulnerability in *Populus* (N.N. Alder and J.S. Sperry, unpublished).

Water relations adjustments

Stomatal adjustment

Because of the large potential energy gradient between the plant and the atmosphere, adjustments in g_s regulate water economy and increase the ability of trees to avoid (postpone) drought. Stomata open and close in response to changes in plants and their environment. Light intensity and wavelength, internal and external CO_2 levels, humidity, ψ, stem conductance, and plant metabolism and ABA levels all influence g_s (Hinckley et al. 1991; Hinckley and Braatne 1994). The ability of seedlings to adjust g_s, both on a daily and seasonal basis, was important for adaptation of seedlings to site and g_s was the physiological parameter that showed the highest correlated with early growth rate of transplanted conifers (Blake and Sutton 1988; Blake and Yeatman 1989).

Seasonal variation in g_s in trembling aspen (*Populus tremuloides*) followed soil water content, while diurnal variations followed changes in solar radiation, vapour pressure deficit (VPD), and air temperature (Iacobelli and McCaughey 1993). Stomata closed and growth slowed in *Populus trichocarpa*, *Populus deltoides*, and their hybrids when ψ_p declined below −1.0 MPa; however, plants reached this level only four times during the growing season (Pezeshki and Hinckley 1982).

Plants adjust to drought long before stomata close; growth rates of fine roots increased while shoot growth declined under drought, which helped postpone dehydration (Blake and Tschaplinski 1992). Once poplars established deep sinker roots to ground water, ψ_{pd} never declined below −0.88 MPa and turgor loss was not observed (Pezeshki and Hinckley 1988). Responses of poplar stomata to declining ψ_{soil} were complex; however, decline in ψ_{pd} lowered g_s of black cottonwood (*P. trichocarpa*), irrespective of vapour pressure gradient. g_s declined when soil ψ declined below −1.0 MPa. As ψ_{soil} declined further, the threshold value for stomatal closure in cottonwood shifted from −1.0 to −0.5 MPa (Pezeshki and Hinckley 1982).

The ability of stomata to close in response to water deficits varied among poplar species and their interspecific hybrids (Blake et al. 1984; Schulte and Hinckley 1987a). Some species required larger changes in leaf and/or root hydration for stomatal closure. Not all poplars can close stomata in response to limiting soil moisture. Stomata of well-hydrated *P. trichocarpa* were unresponsive to vapour density gradient and leaf ψ (Schulte and Hinckley 1987a). Stomata of *P. trichocarpa* were unable to close despite increase in ABA levels, increased membrane leakage, plasmolysis of guard cells and loss of turgor. However, they closed under field conditions, when drought developed slowly, and when drought stress was repeated (Schulte and Hinckley 1987b). In contrast to *P. trichocarpa*, stomata of *P. deltoides* and its hybrids were much more sensitive to reduced root growth and drought lowered g_s long before P started to decline and the response of parental types differed from those of their hybrids (Schulte and Hinckley 1987b).

Stomata on excised *P. trichocarpa* shoots failed to close despite desiccation and wilting. By contrast, stomata of *P. trichocarpa* × *P. deltoides* closed rapidly as soon as excised leaves started to wilt. In *P. trichocarpa*, young, expanding foliage and gradual drought showed the greatest decline in g_s, falling to 20% of maximum values (Hinckley and Braatne 1994).

Osmotic adjustment

Many plant species tolerate drought by maintaining low osmotic potential (π) or through osmotic adjustment, the active accumulation of solutes in response to drought (Morgan 1984). Low osmotic potential enhances the capacity of plants to take up water from a drying soil, and may be as important as root growth in facilitating water uptake (Tyree and Jarvis 1982). Species and hybrids with tolerance mechanisms may have the advantage of maintaining cell turgor and growth under adverse conditions, whereas many dehydration postponement responses (e.g., stomatal closure, leaf abscission, leaf rolling) greatly reduce plant growth rates under stress (Turner 1979). However, Turner (1986) cautioned that there is typically a limit to the degree of full turgor maintenance, and given that C_s and P_n decrease under stress in spite of turgor maintenance, osmotic adjustment in the shoot will be effective in maintaining shoot growth under declining soil water potential only if osmotic adjustment also occurs in the root. The role of drought-induced osmotic adjustment for turgor maintenance and growth under drought has also been questioned. Rieger (1995) provided evidence that the potential turgor maintenance of osmotic adjustments of 0.34 MPa and 1.43 MPa in leaves of citrumelo (*Poncirus trifoliata* Raf. × *Citrus paradisi* Macf.) and olive (*Olea europa* L.) were offset by reductions in hydraulic conductivity. Although turgor maintenance and growth under stress may be limited, osmotic adjustment enables plants to function over a wider range of Ψ, and facilitates the maintenance of tissue integrity during drought and recovery of growth rates after drought relief.

Many poplar hybrids currently used for short-rotation forestry are not considered drought tolerant, although sufficient genetic variability exists to identify drought-tolerant poplar clones. Tyree et al. (1979) screened 21 well-watered poplar clones grown under greenhouse conditions to determine variability of osmotic potential at saturation (π_o) and reported that π_o ranged from -1.18 MPa for *P. alba* L. × *P. tremuloides* Michx. 'AT142' to -1.89 MPa for *P. grandidentata* Michx., suggesting that there is considerable variability within the genus. The assumption that fast growing clones lack drought tolerance may be incorrect. Clones with the lowest π_o were found to be medium to fast-growing clones, whereas slow growing clones had the highest π_o. There was a great range of π_o values, from -1.38 MPa to -2.35 Mpa, when progeny of black cottonwood (*P. trichocarpa* ♀ Torr. & Gray '93–968'), eastern cottonwood (*P. deltoides* ♂ Bartr. 'Ill–129'), two hybrid progeny (53–242 ♂, 53–246 ♀) and 55 of their F_2 progeny were compared (T.J. Tschaplinski, unpublished data).

Faster-growing clones of poplar hybrids were found to be more, rather than less, drought tolerant, displaying lower π_o (Tschaplinski and Blake 1989*a*, *b*). Faster-growing poplar clones (*Populus balsamifera* L. × *P.deltoides* Bartr. 'Jackii 4' and *Populus deltoides* × *P. nigra* L. 'DN2'), maintained higher transpiration rates, and P_n compared with slower-growing clones, Jackii 7 and DN15, at midseason at midday or later (Tschaplinski and Blake 1989*a*). The tendency of faster-growing clones to quickly root from cuttings and allocate a greater proportion of carbon to roots (Tschaplinski and Blake 1989*a*, *c*) and the lower π_o both likely facilitated water uptake resulting in less drought stress than that experienced by slower-growing clones. Photosynthetic capacity was lower in the slower-growing poplar clones and leaf senescence was earlier, presumably as a result of higher plant moisture stress.

Recently-mature foliage of drought tolerant *Populus deltoides* × *P. nigra* and *P. deltoides* × *P. balsamifera* hybrids had the highest concentrations of sucrose and low molecular weight phenolic compounds, including salicyl alcohol, and its glucoside, salicin, and these compounds appeared to contribute most to osmotic adjustment (Tschaplinski and Blake 1989*b*). This study suggested that phenolic compounds and their glucosides may have a role in drought tolerance in hybrid poplar, particularly in expanding foliage, which can have low concentrations of sucrose due to its rapid utilization. In contrast, sucrose is typically an important osmolyte in fully expanded leaves. Sucrose and malic acid were the most abundant organic solutes contributing to π_o in *P. deltoides* 'Ohio Red,' with concentrations of both solutes increasing under drought (Gebre et al. 1994).

The three clones studied all displayed osmotic adjustment ranging from 0.24 to 0.48 MPa, with 'Ohio Red' and 'Wildcat' achieving lower π_o than 'Platte.' This confirmed a previous demonstration of a substantial degree of osmotic adjustment in the poplar hybrid, *P. deltoides* × *P. nigra* 'DN22' (Tschaplinski and Blake 1989*b*). Osmotic potential at saturation of leaves of stressed trees of this hybrid was reduced by 0.55 MPa due to a 4-fold increase in concentrations of all soluble sugars, including sucrose, glucose, fructose, myoinositol, galactose, and salicin.

Drought tolerance of six clones of the black cottonwood and eastern cottonwood pedigree described above, including the parental clones and four hybrid progeny, was investigated by subjecting trees to repeated cyclical stresses of 1- to 2-days' duration over the 14-wk study (Tschaplinski et al. 1994; Tschaplinski and Tuskan 1994). Male clones tended to have a greater degree of drought tolerance than did the ♀ clones, as evidenced by greater osmotic adjustment (up to 0.25 MPa), and maintenance of dry matter production of the main stem under drought. Osmotic adjustment in the leaves of both hybrid 53–242 ♂ and the *P. deltoides* ♂ parent was primarily associated with increases in malic acid, K, sucrose, and glucose, with the same metabolites also increasing in fine roots of the male hybrid 53–242, which was the only clone to display osmotic adjustment in roots. Given that the TRIC ♀ parent did not display osmotic adjustment in either tissue, the hybrids' capacity for adjustment was likely conferred by the *P. deltoides* male parent. Osmotic adjustment in roots may steepen the water potential gradient between the soil and plant, allowing root growth to continue during moderate stress. The capacity for continued root growth afforded by osmotic adjustment would allow the root system of a tree to explore new reserves of water that would be otherwise unavailable during drought.

To determine whether *in vitro* assays can be used as a surrogate for whole plants in the assessment of drought tolerance, six clones of the same pedigree described above, including the parents and 4 hybrid progeny, were regenerated in tissue culture and subjected to 0–20% polyethylene glycol MW 8000 (PEG) in the nutrient medium, providing osmotic potentials that ranged from −0.40 MPa to −1.29 MPa (Tschaplinski et al. 1996). Drought tolerance was assessed by callus growth and solute accumulation under various stress conditions. There were few clonal differences in relative growth rate across PEG concentrations and differences were not related to solute accumulation. In contrast to whole plants, where carbon-based compounds are most abundant, inorganic cations and free amino acids constituted the bulk of the solutes in calli and concentrations of soluble carbohydrates were low. Belanger et al. (1990) also reported that osmotic stress applied to tissue-cultured plantlets generally increased asparagine, glutamine, arginine, alanine, glutamic acid, and proline in five *P. tremuloides* clones. Therefore, results obtained from callus cultures growing under artificially-induced drought regimes did not parallel those obtained from drought studies on whole plants.

In summary, a variety of biochemical, physiological, and morphological adjustments tend to minimize the deleterious effects of water deficits (Dale and Sutcliffe 1986) which are detailed in Chapter 17, Neuman et al. (Stress physiology). The energy (ATP) cost of different of adaptations does not appear to have been compared in poplars or other types of trees. However, stomatal adjustment in drought-stressed plants, were no less costly in terms of energy, compared to osmotic adjustment (Morgan 1984).

Root-to-shoot signals

Stomatal closure is closely correlated with cavitation (sections on 'Xylem cavitation' and 'Osmotic adjustment') and although stomatal closure can aid survival by limiting water loss in a drought, chronic closure not only reduces growth, but may also induce senescence in woody plants (Blake and Sutton 1988). In the absence of other limiting factors, shoot growth continues until root water absorption falls behind the loss of water in transpiration. When this occurs, stomata close and water loss declines until water loss and gain are in balance. Although prolonged imbalance between roots and shoots causes leaf, and ultimately whole plant senescence (Tschaplinski and Blake 1985), a flushing in growth, induced by drought, helps to keep root and shoot growth in a functional balance (Borchert 1975). The question arises as to how soil drought induces the changes in g_s and growth in shoots. Recent evidence suggests that ABA generated in dehydrating roots is carried in the sap stream to the leaves, causing g_s and shoot growth to be inhibited (Tardieu and Davies 1993). When only a portion of the soil volume was dry and adequate water was available to lupins in an adjacent wet zone, root signals did not appear to influence g_s (Gallardo et al. 1994).

Opinions vary greatly on the relative importance for stomatal movements of hydraulic signals such as ψ, and metabolic signals such as ABA. Zhang and Davies (1990) observed that stomatal closure in maize plants followed significant increases in the ABA content of roots, without any corresponding decline in ψ_l. However, Kramer (1988) argued that water relations changes alone cause stomatal closure and suggested the role of ABA in stomatal movements may have been overemphasized. Shoots of many species appear responsive to both hydraulic and metabolic 'signals' (Boyer 1989; Schulze et al. 1988). Leaf water status may control stomatal movement in combination with root produced ABA (Tardieu and Davies 1993). Alternatively, hydraulic changes in drying roots may increase ABA synthesis in roots and, for inhibition of g_s and shoot growth, the sequential action of both hydraulic and metabolic signals is required. The hydraulic and metabolic signals may operate sequentially in plants, so that the two hypotheses may not be mutually exclusive.

Two types of plant responses to drought have been suggested: in 'isohydric' plants, root-sourced ABA may accumulate in shoots, where it reduces g_s and growth, without any prior reduction in ψ_l or turgor. By contrast, in 'anisohydric' plants, changes in ψ_l follow changes in evaporative demand (Tardieu and Davies 1993). In one study, stomatal conductance declined in black spruce after 2 d of low humidity treatment, while significant increase in ABA levels in roots, xylem sap, and leaves were detected after one month's dry-air treatment, i.e., long after stomata closed (Blake et al. 1995).

The sensitivity of stomata to drought-induced ABA varies among poplar clones. The decline in g_s under drought in two Eugenii clones (*P.* ×*euramericana* cv. Eugneii) was unaffected by soil drying until leaf ABA concentrations exceeded 100 ng·g·dw^{-1}. By contrast, Tristis clones (*Populus tristis* × *P. balsamifera* cv. Tristis No. 1) closed stomata when leaf concentrations of ABA reached 10 ng·g·dw^{-1} (Liu and Dickmann 1992).

Application of ABA to roots was variously found to decrease (Blake et al. 1990; Zeevaart and Creelman 1988), stimulate (Abou-Mandour and Hartung 1980), or have no effect (Blake et al. 1990) on root dry weight accumulation. Cell wall relaxation under drought induced tissue shrinkage which helps to alleviate drought (Davies et al. 1989; Blake et al. 1990*b*; Fan et al. 1994) and most studies show that drought causes root elongation to cease in the region of root drying, but root branching was found to increase, which allowed black spruce roots to expand into the deeper regions of the soil profile where moisture was more abundant (Q. Liu and T.J. Blake, unpublished data).

Application of ABA as a root drench quickly closed stomata of three conifer species and caused a temporary bud dormancy (Blake et al. 1990*b*). Root drying experiments showed that ABA, generated by dehydrating root tips, traveled upwards in the sap stream to the shoot where it closed stomata (Davies et al. 1989; Blake et al. 1995). Increased ABA production in roots, xylem sap, and leaves under soil drought (Fan and Blake 1994) and osmotic stress (Tan and Blake 1993) was highly correlated with stomatal closure and reduced shoot growth of black spruce and jack pine. Despite the above evidence, the role of ABA as a root-to-shoot signal has been questioned: (*i*) Stomatal closure, induced by applied ABA, was transitory, lasting hours to days and this was followed by an increase in stomatal conductance after 7 d (Blake et al. 1990*a*). Despite an increase in transpiration, stomata of ABA-treated plants were sensitized and closed more rapidly under a subsequent drought. (*ii*) ABA levels in the sap stream of wheat were insufficient to induce an inhibitory response and stomatal closure still occurred following reapplication of ABA-free sap to leaves (Munns 1990). (*iii*) Concentrations of ABA in roots of sugar maple (*Acer saccharum* Marsh) were higher in May and June, when root elongation was maximal (Cohen et al. 1978).

Higher levels of ABA accumulating under drought were associated with an increase in the membrane injury index in shoots of drought stressed plants (Tan and Blake 1993; Fan and Blake 1994) and an application of ABA to undroughted plants of three woody species increased membrane leakage (Fan and Blake 1994). Progressive killing of the subsidiary cells under drought was observed to open stomata of some species (Mallock and Fenton 1979). Taken together, drought appears to induce the following sequence of events: higher levels of ABA increase in electrolyte leakage and increased leaf senescence, with a subsequent reduction in leaf area. Supporting such a sequence of events, ABA accumulation in the leaves of drought stressed hybrid poplars ('*Eugenei*' and '*Tristis*') resulted in leaf senescence and abscission (Liu and Dickmann 1992).

Species that can lower g_s during midsummer drought without senescing may have a competitive advantage. Black cottonwood (*Populus trichocarpa* Torr. & Gray) reduced osmotic potentials at full saturation and at the turgor loss point and partially closed its stomata as drought increased over the growing season (Pezheshki and Hinckley 1988). By contrast, red alder (*Alnus rubra* Bong.) was unable to make these responses and the resulting water stress caused leaf shedding which depressed growth, relative to black cottonwood (Pezheshki and Hinckley 1988). Although leaf senescence increases the root/shoot ratio and survival of droughted plants in the short term, the consequent reduction in photosynthetic capacity and dry matter production (Araujo et al. 1989) could reduce competitive ability in the longer term.

Water use efficiency (WUE)

WUE is the term used to relate plant production to water consumption in transpiration of leaves and plants or evapotranspiration of whole stands. Whole plant WUE can be assessed from the amount of dry matter produced per unit water loss in transpiration (from weight loss, determined gravimetrically, by multiplying transpiration rate per unit leaf area by the total leaf area per plant). The ratio of net carbon assimilation rate (A) to the transpiration rate (E) provides an instantaneous leaf-level assessment of WUE (A/E). The use of carbon isotope discrimination (Δ) analysis has led to its general acceptance as a surrogate for p_i/p_a (the ratio of intercellular and ambient partial pressures of CO_2, respectively) in C_3 plants (Farquhar et al. 1982; Farquhar 1989). Instantaneous WUE (A/E) can also be determined from p_i and p_a and the water vapor pressure difference between the intercellular spaces of a leaf and the atmosphere according to the following equation:

$$A/E = p_a (1 - p_1/p_a)/1.6\,v$$

where v is the water vapor pressure difference between intercellular spaces of a leaf and the atmosphere and the 1.6 factor is the binary diffusivity of water vapor in air, relative to CO_2 (Farquhar et al. 1989). Caution is required when WUE determined at one scale (e.g., instantaneous leaf-level WUE) is extrapolated to another (e.g., stand-level WUE).

There was a high degree of genetic variation in whole-plant WUE among poplar clones when WUE was measured under conditions of high water availability (Blake et al. 1984). Several poplar clones in the section *Tacamahaca* had higher WUE than clones in the Section *Aegeiros*. However, there was also much variation among clones of the same species, and within clones of the same species. WUE was correlated with several combinations of foliar adaptations that reduce water loss: (*i*) absence of stomata on the adaxial surface, (*ii*) presence of dense hairs on the abaxial surface, (*iii*) presence of cuticular ridges on the stomata coupled with more deeply recessed stomata, (*iv*) earlier stomatal opening in the morning, and (*v*) higher frequency of smaller stomata per unit leaf area. For example, high WUE of two clones *P. maximowiczii* 'M4' and *P. alba* 'A499' was associated with high midday diffusive resistance (low stomatal conductance) on the abaxial leaf surface, which has also been reported in other studies (Ceulemans et al. 1978; Pallardy and Kozlowski 1980). Stomatal resistance accounted for only 40% of the variation in WUE between genotypes. Although mesophyll resistances were not examined, leaf diffusive resistances were higher in water-efficient clones, while other morphological and physiological factors failed to consistently explain the variation observed in WUE (Blake et al. 1984).

Uncertainty has arisen as to whether species with higher WUE have a greater productivity under drought than species with a lower WUE. WUE showed a low, but statistically significant, positive correlation coefficient ($r^2 = 0.30$) with total dry matter production when WUE of 17 poplar clones were compared (Blake et al. 1984). The relationship between WUE and productivity can be established from the components that determine WUE. If variation in WUE is the result of comparatively greater variation in stomatal conductance than in CO_2 assimilation rate, then the highest WUE would be coupled with restricted dry matter productivity. In contrast, if variation in WUE is primarily due to greater variation in CO_2 assimilation rate than in stomatal conductance, as may be the case in well-watered plants, then the correlation between productivity and WUE may be negative. It should be cautioned that conditions under which the determinations of WUE occur may greatly influence the nature of the correlation, or lack thereof, observed.

Stand-level WUE may be more important than whole tree WUE, for productivity of trees grown under short rotation intensive culture. High WUE, if coupled with high dry matter productivity, should reduce stand water use and hence lower irrigation costs while still maintaining productivity. In the absence of irrigation, it would extend the length of time it takes for a stand to deplete available soil water. If polyclonal cultures are established with clones displaying a wide range in WUE, a highly-productive clone with low WUE (but one that exhibits other drought resistance mechanisms) could exploit the available water relative to a high WUE clone, placing the latter clone at a competitive disadvantage. Given the current lack of any consistent correlations between morphological and physiological parameters and WUE, or between WUE and productivity, it cannot be assumed that a high WUE will necessarily increase productivity in plantations where water supply is frequently limiting.

Clustering of adjustments

Multiple adaptations and a "clustering" of strategies appear more important for survival and growth under drought than single adaptations. The significance of any single adjustment for drought tolerance is doubtful until its relative importance has been confirmed. The significance of different morphological and physiological parameters was studied in 3 willow clones by simulating drought in a root misting chamber. Although ψ_x declined to similar levels, stomata of the faster growing *S. viminalis* clones (W559 and W557) were more responsive to drought and closed more rapidly than the slow growing *S. eriocephala* clone ER57, which maintained a lower g_s. The faster growing *S. viminalis* clones (W559 and W557) maintained a higher g_s and lower membrane leakage under simulated drought, compared to the slow growing *Salix eriocephala* (clone ER57), suggesting the *S. Viminalis* clones were more drought tolerant.

Height growth showed significant negative correlations ($r^2 = -0.68$) with membrane injury index, while height and diameter growth showed even stronger, positive correlations with g_s and A ($r^2 > 0.73$; $p < 0.01$). Height also showed a significant positive correlation with ψ ($r^2 = 0.58$). A study of Table 2 also shows that g_s explained much of the variation in transpiration (86%), height (84%), and RCD (73%). Significant correlations were observed between g_s and ψ (76%) and between C_s and membrane leakage (−74%). These results suggest that a high proportion of the variation in g_s and growth in willow can be explained by water relations parameters, as observed previously in poplar species and clones (Tschaplinski and Blake 1989*b*). The faster growth in the two *S. viminalis* clones and in hybrid poplar clones was correlated with higher g_s and A, which, in turn, resulted from their greater drought tolerance and earlier rooting. The fast growing clones suffered less membrane damage under drought and were able to lower g_s and A under water stress (A. Vlainic and T.J. Blake, unpublished data).

Fast growing *Populus deltoides* hybrids also showed earlier rooting of cuttings, slower stomatal closure, postponed dehydration under drought, and a greater tolerance of dehydration, compared to slow growing clones (Tschaplinski and Blake 1989*a*). More vigorous types supported a greater leaf area, higher A rates and photosynthesis continued until later in the season due to delayed leaf senescence (Tschaplinski and Blake 1989*b*). g_s and P_n were greater, over the day, in fast growing *Populus deltoides* × *P. nigra*, clones (DN2 and DN22) and *P. balsamifera* × *P. deltoides* (Jackii 4), and less in slower growing DN15 and Jackii 7 clones (Tschaplinski and Blake 1989).

The positive relationship between g_s, A, drought tolerance, and growth rate observed in poplars and willows was not observed in fast-growing *Eucalyptus*

Table 2. Spearman's rank correlation coefficients and their statistical significance in *Salix viminalis* W557 and W559 (faster growing clones) and *S. eriocephala* ER57 (a slow growing clone).

	A	g_s	H	RCD	Ψ	E	MII
A	1.00	0.76**	0.74**	0.72**	0.73**	0.81	−0.49**
g_s	—	1.00	0.84**	0.73**	0.76**	0.86***	−0.74**
H	—	—	1.00	0.93***	0.58*	0.84***	−0.68**
RCD	—	—	—	1.00	0.50	0.71**	−0.48
Ψ	—	—	—	—	1.00	0.65**	−0.28
E	—	—	—	—	—	1.00	−0.72**
MII	—	—	—	—	—	—	1.00

Note: Net assimilation, A; stomatal conductance, g_s; height increment; root collar diameter, RCD; stem Ψ; transpiration rate, E; and membrane injury index, MII.

$**P \leq 0.01$
$***P \leq 0.001$

grandis clones, which exhibited a greater photosynthetic capacity, despite lower unit rates of A, by virtue of their greater leaf area (Blake and Tschaplinski 1992).

Conclusions

A variety of biochemical, physiological, and growth adjustments either postpone or minimize the importance of water deficits in plants (Dale and Sutcliffe 1986). Although responses to water deficits are usually considered singly, they do not occur independently and a continuous sequence of adjustments coincide in groups or "strategies" characteristic of a species (Hinckley et al. 1991). Although the importance of dehydration tolerance adaptations, e.g., osmotic adjustment, has been demonstrated, experiments in *P. trichocarpa*, *P. deltoides*, and their hybrids indicate that survival may be more closely related to loss of hydraulic conductance under drought, rather than dehydration tolerance *per se*. Recent studies suggest the importance for growth rate of a high stem conductance of water vapor. Although the vulnerability of poplar xylem to cavitation is highly variable, survival may be threatened by the relatively small margin of safety from critical levels of cavitation in some poplar species.

Dehydration tolerance adaptation were, however, important for the more vigorous growth of some genotypes, when these were compared with slow growing types. Although counter-intuitive, more vigorous clones were more drought tolerance, despite their greater leaf area. Stomatal conductance often shows significant, positive correlation with dry matter production, however, the relationship appears complex. Correlations between g_s and dry matter production varied from positive to negative, depending on the species and its ecological zone (Blake and Tschaplinski 1992). A low g_s may increase short-term survival under drought; however prolonged, stomatal closure is a chronic condition that correlates with slow growth (Tschaplinski and Blake 1985; Blake and Tschaplinski 1992). Because more vigorous genotypes exhibited a lower membrane injury index, they were better able to maintain growth during the early stages of a drought and could more rapidly reopen their stomata when released from drought. The following sequence of adjustments under drought promoted growth in more vigorous poplar clones, compared with slow growing types. Faster-growing genotypes exhibited earlier root development on cuttings, increased ψ_x, A, and some fast-growing clones exhibited significantly greater osmotic adjustment (Tschaplinski and Blake 1989a) and lost turgor at lower osmotic potentials (π_{TLP}). Slower growing poplar clones (Tschaplinski and Blake 1989a), black spruce provenances (Tan and Blake 1995) and less vigorous inbred families (Blake and Yeatman 1989) lacked these adaptations and were more drought prone when challenged by drought.

Further work is required to elucidate the relative efficiency of different drought tolerance adaptations. Such correlations will permit the development of biochemical and molecular markers, particularly quantitative trait loci (QTLs). These can be subsequently validated by comparison of a greater number of pedigrees growing under field conditions. Development of screening procedures would greatly reduce time and cost in assessing drought tolerance. Individual poplar species may lack one or more of these adaptations, but still be relatively drought tolerant. For example, the riparian poplar *P. trichocarpa* was resistant to cavitation but had a diminished ability to reduce g_s under drought (see section on 'Xylem transport'). Although further work is required to determine how various adaptations are "clustered," correlations, when established, can be used to produce linkage maps that would be a useful tool for tree breeding and for the selection of the most vigorous poplar clones at an early age. Further work is also required to interrelate water relations in process models to explain genetic differences in growth rate.

Acknowledgements

The writing of this chapter was supported by an Operating Grant (A-7815) to T.J. Blake. from the Natural Sciences and Engineering Research Council of Canada. It was also supported by the Biofuels Feedstock Development Program, U.S. Department of Energy, at Oak Ridge National Laboratory, managed by Lockheed Martin Energy Research Corp. for the U.S. Department of Energy under contract DE-ACO5-96OR22464, publication no. 4594, Environmental Sciences Division, Oak Ridge National Laboratory, Oak Ridge, TN 37831, USA.

References

Abou-Mandour, A.A., and Hartung, W. 1980. The effects of ABA on growth and development of intact roots seedlings, root and callus cultures and stem and root segments of *Phaseolus coccineus*. Z. Pflanzenphysiol. **100**: 25–30.

Alder, N. N., Sperry, J.S., and Pockman, W.T. 1995. Root and stem xylem embolism, stomatal conductance, and leaf turgor in *Acer grandidentatum* populations growing along a soil moisture gradient. Oecologia. In review.

Araujo, M.C., Pereira, J.S., and Pereira, H. 1989. Biomass production by *Eucalyptus globulus*: Effects of climate, mineral fertilisation and irrigation. *In* Proc of 5th E.C. Conf. Biomass for Energy and Industry, Lisbon, Portugal. *Edited by* G. Grassi, G. Gosse, and G. dos Santos. Elsevier Applied Science, Vol. 1: 446–462.

Belanger, R.R, Manion, P.D., and Griffin, D.H. 1990. Amino acid content of water-stressed plantlets of *Populus tremuloides* clones in relation to clonal susceptibility to *Hypoxylon mammatum in vitro*. Can. J. Bot. **68**: 26–29.

Blake, T.J., and Sutton, R.F. 1988. Stomatal responses, the key to adaptation in newly planted jack pine and black spruce. Plant Physiol. (Life Sci. Adv.) **7**: 125–130.

Blake, T.J., and Tschaplinski, T.J. 1992. Water relations. *In* Ecophysiology of short rotation forest crops. *Edited by* C.P. Mitchell, J.B. Ford-Robertson, T. Hinckley, and L. Sennerby-Forsse. Elsevier Applied Science, Essex, UK. pp. 66–94.

Blake, T.J., and Yeatman, C.W. 1989. Water relations, gas exchange and early growth rates of outcrossed and selfed *Pinus banksiana* (Lamb.) families. Can. J. Bot. **67**: 1618–1623.

Blake, T.J., Tschaplinski, T.J., and Eastham, A. 1984. Stomatal control of water use efficiency in poplar clones and hybrids. Can. J. Bot. **61**: 1344–1351.

Blake, T.J., Tan, W., and Abrams, S.R. 1990*a*. Antitranspirant action of abscisic acid and ten synthetic analogs in black spruce. Physiol. Plant. **80**: 365–370.

Blake, T.J., Bevilacqua, E., Hunt, G.A., and Abrams, S.R. 1990*b*. Effects of abscisic acid and its acetylenic alcohol on dormancy, root development and transpiration in three conifer species. Physiol. Plant. **80**: 371–378.

Blake, T.J., Fan, S., Darlington, A., and Halinska, A. 1996. Root to shoot communication and responses of conifer seedlings to soil and air drought. *In* Proceedings of Advances in Tree Development Control and Biotechnique, Inter. Union of For. Res. Organiz. Workshop, Sept. 15–20, 1993. Beijing, China. pp. 172–183.

Borchert, R. 1975. Endogenous shoot growth rhythms and indeterminable shoot growth in oak. Physiol. Plant. **35**: 152–157.

Boyer, J.S. 1989. Water potential and plant metabolism: Comments on Dr. P.J. Kramer's article: Changing concepts regarding plant water relations. Plant Cell Environ. **12**: 213–216.

Braatne, J.H., Hinckley, T.M., and Stettler, R.F. 1992. Influence of soil water on the physiological and morphological components of plant water balance in *Populus trichocarpa, Populus deltoides* and their F$_1$ hybrids. Tree Physiol. **11**: 325–339.

Ceulemans, R., Impens, I., Lemour, R., Moermans, R., and Samsuddin, Z. 1978. Comparative study of the transpiration regulation during water stress situations in four different clones. Ecol. Plant. **13**: 139–146.

Cochard, H., Cruziat, P., and Tyree, M.T. 1992. Use of positive pressures to establish vulnerability curves. Plant Physiol. **100**: 205–209.

Cohen, D.B., Dumbroff, E.B., and Webb, D.P. 1978. Seasonal patterns of abscisic acid in roots of *Acer saccharum*. Plant Sci. Lett. **11**: 35–39.

Crombie, D.S., Hopkins, H.F., and Milburn, J.A. 1985. Gas penetration of pit membranes in the xylem of *Rhododendron* as the cause of acoustically detectable sap cavitation. Aust. J. Plant Physiol. **12**: 445–453.

Dale, J.E., and Sutcliffe, J.F. 1986. Water relations. *In* Plant physiology. A treatise. Vol. IX. Water and solutes in plants. *Edited by* R.C. Steward, J.F. Sutcliffe and J.E. Dale. Academic Press, Toronto, ON. pp. 1–43.

Davies, W.J., Rhizopoulou, S., Sanderson, R., Taylor, G., Metcalfe, J.C., and Shang, J. 1989. Water relations and growth of roots and leaves of woody plants. *In* Biomass production by fast-growing trees. *Edited by* J.S. Pereira and J.J. Landsberg. Kluwer Academic Press Publ., Dordrect, The Netherlands. pp. 13–36.

Elias, T.S. 1980. Trees of North America. Van Nostrand Reinhold Company, New York, NY.

Fan, S., and Blake, T.J. 1994. Abscisic acid induced electrolyte leakage in woody species with contrasting ecological requirements. Physiol. Plant. **90**: 414–419.

Fan, S., Blake, T.J., and Blumwald, E. 1994. The relative contribution of elastic and osmotic adjustments to turgor maintenance of woody species. Physiol. Plant. **90**: 408–413.

Farquhar, G.D. 1989. Models of integrated photosynthesis of cells and leaves. Philos. Trans. R. Soc. Lond., Ser. B, **323**: 357–367.

Farquhar, G.D., O'Leary, M.H., and Berry, J. 1982. On the relationship between carbon isotope discrimination and intercellular carbon dioxide concentration in leaves. Aust. J. Plant Physiol. **9**: 121–137.

Farquhar, G.D., Ehrleringer, J.R., and Hubbick, K.T. 1989. Carbon isotope discrimination and photosynthesis. Annu. Rev. Plant Physiol. Plant Mol. Biol. **40**: 503–537.

Gallardo, M., Turner, N.C., and Ludwig, C. 1994 Water relations, gas exchange and abscisic acid content of *Lupinus consentii* leaves in response to different proportions of the rootsystem. J. Exp. Bot. **45**: 909–918.

Gebre, G.M., Kuhns, M.R., and Brandle, J.R. 1994. Organic solute accumulation and dehydration tolerance in three water-stressed Populus deltoides clones. Tree Physiol. **14**: 575–587.

Hacke, U., and Sauter, J. 1995. Vulnerability of xylem to embolism in relation to leaf water potential and stomatal conductance in *Fagus sylvatica purpurea* and *Populus balsamifera*. J. Exp. Bot. **46**: 1177–1183.

Hinckley, T.M., and Braatne, J.N. 1994 Stomata. *In* Plant and environment interactions. *Edited by* R.E. Wilkinson. Marceldekker Inc. New York, NY. pp. 325–355.

Hinckley, T.M., Ceulemans, R., Dunlap, J.M., Figliola, A., Heilman, P.E., Isebrands, J.G., Scarascia-Mugnozza, G., Schulte, P.J., Smit, B., Stettler, R.F., Van Volkenburgh, E., and Wiard, B.M. 1989. Physiological, morphological and anatomical components of hybrid vigor in *Populus*. *In* Structural and functional responses to environmental stresses. *Edited by* K.H. Kreeb, H. Richter, and T.M. Hinckley. Academic Publishing, The Hague, The Netherlands. pp. 199–217.

Hinckley, T.M., Richter, H., and Schulte, P.J. 1991. Water relations. *In* Physiology of trees. *Edited by* A.S. Raghacendra. Wiley, New York, NY. pp. 137–162.

Hinckley, T.M., Brooks, J.R., Cermak, J., Ceulemans, R., Kucera, J., Meinzer, F.C., and Roberts, J.D.A. 1994. Water flux in a hybrid poplar stand. Tree Physiol. **14**: 1005–1918.

Holbrook, N.M., Burns, M.J., and Field, C.B. 1995. Negative xylem balance pressures in plants: a test of the balance pressure technique. Science (Washington, DC), **270**: 1193–1194.

Iacobelli, A., and McCaughey, J.H. 1993. Stomatal conductance in a northern temperate deciduous forest: Temporal and spatial patterns. Can. J. For. Res. **23**: 245–252.

Jarbeau, J.A., Ewers, F.W., and Davis, S.D. 1995. The mechanism of water-stress induced embolism in two species of chaparral shrubs. Plant Cell Environ. **18**: 189–196.

Jones, H.G., and Sutherland, R.A. 1991. Stomatal control of xylem embolism. Plant Cell Environ. **14**: 607–612.

Kaufmann, M.R., and Troendle, C.A. 1981. The relationship of leaf area and foliage biomass to sapwood conducting area in four subalpine forest tree species. For. Sci. **27**: 477–482.

Kramer, P.J. 1988. Changing concepts regarding plant water relations. Plant Cell Environ. **11**: 565–568.

Liu, Z., and Dickmann, D. 1992. Abscisic acid accumulation in leaves of two contrasting hybrid poplar clones affected by nitrogen fertilization plus cyclic flooding and soil drying. Tree Physiol. **11**: 109–122.

Mallock, K.R.J., and Fenton, R. 1979. Inhibition of stomatal opening by analogues of abscisic acid. J. Exp. Bot. **30**: 1201–1209.

Morgan, J.M. 1984. Osmoregulation and water stress in higher plants. Annu. Rev. Plant Physiol. **35**: 299–319.

Munns, R. 1988. Why measure osmotic adjustment? Aust. J. Plant Physiol. **15**: 717–726.

Munns, R. 1990. Chemical signals moving from roots to shoots: the case against ABA. Br. Soc. Plant Growth Regul. Monogr. **21**: 175–183.

Pallardy, S.G., and Kozlowski, T.T. 1980. Cuticle development in the stomatal region of *Populus* clones. New Phytol. **85**: 363–368.

Pallardy, S.G., and Kozlowski, T.T. 1981. Water relations of *Populus* clones. Ecology, **62**:159–167.

Pallardy, S.G., Cermak, J., Ewers, F.W., Kaufmann, M.R., Parker, W.C., and Sperry, J.S. 1996. Water transport dynamics in trees and stands. *In* Resource physiology of conifers. *Edited by* W.K. Smith. Academic Press, New York, NY. In press.

Panshin, A.J., and de Zeeuw, C. 1980. Textbook of wood technology. 4th ed. McGraw Hill, New York, NY.

Pezeshki, R., and Hinckley, T.M. 1982. The stomatal response of red alder and black cottonwood to changing water status. Can. J. For. Res. **12**: 761–771.

Pezeshki, R., and Hinckley, T.M. 1988. Water relation characteristics of *Alnus rubra* and *Populus trichocarpa*: responses to field drought. Can. J. For. Res. **18**: 1159–1166.

Pickard, W.F. 1981. The ascent of sap in plants. Progr. Biophys. Mol. Biol. **37**: 181–229.

Pockman, W.T., Sperry, J.S., and O'Leary, J.W. 1995. Sustained and significant negative water pressure in xylem. Nature (London), **378**: 715–716.

Rieger, M. 1995. Offsetting effects of reduced root hydraulic conductivity and osmotic adjustment following drought. Tree Physiol. **15**: 379–385.

Scholander, P.F., Hammel, H.T., Bradstreet, E.D., and Hemmingsem, E.A. 1965. Sap pressure in vascular plants. Science (Washington, DC), **148**: 339–346.

Schulte, P.J., and Hinckley, T.M. 1987a. The relationship between guard cell water potential and the aperture of stomata of *Populus*. Plant Cell Environ. **10**: 313–318.

Schulte, P.J., and Hinckley, T.M. 1987b. Abscisic acid relations and the response of *Populus trichocarpa* stomata to leaf water potential. Tree Physiol. **3**: 103–113.

Schulte, P.J., Hinckley, T.M., and Stettler, R.F. 1987. Stomatal response of *Populus* to leaf water potential. Can. J. Bot. **65**: 255–260.

Schulze, E.D., Streudle, E., Gollant, R., and Schurr, E. 1988. Response to Dr. P.J. Kramer's article: Changing concepts regarding plant water relations. Plant Cell Environ. **11**: 573–576.

Sperry, J.S., and Saliendra, N.Z. 1994. Intra- and inter-plant variation in xylem cavitation in *Betula occidentalis*. Plant Cell Environ. **17**: 1233–1241.

Sperry, J.S., and Sullivan, J.E.M. 1992. Xylem embolism in response to freeze-thaw cycles and water stress in ring-porous, diffuse-porous, and coniferous species. Plant Physiol. **100**: 605–613.

Sperry, J.S., Perry, A., and Sullivan, J.E.M. 1991. Pit membrane degradation and air-embolism formation in aging xylem vessels of *Populus tremuloides* Michx. J. Exp. Bot. **42**: 1399–1406.

Sperry, J.S., Nichols, K.L., Sullivan, J.E.M., and Eastlack, S.E. 1994. Xylem embolism in ring-porous, diffuse-porous, and coniferous trees of northern Utah and interior Alaska. Ecology, **75**: 1736–1752.

Sperry, J.S., Saliendra, N.Z., Pockman, W.T, Cochard, H., Cruziat, P., and Tyree, M.T. 1996. New evidence for large negative xylem pressures and their measurement by the pressure chamber method. Plant Cell Environ. **19**: 427–436.

Tan, W., and Blake, T.J. 1993. Drought tolerance, abscisic acid and electrolyte leakage in fast- and slow-growing black spruce (*Picea mariana*) progenies. Physiol. Plant. **89**: 817–823.

Tardieu, F., and Davies, W.J. 1993. Root-shoot communication and whole plant regulation of water flux. *In* Water deficits — Plant responses from cell to community. Bios Scientific Publishers Ltd., Aransas Pass, TX.

Tschaplinski, T.J., and Blake, T.J. 1985. Effects of root restriction on growth correlations, water relations and senescence of alder seedlings. Physiol. Plant. **64**: 167–176.

Tschaplinski, T.J., and Blake, T.J. 1989*a*. Water relations and photosynthetic capacity as determinants of productivity in hybrid poplar cultivars. Can. J. Bot. **67**: 1689–1697.

Tschaplinski, T.J., and Blake, T.J. 1989*b*. Water-stress tolerance and late-season organic solute accumulation in hybrid poplar. Can. J. Bot. **67**: 1681–1688.

Tschaplinski, T.J., and Blake, T.J. 1989*c*. Correlation between early root production, carbohydrate metabolism and subsequent dry matter production in hybrid poplar. Can. J. Bot. **67**: 2168–2174.

Tschaplinski, T.J., and Tuskan, G.A. 1994. Water-stress tolerance of black cottonwood and eastern cottonwood clones and four of their hybrid progeny. II. Metabolites and inorganic ions that constitute osmotic adjustment. Can. J. For. Res. **24**: 681–687.

Tschaplinski, T.J., Tuskan, G.A., and Gunderson, C.A. 1994. Water-stress tolerance of black cottonwood and eastern cottonwood clones and four of their hybrid progeny. I. Growth, water relations and gas exchange. Can. J. For. Res. **24**: 346–371.

Tschaplinski, T.J., Gebre, G.M., Dahl, J.E., Roberts, G.T., and Tuskan, G.A. 1996. Growth and solute adjustment of calli of *Populus* clones cultured on nutrient media containing polyethylene glycol. Can. J. For. Res. **25**: 1425–1433.

Turner, N.C. 1979. Drought resistance and adaptation to water deficits in crop plants. *In* Stress physiology in crop plants. *Edited by* H. Mussell and R.C. Staples. Wiley-Interscience, New York, NY. pp. 181–194.

Turner, N.C. 1986. Adaptation to water deficits: A changing perspective. Aust. J. Plant Physiol. **13**: 175–190.

Tyree, M.T., and Jarvis, P.G. 1982. Water in tissues and cells. *In* Physiological plant ecology II. Water relations and carbon assimilation. *Edited by* O.L. Lange, P.S. Nobel, C.B. Osmond, and H. Zeigler. Springer-Verlag, Berlin. pp. 35–77.

Tyree, M.T., and Sperry, J.S. 1988. Do woody plants operate near the point of catastrophic xylem dysfunction caused by dynamic water stress? Answers from a model. Plant Physiol. **88**: 574–580.

Tyree, M.T., MacGregor, M.E., and Cameron, S.I. 1979. Physiological determinants of poplar growth under intensive *culture. In* Poplar research, management and utilization in Canada. *Edited by* D.C.F. Fayle, L. Zsuffa, and H.W. Anderson. Ont. Min. Natur. Resourc. Inform. Pap. No. 102.

Tyree, M.T., Alexander, J., and Machado, J.L. 1992. Loss of hydraulic conductivity due to water stress in intact juveniles of *Quercus rubra* and *Populus deltoides*. Tree Physiol. **10**: 411–415.

Tyree, M.T., Kolb, K.J., Rood, S.B., and Patino, S. 1994. Vulnerability to drought-induced cavitation of riparian cottonwoods in Alberta: a possible factor in the decline of the ecosystem? Tree Physiol. **14**: 455–466.

Wang, J., Ives, N.E., and Lechowicz, M.J. 1992. The relation of foliar phenology to xylem embolism in trees. Funct. Ecol. **6**: 469–475.

Waring, R.H., Gholz, H.L., Grier, C.C., and Plummer, M.L. 1977. Evaluating stem conducting tissue as an estimator of leaf area in four woody angiosperms. Can. J. Bot. **55**: 1474–1477.

Zeevart, J.A.D., and Creelman, R.A. 1988. Metabolism and physiology of abscisic acid. Annu. Rev. Plant Physiol. Plant Mol. Biol. **39**: 439–473.

Zhang, J., and Davies, W.J. 1990 Changes in the concentration of ABA in xylem sap as a fuction of changing soil water status, can account for changes in leaf conductance and growth. Plant Cell Environ. **13**: 227–285.

Zimmermann, M.H. 1978. Hydraulic architecture of some diffuse-porous trees. Can. J. Bot. **56**: 2286–2295.

Zimmermann, M.H. 1983. Xylem structure and the ascent of sap. Springer-Verlag, New York, NY. pp. 169–216.

Zimmermann, M.H, and Jeje, A.A. 1981. Vessel-length distribution in some American woody plants. Can. J. Bot. **59**: 1882–1892.

Zimmermann, U., Meinzer, F.C., Zhu, J.J., Schneider, H., Goldstein, G., Kuchenbrod, E., and Haase, A. 1994. Xylem water transport: is the available evidence consistent with the cohension theory? Plant Cell Environ. **17**: 1169–1181.

CHAPTER 17
Stress physiology — abiotic

Dawn S. Neuman, Michael Wagner,
Jeffrey H. Braatne, and Jon Howe

Introduction

Poplars display well-developed adaptations to environmental extremes. Yet, there has been no complete characterization of how any one stress factor influences leaf growth, how biomass is accumulated, or how roots function. As a result, we draw our information from several areas of stress physiology to describe the physiological responses of poplars to environmental stress from the molecular to the whole plant levels of organization. The field of stress physiology is concerned with not only understanding how growth is regulated, but also how growth responds to changes in the environment, and how final productivity is affected. In the last two decades, the field of stress physiology has grown beyond descriptive experiments and has started to approach an understanding of the underlying mechanisms of how plants respond to the environment. Here, we will attempt to detail many of the important studies that have provided a picture of how poplars respond to changes in their environment.

This chapter will focus on the responses and resistance mechanisms used by poplars for surviving a variety of environmental stresses such as flooding, salinity, cold, and atmospheric pollutants (specifically ozone). Specific topics addressed in this chapter relate to the historical focus of many researchers involved in attempting to understand poplar physiology. The response of poplars to decreased water availability can be found in Chapter 16 (Blake, Sperry, Tschaplinski, and Wang) and will not be addressed in this chapter, although mechanisms of drought response are clearly integrated with other abiotic stresses such

D.S. Neuman. Biological Sciences, University of Nevada, 4505 Maryland Parkway, Las Vegas, NV 89154-4004, USA.
M. Wagner and J. Howe. School of Forestry, Northern Arizona University, P.O. Box 15018, Flagstaff, AZ 86011, USA.
J.H. Braatne. College of Forest Resources, University of Washington, Seattle, WA 98195, USA.
Correct citation: Neuman, D.S., Wagner, M., Braatne, J.H., and Howe, J. 1996. Stress physiology — abiotic. *In* Biology of *Populus* and its implications for management and conservation. Part II, Chapter 17. *Edited by* R.F. Stettler, H.D. Bradshaw, Jr., P.E. Heilman, and T.M. Hinckley. NRC Research Press, National Research Council of Canada, Ottawa, ON. pp. 423–458.

as salinity and cold. In this chapter, we examine the following topics: flooding stress on poplars; the effects of salinity on poplar physiology; responses to cold and hardening of plant tissue; the specific effects of ozone on the physiology and growth of poplars; and some perspectives on the direction for future research for increased understanding of poplar responses to a changing environment.

Flooding stress

Plants that are not adapted to wet or flooded soil conditions, soil flooding reduces shoot and root growth. This results from the disruption of many physiological functions of the plant caused by inadequate oxygen supply to roots in flooded soils. Soils at field capacity have 10–30% air-filled spaces. Percentages of air-filled spaces are decreased as they fill with water resulting in an anaerobic or hypoxic root environment. Consequently, oxygen availability to submerged plant tissues can be very low. An inadequate oxygen supply promotes fermentation. When aerobic activity in root tissues is decreased, the production of energy producing compounds also becomes limited because mitochondrial respiration approaches zero. Although some anaerobic respiration may occur in waterlogged roots, the decrease in energy (Davies et al. 1987; Kennedy et al. 1992) when oxygen is limited is most likely insufficient to maintain physiological process without special adaptations, such as those found in rice (Setter et al. 1987; reviewed by Jackson and Pearce 1991).

Although it is not known exactly how decreasing oxygen availability results in cellular death, it has been reported that death results from cytoplasmic acidification. The ability of cells to pump protons out of the cytoplasm becomes impaired and ATP production is reduced in oxygen deficient environments (Roberts et al. 1985). In addition, root tips of flooding-susceptible species accumulate lactic acid, which also may increase cytosolic acidification (Roberts et al. 1984; Xia and Saglio 1992). Another possible explanation for cellular death in oxygen deficient environments may be due to the increase in ethanol concentration within tissues brought about by fermentation processes (Jackson et al. 1982). This explanation has been questioned however, because ethanol may not be as toxic to plant tissues as suggested by early reports (Kennedy et al. 1992; Ricard et al. 1994).

Finally, although extensive research has been devoted to the study of plant metabolism under oxygen deficits, it is becoming clear there is a considerable amount of metabolic flexibility concerning the responses of plants to "flooding stress." Reports range from classical views of stress responses detailing changes in glycolytic pathways and the mitochondrial TCA cycle, to attempts to integrate changes in cytoplasmic pH to translations of specific mRNAs and descriptions of altered polypeptide profiles in tissues exposed to low concentrations of oxygen. For very fine reviews of this topic see Kennedy et al. (1992) and Armstrong et al. (1994).

Physiological responses to flooding in poplars

Several studies have shown that woody plants have a wide range of metabolic responses to reduced oxygen in the root zone (Kozlowski 1984). Among the general symptoms associated with flooding are: (*i*) yellowed leaves, (*ii*) leaf epinasty, (*iii*) formation of adventitious roots, and (*iv*) wilting in some plants (reviewed in Kozlowski 1984). Poplars are generally considered to be fairly tolerant of excess soil moisture (Harrington 1987), although lack of soil aeration will reduce growth in some species (Smit et al. 1989; Lui and Dickmannn 1992*a*). Unlike drought stress, flooding slows root growth (Lui and Dickmannn 1992*a*). In addition, flooding also causes decreased uptake of nutrients (Harrington 1987). Some poplars are more flood tolerant than others and there are discrepancies in the literature between closely related clones. For example, Harrington (1987) has shown that 20 d of flooding did not affect the biomass accumulation of black cottonwood. This observation is in opposition to a report by Smit et al. (1989) where flooding was observed to reduce biomass in the same species. Most likely, the discrepancy results from extensive variability in stress responses found among the many poplar clones.

Contrasting responses to flooding suggest that some clones may function as distinct hybrid genotypes. Lui and Dickmann (1992*a*, 1993) made extensive comparisons of two hybrid clones (*P.* ×*euramericana* Eugenei, and *P. tristis* × *P. balsamifera* Tristis) under flooding and soil drying. In their 1992*b* study, repeated flood stress was imposed on Eugenei and Tristis under two nitrogen levels. Flooding led to smaller leaves in Tristis while only a small response was found in Eugenei. Differences between these hybrids were not observed when other physiological responses were assessed. In their 1993 study, gas exchange variables were measured over 18 d of flooding. Neither of the clones displayed any reduction in photosynthesis during the first few days, although photosynthesis and stomatal conductance decreased after 18 d of flooding. Interestingly, these responses can be ameliorated by the addition of nitrogen. Supplemental N enabled plants to resist the negative effects of flooding on both photosynthesis and conductance (Lui and Dickmann 1993). Some studies have shown that flooding causes a decline in the concentration of foliar N in many woody species (see Kozlowski 1984).

Why do stomata close in leaves of flooded poplars?

In many species stomatal closure occurs within hours of root inundation (reviewed in Kozlowski 1984). Flooding causes symptoms of water stress in some species (Kramer 1951; Kozlowski and Pallardy 1984) but not in others (Regehr et al. 1975; Pereira and Kozlowski 1977). In some species, this stomatal closure has been shown to occur in spite of adequate turgor (Zhang and Davies 1986), although measurements of bulk leaf water potential may not be sensitive enough to detect small changes in water status.

The plant hormone, abscisic acid (ABA), has been shown to have a role in stomatal closure and has been suggested as an important hormonal regulator during flooding (Zhang and Davies 1987). Wright and Hiron (1972) first reported that flooding, like soil drying, promoted ABA accumulation in leaves of *Phaseolus vulgaris* after several days of waterlogging; similar results have been found in other species (Zhang and Davies 1987). The source of increased ABA in flooded plants is unclear. As with soil drying, several investigators have suggested that ABA might be supplied to leaves from oxygen-deficient roots through the xylem (Davison and Young 1973; Walton et al. 1976; Cornish and Zeevart 1985). Davies et al. (1987) found that ABA levels in roots increased within 24 h of waterlogging, while ABA levels in leaves did not increase until 36 h. Endogenous levels of bulk leaf ABA however, were poorly correlated with the stomatal conductance, which decreased at least 20 h before ABA levels were increased. Although these results do not point to a role for ABA in stomatal closure during flooding, several authors have suggested that measurements of bulk leaf ABA might not represent changes in the ABA concentration at the stomata.

There are several studies showing that flooding causes stomatal closure in poplars (Fig. 1), although the timing and degree of closure depends upon the length of the flooding event and clonal variety. For example, Regehr et al. (1975) found that inundation of the roots reduced stomatal conductance and photosynthesis by 50% in a clone of *P. deltoides* Marsch. Alternatively, Harrington (1987) found that flooding initially decreased leaf conductance in *P. trichocarpa* Torr. and Gray, but stomata did not stay closed. Historically, flood induced stomatal closure has been attributed to a decrease in root hydraulic conductance, followed by decreasing leaf water potential, which in turn, causes the stomatal closure associated with flooding. Although, flooding appears to cause a reduction in root hydraulic conductivity in poplars, there is little indication of any significant associated decrease in leaf water potential (Harrington 1987; Smit and Stachowiak 1988; Smit et al. 1990; Lui and Dickmann 1992a). Specifically, the mechanism by which flooding causes stomatal closure is still unclear. Under nonflooded conditions, changes in leaf water potential of poplars are correlated with leaf conductance (Schulte et al. 1986), but this does not appear to be the case during flooding. Instead, flooding either has no effect on leaf water potential or actually increases leaf water potential (Smit et al. 1990). The latter would suggest that stomatal closure acts to maintain water balance in leaves of plants with flooded roots.

Because stomatal closure is not always associated with changes in leaf water status, it has been suggested that stomata may be closing in response to changes in the concentration or flow of some chemical messenger that is moving between roots and shoots (Davies et al. 1987). Different kinds of stresses in the root zone may affect the biosynthesis of hormones from roots to shoots and influence growth. Drought, temperature, waterlogging, and flooding have been shown to

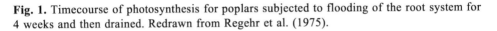

Fig. 1. Timecourse of photosynthesis for poplars subjected to flooding of the root system for 4 weeks and then drained. Redrawn from Regehr et al. (1975).

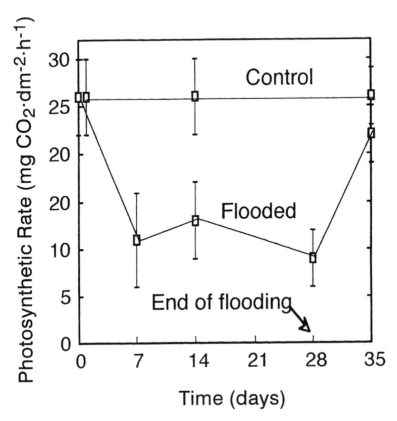

change the transport of root-sourced hormones to the shoots (Itai and Vaadia 1965; Skene and Kerridge 1967; Burrows and Carr 1969; Skene 1975). Railton and Reid (1973) found that applications of cytokinins can substitute for roots in inducing shoot growth, and that applications of cytokinins can be used to mitigate a stressful root environment. In addition, correlations can be made between leaf physiology and hormone content in the sap or root (Itai and Vaadia 1965; Zhang and Davies 1987). Therefore, hormones are likely candidates to exert some regulatory function in root–shoot interactions.

Based on the idea that roots and leaves might be connected by the flow of plant hormones, several suggestions have been given to explain how flooding could cause stomatal closure (Bradford and Hsiao 1982; Jackson et al. 1988). For example, it has been suggested: (*i*) that flooded roots export factor(s) such as ABA to leaves that bring about stomatal closure, or (*ii*) that flooded roots fail to export some factor(s) such as a cytokinin required to maintain stomatal opening, or (*iii*) that flooding causes a reduction in phloem transport down to roots which results in the build-up of some factor such as ABA leading to stomatal closure. Neuman and Smit (1990) and Smit et al. (1990) have examined the first two cases using a *P. trichocarpa* × *deltoides* hybrid. To test the first possibility, they measured the concentration of ABA in leaves and xylem sap of plants in which the root system was made hypoxic by gassing with 100% nitrogen gas in hydroponic cultures. Although leaf ABA concentration increased two-fold early in the experiment, concentrations of ABA in leaves returned to control values by the second day, in spite of the fact that stomata remained closed. This is consistent with the work of Lui and Dickmann (1992*b*) where long-term flooding (weeks) did not cause any significant increase in leaf ABA concentration. While there appears to be a role for ABA in regulating the aperture of stomata in poplars exposed to drought, this does not appear to be the mechanism by which stomatal closure occurs during flooding. Nevertheless, small changes in ABA that might be enough to trigger the closure of stomata may not be detected in measurements of bulk leaves.

Neuman and Smit (1990) also tested the second hypothesis which suggests that flooded roots fail to export cytokinins to leaves. In this study, they measured the concentration of zeatin riboside (ZR), a transported form of cytokinin which has been reported to counteract the effects of ABA-induced stomatal closure in some plants. They found that although root hypoxia does reduce the fluxes of ZR coming out of roots, there were no differences in bulk leaf concentrations between leaves of poplars with aerated and hypoxic roots. In addition, applications of cytokinins do not reopen stomata closed by either root hypoxia or applied ABA (D.S. Neuman, unpublished).

From these studies, we conclude that additional information is required on the underlying mechanisms by which hormones mediate responses to the environment. Specific information on the site and mechanism of cytokinin action, and the ways in which cytokinins are compartmentalized within plant cells will be required to understand the physiological significance of reduced cytokinin transport in the transpiration stream in flooded poplars.

Flooding and leaf growth of poplars

In addition to stomatal closure, flooding can also have a rapid and profound effect upon leaf expansion. To understand how flooding slows leaf growth, we need to review how leaves grow (see Chapter 12, Van Volkenburgh and Taylor, for a complete presentation). Essentially, there are two components controlling enlargement of leaf cells: water uptake and wall metabolism. Both of these controlling aspects can be under metabolic regulation. The hydraulic aspect of water uptake during cellular growth has been defined by the following relationship, $(1/V)\,dV/dt = L\,(\sigma\,\Delta\pi - P)$ (reviewed by Cosgrove 1986) where $(1/V)\,dV/dt$ is the rate of volume increase, L is cellular hydraulic conductance, σ is the solute reflection coefficient, $\Delta\pi$ is the difference in osmotic pressure between the cell and its surroundings, and P is the turgor pressure. For growth to occur, cells must take up water. The driving force for water uptake increases when $\Delta\pi$ increases or when P decreases (but still remains above the threshold required for growth). In intact plants, water and solute transport may be important growth-controlling factors. Water potential gradients between growing tissues and the xylem water supply suggest that growth could be affected by hydraulic conductivity (Michelena and Boyer 1982; Cosgrove and Cleland 1983). The capacity for water transport of the cells near the xylem probably determines water flux to nearby expanding cells.

The examination of water relations during leaf expansion provides evidence for how closely leaf growth is integrated into whole plant function. Leaf expansion is highly dependent on water uptake. Hsiao and Jing (1987) found that the rate of elongation of growing maize leaves almost stopped if the water potential of the solution around the roots was lowered by as little as 0.2 MPa. This effect could be reversed if the water potential was raised to the original value. Large decreases in water availability may reduce cell enlargement due to the reduction in turgor which is inadequate to maintain stretching and growth of the cell walls. Although water availability can clearly influence leaf growth, there are also cases in which leaf growth is slowed in spite of sufficient turgor. For example, the leaves of bean and well-watered sunflower have been shown to grow faster in the day than at night even though turgor was lower during the day (Davies and Van Volkenburgh 1983; Matthews et al. 1984). This suggests that growth is regulated through more than one biophysical or biochemical pathway.

Another way growth may be controlled is through cell wall properties. The rate of growth has been described as a function of the turgor pressure (P), wall extensibility (m) and wall yield threshold (Y) (e.g., turgor pressure sufficient to cause irreversible wall effects): $dV/dt = m(P - Y)$ (Lockhart 1965; Van Volkenburgh and Cleland 1984; Ray 1987). In some cases, extensometric measurements of cell walls show that growth is related to the ability of cells to extend (Cleland 1987; Van Volkenburgh 1987).

Smit et al. (1989) detailed the effects of root hypoxia on leaf growth of a *P. trichocarpa* × *deltoides* hybrid. In this study, plants were subjected to hypoxic root conditions in solution culture by gassing with nitrogen. Root hypoxia caused the rate of leaf expansion to decrease within 8 h and was generally detectable within 1 h. Leaf elongation was suppressed for the duration of the treatment and final leaf size was reduced by 35–60% compared to plants with well-aerated roots. Epidermal cell size and number were found in this study to be reduced, depending upon the developmental stage of the leaf at the start of the stress, and on the duration of the treatment. Like other studies using various poplar clones, no differences in bulk leaf water potential were measured between the plants with hypoxic and aerated roots. Estimates of turgor potential were higher in leaves of plants with hypoxia-treated roots than the turgor of leaves of plants with aerated roots.

So, why do leaves of poplars with flooded roots stop growing? Apparently reduced leaf growth may be due to changes in the nature of cell walls. Smit et al. (1989) found that cell wall extensibility was lower in leaves of plants with hypoxia-treated roots than in leaves of plants with aerated roots. These data suggest that leaf growth of hypoxia-stressed plants is limited by cell wall extensibility. Extensibility is a poorly understood property of plant cell walls and the mechanism by which the root stress induces changes in leaf cell wall characteristics is not known. The complexity of cell walls (Biggs and Fry 1987; Carpita 1987; Frost and Taylor 1990) and the variety of possible loosening mechanisms (Cleland 1987) suggest multiple ways in which walls are regulated.

Root-to-shoot messages and changes in gene expression during flooding

There is evidence that flooding of the root system induces changes in gene expression. Neuman and Smit (1993) found substantial changes in the pattern of translatable mRNAs in poplar leaves during root hypoxia. They used the pattern of *in vitro* translatable mRNAs from leaves on plants with oxygen-deprived (hypoxic) roots as a molecular phenotype to test for the action and specificity of putative root-to-shoot signal molecules. In their study, root hypoxia caused a decrease in the abundance of a group of leaf mRNAs which translated to low molecular weight (LMW) proteins (12–28 kDa). Root excision induced a similar reduction in a subset of the same LMW leaf mRNAs. This suggests that the effects of root hypoxia on leaves are physiologically related to the effects of root excision. Feeding ABA to leaves did not mimic the effects of root aeration or hypoxia. Applications of a cytokinin known to be the transportable species (zeatin riboside, ZR), to leaves of plants with hypoxic roots duplicated the effects of root aeration on the LMW translation products. The specificity of the leaf response to ZR was tested by feeding KCl and sucrose to leaves of root excised plants. Applications of either KCl or sucrose also increased the abundance of the LMW translation products. Although the results reported in this paper are consistent with the concept that shoot responses to root stress may be modulated by substances moving in the xylem, a more complete understanding of how hormones control cellular responses needs to be reached before the mechanism of root–shoot integration can be understood.

Summary

The reduced ability of poplars to grow when subjected to hypoxic conditions has implications for plant productivity. Available information indicates that in plants with oxygen-deficient roots, leaf cell expansion and stomatal closure are limited by hydraulic and chemical factors. Initial effects on leaf growth and stomatal conductance may be due to very small changes in local turgor.

Subsequent events in leaves of plants with oxygen-deficient roots, however, may be metabolically regulated. These events could be regulated by one or more of the plant hormones and could involve changes in wall synthesis and/or cell wall stiffening. Over the long term, the initial slowing of growth might be transduced in some way into modifications of wall properties, such as wall extensibility or changes in wall chemistry. This effect may be responsible for reductions in leaf growth if exposure to the root hypoxia is prolonged (Smit et al. 1989). The relative importance of how much extensibility limits cell enlargement in comparison with other factors during episodes of root hypoxia is difficult to determine.

Salinity stress

Irrigation often improves biomass production of poplars under many climactic conditions (Rawitz et al. 1966; Hansen 1978, 1983, 1988). In arid regions, however, sustainable irrigation systems are difficult to maintain because waterborne salts tend to accumulate in rooting zones (Tanji 1990). Salt accumulations can have strong negative effects on soil structure (Shainberg and Letey 1984; Goldberg et al. 1988), resulting in soils with reduced water infiltration, poor aeration, poor water retention, and increased resistance to root expansion. Like flooding increased salinity can also impact the overall physiology of many woody species, including poplars. For example, salinity has been shown to alter plant water relations and enzyme activity, as well as nutritional and hormonal physiology (Pearson 1960; Bernstein 1975; Casey 1972; Maas 1986; Rengel 1992; Catalan et al. 1994). Though the growth of *Populus* spp. generally benefits from supplemental water, this genus is also more sensitive to elevated salinity than many commonly irrigated tree species (Firmin 1968).

Salinity is a term used to describe the total concentration of dissolved mineral salts in water or soil. For soil structure and plant nutrition, the most important salts are cations such as sodium, calcium, magnesium, boron, and potassium or anions such as chlorine, sulfate, carbonate, and nitrate (Tanji 1990). Of these, calcium, magnesium, sodium, sulfate, and chlorine are most often studied. Soils that are characterized by high concentrations of sodium salts are known as sodic, while soils that have high relative concentrations of other salts are known as saline. Salinity in the soil is typically quantified by electroconductivity (EC; deciseimens/metre). Electroconductivity is a measure of the ability of a soil saturation extract to conduct electricity and is correlated with the total number of ions or salts in the soil solution, although EC does not quantify the relative concentrations of specific salts. Where sodium is the problem, exchangeable sodium percentage (ESP; percent sodium ions in solution) may also be measured because of the unique effects of sodium on soil structure and plant physiology (USDA 1954).

The presence of mixtures of different salts often results in specific interactions between ions. For example, although high concentrations of calcium can be very toxic to cells, several studies have shown that excess calcium can lessen the negative effects of excess sodium on root and shoot development (Cramer et al. 1986; Nakamura et al. 1990; Subbarao et al. 1990; Ilyas et al. 1993). Calcium improves the uptake of potassium (Cramer et al. 1987; Nakamura et al. 1990), phosphate (Cramer et al. 1986), and other nutrients in sodic soils (Burgos et al. 1993) and is important for maintaining cell integrity. The presence of calcium in sodic soils may act to prevent potassium leakage from tissues by reducing sodium binding to cell walls and plasma membranes (Cramer et al. 1985, 1986, 1987, 1990; Rengel 1992). This idea is supported by the observation that salt-tolerant genotypes are more positively affected than salt-sensitive genotypes by the mitigating effects of calcium (Subbarao et al. 1990).

Neither EC nor ESP alone provides a complete picture of the effects of salinity on plant productivity. Interactions between soil type and sodium concentrations (ESP) relative to other salts (EC) have long been known to be important factors in determining soil water potential and structural stability (Frenkel et al. 1977; Shainberg et al. 1980; Levy et al. 1988). In addition to the known effects of sodium on the biology of plants, other ions, especially calcium, have been shown to interact in sodic soils to influence plant growth, specifically seed germination and root growth (Ben-Hayyim et al. 1987; Cramer et al. 1987; Greive and Maas 1988; Nakamura et al. 1990; Colorado et al. 1991). Although many ionic inter-actions may occur in particular soils to limit plant growth, tolerance to salinity is still most often reported in reference to the broad measure, EC.

Salinity impacts

High salinity often limits plant growth by inducing water stress through a decrease in the osmotic potential of soil solutions (Casey 1972; Bernstein 1975; Rhoades 1989; Tanji 1990; Rengel 1992). Even at field capacity, salt-sensitive plants may become water stressed in saline environments (Staples and Toenni-essen 1984; Armitage 1985). Although, the osmotic component of water stress is not strongly dependent on the relative concentrations of particular ions, a high relative concentration of sodium ions weakens soil structure and increases soil particle dispersion (Frenkel et al. 1977; Goldberg et al. 1988). This has the effect of reducing water infiltration and gas exchange within the soil (Shainberg and Letey 1984). As a result, the viability of even the most salt-tolerant plants may be reduced by the presence of high relative concentrations of sodium (Mass 1986).

Sodium can also negatively affect growth independently of soil effects. High relative concentrations of sodium can disturb the nutrient balance in roots such that relative concentrations of important nutrients (phosphorous, potassium, calcium, and magnesium) are decreased in plant tissues (Cramer et al. 1986; Naka-mura et al. 1990; Garcia and Charbaji 1993). In addition, the transport of increased concentrations of sodium from roots to leaves has been shown to decrease chlorophyll concentration and stomatal conductance. As a result, net photosynthesis is reduced (Greenway 1962; Morgan 1984; Weinberg 1988; Nieves et al. 1991; Garcia and Charbaji 1993). Lack of photosynthates has a strong effect on roots. Although root-to-shoot ratio (Greenway 1962; Cramer et al. 1986) and root density are increased (Alva and Syvertsen 1991) in sodic soils, overall root length declines (Cramer et al. 1986; Nakamura et al. 1990; Silberbush and Lips 1991), as does root hydraulic conductivity (Joly 1989).

Biological responses to salinity tend to be species specific. A common approach to studying salt tolerance in plants is to use tissue cultures. This technique has limitations because salt tolerance is a whole-plant response and individual cells do not serve as good models (Miles 1991). The paucity of information about responses of particular poplar genotypes to increasing soil salinity makes generalizations difficult.

Salt tolerance of plants

Most plants have evolved some mechanisms to cope with excess of salts in the rooting environment; although the mechanisms by which tolerance is achieved are species specific and not clearly understood at this time. If increases in salinity are endured without large losses in productivity, then plants are considered to be salt tolerant. Deciduous fruits, nuts, citrus, and avocado exhibit sodium toxicity symptoms at ESP levels of 2–10% and an EC of around 4 dS·m^{-1}, while clover, oats, and rice do not experience similar symptoms until an ESP of 20–40% and an EC of 8–10 dS·m^{-1}. Wheat, cotton, and beets can tolerate ESP levels above 50% and an EC of 10–16 dS·m^{-1} (Withers and Vipond 1980). *Acacia* spp. and *Eucalyptus* spp. can be grown without severe declines in growth in the 10–25 dS·m^{-1} range. Well-known salt-tolerant plants, like *Tamarix* spp., *Casuarina* spp., and *Prosopis* spp., can maintain reasonable growth rates under conditions with an EC above 35 dS·m^{-1}. Poplars are much less salt tolerant and show the best growth when the EC is in the 1–5 dS·m^{-1} range (Firmin 1968; Armitage 1985) (Fig. 2). Adams et al. (1979) ranked several species of *Populus* for salt tolerance as follows: *P. euphratica* (5 dS·m^{-1}), *P. bolleana* (4.5 dS·m^{-1}), *P. oblega* (2 dS·m^{-1}), *P. ×euramericana* (1.0 dS·m^{-1}), *P. thevestina* (1.0 dS·m^{-1}). *P. diversifolia* is reported to be an indicator of saline conditions in the major Asian deserts (e.g., Gobi, Mongolia, Takla-makam, Tibet; and Turkestan, Russia). As a consequence of the considerable variability of salinity tolerances within *Populus*, there may be opportunities for breeding improvement.

The ability to tolerate drought also plays a role in how sensitive a specific species might be to increasing salinity. Species that are tolerant of drought tend to also be more tolerant of saline conditions. Poplars are relatively sensitive to

Fig. 2. Relative crop yield as a function of soil salinity for woody plants of different salt tolerance. Modified from Maas (1986), Armitage (1985), and Firmin (1968).

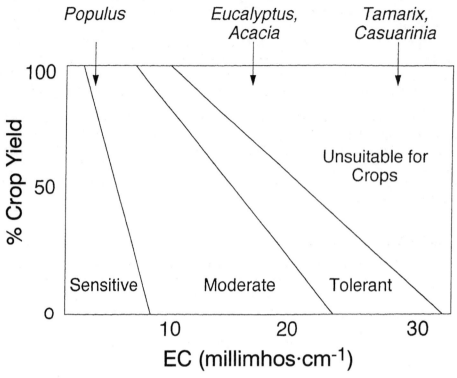

drought and grow best at soil moistures around field capacity (Hansen 1988, 1983). Productivity is fairly linear with increased water content of the soil. Optimal growth appears to occur between soil water potentials of around -0.07 to -0.03 mPa (Rawitz et al. 1966; Hansen 1988).

The osmotic component of water potential is linearly proportional to the EC of the soil solution over the range at which plants typically grow. For example if the EC of the soil solution increases from 1 to 3 dS·m^{-1} the osmotic potential will decrease from -0.3 to -1 MPa (USDA 1954). Increased salinity, therefore, will negatively affect the growth of poplars even if the soil water content is nearly optimal for growth. Continuous irrigation with wastewaters can exacerbate this problem. Juwarkar and Subrahmanyan (1987) found that the EC of the soil solution can easily increase 4 fold within 3 years in saline/sodic wastewater irrigation systems.

In the unique situation where wastewater disposal is used in an irrigation system, flood tolerant poplars may have an initial advantage over more drought tolerant species (Hansen et al. 1980). Often in these situations, wastewater application rates and consumption are maximized and soil moisture is around field capacity. The initial high growth rates enjoyed by poplars in these situations, however, may be short-lived as salts accumulate over time. Saline conditions which develop may reduce initial high growth rates. Also, as sodium accumulates, nutrient imbalances and toxic effects may begin to occur. As woody plants age and sapwood is converted to heartwood, sodium sequestered in living tissue may be released, thereby increasing internal sodium concentrations in other living tissues which may cause leaf burn (Maas 1986) or interfere with plant enzyme systems (Bernstein 1975; Davis and Jaworski 1979).

Salinity impacts on poplars

For poplars in general, salinity has been well-documented to inhibit growth (Kearney and Scofield 1936; Butijn 1954; Dimitri 1973; Sucoff 1975). Little information, however, is available on the specifics of poplar physiology during exposure to high salinity. Not surprisingly, there is some evidence that clonal variability exists with respect to salinity sensitivity. For example, in a study of growth and survival among 32 tree species (including 6 genotypes of poplar), irrigated with pulp and wastewater (EC = 3.21 mmhos·cm^{-2}), Aw and Wagner (1993) found that even within specific hybrids (*P. deltoides* × *P. nigra*) substantial variation in salinity tolerance exists. In addition, response variability of poplars to salt stress may depend upon prior growth conditions. For example, Luo and Zhou (1991) found that poplar clones irrigated with salinized waters during their first year of growth had higher second year survival than controls when exposed to subsequent high salinity. This study points to the increased ability of preconditioned plants to survive exposure to increased salinity.

From the perspective of organismal physiology, poplars appear to respond similarly to herbaceous species when exposed to increasing salinity. For example, Bray et al. (1991) used small plants of *P. trichocarpa* × *deltoides* to study the effects of NaCl treatments on organic solutes and growth. Plants were grown on gelled Murashige and Skoog's medium supplemented with varied amounts of NaCl to simulate different levels of salinity stress. Bray and colleagues found that increasing the NaCl around roots results in leaf necrosis, growth inhibition, and a general decrease in relative water content, with the youngest tissues being most affected. The formation of new roots was halved at the lowest salt concentration (40 mM) and mostly stopped at the highest salt concentrations (200 mM). Furthermore, poplars did not accumulate glycine betaine and/or proline, which is a common response to salinity. Trigonelline, a methylated form of nicotinic acid, however, increased in the plants grown in 100 mM NaCl. The role of this compound in the salinity responses is unclear and the increase observed in the Bray et al. (1991) study may have been due to the presence of nicotinic acid used in their *in vitro* system.

Inherent variability in sodium uptake by poplars

The ability of plants to cope with excessive ions also involves the capacity to exclude ions or to compartmentalize them in selected tissues. Among poplars, different genotypes have different inherent abilities to take up salts without adverse effects on growth. Aw (1994) found substantial variation in sodium uptake among seven poplar genotypes ranging from 0.15 g·kg^{-1} for *P. alba* × *P. grandidentata* to 3.48 g·kg^{-1} for *P.* × *euramericana* Robusta. In addition, though two genotypes may remove similar amounts of sodium from the soil, there is variability in how much is then translocated from the root to aerial tissues (Smith et al. 1978) and in which tissues incorporation of the element ultimately occurs. For instance, *P. deltoides* Bartr. ex Marsh clone ANV 70/51 under municipal wastewater irrigation retains 17% of its sodium uptake in roots, 10% in stem wood, and 13% in leaves. Under the same conditions, *P. deltoides* × *P. nigra* L. ANV 65/31 clone retains 31% of its sodium uptake in roots, 9% in stem wood, and 2% in leaves (Stewart et al. 1990).

Among the plants that have the ability to store ions in permanent tissues are *Eucalyptus grandis*, *E. saligna*, *Phalaris aquatica*, and some species of corn. Other plants such as *Tamarix* and *Atriplex* do not store ions in permanent tissues, but instead ions are transported to salt glands on leaf surfaces where they are temporarily stored and are not harmful to the physiology of the plant. Some of the more salt-sensitive plants such as poplars, *Pinus radiata*, and *Medicago sativa* exclude ions from roots; although, there is great interspecific variation in this regard (Greenway and Munns 1980; Sherrell 1984; Staples and Toenniessen 1984; Mills et al. 1985; Binzel et al. 1988; Hopmans et al. 1990; Stewart et al. 1990). Not only is interspecific variation common, there is also variation within species (Greenway 1962; Kingsbury and Epstein 1986).

Mechanisms that plants employ to acquire important nutrients from the soil are selective, resulting in the uptake of nutritionally important cations such as potassium, calcium, and magnesium (Epstein 1961; Davis and Jawarski 1979; Greenway and Munns 1980; Cramer et al. 1985; Kingsbury and Epstein 1986; Schatchtman et al. 1991). Because sodium is excluded, this results in low relative concentrations of sodium in plant tissues when compared to soil concentrations (Davis and Jawarski 1979). Plants may even actively exude sodium back into the soil after it has been absorbed (Ratner and Jacoby 1976; Davis and Jawarski 1979; Mills et al. 1985). The exclusion of sodium by roots concentrates sodium in the rooting zone and increases water stress (Howe and Wagner 1995).

Some plants adjust to increases in soil salinity and sodicity through a process called osmotic adjustment in which solutes are concentrated in tissues, thereby reducing internal water potentials. If saline conditions develop gradually, osmotic adjustment reduces, but does not eliminate, the negative effects on plant–soil water relations generally induced by saline and sodic conditions (Hayward and Spurr 1943; Bernstein 1961; Cooper and Dumbroff 1973).

Mitigating salinity

Leaching sodium and salts out of the rooting zone is one of the most common and effective methods for controlling salinity and sodicity in irrigated agriculture (O'Conner 1972a, b; Miller and Scifres 1988; Armstrong and Tanton 1992; Ilyas et al. 1993). Leaching, water infiltration, and soil structure can be improved in a wide range of soils by applying calcium amendments (Fletcher et al. 1984; Levy et al. 1988; Ilyas et al. 1993). Sodium and other cations displaced from cation exchange sites by calcium are leached out of the rooting zone by heavy water application, reducing their specific detrimental effects on soil structure and plant growth (Miller and Scifres 1988; Singh and Bajwa 1991; Armstrong and Tanton 1992; Ilyas et al. 1993). Gypsum in various forms is typically used as a source of calcium (Keren and Shainberg 1981; Kazman et al. 1983). The application of calcium carbonate, decaying organic matter, or other acid forming materials in calcareous soils may be employed as well to increase calcium ion activity (Robbins 1986; Sekhon and Bajwa 1993; Ilyas et al. 1993).

The productivity of one-third of all irrigated land, or about 70 million ha worldwide, is negatively affected to some degree by salinity (Armitage 1985), and abandonment of agricultural land due to increasing salinity is recognized as a significant problem (Grainger and Tinker 1982). Historically, many promising irrigation projects have failed as a result of sodium accumulation over time (Tanji 1990). Salt loading and structural damage to soil are difficult to reverse. Control of salinity, therefore, should occur before soil damage has developed and plant productivity is reduced.

Freezing stress

Susceptibility to injury by cold is one of the most important factors governing plant distribution. A considerable amount of research has been done in this area because of the importance of understanding the impacts of damage by cold temperatures. In much of the literature, plants have been characterized according to their responses to low temperature. Species have been categorized as sensitive or tolerant to freezing or chilling (Levitt 1980; Larcher 1981; Kozlowski et al. 1991). For species comparisons, these categories are becoming difficult to use because temperature thresholds and experimental approaches in freezing and chilling studies have not been standardized. In addition, the minimum temperatures under which a plant continues to function, may be mostly a reflection of its previous exposure to low temperatures, thus complicating the evaluation of research results.

The development of resistance to low temperatures occurs in step with changes in environmental temperature. This is known as acclimation, hardening, or frost hardening, and is well-illustrated by alpine plants (Larcher 1981). In winter, when temperatures are low, these plants have developed the ability to survive freezing to below –30°C. During the growing season, the same plants are killed by temperatures between –5 and –7°C. For poplars and many other woody species, tolerance to cold temperatures, and eventually freezing temperatures can change with the season or with other changes in environmental conditions. Poplars can develop resistance to the cold when exposed to low temperatures. Sakai and Yoshida (1968) found that poplars held at 15°C increased in tolerance to the cold from a killing temperature of –2 to –30°C over a 2-month period. There are even reports of survival at much lower temperature exposures. Tumanov (1969) found that poplar twigs survived temperatures as low as –50°C if cooled quickly and –60°C if cooled gradually.

The capacity to withstand cold temperatures, therefore can vary greatly even within a genetically pure strain depending on the stage of development, previous exposure, and other environmental factors (e.g., temperature, light, water, and mineral availability). Within this section, we will focus on several points that are relevant to the physiological basis of damage caused by low temperatures, including: changes in carbohydrate metabolism, membrane chemistry, and gene expression.

Physiological basis of damage imposed by low temperatures

Many plants are sensitive to low nonfreezing temperatures, resulting in changes in metabolism and signs of injury. These responses to low temperatures are commonly influenced by stage of development in the life cycle. For example, seeds are sensitive to cold during inbibition (Wolk and Herner 1982). During vegatative and reproductive stages, leaf growth of many crop species, such as cotton, corn, and rice, is reduced by low temperatures with symptoms such as wilting and yellowing of leaves (Wang 1982; Wolk and Herner 1982). From a cellular and biochemical perspective, there are few processes that are not altered by cold temperatures. The main cellular-biochemical processes which appear to be most influenced include respiratory pathways (Lyons and Breidenbach 1990), photosynthetic processes (Peeler and Naylor 1988), protein synthesis (Cooper and Ort 1988) and membrane permeability (Minorsky 1985).

There is general agreement that the plasma membrane is a primary site of injury during exposure to low temperatures (Steponkus 1990). In addition, a number of studies have shown that freezing temperatures can cause severe water loss from tissues when ice forms in the cell wall, causing a freeze-induced concentration of the solutes outside the protoplast. Concentration of the apoplastic solution results in the movement of osmotically active water out of the protoplast, causing cytoplasmic dehydration. For the plant to survive, membranes must remain stable in spite of the presence of high solute concentrations and an increase in the rate of water flux out of the protoplast (Steponkus 1990). Exactly how the plasma membrane maintains its integrity is unclear. There is, however, circumstantial evidence which suggests that membrane lipid composition determines the resistance of a particular species to low temperatures (Christiansen 1983; Lynch 1990).

The membrane as a site of cold damage was suggested as early as 1929 (Ivanov 1929). One of the major beliefs related to the membrane damage hypothesis is that the lipid composition of chilling-sensitive and chilling-resistant plants will be different. For poplars, as with many other species, there are seasonal changes in phospholipid (maximum in winter) and triglyceride concentrations (minimum in winter) in tissues exposed to cold temperatures. Yoshida (1973) showed that phosphoglycerides (mostly phosphatidylcholine) and phospholipids in the cortex and xylem of *Populus* ssp. (Gelrica) changed with the development of hardiness in conjunction with an increase in free sterols in the cortex. The same free sterols in the xylem do not appear to change seasonally. Such associations point to a role of membrane modifications in the development of hardiness for poplars. It will be difficult to determine which of the changes in lipid composition are associated with acclimation ability. Instead, it is possible that changes in lipid composition may only reflect fluidity adjustments at reduced temperatures.

Although the membrane damage hypothesis has significance, others have suggested alternative explanations for how low temperatures cause injury and how resistance may be developed. For example, changes in lipid composition may be important for the development of cold tolerance through specific interactions between enzyme/protein chemistry and certain lipids (Lynch 1990). Others have suggested that phosphatidylglycerol plays a major role in protein–lipid interactions by changing the phase properties of membranes (Murata 1983; Orr and Raison 1990).

The symptoms of cold injury are numerous and appear to reflect a range of metabolic dysfunctions leading to cell injury or death. Relatively few studies have detailed the specific effects of low temperatures on the metabolism of poplars. In a study of cold on the kinetics of specific enzymes of the glucose pathways, Sagisaka (1985) found that low temperatures decreased activity of glucose pathway enzymes in acclimatized poplar twigs kept at $-10°C$ (Table 1). He suggests that in extreme cold the supply of G6P will be reduced to the point where NADPH will be limited. The ability of poplars to overwinter may be determined by the G6P supply as has been found in many cereals. Within the membrane, diminishing reducing power would conceivably cause peroxidation of membrane lipids (Christophersen 1969; Sagisaka 1974c, 1985). Because the relationship between peroxide activity and cold damage remains unclear, it may be that peroxidase activity may have nothing to do with the development of low temperature tolerance. Many factors could induce changes in peroxidase activity in plants and not be part of the specific mechanism(s) associated with the

Table 1. Changes in substrate concentrations after exposure to freezing temperatures.

Substrate tissue		Substrate concentration ($\mu mol \cdot g^{-1} \cdot dwt$)	
		Control	Frozen
G6P	LB	0.30	0.05
	X	0.30	0.04
Glutathione:			
GSH	LB	0.66	0
	X	0.31	0
GSSG	LB	0.15	0.17
	X	0.16	0.24

Note: LB, living bark; X, xylem. Twigs of poplar were stored in plastic bags at –10°C. Data are from Sagisaka (1985).

development of cold or freezing tolerance (Stuber and Levings 1969; Lavee and Galston 1989).

In addition to changes in enzyme chemistry, qualitative and quantitative changes in water-soluble carbohydrates have been observed during cold treatments. Levitt (1980) suggested that the accumulations of such carbohydrates may be related to the development of cold tolerance (Levitt 1980). On the other hand, changes in the carbohydrate profile may not be part of a mechanism associated with the development of cold tolerance. Instead, the accumulation of carbohydrates observed during cold treatments could be related to reduced growth rate or some other side effect of low temperature metabolism (Pollock et al. 1988; Susuki 1989). This may be the case for poplars. Sagisaka (1974*a*, *b*) found that excessive cold induces poplars to form sucrose from glucose and fructose. This would interfere with other carbon pathways because the resultant glucose concentrations will be too low for the hexokinase reaction, even though ATP concentration in the cells may be unchanged. Additional work is needed to understand the complexities of cold damage in poplars and other woody species.

Low temperature acclimation in poplars — metabolic adjustments

Hardy plants must be able to maintain metabolism during exposure to low temperatures or freezing conditions. The first stages of cold acclimation can be induced by short-day photoperiods (Wareing 1956; Weiser 1970). After growth slows, temperature is the main factor responsible for the development of hardiness (Levitt 1980). For survival of plants at low temperatures several metabolic adjustments must occur. Exactly how these metabolic changes increase hardiness is not yet known. Mounting evidence from work with several species, including poplars, suggests that considerable variety exists for this trait even among closely related plants. In a study of inheritance of cold hardiness traits in poplars, Drew and Chapman (1992) used progeny from combinations of *P. trichocarpa* from Alaska crossed with the same species from a Montana population. The resulting hybrid was crossed with *P. deltoides* from Minnesota. Progeny from crosses like this will usually have hardiness capabilities intermediate between parents. Instead, Drew and Chapman (1992) found a range of responses where the temperature adaptation of the interspecific hybrid resembled the *P. deltoides* parent, but conductance characteristics were unlike either parent. Apparently, the potential to develop hardiness in poplars involves multiple characteristics with both additive and nonadditive effects. There is little information on how cold tolerance is inherited or the number of genes involved. Clearly, additional information concerning the genetics of cold tolerance is needed.

Endogenous factors such as hormones also appear to be involved in the development of cold hardiness (Lyons 1973; Rikin et al. 1975; Chen et al. 1983). In several species, ABA may act to mediate responses to cold temperatures. Applications of ABA induce chilling tolerance in several species (Rikin et al. 1975; Chen et al. 1983) and has been shown to rise in cold-treated plant tissues (Daie and Campbell 1981; Chen et al. 1983). This may be the case for many woody species, but we are unaware of any such studies detailing the role of ABA in mediating this response in poplars. There is evidence, however, suggesting that another group of hormones, cytokinins, may be involved in metabolic adjustments of poplars during acclimation to low temperatures. For example, cold-induced dormancy is overcome when growth promoting hormones, such as cytokinins, increase in sufficient concentrations to overcome the influence of endogenous inhibitors (Domanski and Kozlowski 1968; using *P. balsamifera*). This idea is supported by the observation that during low temperature acclimation in poplars, there is increased concentration in tissues of a compound with kinetin-like activity. This compound then decreases in conjunction with bud opening as weather warms (Domanski and Kozlowski 1968).

In poplars, as with many other woody species, there appear to be large changes in the carbohydrate profile during adaptation to cold. This appears to involve the conversion of both complex carbohydrates and proteins into lower molecular weight products (Levitt 1966). The development of cold hardiness occurs in conjunction with changes in the pools of starch and water soluble sugars (sucrose, raffinose, glucose, and fructose). Bonicel and Veercosa de Medeiros Raposo (1990) measured the concentrations of various branch and bud scale carbohydrates during the vegetative rest period using *P. trichocarpa* × *P. deltoides*, cv. Raspalje. They found that deep dormancy (when no budding is possible, even in optimal environments) was characterized by low concentrations of carbohydrates. In contrast, postdormancy (when budding is possible) was characterized by declining starch levels and increasing levels of soluble sugars (mostly represented by increased concentrations of raffinose and sucrose).

The development of cold adaptation may be associated with large changes in the profile of various carbohydrates. Details of the sugar and starch conversions that occur in association with the development of cold adaptation in poplars

(*P. canadensis* Moench Robusta) have been published by Witt and Sauter (1994). At the beginning of the dormant season, they found that starch in woody tissues is converted to various soluble sugars including sucrose, raffinose, stachyose, and maltose. The activity of starch degrading enzymes also change. Initial steps in the degradation of starch to various sugars are catalyzed by alpha amylose, beta amylose, or starch phosphorylase. Still other enzymes are needed to finish the degradation process. In poplars, increased activities of several starch degrading enzymes occurs during cold acclimation (e.g., alpha-amylase, EC 3.2.1.1; starch phosphorylase, EC 2.4.1.1; debranching enzyme, EC 3.2.1.4 1; D-enzyme, EC 2.4.1.25; maltose phosphorylase, EC 2.4.1.8; starch synthase, EC 2.4.1.21 and ADPGlcPPase, EC 2.7.7.27). Enzyme activities were highest during October, a period associated with decreasing concentrations of starch. This occurs after leaf abscission and appears to be related to declining temperature. The degree of starch hydrolysis (degradation) is related to the extent of cold temperature exposure. Sauter and Van Cleve (1991) found that total sugar content in wood increased from 27 $\mu g \cdot mg^{-1}$ DW (in hexose units) at 0°C to 35 $\mu g \cdot mg^{-1}$ DW when plants were held at −5°C. Interestingly, exposure to increasingly cold temperatures is also associated with the formation of a large maltose pool (Sauter 1986; Sauter and Van Cleve 1991; Sauter and Neumann 1994). Maltose seldom accumulates to any appreciable extent in plant tissues because maltose is usually hydrolyzed to glucose by alpha-amylase or maltase. The resulting maltose is then converted into sucrose and its galactosides.

In addition to changes associated with carbohydrate composition, colder temperatures also induce modifications in other compounds required for growth and survival. Sauter and Van Cleve (1991) found that low temperatures result in decreased fat concentration in cold-acclimated tissues while the concentration of glycerol increased. Parallel to modifications in the composition of secondary plant products, cold also causes other changes. Protoplasts become enriched with tubular and vesicular smooth endoplasmic reticulum (ER) cisternae, and structural changes occur in protein bodies that may be related to protein translocation within the cell. Some of these changes may be related to membrane metabolism. For example, the ER of all cells can convert fatty acids into phospholipids needed for growth of the ER itself and other cellular membranes (Moore 1982). These processes may be important in formation and stabilization of membranes formed during cold acclimation (Sauter and Van Cleve 1991).

Gene expression and photoperiod

In addition to the biochemical and physiological changes mentioned above, the combination of low temperatures and short-day photoperiods appear to induce alterations of gene expression. In a study using woody tissues from *P. ×canadensis* Moench Robusta, Van Cleve and Apel (1993) found storage proteins which accumulate during the winter and disappears with the development of terminal buds. Langheinrich and Tischner (1991) found two small proteins (a 32-kDa and a 36-kDa polypeptide), also considered to be storage proteins, in bark wood and root tissues of several clones. These polypeptides were abundant in winter and not detectable in summer. The authors have suggested that the two polypeptides are glycoforms differing only in the extent of glycosylation. Protein accumulation was preceded by an associated increase in the level of corresponding mRNA. The occurrence of these proteins, however, did not appear in all clones tested. Van Cleve and Apel (1993) studied the expression of these proteins in *P.* ×*canadensis* Moench. They found these proteins could be induced by either: (*i*) a shift from long-day to short-day conditions which may be a classical phytochrome-mediated response (Coleman et al. 1991, 1993) or, (*ii*) by a low-temperature treatment or, (*iii*) by applications of nitrogen under continuous long-day conditions. The synthesis of these possible storage proteins did not depend on the cessation of growth or the formation of a terminal bud. The specific function these proteins perform are unknown and it is unclear if they are actually related to the development of cold tolerance in poplars.

Ozone stress

Although air pollution has many components, ozone has been recognized as the primarily adverse agent of air pollution on vegetation (Darrell 1989; Treshow and Anderson 1991; Retzlaff et al. 1991). This molecule, made up of three oxygen atoms, is formed when nitrogen dioxide (NO_2) is converted to nitric oxide (NO) by the action of sunlight. Ozone is formed by the reaction of the freed oxygen atom with oxygen molecules in the atmosphere. NO_2 and other hydrocarbons are produced primarily by the combustion of fossil fuels in motor vehicles and other sources of high-temperature combustion found in a modern world (see Chapter 12, Van Volkenburgh and Taylor, for additional information on ozone impacts on leaf growth).

Ozone and growth responses of poplars

There have been numerous studies of the effects of ozone on woody plants (Reich and Amundsen 1985; Reich 1987; Retzlaff et al. 1991). Collectively, these studies have shown that exposure of plants to elevated ozone concentrations reduces net photosynthesis (Reich and Amundsen 1985; Reich 1983, 1987). This reduction in photosynthetic activity has been related to decreases in stomatal conductance (Coyne and Bingham 1982; Reich and Lassoie 1984; Reich 1987; Wang et al. 1995), increases in maintenance respiration (Amthor 1988), and/or damage to biochemical processes, such as oxidative action on chloroplasts (Landry and Pell 1993; Violin et al. 1993). Interspecific differences in ozone response have been commonly related to differences in stomatal conductance (Reich 1987). Typically species with high values of leaf conductance show a greater reduction in photosynthetic activity upon exposure to ozone than those with lower leaf conductances. Furthermore, biochemical changes within the photosynthetic apparatus lead to significant changes in: (*i*) quantum yield, (*ii*) ribulose 1,5-biphosphate (RuBP) regeneration, and (*iii*) RuBP carboxylase-oxygenase (Rubisco) activity (Black et al. 1982; Pell and Pearson 1983; Reich 1987; Sasek and Richardson 1989; Farage et al. 1991; Violin et al. 1993).

Ozone-induced reductions in net photosynthesis are closely related to declines in biomass production (Reich and Amundsen 1985; Amundsen et al. 1986; Amundson et al. 1987; Reich 1987; Retzlaff et al. 1991). Long-term exposure to ozone commonly leads to the appearance of scattered flecks and lesions on leaf surfaces and significant reductions in growth (Amundson et al. 1986, 1987; Patton and Garraway 1986; Reich 1987; Wang et al. 1985; Taylor and Frost 1992). Interestingly, ranking of the susceptibility of species to ozone based on foliar damage does not always correlate with rankings according to the growth responses of woody plants (Heagle 1979; Kress and Skelly 1982). Based upon the appearance of visible symptoms poplars are considered to be one of the more sensitive woody plants, along with several pines and a few oaks and ashes.

The relative sensitivity of poplars to ozone has resulted in numerous studies of their response to elevated ozone concentrations (Reich and Lassoie 1984; Berrang et al. 1991; Gupta et al. 1991; Karnosky et al. 1992*a*, *b*; Tjoelker et al. 1993; Violin et al. 1993). As in responses to flooding and salinity, the general consensus is that there are considerable variability in the response of poplars to higher ozone concentrations although variation in study results may be due to genetic differences in ozone sensitivity or differences in experimental design. Berrang et al. (1991) studied ozone sensitivity among several populations of *P. tremuloides* Michx. from 15 locations in the United States. Clones were selected from each population, greenhouse grown, fumigated with 150 ppb ozone for 6 h, and evaluated for visible injury. They found that the most tolerant populations were obtained from the West Coast, the northeast, and the industrialized portions of the Great Lakes. These resistant populations, which were from areas with the highest concentrations of ozone, had significantly less injury than populations from areas with lower ozone concentrations. In a related study focused on clonal variability, Karnosky et al. (1992*a*) evaluated ozone sensitive (symptomatic) and ozone tolerant (asymptomatic) *P. tremuloides* clones over two seasons. In the first season, there was no effect of ozone on the biomass accumulation of 18 ozone-tolerant clones compared to a decrease of 46.4% for 18 ozone-sensitive clones. During the second season, however, there was a small decrease in biomass accumulation in the tolerant clones (5% decreased biomass) compared to a 74% decrease in biomass for the sensitive clones. Interestingly, the results of these studies suggest that high ambient levels of ozone may eventually elimi-nate the presence of the ozone-sensitive clones from natural populations (Berrang et al. 1991; Karnosky et al. 1992*a*, *b*).

Much of the variability reported in the literature may be due to different experi-mental design used by various researchers. Matyssek et al. (1993) found that the biomass of clones of hybrid *Populus* (*Populus* ×*euramericana*) exposed to ozone in filtered air (0 = control, 0.05, 0.10 $\mu^{-1}\cdot L^{-1}$) or in ambient air (mean = 0.03 $\mu^{-1}\cdot L^{-1}$) was reduced only at the highest concentration of ozone. This was related to leaf loss rather than reduced leaf formation. Ozone was also found to cause a reduction in starch concentration, ratio of stem weight to length, and root-to-shoot biomass. Specific responses to ozone can also vary depending upon which tissues are considered (Table 2). Büecker and Ballach (1992) found that ozone effects were strongest in older leaves, with little or no effect in petioles and roots (*P. nigra* L. cv. Loenen and *P. maximowiczii* Henri × *P. nigra* L. cv. Rochester).

Ozone also has other effects on the physiology of poplars. In studies that exposed poplars to high concentrations of ozone, responses included earlier leaf abscission and decreases in stem basal diameter, stem mass, internode length, shoot height, and leaf area (Woodbury et al. 1994, using DN34, *P. deltoides* × *P. nigra*; Tenga et al. 1993, using *P. deltoides* × *P. nigra*). Unlike flooding responses, there appears to be no effect of ozone on the biophysical parameters associated with cell elongation such as wall extensibility or turgor (Taylor and Frost 1992). There is one report, however, which suggests that lengthy exposure to ozone can alter water relations in leaves of poplars. The change in leaf water status is associated with cell collapse, accompanied by water loss and an increase in air space. In stems, ozone has been associated with reduced radial width in xylem tissue (Landolt et al. 1994).

Collectively, these results suggest that poplars will respond in many ways to increased ozone concentrations in the environment. The response appears to depend upon the tissues studied, the clones tested, and the concentration of ozone used. When overall biomass is decreased in response to ozone, the growth depression does not appear to be directly linked to effects of ozone on the photosynthetic apparatus, although ozone can limit photosynthesis and growth via stomatal responses in some clones (see Tjoelker et al. 1993). At the whole plant level, the major effect of increased ozone concentration, however, appears to be associated with early senescence and decreased productivity (Reich 1987; Tjoelker et al. 1993; Volin et al. 1993).

Causes of ozone damage at the cellular and subcellular levels of organization

Ozone has been found to have profound influences at several levels of physiological organization in poplars. At the level of membrane organization, ozone can influence chloroplasts by mechanisms that alter oxidative properties of membranes (Pechak et al. 1986). Other indications of membrane changes are the accumulation of serine and raffinose. These compounds are components of the hydrophilic headgroups of membrane phospho- and galactolipids and may reflect

Table 2. Weights of leaves and roots of two cultivars of *Populus* after a 6-week exposure to air pollutants containing ozone.

Cultivar	Exposure	Leaves (g) (fresh weight)	Roots (g) (dry weight)
Loenen	CON	109 ± 1	4.7 ± 0.2
	CON	108 ± 11	4.2 ± 0.5
	CON	89 ± 3	3.6 ± 0.3
	CON	93 ± 9	3.6 ± 0.3
Rochester	CON	140 ± 7	4.0 ± 0.1
	CON	116 ± 4	3.2 ± 0.4
	CON	134 ± 19	3.1 ± 0.6
	CON	122 ± 11	2.9 ± 0.3

Note: CON = charcoal filtered ambient air (control); S/N = CON + SO_2 + NO_x; O/N = nonfiltered ambient air; O/S/N = O/N + S/N. Data are from Büecker and Ballach (1992).

changes in lipid esterification (Büecker and Guderian 1993). Cellular antioxidants such as glutathione and superoxide dismutase have also been found to accumulate in poplar leaves following short exposures to elevated ozone concentrations (Gupta et al. 1991).

At the leaf level, exposure of mature leaves can result in large depressions in photosynthesis (Reich 1983; Reich and Amundsen 1985; Reich 1987). The decreased photosynthesis has been attributed to reduced carboxylation efficiency associated with decreased regeneration of ribulose 1,5-bisphosphate (Sasek and Richardson 1989). Of all plant reactions, photosynthesis taking place in the chloroplasts must be regarded as most critical. Exposure of isolated chloroplasts to even low concentrations of ozone for brief periods inhibits evolution of oxygen and changes in electron transport (Landry and Pell 1993).

Landry and Pell (1993) have also studied the structural changes in Rubisco using an ozone-stressed hybrid poplar (*P. maximowiczii ×trichocarpa,* clone 245) in leaf extracts and isolated chloroplasts. Exposing the hybrid to ozone resulted in the visual symptoms of senescence and a decrease in the activity and quantity of ribulose 1,5-bisphosphate carboxylase/oxygenase (Rubisco). The results of their biochemical analysis showed that ozone decreased total Rubisco activity and binding of the enzyme's transition-state analog, 2-carboxyarabinitol bisphosphate, and promoted the loss of Rubisco large subunit (LSU). The 55 kDa Rubisco LSU appeared to aggregate (visualized by immunoblotting).

Ozone also causes changes in carbohydrate pool sizes. In addition, polyols and phenolics in leaf-lamina, petiole, shoot-axis, and roots of several poplar clones are altered by exposure to ozone (Büecker and Ballach 1992; Büecker et al. 1993). Ozone can increase or decrease the concentrations of soluble sugars and carbohydrates depending upon which tissue is measured. Landolt et al. (1994)

showed that by exposing hybrid *Populus* (*P.* ×*euramericana* var. Dorskamp) to ozone for three months, starch concentrations increased in leaves, stem bark, and bundle sheath cells located along small leaf veins. At the same time, sucrose and inositol content increased in the leaves. The significance of these changes to the causes of ozone damage at a cellular level remains unclear as a comprehensive series of such studies has not been attempted for a single clone. Clearly, more research is needed in this area to understand the complexities of plant responses to increased ozone in the environment.

Ozone and plant nutrition

We have discussed several conflicting responses of poplars to ozone. Nutritional status may play an important role in influencing the responses of woody species to environmental pollutants like ozone. Perhaps some of the confusion may be due to the influence of continuously variable soil factors such as pH, water relations, soil type, or nutrient availability. Karnosky et al. (1992a) studied the effects of genotype on the response of *P. tremuloides* Michx. to ozone and nitrogen deposition. Using *P. tremuloides* clones they evaluated growth under three ozone treatments: charcoal-filtered air, nonfiltered air and ozone (added at the rate of 80 ppb for 6 $h \cdot d^{-1}$, 3 days per week), and four N deposition levels (0, 10, 20, and 40 $kg \cdot ha^{-1} \cdot yr^{-1}$). Applications of nitrogen were found to reduce the negative effects of ozone on growth.

Clearly, any change in proportional allocation of carbon in response to nutrient deficiency will change the physiological status of the plant. In conjunction with air pollution stress, nutrient deficiency may further alter carbon allocation. The final result may be that the effects of air pollution stress could be mitigated (or exacerbated) when certain nutrients are abundant or deficient. This suggests that in experiments designed to address the effects of atmospheric pollutants on plants, nutritional status should be considered along with other environmental stresses (Greitner et al. 1994).

Ozone and ethylene

Ethylene acts as a plant hormone, causing distortion of plant parts, decreasing apical dominance, stimulating lateral buds, and hastening the aging process (Abeles 1973). Many of these responses are similar to those seen when plants are exposed to increasing concentrations of ozone. Kargiolaki (1989) and Kargiolaki et al. (1991) studied the release of ethylene from several clones of *Populus* under increased concentrations of ozone (*P. deltoides* var. missiouriensis Marsh., *P. nigra* cv. Italica L., and the hybrids *P. nigra* cv. Italica × *P. deltoides* (He-X/3) and *P. nigra* cv. Italica × *P. nigra* cv. Serres (He-K/7)). They found that ozone caused changes in the anatomical characteristics, including the formation of aerenchyma in stem lesions and the sloughing of superficial cells from the injured areas. These responses were correlated with the level of pollutant-induced ethylene evolution from leaves. In addition, there were clonal differences in ethylene emissions in response to ozone. Ethylene emission occurs under several environmental stresses and may be an indicator of ozone damage. Interestingly, in the Kargiolaki et al. (1991) study, plants that showed the lowest ethylene production were also the least susceptible to ozone damage.

Perspectives

Plants experience many types of environmentally induced stresses. The cumulative effects of stress determine distribution, reproductive success, and productivity. Experimentally, as long as no integrative approach is available for attempting to understand abiotic stresses, we continue to target each stress individually. While a significant amount of information concerning mechanisms of stress tolerance has been accumulated, there is no example where all parts of a specific mechanism of tolerance have been completely characterized. An understanding of the mechanisms of stress responses is important, because without this information we cannot predict the impact of changes in the environment on plants.

We do not yet know the genetic makeup that distinguishes some *Populus* varieties from others with respect to stress tolerance or resistance. Nor do we know why some clones survive conditions that others do not. We do not even understand the significance of observed responses. What does our knowledge mean in the big picture? On one hand, abiotic stresses may result in a new metabolic equilibrium that confers tolerance. On the other hand, stress responses may possibly be the result of accelerated development, or senescence, or simply a centralized response to injury. Of course, these views are all speculative. So far we do not know if changes in any cellular process are related quantitatively to the degree of stress. Some general questions that remain to be answered follow:

- What is the basis of stress tolerance or resistance? Do susceptible and nonsusceptible plants have prominent differences in metabolism in response to various stresses? When plants do respond to a particular stress, is there a centralized stress response system, whereby the same general metabolic responses occur to other types of environmental stress? What occurs when plants are exposed to multiple stresses (e.g., nutrients, ozone, and drought interactions)?

- What is the site of perception for environmental stress? Does the response start at the membrane? If so, how is recognition of a stress event transmitted throughout the plant? Is transmission a physiochemical event or is it a change in metabolism related to proteins or carbohydrates?

- What role do hormones play in stress responses? Which hormones are involved? Are there specific receptors and hormones related to specific stresses, or is there a centralized stress response system?

- Finally, to what extent do changes in gene expression confer any ecological advantage? Will we ever be able to understand stress responses at a molecular level? If so, will we be able to genetically engineer more stress resistant plants?

References

Abeles, F.B. 1973. Ethylene in plant biology. Academic Press, New York, NY. 302 p.

Alva, A.K., and Syverston, J.P. 1991. Irrigation water salinity affects soil nutrient distribution, root density and leaf nutrient levels of citrus under drip irrigation. J. Plant Nutr. **14**: 715–727.

Amthor, J.S. 1988. Growth and maintenance respiration in leaves of bean exposed to ozone in open-top chambers in the field. New Phytol. **110**: 319–325.

Amundson, R.G., Raba, R.M., Schoettle, A.W., and Reich, P.B. 1986. Response of soybean to low concentrations of ozone. II: Effects on growth, biomass allocation and flowering. J. Environ. Qual. **15**: 161–167.

Amundson, R.G., Kohut, R.J., Schoettle, A.W., Raba, R.M., and Reich, P.B. 1987. Ozone induces correlative reductions in whole-plant photosynthesis and yield of winter wheat. Phytopathology, **77**: 75–79.

Armitage, F.B. 1985. Irrigation forestry in arid and semi-arid lands. International Development Research Ctr., Ottawa, ON. 160 p.

Armstrong, A.S.B., and Tanton, T.W. 1992. Gypsum applications to aggregated saline-sodic clay topsoils. J. Soil Sci. **43**: 249–260.

Armstrong, W., Brandle, R., and Jackson, M.B. 1994. Mechanisms of flood tolerance in plants. Acta Bot. Neerl. **43**: 307–358.

Aw, M. 1994. Saline pulp and paper mill wastewater reclamation using woody plant species. M.Sc. thesis, Northern Arizona University, Flagstaff, AZ. 101 pp.

Aw, M., and Wagner, M.R. 1993. Suitable woody species for a land application alternative to pulp and paper mill wastewater disposal. *In* Proceedings of the First Biomass Conference of the Americas: Energy, environment, agriculture, and industry. Aug./Sept. 1993. Burlington, VT. National Renewable Energy Lab Publication DE93010050. pp. 1634–1639.

Ben-Hayyim, G., Kafkafi, U., and Ganmore-Neuman, R. 1987. Role of internal potassium in maintaining growth of cultured Citrus cells on increasing NaCl and $CaCl_2$ concentrations. Plant Physiol. **85**: 434–439.

Bernstein, L. 1961. Osmotic adjustment of plants to saline media. I. Steady state. Amer. J. Bot. **48**: 909–917.

Bernstein, L. 1975. Effects of salinity and sodicity on plant growth. US Salinity Laboratory. USDA. Annu. Rev. Phytopathol. **13**: 295–312.

Berrang, P., Karnosky, D.F., and Bennett, J.P. 1991. Natural selection for ozone tolerance in *Populus tremuloides*: an evaluation of nationwide trends. Can. J. For. Res. **21**: 1091–1097.

Biggs, K.J., and Fry, S.C. 1987. Phenolic cross-linking in the cell wall. *In* Physiology of cell expansion during plant growth. *Edited by* D.J. Cosgrove and D.P. Knievel. *In* Proceedings of the Second Annual Penn State Symposium in Plant Physiology. American Society of Plant Physiologists, MD. pp. 46–57.

Binzel, M.L., Hess, F.D., Bressan, R.A., and Hasegawa, P.H. 1988. Intracellular compartmentation of ions in salt adapted tobacco cells. Plant Physiol. **86**: 607–614.

Black, V.J., Ormrod, D.P., and Unsworth, M.H. 1982. Effects of low concentrations of ozone, singly and in combination with sulphur dioxide on net photosynthesis rates of *Vicia faba*. J. Exp. Bot. **33**: 1302–1311.

Bonicel, A., and Vercosa De Medeiros Raposo, N. 1990. Variation of starch and soluble sugars in selected sections of poplar buds during dormancy and post-dormancy. Plant Physiol. Biochem. **28**: 577–586.

Bradford, K.J., and Hsaio, T.C. 1982. Stomatal behavior and water relations of waterlogged tomato plants *Lycopersicon esculentum*. Plant Physiol. **70**: 1508–1513.

Bray, L., Chriqui, D., Gloux, K., Le-rudulier, D., Meyer, M., and Peduzzi, J. 1991. Betaines and free amino acids in salt stressed vitroplants and winter resting buds of *Populus trichocarpa* × *Populus deltoides*. Physiol. Plant. **83**: 136–143.

Büecker, J., and Ballach, H.J. 1992. Alterations in carbohydrate levels in leaves of *Populus* due to ambient air pollution. Physiol. Plant. **86**: 512–517.

Büecker, J., and Guderian, R. 1993. Marked increases in raffinose in leaves of *Populus* due to ambient air pollution. J. Plant Physiol. **141**: 654–656.

Büecker, J, Guderian, R., and Mooi, J. 1993. A novel method to evaluate the phytotoxic potential of low ozone concentrations using *Populus* cuttings. Water Air Soil Pollut. **66**: 193–201.

Burgos, P.A., Roldau, M., Agui, I., and Donaire, J.P. 1993. Effect of sodium chloride on growth, ion content, and hydrogen extrusion activity of sunflower and jojoba roots. J. Plant Nutr. **16**: 1047–1058.

Burrows, W.J., and Carr, D.J. 1969. Effects of flooding the root system of sunflower plants on the cytokinin content of the xylem sap. Physiol. Plant **22**: 1105–1112.

Butijn, J. 1954. The salt susceptibility of wind screens and fruit trees on saline spots in New Zealand. Neth. Dir. Van De Tuinbow Meded. **17**: 821–824.

Carpita, N.C. 1987. The biochemistry of "growing" cell walls. *In* Physiology of cell expansion during plant growth. *Edited by* D.J. Cosgrove and D.P. Knievel. Proceedings of the Second Annual Penn State Symposium in Plant Physiology. American Society of Plant Physiologists, MD. pp. 28–45.

Casey, H.E. 1972. Salinity problems in arid lands irrigation. Literature Review. U.S. Dep. of the Interior. Office of Water Research. Washington, DC.

Catalan, L., Balzarini, M., Talesknik, E., Sereno, R., and Karlin, U. 1994. Effects of salinity on germination and seedling growth of *Prosopis flexuosa* D.C. For. Ecol. Manage. **63**: 347–357.

Chen, H.-H., Li, P.H., and Brenner, M.L. 1983. Involvement of abscisic acid in potato cold acclimation. Plant Physiol. **71**: 362–365.

Christiansen, M.N. 1983. Temperature stress and lipid modification. *In* Phytochemical adaptations to stress. *Edited by* B. Timmermann, C. Steelink, and F.A. Loewus. Plenum Press, New York, NY. pp. 177–196.

Christophersen, B.O. 1969. Reduction of linolenic acid hydroperoxide by a glutathione peroxidase. Biochem. Biophys. Acta, **176**: 463–470.

Cleland, R.E. 1987. The mechanism of wall loosening and wall extension. *In* Physiology of cell expansion during plant growth. *Edited by* D.J. Cosgrove and D.P. Knievel. *In* Proceedings of the Second Annual Penn State Symposium in Plant Physiology. American Society of Plant Physiologists, MD. pp. 18–27.

Coleman, G.D., Chen, T.H.H., Ernst, S.G., and Fuchigami, L. 1991. Photoperiod control of *Populus* bark storage protein accumulation. Plant Physiol. **102**: 5359.

Coleman, G.D., Englert, J.M., Chen, T.H.H., and Fuchigami, L.H. 1993. Physiological and environmental requirements for *Populus* (*Populus deltoides*) bark storage protein degradation. Plant Physiol. **102**: 53–59.

Colorado, P., Nicholas, G., and Rodriguez, A. 1991. Calcium dependence of the effects of abscisic acid on RNA synthesis during germination of *Cicer arietinum* seeds. Physiol. Plant. **83**: 457–462.

Cooper, A.W., and Dumbroff, E.B. 1973. Plant adjustment to osmotic stress in balanced mineral nutrient media. Can. J. Bot. **51**: 763–773.

Cooper, P., and Ort, D.R. 1988. Changes in protein synthesis induced in tomato by chilling. Plant Physiol. **88**: 454–461.

Cornish, K., and Zeevaart, J.A.D. 1985. Abscisic acid accumulation by roots of *Xanthium strumarium* L. and *Lycopersicon esculentum* Mill. in relation to water stress. Plant Physiol. **81**: 653–658.

Cosgrove, D.J. 1986. Biophysical control of plant cell growth. Annu. Rev. Plant Physiol. **37**: 377–405.

Cosgrove, D.J., and Cleland, R.E. 1983. Osmotic properties of growing pea epicotyl segments. Planta, **153**: 343–350.

Coyne, P.I., and Bingham, G.E. 1982. Variation in photosynthesis and stomatal conductance in an ozone-stressed ponderosa pine stand: light response. For. Sci. **28**: 257–273.

Cramer, G.R., Lauchli, A., and Polito, V.S. 1985. Displacement of Ca^{2+} by Na^+ from the plasmalemma of root cells. Plant Physiol. **79**: 207–211.

Cramer, G.R., Lauchli, A., and Epstein, E. 1986. Effects of NaCl and $CaCl_2$ on ion activity in complex nutrient solutions and root growth of cotton. Plant Physiol. **81**: 792–797.

Cramer, G.R., Lynch, J., Lauchli, A., and Epstein, E. 1987. Influx of Na^+, K^+, and Ca^{2+} into roots of salt stressed cotton seedlings. Plant Physiol. **83**: 510–516.

Cramer, G.R., Epstein, E., and Lauchli, A. 1990. The effect of sodium, potassium and calcium on salt stress in barley. I. Growth analysis. Physiol. Plant. **80**: 83–88.

Daie, J., and Campbell, W.F. 1981. Response of tomato plants to stressful temperatures. Plant Physiol. **67**: 26–29.

Darrell, N.M. 1989. The effect of air pollutants on physiological processes in plants. Plant Cell Environ. **12**: 1–30.

Davies, D.D., Kemnworthy, P., Mocquot, B., and Roberts, K. 1987. The effects of anoxia on the ultrastructure of pea roots. *In* Plant life in aquatic and amphibious habitats. *Edited by* R.M.M. Crawford. Br. Ecol. Soc. Publ. No. 5. Blackwell Scientific Publications, Oxford, England. pp. 265–277.

Davies, W.J, and Van Volkenburgh, E. 1983. The influence of water deficit on the factors controlling the daily pattern of growth of *Phaseolus trifoliates*. J. Exp. Bot. **34**: 987–999.

Davies, W.J., Metcalf, J.C., Schurr, U., Taylor, G., and Zhang, J. 1987. Hormones as chemical signals involved in root to shoot communication of effects of changes in the soil environment. *In* Hormone action in plant development, a critical appraisal. *Edited by* G.V. Hoad, J.R. Lenton, M.B. Jackson, and R.K. Atkin. Butterworth and Co. Publishers, London, England. pp. 201–216.

Davis, R.F., and Jaworski, A.Z. 1979. Effects of low temperature on the sodium efflux pump in excised corn roots. Plant Physiol. **63**: 940–946.

Davison, R.M., and Young, H. 1973. Abscisic acid content of xylem sap. Planta, **109**: 95–98.

Dimitri, L. 1973. The salt tolerance of various *Populus* and willow species and clones in the laboratory and field. [*Populus, Salix*]. Eur. J. For. Pathol. **3**: 24–38. (Ref. English summary.)

Domanski, R., and Kozlowski, T.T. 1968. Variations in kinetin-ike activity of buds of *Betula* and *Populus* during release from dormancy. Can. J. Bot. **46**: 397–403.

Drew, A.P., and Chapman, J.A. 1992. Inheritance of temperature adaptation in intra- and interspecific *Populus* crosses. Can. J. For. Res. **22**: 62–67.

Epstein, E. 1961. The essential role of calcium in selective cation transport by plant cells. Plant Physiol. **36**: 437–444.

Farage, P.K., Long, S.P., Lechner, E.G., and Baker, N.R. 1991. The sequence of change within the photosynthetic apparatus of wheat following short-term exposure to ozone. Plant Physiol. **95**: 529–535.

Firmin, R. 1968. Forestry trials in Kuwait. *In* Saline irrigation for agriculture and forestry. *Edited by* H. Boyko. W. Junk Publ., The Hague, The Netherlands. pp. 7–42.

Fletcher, P., Sposito, G., and LeVesque, C.G. 1984. Sodium-calcium-magnesium exchange reactions on montmorillinite soil: I. Binary reactions. Soil Sci. Soc. Amer. **48**: 1016–1021.

Frenkel, H., Goertzen, J.O., and Rhoades, J.D. 1977. Effects of clay type and content, exchangeable sodium percentage, and electrolyte concentration on clay dispersion and soil hydraulic conductivity. Soil Sci. Soc. Amer. **42**: 32–39.

Frost, D.L., and Taylor, G. 1990. Wall properties, turgor pressure and leaf growth of hybrid *Populus* (*Populus deltoides* × *P. nigra*). *In* Proceedings of a meeting, September 13–15, 1989, Lancaster, England. Importance of root to shoot communication in the responses to environmental stress. *Edited by* W.J. Davies and B. Jeffcoat. Monogr. Br. Plant Growth Regul. Group. Bristol: The Group, **21**: 28–34.

Garcia, M., and Charbaji, T. 1993. The effect of sodium chloride salinity on cation equilibrium in grapevine. J. Plant Nutr. **16**: 2223–2237.

Goldberg, S., Suarez, D.L., and Glaubig, R.A. 1988. Factors effecting clay dispersion and aggregate stability in arid zone soils. Soil Sci. **146**: 317–326.

Grainger, A., and Tinker, J. 1982. Desertification. International Institute for Environmental Development. London, Washington, DC. 94 p.

Greenway, H. 1962. Plant response to saline substrates. J. Aust. Biol. Sci. **15**: 16–38.

Greenway, H., and Munns, R. 1980. Mechanisms of salt tolerance in nonhalophytes. Annu. Rev. Plant Physiol. **31**: 149–190.

Greitner, C.S., Pell, E.J., and Winner, W.E. 1994. Analysis of aspen foliage exposed to multiple stresses, ozone, nitrogen deficiency and drought. New Phytol. **127**: 579–589.

Greive, C.M., and Maas, E.V. 1988. Differential effects of sodium/calcium ratio on sorghum genotypes. Crop Sci. **28**: 659–665.

Gupta, A.S., Alscher, R.G., and McCune, D. 1991. Response of photosynthesis and cellular anti-oxidants to ozone in *Populus* leaves. Plant Physiol. **96**: 650–655.

Hansen, E.A. 1978. Forest irrigation — a review. *In* Proceedings of soil moisture and site productivity. *Edited by* W.E. Palmer. Nov. 1–3, 1977, Myrtle Beach, SC. pp. 126–134.

Hansen, E.A. 1983. Irrigating forest plantations. *In* Intensive plantation culture. U.S. Dep. Agric. For. Serv. Gen. Tech. Rep. N9-91: 46–52.

Hansen, E.A. 1988. Irrigating short rotation intensive culture hybrid *Populus*. Biomass **16**: 237–250.

Hansen, E.A., Dawson, D.H., and Tolsted, D. 1980. Irrigation of intensively cultured plantations with paper mill effluent. *In* Tech. Bull. 459 5/15/85, National Counc. Paper, TAPPI. **63**: 139–142.

Harrington, C.A. 1987. Responses of red alder and black cottonwood seedlings to flooding. Physiol. Plant. **69**: 35–48.

Hayward, H.E., and Spurr, W.B. 1943. Effects of osmotic concentration of substrate on the entry of water into corn roots. Bot. Gaz. **105**: 152–164.

Heagle, A.S. 1979. Ranking of soybean cultivars for resistance to ozone using different ozone doses and response measures. Environ. Pollut. **19**: 1–10.

Hopmans, P., Stewart, H.T.L, Flinn, D.W., and Hillman, T.J. 1990. Growth, biomass production and nutrient accumulation by seven tree species irrigated with municipal effluent at Wodonga, Australia. For. Ecol. Manage. **30**: 203–211.

Howe, J., and Wagner, M. 1995. The effect of papermill wastewater and organic amendments on sodium accumulation in a potted cottonwood soil system. Environ. Pollut. In press.

Hsiao, T.C., and Jing, J. 1987. Leaf and root expansive growth in response to water deficits. *In* Proceedings of the Second Annual Penn. State Symposium in Plant Physiology. Physiology of cell expansion during plant growth. *Edited by* D.J. Cosgrove and D.P. Knievel. American Society of Plant Physiologists, MD. pp. 180–192.

Ilyas, M., Miller, R.W., and Qureshi, R.H. 1993. Hydraulic conductivity of saline-sodic soil after gypsum application and cropping. Soil Sci. Soc. Amer. **57**: 1580–1585.

Itai, C., and Vaadia, Y. 1965. Kinetin-like activity in root exudate of water stressed sunflower plants. Physiol. Plant. **18**: 941–944.

Ivanov, S.M. 1929. Dependence of the chemical composition of oil-containing plants on the climate. Chem. Zbl. **1928**: 1971 (abstract).

Jackson, M.B., and Pearce, D.M.E. 1991. Hormones and morphological adaptations to aeration stress in rice. *In* Plant life under oxygen stress. *Edited by* M.B. Jackson, D.D. Davies, and H. Lambers. SPB Academic Publishing bv, The Haugue, The Netherlands. pp. 47–67.

Jackson, M.B., Herman, B., and Goodenough, A. 1982. An examination of the importance of ethanol in causing injury to flooded plants. Plant Cell Environ. **5**: 163–172.

Jackson, M.B., Young, S.F., and Hall, K.C. 1988. Are roots a source of abscisic acid for the shoots of flooded pea plants? J. Exp. Bot. **39**: 1631–1637.

Joly, R. 1989. Effects of sodium chloride on hydraulic conductivity of soybean root systems. Plant Physiol. **91**: 1262–1265.

Juwarkar, A.S., and Subrahmanyam, P.V.R. 1987. Impact of pulp and paper mill wastewater on crop and soil. Water Sci. Tech. **19**: 693–700.

Kargiolaki, H. 1989. Cell separation events in *Populus* in response to sulphur dioxide and ozone: involvement of ethylene. NATO ASI (Adv. Sci. Inst.) Ser. Ser. H Cell Biol. **35**: 405–419.

Kargiolaki, H., Osborne, J., and Thompson, F.B. 1991. Leaf abscission and stem lesions (intumescences) on poplar clones after sulfer dioxide and ozone fumigation: A link with ethylene release. J. Exp. Bot. **42**: 1189–1198.

Karnosky, D.F., Gagnon, Z.E., Reed, D.D., and Witter, J.A. 1992*a*. Effects of genotype on the response of *Populus tremuloides* Michx. to ozone and nitrogen deposition. Water Air Soil Pollut. **62**: 189–199.

Karnosky, D.F., Gagnon, Z.E., Reed, D.D., and Witter, J.A. 1992*b*. Growth and biomass allocation of symptomatic and asymptomatic *Populus tremuloides* clones in response to seasonal ozone exposures. Can. J. For. Res. **22**: 1785–1788.

Kazman, Z., Shainberg, I., and Gal, M. 1983. Effect of low levels of sodium and applied phytogypsum on the infiltration rate of various soils. Soil Sci. **135**: 184–192.

Kearney, T.H., and Scofield, C.S. 1936. The choice of crops for saline land. U.S. Dep. Agric. Circ. 404. 24 p.

Kennedy, R.A., Rumpho, M.E., and Fox, T.C. 1992. Anaerobic metabolism in plants. Plant Physiol. **100**: 1–6.

Keren, R., and Shainberg, I. 1981. Effect of dissolution rate on the efficiency of industrial and mined gypsum in improving infiltration of a sodic soil. Soil Sci. Soc. Amer. **45**: 103–107.

Kingsbury, R., and Epstein, E. 1986. Salt sensitivity in wheat. Plant Physiol. **80**: 651–654.

Kozlowski, T.T. 1984. Flooding and plant growth. Academic Press, New York, NY. 356 p.

Kozlowski, T.T., and Pallardy, S. 1984. Effect of flooding on water, carbohydrate and mineral relations. *In* Flooding and plant growth. *Edited by* T.T. Kozlowski. Academic Press, New York, NY. pp. 165–188.

Kozlowski, T.T., Kraemer, P.J., and Pallardy, S.G. 1991. The physiological ecology of woody plants. Academic Press, New York, NY. 657 p.

Kramer, P.J. 1951. Causes of injury resulting from flooding of the soil. Plant Physiol. **26**: 722–726.

Kress, L.W., and Skelly, J.M. 1982. Response of several eastern forest tree species to chronic doses of ozone and nitrogen dixoide. Plant Dis. **66**: 1149–1152.

Landolt, W., Gunthardt-Goerg, M., Pfenninger, I., and Scheidegger, C. 1994. Ozone-induced microscopical changes and quantitative carbohydrate contents of hybrid *Populus* (*Populus* ×*euramericana*). Trees (Berlin), **8**: 183–190.

Landry, L.G., and Pell, E.J. 1993. Modification of rubisco and altered proteolytic activity in ozone-stressed hybrid *Populus* (*Populus maximowiczii* ×*trichocarpa*). Plant Physiol. **101**: 1355–1362.

Langheinrich, U., and Tischner, R. 1991. Vegetative storage proteins in *Populus*: Induction and characterization of a 32- and a 36-kilodalton polypeptide. Plant Physiol. **97**: 1017–1020.

Larcher, W. 1981. Effects of low temperature stress and frost injury on productivity. *In* Physiological processes limiting plant productivity. *Edited by* C.B. Johnson. Butterworth & Co., London, England. pp. 253–269.

Lavee, S., and Galston, A.W. 1989. Hormonal control of peroxidase activity in cultured *Pelargonium* pith. Amer. J. Bot. **55**: 890–893.

Levitt, J. 1966. Cryochemistry of plant tissue. Cryobiology, **13**: 243–242.

Levitt, J. 1980. Responses of plant to environmental stress, Vol 1. Chilling, freezing and high temperature stresses. 2nd ed. Academic Press, New York, NY. 497 p.

Levy, G.J., and Van Der Watt, H.V.H., and Du Plessis, H.M. 1988. Effect of sodium-magnesium and sodium calcium systems on soil hydraulic conductivity and infiltration. Soil Sci. **146**: 303–310.

Lockhart, J.A. 1965. An analysis of irreversible plant cell elongation. J. Theor. Biol. **8**: 264–275.

Lui, Z., and Dickmannn, D.I. 1992*a*. Responses of two hybrid *Populus* clones to flooding, drought, and nitrogen availability: I. Morphology and growth. Can. J. Bot. **70**: 2265–2270.

Lui, Z., and Dickmannn, D.I. 1992*b*. Abscisic acid accumulation in leaves of two contrasting hybrid *Populus* clones affected by nitrogen fertilization plus cyclic flooding and soil drying. Tree Physiol. **11**: 109–122.

Lui, Z., and Dickmannn, D.I. 1993. Responses of two hybrid *Populus* clones to flooding, drought, and nitrogen availability: II. Gas exchange and water relations. Can. J. Bot. **71**: 927–938.

Luo, B., and Zhou, S. 1991. Study on the salt-resistance of *Populus euphratica* under water-culture. For. Res. **4**: 486–591.

Lynch, D.V. 1990. Chilling injury in plants: The relevance of membrane lipids. *In* Environmental injury to plants. *Edited by* F. Katterman. Academic Press, New York, NY. pp. 17–34.

Lyons, H.M. 1973. Chilling injury in plants. Annu. Rev. Plant Physiol. **24**: 445–466.

Lyons, J.M., and Breidenbach, R.W. 1990. Relation of chilling stress to respiration. *In* Chilling injury of horticultural crops. *Edited by* C.Y. Wang. CRC Press, Boca Raton, FL. pp. 223–233.

Maas, E.V. 1986. Salt tolerance of plants. U.S. Salinity Laboratory, USDA Agricultural Research Service, Riverside, CA. Appl. Agric. Res. **1**: 12–26.

Matthews, M.A., Van Volkenburgh, E., and Boyer, J.S. 1984. Adaptation of sunflower leaf growth to water deficits. Plant Cell Environ. **7**: 199–206.

Matyssek, R., Gunthardt-Goerg, M.S., and Landolt, W., and Keller, T. 1993. Whole-plant growth and leaf formation in ozonated hybrid *Populus* (*Populus* ×*euramericana*). Environ. Pollut. **81**: 207–212.

Michelena, V., and Boyer, J.S. 1982. Complete turgor maintenance at low water potentials in the elongating region of maize leaves. Plant Physiol. **69**: 1145–1149.

Miles, D. 1991. Increasing salt tolerance of plants through cell culture requires greater understanding of tolerance mechanisms. Aust. J. Plant Physiol. **18**: 1–15.

Miller, W.P., and Scifres, J. 1988. Effect of sodium nitrate and gypsum on infiltration and erosion of a highly weathered soil. Soil Sci. **145**: 304–309.

Mills, D., Robinson, K., and Hodges, T.K. 1985. Sodium and potassium fluxes and compartmentation in roots of *Atriplex* and oat. Plant Physiol. **78**: 500–509.

Minorsky, P.V. 1985. An heuristic hypothesis of chilling injury in plants: a role for calcium as the primary physiological transducer of injury. Plant Cell Environ. **8**: 75–94.

Moore, T.S. 1982. Phospholipid biosynthesis. Annu. Rev. Plant Physiol. **33**: 235–239.

Morgan, J.M. 1984. Osmoregulation and water stress in higher plants. Annu. Rev. Plant Physiol. **35**: 299–319.

Murata, N. 1983. Molecular species composition of phosphatidylglycerols from chilling-sensitive and chilling-resistant plants *Luffa cylindrica*. Plant Cell Physiol. **24**: 81–86.

Nakamura, Y., Tanaka, T., Ohta, E., and Sakata, M. 1990. Protective effect of external Ca^{2+} on elongation and the intracellular concentration of K^+ in intact mung bean roots under high NaCl stress. Plant Cell Physiol. **3**: 815–821.

Neuman, D.S., and Smit, B.A. 1990. Does cytokinin transport from root-to-shoot in the xylem sap regulate leaf responses to root hypoxia? J. Exp. Bot. **41**: 1325–1333.

Neuman, D.S., and Smit, B.A. 1993. Root hypoxia-induced changes in the pattern of translatable mRNAs in *Populus* leaves. J. Exp. Bot. **44**: 1781–1786.

Nieves, M.A., Cerda, G., and Botella, M. 1991. Salt tolerance of two lemon scions measured by leaf chloride and sodium accumulation. J. Plant Nutr. **14**: 623–636.

O'Conner, G.A. 1972*a*. Reclamation of a sodium affected soil with gypsum. N.M. Agric. Exp. Stn. Res. Rep. 290.

O'Conner, G.A. 1972*b*. Reclamation of a sodium affected soil with gypsum. N.M. Agric. Exp. Stn. Res. Rep. 292.

Orr, G.R., and Raison, J.K. 1990. The effect of changing the composition of phosphatidylglycerol from thylakoid polar lipids of oleander and cucumber on the temperature of the transition related to chilling injury. Planta, **181**: 137–143.

Patton, R.L., and Garraway, M.O. 1986. Ozone-induced necrosis and increased peroxidase activity in hybrid poplar leaves. Environ. Exp. Bot. **26**: 137–141.

Pearson, G.A. 1960. Tolerance of crops to exchangeable sodium. U.S. Dep. Agric. Agric. Info. Bull. **216**. 4 p.

Pechak, D.G., Noble, R.D., and Dochinger, L. 1986. Ozone and sulfur dioxide effects on the ultrastructure of the chloroplasts of hybrid *Populus* leaves. Bull. Environ. Contam. Toxicol. **36**: 421–428.

Peeler, T.C., and Naylor, A.W. 1988. A comparison of the effects of chilling on leaf gas exchange in pea (*Pisum sativum* L.) and cucumber (*Cucumis sativus* L.). Plant Physiol. **86**: 143–146.

Pell, E., and Pearson, N.S. 1983. Ozone-induces reduction in quantity of Ribulose-1,5-biphosphate Carboxylase in alfalfa foliage. Plant Physiol. **73**: 185–187.

Pereira, J.S., and Kozlowski, T.T. 1977. Variation among woody angiosperms in response to flooding. Physiol. Plant. **41**: 184–192.

Pollock, C.J., Eagles, C.F., and Sims, I.M. 1988. Effect of photoperiod and irradiance changes upon development of freezing tolerance and accumulation of soluble carbohydrates in seedlings of *Lolium perenne* grown at 2°C. Ann. Bot. **62**: 95–100.

Railton, I.D., and Reid, D.M. 1973. Effects of benzyladenine on the growth of waterlogged tomato plants. Planta, **111**: 261–266.

Ratner, A., and Jacoby, B. 1976. Effect of K its counter anion and pH on sodium efflux from barley root tips. J. Exp. Bot. **27**: 843–852.

Rawitz, E., Karschon, R., and Mitrani, K. 1966. Growth and consumptive water use of two Poplar clones under different irrigation regimes. Isr. J. Agric. Res. **16**: 77–88.

Ray, P.M. 1987. Principals of plant cell growth. *In* Proceedings of the Second Annual Penn State Symposium in Plant Physiology. Physiology of cell expansion during plant growth. *Edited by* D.J. Cosgrove and D.P. Knievel. American Society of Plant Physiologists, MD. pp. 1–17.

Regehr, D.L., Bazzaz, F.A., and Boggess, W.R. 1975. Photosynthesis, transpiration, and leaf conductance of *Populus deltoides* in relation to flooding and drought. Photosynthetica, **9**: 52–61.

Reich, P.B. 1983. Effects of low concentrations of O_3 on net photosynthesis, dark respiration and chlorophyll contents in aging hybrid poplar leaves. Plant Physiol. **73**: 291–296.

Reich, P.B. 1987. Quantifying plant response to ozone: a unifying theory. Tree Physiol. **3**: 63–91.

Reich, P.B., and Amundson, R.G. 1985. Ambient levels of ozone reduce net photosynthesis in tree and crop species. Science (Washington, DC), **230**: 566–570.

Reich, P.B., and Lassoie, J.P. 1984. Effect of low level ozone and/or sulfur dioxide on leaf diffusive conductance and water-use efficiency in hybrid poplar. Plant Cell Environ. **7**: 661–668.

Rengel, Z. 1992. The role of Ca^{+2} in salt toxicity. Plant Cell Environ. **15**: 625–632.

Retzlaff, W.A., Williams, L.E., and Dejong, T.M. 1991. The effect of different atmospheric ozone partial pressures on photosynthesis and growth of nine fruit and nut tree species. Tree Physiol. **8**: 93–105.

Rhoades, J.D. 1989. Effect of salts on plants. *In* Proceedings of National Water Conference. July 17–20, 1989. IR and WR Div/ASCE, Newark, DE.

Ricard, B., Courée, I., Raymond, P., Saglio, P., Saint-Ges, V., and Pradet, A. 1994. Plant metabolism under hypoxia and anoxia. Plant Physiol. Biochem. **32**: 1–10.

Rikin, A., Waldman, M., Richmond, A.E., and Dovrat, A. 1975. Hormonal regulation of morphogenesis and cold-resistance. J. Exp. Bot. **26**: 175–183.

Robbins, C.W. 1986. Sodic calcareous soil reclamation as affected by different amendments and crops. Agron. J. **78**: 916–920.

Roberts, J.K.M., Callis, J., Wemmer, D., Walbot, V., and Jardetzky, O. 1984. Mechanism of cytoplasmic pH regulation in hypoxic maize root tips and its role in survival under hypoxia. Proc. Nat. Acad. Sci. U.S. **81**: 3379–3383.

Roberts, J.K.M., Andrade, F.H., and Anderson, I.C. 1985. Further evidence that cytoplasmic acidosis is a determinant of flooding tolerance in plants. Plant Physiol. **77**: 492–494.

Sagisaka, S. 1974*a*. Effect of low temperature on amino acid metabolism in wintering *Populus*. Plant Physiol. **53**: 319–322.

Sagisaka, S. 1974*b*. Transition of metabolism in living *Populus* bark from growing to wintering stages and visa versa. Plant Physiol. **54**: 544–549.

Sagisaka, S. 1974*c*. The occurrence of peroxide in a perennial plant, *Populus gelrica* Plant Physiol. **57**: 308–309.

Sagisaka, S. 1985. Limited survival period of cold acclimatized trees in frozen ambient temperatures. Plant Cell Physiol. **27**: 1209–1212.

Sakai, A., and Yoshida, S. 1968. The role of sugar and related compounds in variations of freezing resistance. Cryobiology, **5**: 160–174.

Sasek, T.W., and Richardson, C.J. 1989. Effects of chronic doses of ozone on loblolly pine: photosynthetic characteristics in the third growing season. For. Sci. **3**: 745–755.

Sauter, J.J. 1986. Temperature induced changes in starch and sugars in the stem of *Populus* ×*canadensis robusta*. J. Plant Physiol. **132**: 608–612.

Sauter, J.J., and Van Cleve, B. 1991. Biochemical and ultrastructural results during starch-sugar conversion in ray parenchyma cells of *Populus* during cold adaptation. J. Plant Physiol. **139**: 19–26.

Sauter, J.J., and Neumann, U. 1994. The accumulation of storage materials in ray cells of *Populus* wood (*Populus* ×*canadensis robusta*): Effect of ringing and defoliation. J. Plant Physiol. **143**: 21–26.

Schatchtman, D., Tyermann, S., and Terry, B. 1991. The K^+/Na^+ selectivity of cation channel in the plasma membrane of root cells differ in salt tolerant and salt sensitive species. Plant Physiol. **97**: 598–605.

Schulte, P.J., Hinckley, T.M., and Stettler, R.F. 1986. Stomatal responses of *Populus* to leaf water potential. Can. J. Bot. **65**: 255–260.

Sekhon, B.S., and Bajwa, M.S. 1993. Effect of organic matter and gypsum in controlling soil sodicity in rice-wheat-maize system irrigated with sodic waters. Agric. Water Manage. **24**: 15–25.

Setter, T.L., Kapanchanakul, T., Kapanchanakul, K., Bhekasut, P., Wiengweera, A., and Greenway, H. 1987. Concentrations of CO_2 and O_2 in floodwater and internodal lacunae of floating rice growing at 1–2 metre water depths. Plant Cell Environ. **10**: 767–776.

Shainberg, I., and Letey, J. 1984. Response of soils to saline and sodic conditions. Hilgardia, **52**: 1–57.

Shainberg, I., Rhoades, J.D., and Prather, R.J. 1980. Effect of low electrolyte concentration on clay dispersion and hydraulic conductivity of sodic soil. Soil. Sci. Soc. Amer. **45**: 273–277.

Sherrell, C.G. 1984. Sodium concentration in *Lucerne*, *Phalaris* and a mixture of the two species. N.Z. J. Agric. Res. **27**: 157–160.

Silberbush, M., and Lips, S.H. 1991. Potassium, nitrogen ammonium/nitrate ratio and sodium chloride effects on wheat growth. J. Plant Nutr. **14**: 751–764.

Singh, H., and Bajwa, M.S. 1991. Effect of sodic irrigation and gypsum on the reclamation of sodic soil and growth of rice and wheat plants. Agric. Water Manage. **20**: 163–171.

Skene, K.G.M. 1976. Cytokinin production by roots as a factor in the control of plant growth. *In* The development and function of roots. *Edited by* J.G. Torrey and D. Clarkson. Academic Press, New York, NY. pp. 365–395.

Skene, K.G.M., and Kerridge, G.H. 1967. Effects of root temperature on cytokinin activity in root exudate of *Vitus vinifera*. Plant Physiol. **42**: 1131–1139.

Smit, B.A., and Stachowiak, M. 1988. Effects of hypoxia and elevated carbon dioxide concentration on water flux through *Populus* roots. Tree Physiol. **4**: 153–165.

Smit, B.A., Stachowiak, M., and Van Volkenburgh, E. 1989. Cellular processes limiting leaf growth in plants under hypoxic root stress. J. Exp. Bot. **40**: 89–94.

Smit, B.A,. Neuman, D.S., and Stachowiak, M. 1990. Root hypoxia reduces leaf growth. Role of factors in the transpiration stream. Plant Physiol. **92**: 1021–1028.

Smith, G.S., Middleton, K.R., and Edmonds, A.S. 1978. A classification of pasture and fodder plants according to their ability to translocate sodium from their roots into aerial parts. N.Z. J. Exp. Agric. **6**: 183–188.

Staples, R.C., and Toenniessen, G.H. 1984. Salinity tolerance in plants. John Wiley & Sons, New York, NY. 443 p.

Steponkus, P.L. 1990. Cold acclimation and freezing injury from a perspective of the plasma membrane. *In* Environmental injury to plants. *Edited by* F. Katterman. Academic Press, New York, NY. pp. 1–16.

Stewart, H.T.L., Hopmans, P., Hillman, T.J., and Flinn, D.W. 1990. Nutrient accumulation in trees and soil following irrigation with municipal effluent in Australia. Environ. Pollut. **63**: 155–177.

Stuber, C.W., and Levings, C.S. III. 1969. Auxin induction and repression of peroxidase isozymes in oats (*Avena sativa* L). Crop Sci. **9**: 415–416.

Subbarao, G.V., Johansen, C., Jana, M.K., and Kumar Rao, J.V.D.K. 1990. Effects of sodium/calcium ratio in modifying salinity response of Pigeon Pea (*Cajanus cajan*). J. Plant Physiol. **136**: 439–443.

Sucoff, E. 1975. Effects of deicing salts on woody vegetation along Minnesota roads. Minn. Agric. Exp. Stn. Bull. **303**: 49.

Susuki, M. 1989. Fructans in forage grasses with varying degrees of cold hardiness. J Plant Physiol. **134**: 224–231.

Tanji, K.K. (*Editor*). 1990. Agricultural salinity assessment and management. Amer. Soc. Civil Eng., New York, NY.

Taylor, G., and Frost, D.L. 1992. Impact of gaseous air pollution on leaf growth of hybrid *Populus*. For. Ecol. Manage. **51**: 151–162.

Tenga, A.Z., Hale, B., and Ormrod, D.P. 1993. Growth responses of young cuttings of *Populus deltoides* × *Populus nigra* to ozone in controlled environments. Can. J. For. Res. **23**: 854–858.

Tjoelker, M.G., Violin, J.C., Oleksyn, J., and Reich, P.B. 1993. Light envrionment alters response to ozone stress in seedlings of *Acer saccharum* and hybrid *Populus*: I. *In situ* net photosynthesis, dark respirations and growth. New Phytol. **124**: 627–636.

Treshow, M., and Anderson, F.K. 1991. Plant stress from air pollution. The Bath Press, Bath, UK. 283 pp.

Tumanov, I.I. 1969. Physiology of plants not killed by frost. Izv. Akad. Nauk. USSR Ser. Biol. **4**: 469–480.

USDA. 1954. Salinine and alkali soils. USDA Handbook No. 60.

Van Cleve, B., and Apel, K. 1993. Induction by nitrogen and low temperature of storage-protein synthesis in *Populus* trees exposed to long days. Planta, **189**: 157–160.

Van Volkenburgh E. 1987. Regulation of dicotyledonous leaf growth. *In* Proceedings of the Second Annual Penn State Symposium in Plant Physiology. Physiology of cell expansion during plant growth. *Edited by* D.J. Cosgrove and D.P. Knievel. American Society of Plant Physiologists, MD. pp. 193–201.

Van Volkenburgh, E., and Cleland, R.E. 1984. Control of leaf growth by changes in cell wall properties. What's new. Plant Physiol. **15**: 25–28.

Volin, J.C., Tjoelker, M.G., Oleksyn, J., and Reich, P.B. 1993. Light environment alters response to ozone stress in seedlings of *Acer saccharum* Marsh. and hybrid *Populus* L.: II. Diagnostic gas exchange and leaf chemistry. New Phytol. **124**: 637–64.

Walton, D.C., Harrison, M.A., and Cote, P. 1976. The effects of water stress on abscisic acid levels and metabolism in roots of *Phaseolus vulgaris* L. and other plants. Planta, **131**: 141–144.

Wang, C.Y. 1982. Physiological and biochemical responses of plants to chilling stress Hort-Science, **17**: 173–186.

Wang, D., Karnosky, D.F., and Borman, F.H. 1985. Effects of ambient ozone on the productivity of *Populus tremuloides* grown under field conditions. Can. J. For. Res. **16**: 47–55.

Wang, D., Hinckley, T.M., Cumming, A.B., and Braatne, J. 1995. A comparison of measured and modeled ozone uptake into plant leaves. Environ. Pollut. **89**: 247–254.

Wareing, P.F. 1956. Photoperiodism in woody plants. Annu. Rev. Plant Physiol. **7**: 191–214.

Weinberg, R. 1988. Modification of foliar solute concentrations by potassium chloride. Physiol. Plant. **73**: 418–425.

Weiser, C.J. 1970. Cold resistance and injury in woody plant. Science (Washington, DC), **169**: 269–1278.

Withers, B., and Vipond, S. 1980. Irrigation design and practice. Cornell University Press, Ithaca, NY.

Witt, W., and Sauter, J.J. 1994. Enzymes of starch metabolism in *Populus* wood during fall and winter. J. Plant Physiol. **143**: 625–631.

Wolk, W.D., and Herner, R.C. 1982. Chilling injury of germinating seeds and seedlings. Hort-Science, **17**: 169–173.

Woodbury, P.B., Laurence, J.A., and Hudler, G.W. 1994. Chronic ozone exposure alters the growth of leaves, stems and roots of hybrid *Populus*. Environ. Pollut. **85**: 103–108.

Wright, S.T.C., and Hiron, R.W.P. 1972. The accumulation of abscisic acid in plants during wilting and under other stress conditions. *In* Plant growth substances, 1970. *Edited by* D.J. Carr. Springer-Verlag, Berlin and New York. pp. 291–298.

Xia, J.H., and Saglio, P.H. 1992. Lactic acid efflux as a mechanism of hypoxic accumulation of maze root tips to anoxia. Plant Physiol. **93**: 453–459.

Yoshida, S. 1973. Seasonal changes in lipids and freezing resistance in *Populus* trees. Low Temp. Sci. **31**: 9–20.

Zhang, J., and Davies, W.J. 1986. Chemical and hydraulic influences on the stomata of flooded plants. J. Exp. Bot. **37**: 1471–1491.

Zhang, J., and Davies, W.J. 1987. ABA in roots and leaves of flooded pea plants. J. Exp. Bot. **37**: 649–659.

CHAPTER 18
Production physiology

P.E. Heilman, T.M. Hinckley, D.A. Roberts, and R. Ceulemans

Introduction

Production physiology is concerned with the factors that influence and control the productivity of plants grown in groups or stands. It seeks to understand the process of capture of solar energy and its conversion to plant growth. It is also based upon the relationships between environmental factors and yield and how plants can be managed and genetically improved to increase yields. It may also be concerned with how site factors can be modified in specific cases to obtain maximum yields.

Stress physiology, in contrast, focuses on the reaction of plants' organelles and organs to abiotic and biotic factors and is concerned with plant survival and growth under natural growing conditions (see Neuman et al., Chapter 17). It is based upon understanding mechanisms used by plants to resist the suite of stresses typically encountered under natural conditions. Similar to production physiology, stress physiology addresses how plants, together with management

Abbreviations: A_c, canopy photosynthesis; A_m, maximum or light saturated net photosynthesis of a leaf; AVIRIS, Airborne Visible/Infrared Imaging Spectrometer; D, *P. deltoides*; IPAR, intercepted photosynthetically active radiation, k, light extinction coefficient; LAI, leaf area index; L, leaf area index; MJ, megajoule; N, *P. nigra*; NPP, net primary production; PAR, photosynthetically active radiation; SRF, short rotation forestry; T, *P. trichocarpa*; ε, energy conversion efficiency.

P.E. Heilman. Washington State University, WSU Puyallup Research and Extension Center, Puyallup, WA 98371, USA.
T.M. Hinckley. University of Washington, College of Forest Resources, Seattle, WA 98195, USA.
D.A. Roberts. University of California, Department of Geography, Santa Barbara, CA 93106, USA.
R. Ceulemans. University of Antwerpen (UIA), Department of Biology, B-2610 Wilrijk, Belgium.

Correct citation: Heilman, P.E., Hinckley, T.M., Roberts, D.A., and Ceulemans, R. 1996. Production physiology. *In* Biology of *Populus* and its implications for management and conservation. Part II, Chapter 18. *Edited by* R.F. Stettler, H.D. Bradshaw, Jr., P.E. Heilman, and T.M. Hinckley. NRC Research Press, National Research Council of Canada, Ottawa, ON. pp. 459–489.

practices, can be modified to increase their capacity to be productive under stressful conditions.

In our work with hybrid poplars, we continue to be fascinated by the rate at which these trees grow on favorable sites in the Pacific Northwest (PNW) of North America. This fascination is carried by two questions: (1) what are the physiological and morphological attributes of these trees that allow them to grow so well, and (2) what are the key abiotic factors that favor the growth of poplar? With better understanding of these two questions, we think it is possible to obtain even higher yields from poplar plantations in the PNW as well as in other regions and environments. These elements are developed in this chapter through an examination of above- and below-ground features associated with the interception of light, the uptake of carbon dioxide, and the uptake of water and nutrients. We then examine the environmental factors most frequently associated with stand productivity, namely, climate, water, and nutrients. Our chapter relies heavily on the information presented and discussed in earlier chapters in this book.

Aboveground components and production

Rather than focus on the integrated organism in our discussion on production physiology, we have chosen to present first the aboveground components of production and then the belowground. Then we continue with the largely abiotic factors that affect the expression of maximum production.

Net primary or biomass production in agricultural crops has been proposed to be directly related to the radiant energy interception by the foliage (Monteith 1977, 1981, 1994). The empirical model is frequently expressed as, NPP = IPAR ε, where IPAR is the intercepted photosynthetically active radiation and ε is the energy conversion efficiency which ranges from 1 to 5 g of dry matter produced per megajoule of energy absorbed.

Linder (1985) first demonstrated that a linear relationship between solar radiation capture and biomass production also exists for forest stands. The relationship is also evident for poplar clones in experimental field plantations in the Pacific Northwest (Scarascia-Mugnozza et al. 1989), in Scotland, UK (Cannell et al. 1988), and in Flanders, Belgium (Impens et al. 1990; Ceulemans et al. 1992b). However, variability in crown and canopy architecture among plant genotypes can strongly influence the efficiency of conversion of solar energy into biomass production (Scarascia-Mugnozza et al. 1989; Ceulemans et al. 1990; Chen et al. 1994). In spite of the attractiveness of this empirical model, it has been criticized (see Demetriades-Shah et al. 1994). At a minimum, most investigators find it necessary to add environmental or phenological modifiers that act on an optimal ε because optimal ε itself may be age- and genotype-specific.

The conversion efficiency ε (also called light use efficiency) is defined as the ratio of biomass produced per unit of solar radiation energy intercepted by the foliage or per unit of absorbed photosynthetically active radiation, and can be represented as the slope of the abovementioned linear regression line. Under conditions of adequate moisture and nutrients, the conversion efficiency remains rather constant throughout the entire growing season. Table 1 summarizes some of the values found in the literature for fast growing, short rotation forestry plantations of poplar. All conversion efficiency values in Table 1 are based on intercepted PAR and need to be halved to estimate the ε value based on total solar radiation. A wide variety of ε values have been reported for poplar SRF plantations, ranging from 0.32 to as much as 3.14 g·MJ^{-1}. Conversion efficiency values as high as 3 g·MJ^{-1} were reported for intensive experimental trials in Scotland, but much lower ε values were reported for an irrigated and fertilized clonal stand of *Populus* in Pennsylvania, USA, where the conversion efficiency was on the order of 1.4 g·MJ^{-1} (Landsberg and Wright 1989). In Flanders and the Pacific Northwest the seasonal mean solar conversion efficiency of intercepted radiation to aboveground dry mass ranged on average from 0.72 g·MJ^{-1} for the parental species *P. trichocarpa* and *P. deltoides* to 0.90 g·MJ^{-1} for the interspecific hybrid clones (Table 1). For total aboveground dry mass production, including branches and leaves, ε values ranged from 0.72 g·MJ^{-1} for an Euramerican clone to 1.10 g·MJ^{-1} for *P. trichocarpa* × *P. deltoides* clone Beaupré (Table 1).

From Table 1 it should be noted that SRF poplar plantations have seasonal mean conversion efficiencies for total aboveground biomass production that are very similar to the ε values measured for various C_3 agricultural crops (Cannell 1989; Sinclair and Horie 1989). The values in Table 1 are for SRF systems that are managed as either coppiced or noncoppiced systems. Using containerized poplars and willows, similar values were reported for coppiced and noncoppiced plantations (Cannell et al. 1987; Cannell 1989). However, it might be anticipated that the ε values would be higher in the early stages after coppicing because of (1) the use of carbon reserves in stems and roots stored from previous periods of photosynthesis, (2) higher rates of photosynthesis in the new sprouts, and (3) little selfshading in the foliage. The large ε values of the high-performing and efficiently light-intercepting *P. trichocarpa* × *P. deltoides* clones as compared with those in the Euramerican hybrids or parental species have been attributed to a higher maximum photosynthetic performance (see Ceulemans and Isebrand, Chapter 15). In addition, a number of canopy structural traits, including greater LAI, longer individual and total leaf-area duration, increased clumping

Table 1. Conversion efficiency of absorbed PAR (photosynthetically active radiation) into aboveground dry mass (and/or total dry mass) and maximum leaf area index (LAI) for high density, intensively managed poplar stands from the available literature data. The conversion efficiency (ε) is expressed in $g \cdot MJ^{-1}$. Equivalent values of energy use efficiency based on intercepted total solar radiation will be approximately half the ε values listed. Specifications on the experimental set-up (such as age of the stand, place, length of the growing season) have also been indicated when available. (Table with input from Ruimy et al. 1994).

Treatment	Locality	Age (years)	Growing season (months)	Conversion efficiency (ε) $(g \cdot MJ^{-1})$	Maximum LAI $(m^2 \cdot m^{-2})$	References*
Not irrigated, not fertilized	Flanders, Belgium	5	6	0.9–1.9	4	5
Five clones, irrigated	Flanders, Belgium	2	6	0.72–1.10	6.1–9.4	2, 3
Irrigated, fertilized	Scotland, UK	1	6.5	2.11 (3.14 total)	3.5	1
Three clones, containerized plants, fertilized	Scotland, UK	1	6.5	1.18 (3.06 total)	5.1–5.6	6
Different treatments	Pennsylvania, USA	1	6	0.32	6	4
Different treatments	Pennsylvania, USA	2	6	1.4	7	4
Irrigated, fertilized	Wisconsin, USA	1–3	5	1.16–1.28	8	4

*1, Cannell (1989); 2, Ceulemans et al. (1992*b*); 3, Impens et al. (1990); 4, Landsberg and Wright (1989); 5, Lemeur and Impens (1981); 6, Milne et al. (1992).

of foliage, and more effective leaf display, may play positive roles in performance. These traits are discussed further in the next sections.

The effect of phenology and length of growing season

The field performance of poplar clones is closely tied to the time of bud set and bud break. Pauley and Perry (1954), for example, found that the timing of bud set in several *P. trichocarpa* and *P. deltoides* clones was directly correlated to the length of the frost-free season in the clone's native habitat. It is evident that the length of the growing season of a clone native to northern British Columbia or to the northern plains of Saskatchewan is entirely different from that of a clone that originated in the more temperate climate of France or Italy.

Such differences in bud set date in different clones are clearly attributed to their genetic makeup (Weber et al. 1985; Michael et al. 1988). Genetic differences in phenology and the influence of these differences on growth and productivity have been illustrated by a study of five poplar clones in an experimental plantation in the state of Washington. Phenological characteristics, including dates of bud burst, bud set, leaf development early in the season, and leaf abscission were recorded during three growing seasons (Ceulemans et al. 1992a). The seasonal patterns of leaf retention and leaf loss varied among clones and species. *Populus trichocarpa*, native to the western United States, retained its leaves the longest into the autumn, while *P. deltoides* (native to southeastern United States) exhibited earlier and very rapid rates of leaf abscission. From the five clones in this experimental plantation *P. trichocarpa* clone 12–106 (from southern Oregon) was the first to break bud and had the longest growing period, i.e., 195–216 growing days, depending on the season. Nearly all clones had broken bud by early April with exception of the *P. deltoides* clone. By early September nearly all lateral branches had set bud in this experimental plantation, while buds on the current terminal generally were set 1–3 weeks later (Stettler et al. 1988; Ceulemans et al. 1992a). Associated with bud set is the cessation of elongation, but the continuation of photosynthesis and a major shift in carbon allocation towards storage, the lower stem, and the root system (Hinckley et al. 1989, Scarascia-Mugnozza 1991).

The environment, in addition to genotype, is a predominant determinant of phenology. Common garden studies of a well-defined set of *P. trichocarpa* and interspecific hybrid genotypes have illustrated that fall coloration, leaf abscission, bud set, and bud break show important differences between geographically dispersed plantations with different soil and/or climatic conditions (Weber et al. 1985; Dunlap et al. 1992; Dunlap and Stettler 1996). Measurements of phenological characteristics are an important way to define whole-tree leaf-area fluctuations during the growing season as well as leaf-area duration. This information on total leaf area and the development of LAI is central to most light interception and growth models (see Ceulemans and Isebrands, Chapter 15). Not only is it possible to record the development of green tissue, but also its senescence, and its losses due to herbivory and disease (e.g., Dunlap and Stettler 1996). However, we still need more information on the time-course of leaf-area development for SRF crops — combinations of remote sensing and ground verification may rapidly close this gap (see section on 'Canopy information from remote sensing').

Leaf area index and light interception

As mentioned previously, production is closely and linearly related to the quantity of radiation absorbed. Because foliage absorbs most of the radiation in young plants, the quantity of leaf area is a critical determinant of production. Rapid canopy closure and development of canopy leaf area are essential for the successful establishment and growth of SRF hybrid poplar plantations (Heilman and Xie 1994).

Crown architecture, the combination of total leaf area, leaf-area distribution within the crown, leaf and branch morphology, and orientation, plays a major role in plant productivity as it influences not only the interception of solar radiation, but also its conversion into biomass (Isebrands et al. 1983). The quantity and the display of the leaf area present are perhaps the two most important factors determining stand productivity. Because significant clonal variation exists in crown structure, canopy architecture, branching patterns, leaf orientation, and demography in poplar (Isebrands and Michael 1986; Hinckley et al. 1989; Ceulemans 1990; Ceulemans et al. 1990, Hinckley et al. 1992), it follows that considerable differences in light interception and LAI can be found.

During the establishment year of five different poplar clones in a field study in the state of Washington, maximum light interception for a *P. deltoides* clone from Illinois (Ill–005) was only 75%, a value much lower than that of the other clones (Fig. 1). Note that in Fig. 1 mean penetration values are given for clones of *P. deltoides, P. trichocarpa* and their interspecific hybrids. The highest light interception value of 95% was reached by one of the *P. trichocarpa* × *P. deltoides* clones, i.e., clone 11–11, at the end of September of the first year, whereas an intermediate value of 85% was attained by the *P. trichocarpa* clones and *P. trichocarpa* × *P. deltoides* hybrid clone 44–136. Light absorption by any clone was directly related to its total leaf area (Fig. 1). A rather accurate but destructive method was used to estimate the stand LAI. All leaves from one or two whole trees were harvested, the specific leaf area (i.e., leaf area to dry weight ratio) was estimated for a number of samples, and the leaf area of the whole tree was calculated from total-leaf dry-mass measurements (Ceulemans et al. 1992*a*). The relationship between total leaf area per tree and stem volume (Ridge et al. 1986) could then be used to estimate the total leaf area for the entire plantation or stand, and thus the LAI.

In that study the *P. trichocarpa* × *P. deltoides* hybrid clone 11–11 reached an LAI value of 2.9 during the establishment year, and LAI further increased to 8 and more than 11 in the second and third year, respectively (Fig. 2). LAI kept increasing during the third growing season for all five clones, and by mid-August values were 6.3, 6.2, and 11.2 for *P. trichocarpa* clone 1–12 (from Chilliwack, British Columbia), *P. deltoides* clone Ill–005 (from Illinois) and interspecific hybrid clone 11–11, respectively. Hybrid clone 44–136 showed the largest single year increase in LAI, i.e., from 5.7 in the second year to 11.6 in the third growing season. The increase in LAI, however, leveled off during the fourth growing season in the densely planted (1 m × 1 m) *P. deltoides* and *P. trichocarpa* clones, and even began decreasing in the *T* × *D* hybrid clones 11–11 and 44–136 (Fig. 2). Marked differences in the evolution of LAI with plant age were observed between these hybrid clones (i.e., 11–11 vs. 44–136, with different parentage), and may have important implications for determining the optimum rotation age of intensive SRF plantations. Ranking of the clones for biomass

Fig. 1. Application of the Lambert–Beer law to 1-year-old SRF poplar stands. Logarithm of radiation penetration (ln I/I_0, left ordinate axis) and percentage of radiation penetration (%, right ordinate axis) versus leaf area index (LAI, abscissa) for *P. deltoides* clone Ill–005 (■), *P. trichocarpa* clone 1–12 (△) and *P. trichocarpa* × *P. deltoides* hybrid clone 11–11 (+) during their establishment year in an experimental field plantation in the state of Washington. Dotted regression lines illustrate different radiation extinction coefficients using the Lambert–Beer law (see inset) for radiation interception by a homogeneous canopy, but were not calculated through the experimental data points.

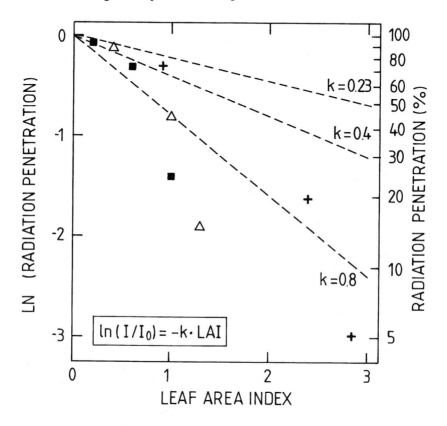

production during their establishment year and later was similar to that for light interception and LAI (Scarascia-Mugnozza et al. 1989).

The LAI values of 10 and more that have been reported for poplar in various studies (Heilman and Xie 1994; see also Fig. 2) are comparable to values reported for several evergreen forest species (such as *Pseudotsuga menziesii, Picea sitchensis, Abies* spp.) and are higher than those noted for agricultural crops as corn, sunflower, etc. Clearly certain *Populus* species and, in particular, the *P. trichocarpa* × *P. deltoides* hybrids, are capable of producing large quantities of leaves at a very young age, that result in their very impressive growth performance (see Van Volkenburgh and Taylor, Chapter 12).

Fig. 2. Leaf area index (LAI) as a function of plant age for *P. deltoides* clone Ill–005 (from Illinois, □), *P. trichocarpa* clone 1–12 (from Chilliwack, BC, ●) and *P. trichocarpa* x *P. deltoides* hybrid clones 11–11 (○) and 44–136 (△) grown under intensive culture in an experimental field plantation in the Pacific Northwest.

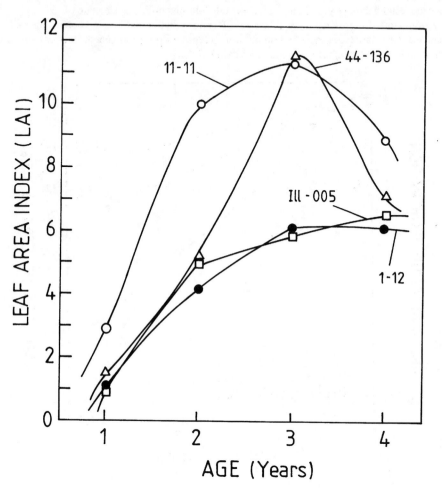

Photosynthesis at the stand or canopy level

The photosynthetic rate of a canopy (A_c) is the sum or integration of the photosynthetic rates of all the leaves that form the canopy (Ceulemans and Saugier 1990). So, A_c depends on the total leaf area, on the orientation of the leaves toward the light at any particular time, on the leaf photosynthetic activity and on different environmental factors, especially the radiation environment. It is well known that the photosynthetic response of a canopy to light is more linear than that of a single leaf (see Fig. 3) (Larcher 1980; Saugier 1986). This response occurs because leaves in the lower canopy are rarely light saturated and will

Fig. 3. Net CO$_2$ exchange at the leaf and the canopy level as a function of photosynthetic photon flux density for a hybrid poplar clone. Simulation by a simplified Monteith-type canopy photosynthesis model describing a nonrectangular hyperbola with LAI = 6 and with 60% erect leaves.

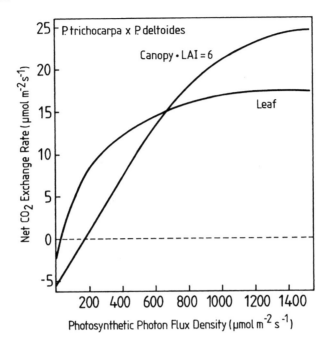

benefit from an increase in light. Different photosynthetic parameters, such as photosynthetic quantum efficiency, light saturated net photosynthesis (A_m), dark respiration rate and light compensation point, for leaves from different canopy positions can be modeled. Several models have already been built to calculate light penetration in a canopy and whole canopy photosynthesis (reviewed by Norman 1980 and Myneni et al. 1986).

As an example of such a modeling exercise, A_c was calculated using a simplified version of a Monteith–canopy photosynthesis model (Monteith 1965) and the output is shown in Fig. 3. In this example, LAI of a poplar stand equaled 6 and crown architecture was chosen in such a way that a horizontal layer with LAI = 1 transmits 60% of the incident light (the next layer would transmit 60% of the remaining 40%, etc.). The ratio between A_c (per unit ground area) and A_m may vary from 1.7 to more than 3.2 in a canopy with high LAI and tree species whose leaf photosynthesis reaches a plateau at relatively low radiation levels (Thornley 1976; Ceulemans and Saugier 1990). A six- to seven-fold difference in A_m at the leaf level between species or clonal genotypes is reduced to a factor of only 3–4 at the canopy flux level. With lower A_m rates differences in this ratio become larger.

To estimate the diurnal change in CO_2 uptake by a hybrid poplar canopy, another simple model using the well-known rectangular hyperbola was applied (Fig. 4). For this simulation a random leaf distribution made up of eight horizontal layers (each with LAI = 0.5) was used. The choice of a total LAI = 4 was based on maximum values observed in a widely spaced experimental field plantation in Flanders, Belgium (Lemeur and Impens 1981). The model assumes an exponential-type light attenuation with depth in the canopy, the extinction coefficient (see further below) that is dependent on height of direct solar radiation. For the calculation of the extinction coefficients an average leaf inclination of 40° was used. This value corresponds with the weighted value of the vertical distribution of the leaf inclination. Light interception was corrected for the fractions of photosynthetically active radiation transmitted through the leaves by decreasing the interception by 10%. Net photosynthesis was calculated for each canopy layer, and the total CO_2 uptake of the poplar canopy per unit of soil surface was obtained by summing the contributions of the separate layers and by multiplying the result by canopy LAI. This procedure was repeated for 10-min steps in order to simulate the course of net photosynthesis as a function of the daytime variation of incoming radiation. The simulated result is shown in Fig. 4. The total CO_2 uptake calculated for the simulated summer day was 61.7 $g \cdot m^{-2}$ (Lemeur and Impens 1981).

In contrast to some agronomic crops, where rather accurate measurements of gas exchange of entire vegetations or canopies can be made (using for example micrometeorologial or whole plant enclosure techniques), the size and dimensions of a tree stand make accurate gas exchange measurements at the canopy or stand level very difficult. A limited number of studies, however, have successfully estimated the CO_2 fluxes of forest stands using different methods (reviewed by Ceulemans and Saugier 1990). Values of CO_2 fluxes above canopies may range from 12 $\mu mol \cdot s^{-1} \cdot m^{-2}$ of soil surface area for *Pinus ponderosa* (Denmead and Bradley 1985) to 40 $\mu mol \cdot s^{-1} \cdot m^{-2}$ for young stands of rubbertree (Monteny 1989). However, very few measurements of canopy photosynthetic rates have been made on poplar stands thus far (Myneni et al. 1986).

The arrangement of leaves within the stand makes the stand more efficient than the individual tree in converting light into dry matter. Literature values for canopy photosynthesis rates of tree species and forest stands are only slightly lower than those reported for crops such as sunflower, rice, and maize. Because of differences in individual crown architecture (Michael et al. 1988; Ceulemans et al. 1990) and competition, as well as in shape and structure of the canopy, isolated trees and trees in a closed stand (for example under SRF) differ in growth and productivity. Not only is their carbon uptake (i.e., photosynthesis per unit of leaf area × total LAI) different, but also the quantity of light that is intercepted. Much potential light interception is wasted in the isolated tree when compared to groups of plants arranged in a canopy.

Fig. 4. Simulated daytime variation of incoming photosynthetic photon flux density (I_0) and the corresponding CO_2 exchange rate of a closed hybrid poplar canopy with LAI = 4. Average leaf inclination = 40° and leaf transmission = 10% (adapted from Lemeur and Impens 1981).

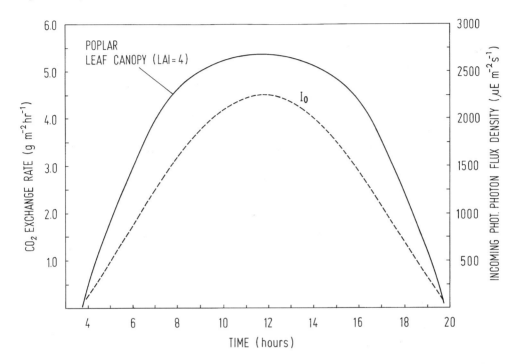

Extinction of light by the canopy

Theoretically the extinction coefficient, k, can be calculated from Lambert–Beer's law (Liou 1980), $I / I_o = e^{-k\,\text{LAI}}$ where I_o is the incident solar radiation above the canopy, I is the solar radiation below the canopy, k is the extinction coefficient, and LAI is the leaf area index. Low canopy extinction coefficients allow in principle a high LAI before full interception is attained, while canopies with an extinction coefficient close to one will have a low LAI (or leaves with an inclination angle close to horizontal) in order to get nearly complete radiation interception (Fig. 1). From Beer's law it follows that 95% interception is obtained when the product $k\,\text{LAI}$ equals three (Monteith 1965; Larcher 1980; Thornley and Johnson 1990). Assuming a minimum k of 0.23 and taking into account the leaf inclination angles of poplar (see Ceulemans and Isebrands, Chapter 15), 95+% interception can be obtained under temperate latitude conditions at LAIs of 13 for the *P. trichocarpa* × *P. deltoides* hybrid clones and of 8 for the *P. trichocarpa* clones. There seems to be a consensus in the literature for a need to describe the clumpiness of the foliage to properly

469

apply Lambert–Beer's law (Gower and Norman 1991; Fassnacht et al. 1994; Stenberg et al. 1995). However, there is very little information regarding the quantification of clumpy foliage in SRF poplar plantations, with the exception of some theoretical modelling studies (Chen et al. 1993).

In a field study in Flanders, Belgium, seasonal average extinction coefficients for solar radiation ranged from ca. 0.23 for *P. trichocarpa* × *P. deltoides* clone Beaupré with a more erectophile foliage (Figs. 5 and 6), to 0.37 for *P. trichocarpa* clone Fritzi Pauley (Ceulemans 1990; Ceulemans et al. 1992*b*). Clonal differences in extinction coefficient can be explained by their different foliar display (Figs. 5 and 6) and branching pattern (Ceulemans et al. 1990). In an experimental field plantation in the Pacific Northwest, USA, *k* values of 0.48 (third year) to 0.63 (fourth year) have been estimated (Heilman and Xie 1994). Fertilization had no effect on *k* in both years of the study (third and fourth year). These *k* values are comparable to data from other deciduous broadleaved tree canopies (Larcher 1980; Thornley and Johnson 1990).

Leaf orientation plays an important role in canopy architecture and radiation extinction (Fig. 5). Distribution of leaf angles within a canopy affects light penetration and its repartitioning between different levels or layers of leaves at different depths (Ledent 1978). Crops with erect leaves have a considerable yield advantage over those with horizontal leaves at high LAIs (Fig. 6), but leaf angle seems to be less important at low LAIs and low solar elevations (Isebrands and Michael 1986). In agricultural plant breeding the ideal type of cereal plants (wheat, rice, etc.) has been determined as having erect upper leaves (Ledent 1978), based on theoretical studies and mathematical models of light interception, as well as on field studies with real crops. Similarly, experimental data from hybrid poplar plantations with high LAIs have indicated the importance of leaf orientation for light interception by the poplar canopy (Figs. 2 and 6). Moreover, the spatial orientation of a single leaf in relation to the orientation of the sun affects the photosynthetic irradiance it receives. Thus leaf angle distributions are important for determining light interception (Fig. 6) and photosynthetic rates of canopies (see Ceulemans and Isebrands, Chapter 15). Furthermore, leaf angle distribution also affects the distribution of leaf temperatures, which in turn affects transpiration and to a lesser extent photosynthesis.

The extinction coefficient calculated for the middle part of the day (i.e., solar noon ± 4 h) remains quite constant during the growing season (Ceulemans et al. 1992*b*). Near the end of the growing season, extinction coefficients slightly increase due to a relatively large proportion of the radiation being intercepted by stems and branches rather than by leaves. In intensively cultured SRF stands these interception values by stems and branches in the winter season may be as high as 45–50% (Zavitkovski 1982; Scarascia-Mugnozza et al. 1989).

Fig. 5. Leaf orientation in poplar can be quantified by measuring three different angles. The leaf lamina angle gives an indication of the directional tilt of the leaf lamina. Its measurement on a selected *P. deltoides* clone from Mississippi is shown here.

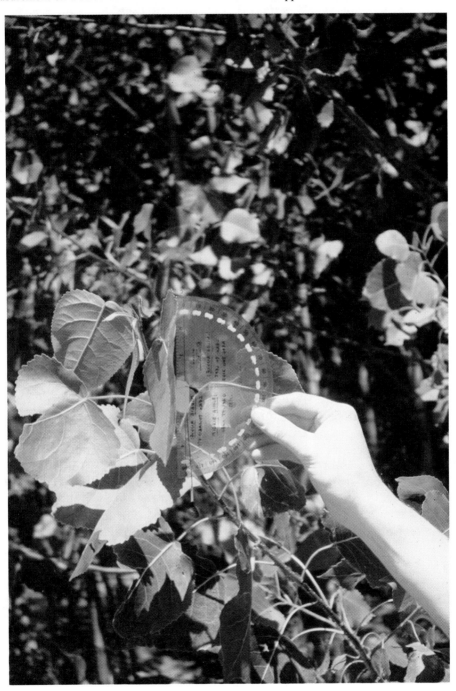

Fig. 6. Leaf inclination angle (determined as deviation from horizontal) versus solar radiation extinction coefficient for five poplar clones grown under intensive culture in an experimental field plantation in Flanders, Belgium. *P. trichocarpa* clones Fritzi Pauley (from Washington state) and Columbia River (from Oregon state), *P. deltoides* × *P. nigra* clone Robusta and *P. trichocarpa* × *P. deltoides* clones Beaupré and Raspalje.

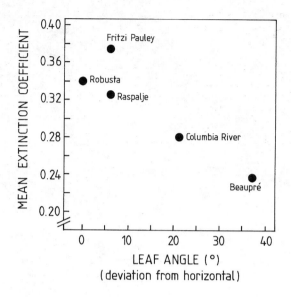

The way vertical distribution of leaf size and leaf area affects the penetration of solar radiation into the canopy, and thereby canopy photosynthesis, can be well illustrated by the approach applied to a *P. trichocarpa* × *P. deltoides* canopy and visualized in Fig. 7. Three different model canopy types were defined, having entirely different leaf size distributions but similar leaf area and leaf orientation distributions. The input data (i.e., leaf area, leaf orientation, cumulative LAI), summarized in Fig. 7, were experimentally observed on 8-year-old hybrid poplar trees in the field (model type I), while for model types II and III theoretical and abstract leaf size distributions were made up as shown. Model type II has the inverse leaf size distribution of the one observed in the field, while model type III assumed a unique and similar leaf size in all layers of the canopy. In all cases the total leaf area of the entire tree was 30 m^2. The procedural approach described by Myneni and Impens (1985) was used to quantify the penetration and absorption of parallel direct solar radiation at different levels within the canopy. A straightforward photosynthesis–light response function was then used to calculate gross photosynthesis (A, in Mg·m^{-2}·s^{-1}) as, $A = (\alpha I A_m)/(\alpha I + A_m)$, where α is the initial photosynthetic quantum efficiency (equal to 0.002 Mg CO$_2$·µmol^{-1} of photons), A_m is the photosynthetic rate at saturating light intensities (equal to 0.5 Mg CO$_2$·m^{-2}·s^{-1}), and I is the light flux density (µmol·m^{-2}·s^{-1}).

Fig. 7. Cumulative leaf area index, percentage of total tree leaf area, leaf inclination angle
(■) and average individual leaf area (cm^2 per leaf) as a function of tree height from the
crown base (m) (panel a). Effect of vertical leaf size distribution on the penetration of direct
solar radiation and canopy photosynthesis in an 8-year-old *P. trichocarpa* × *P. deltoides*
canopy (panels c and d, respectively, model type I). Average leaf area distribution for model
type I was obtained from experimental observations in the field, while model types II and III
represent abstract, hypothetical distribution functions (panel b). The total leaf area of an
individual tree was 30 m^2 in all three cases.

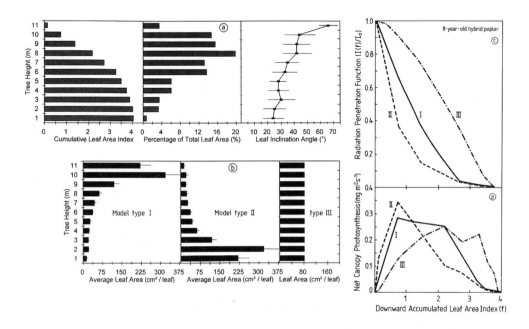

Calculated total canopy photosynthetic rates were 0.641 Mg $CO_2 \cdot s^{-1}$ for model
type I, 0.524 Mg $CO_2 \cdot s^{-1}$ for model type II and 0.685 for model type III. These
results showed that a canopy with a model type I leaf size distribution had a
total (canopy) photosynthesis of 22% more than model type II, and 7% less than
with an even distribution of leaves of one single leaf area (model type III). So,
the leaf size and leaf-orientation distributions observed in the field proved to be
a desirable attribute for optimal light interception and canopy photosynthesis.

Canopy information from remote sensing

A common problem in physiological ecology and production physiology involves scaling up physiological measurements that may be intensive, local, and
of short duration, to larger spatial and longer time scales. One viable means for
scaling up is remote sensing (Waring et al. 1995*b*). Remote sensing is attractive

because it provides a spatial scale ranging from a few meters to tens of kilometers at a temporal scale ranging from multiple-daily to monthly coverage. Remote sensing can be used to derive indirect or direct measures of canopy structure and physiology, and this information can be used to estimate vegetation cover, LAI, IPAR, leaf chemistry, canopy roughness, and canopy gas exchange and to map and classify vegetation (Adams et al. 1989; Baret and Guyot 1991; Chen et al. 1995; Gamon et al. 1993, 1995; Gong et al. 1995; Roberts et al. 1993; Running et al. 1995; Sellers et al. 1992; Spanner et al. 1990; Ustin et al. 1993; Waring et al. 1995*a*, 1995*b*).

Figure 8 is an AVIRIS Image (Airborne Visible/Infrared Imaging Spectrometer: provides a 224 band spectrum at a spatial resolution of 17.4 m) of a group of *Populus* plantations located near Wallula, Washington, USA. These plantations are clonal blocks of architecturally distinct clones: a planophile *P. trichocarpa* × *P. deltoides* (*T* × *D*) hybrid and a more erectophile *P. deltoides* × *P. nigra* (*D* × *Na* or *D* × *Nb*) hybrid. The AVIRIS data were collected on September 22, 1994 and ground truth data were collected a year later under phenologically similar conditions. Architecture at the subpixel scale was assessed using spectral mixture analysis to model canopies as mixtures of green leaves, nonphotosynthetic vegetation (e.g., branches, stems), exposed soil and architecturally derived shade and shadows. Larger scale canopy roughness was assessed by calculating interpixel variance for shade and green leaves for each stand.

The role of leaf angle distribution in modifying canopy reflectance is readily illustrated by comparing mixture models for the *T* × *D* (planophile) and *D* × *N* (erectophile) clones. The more horizontal leaves of *T* × *D* produced a significantly higher green leaf fraction and lower shade fraction than *D* × *N*. A comparison of AVIRIS spectra illustrates how architecture influenced the entire spectrum (Fig. 9). Canopy reflectance was high in *Populus* clones and the clone *T* × *D* had a higher reflectance than *D* × *N*. However, when individual leaves were removed from three canopy positions in the two clones and their individual reflectance spectra evaluated under laboratory conditions, no significant differences could be detected. Therefore, canopy level differences were related to branch and canopy level architectural features such as leaf angle.

The presence of large, single-clone plantations allows for many different tests of accuracy, repeatability, and power to be conducted at the leaf, branch, and stand scales. These tests will provide the foundation for interpreting remotely sensed images (or other scaling techniques) in much more heterogeneous stands. Traditional ground measures of production physiology when coupled with remote sensing offer a unique opportunity to understand process and production at much larger scales.

Fig. 8. AVIRIS (Airborne Visible/Infrared Imaging Spectrometer) image of *Populus* plantations near Wallula, WA. The circles are agricultural fields irrigated with a center pivot system. Spectra are provided for two different clones ($T \times D$: *P. trichocarpa* × *P. deltoides* and $D \times Nb$: *P. deltoides* × *P. nigra*) and for two different aged stands of one clone ($D \times Nb$).

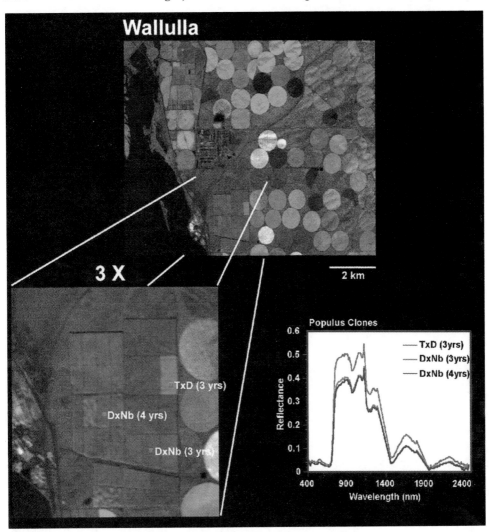

Belowground components and production

Although foliage and its display can be clearly linked to production, the below-ground canopy, although equally important, is not as amenable to study (see Pregitzer and Friend, Chapter 14). In spite of this difficulty, specific components of root growth and morphology have been linked to production. Poplars have an extensive root system that appears to be very effective in tapping soil resources.

Fig. 9. AVIRIS (Airborne Visible/Infrared Imaging Spectrometer) spectra for two *Populus* clones (*T × D*: *P. trichocarpa × P. deltoides* and D × *Na*: *P. deltoides × P. nigra*) located at Wallula, WA., and two stands of *Abies amabilis* (2nd: 47-year-old second growth, and Old: 225-year-old old-growth) located in the city of Seattle's Cedar River Watershed, WA.

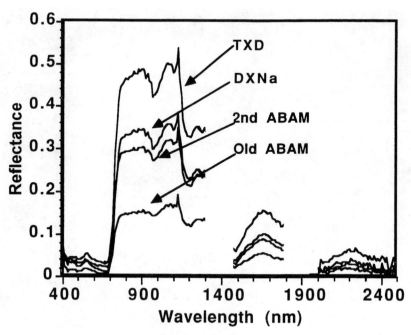

Considerable variation has been observed among clones of poplar species and hybrids in the relative quantity of carbon allocated to root systems, in the rate and pattern of development of roots on cuttings, in tolerance to flooding, and in root system configuration (Dickmann et al. 1988; Smit 1988; Heilman et al. 1994*a*, Hinckley et al. 1989, 1992). The size and arrangement of roots on cuttings, however, seems to exhibit several similarities among clones. For instance, the relationship of root orientation to position on the cutting is generally consistent among clones. Poplar clones seem primarily to differ in number and size of roots, depth, and orientation of growth. Significant differences among families and among both male and female parentage have been observed in the amount of vertical rooting (Heilman et al. 1994*a*, 1994*b*). In a study of a number of 4-year-old poplar clones, we found that mean fine root length density in the top 18 cm of soil averaged 6 cm·cm^{-3}. The $T × D$ clone with the most roots in that study had an estimated 296 000 km of roots per hectare to a depth of 3.2 m, a value far above most plant communities, except for grasslands.

In an experimental plantation study on the west side of the Cascades in the state of Washington, root systems of 2-year-old $T × D$ hybrid clones were characterized by long (3–4 m maximum length) and thin roots with an approximate horizontal orientation. As a result, most of the roots were found in the top 50 cm layer of soil when trees were excavated and horizontal as well as vertical distribution of coarse and fine roots were examined (Friend et al. 1991; Heilman et al. 1994a; Hinckley et al. 1989). Some roots, usually originating from the base of the original cutting, showed a more vertical orientation and could go as deep as 1–1.5 m in the first growing season.

The very low wood density of the coarse roots of poplar together with both the succulence and fineness of the fine roots (<0.06 mm in diameter) means that the root systems are developed with minimal carbon allocation. Furthermore, in contrast to other tree species, fine roots of poplars may be relatively long-lived (Dickmann and Pregitzer 1992). Such characteristics appear to favor allocation of carbon to aboveground production without sacrificing the effectiveness of the root system. A negative correlation of root:shoot ratio with first-year aboveground weight ($R^2 = 0.21$) for $T \times D$ hybrids indicates high aboveground production can occur with low root:shoot ratios (Heilman et al. 1994b). On the other hand, fast-growing clones with low root:shoot ratios can be more susceptible to windthrow (Harrington and DeBell 1996).

In terms of patterns of allocation of carbon to the roots, a progressive shift in allocation of ^{14}C-labeled photosynthate from the upper stem and leaves to the lower stem and to the roots has been observed during the first two growing seasons (Scarascia-Mugnozza 1991, see Fig. 11, Chapter 15). For the five clones in this study, leaves on the main stem exported more labeled ^{14}C than did leaves on branches. Definite differences, however, existed between clones. Hybrid clone 44–136 ($T \times D$), for example, maintained a high allocation to components of belowground biomass (cutting, coarse, and fine roots) throughout the entire growing period, while others only allocate to roots in the latter part of the season (Stettler et al. 1988; Hinckley et al. 1989, 1992; Scarascia-Mugnozza 1991). Although the root systems of hybrid clones 11–11 and 44–136 were not greatly different, that of clone 44–136 had longer and less fibrous roots. Root:shoot ratios at the end of the first growing season were 0.59 and 0.27 in hybrid clones 44–136 and 11–11, respectively. By the end of the second growing season, these root:shoot ratios had declined in both clones and approached 0.18.

In the above study, branches contributed most of their exported photosynthates to the portion of the stem below their attachment points and to belowground components. Only a small fraction of the label appeared in the stem section above the point of insertion of the branch. These components received, from the beginning to the end of the second growing season, proportions ranging from 20 to 55% of the exported assimilates. The remainder was either retained within the labeled foliage and branch section, or respired (Hinckley et al. 1989; Scarascia-Mugnozza 1991).

Finally, another illustration of the considerable clonal variation in root:shoot ratios has been given by Dickmann et al. (1988) and Michael et al. (1988). When comparing the poplar clones Tristis (*P. tristis* × *P. balsamifera*) and Eugenei (*P. ×euramericana*, Table 1) these researchers found two completely different adaptive strategies with different root:shoot ratios and carbon allocation patterns. The indeterminate clone Eugenei was quite plastic in its aboveground responses to environmental changes, but very nonplastic below ground. The determinate clone Tristis was just the opposite, showing few changes above ground, but major changes below ground in response to environment.

Clonal differences in morphology, developmental pattern, and sink strength of roots play a key role in the genotypic interaction between roots under stress and the response of the shoot. Survival, drought-avoidance capability and adaptation to different sites are for each individual poplar clone important aspects resulting from this root × shoot interaction. Therefore, the strong genetic control of phenology and carbon allocation to the roots have important implications for genotype × environment interactions in poplar breeding programs.

One aspect of poplar root systems that has not received much attention in relation to short rotation culture is the role of mycorrhizae. It is known that under natural forest conditions, poplar roots are mycorrhizal. Like most trees in the *Salicaceae* family, they normally form ectomycorrhizae (Marx 1972). However, where roots are frequently flooded and in their early years on gravelly, low-fertility sites, poplars show only vesicular arbuscular mycorrhizae (VAM, J.M. Trappe, personal communication). With advancing stand age and the associated buildup of soil organic matter, infection shifts to ectomycorrhizal species. Those circumstances where the roots are nonmycorrhizal are in water courses where the roots are under more or less permanent inundation, and probably in very fertile soils where ectomycorrhizal development is usually inhibited (Harley 1969).

Mycorrhizal fungi are supported by their symbiotic partners for their carbon nutrition and energy supply (Söderström 1991). Carbon used by these fungi can be as much as 15–20% of the carbon assimilated by the plant (Söderström 1991). Because of the benefits to the plant in terms of increased availability of soil N, P, and water, the carbon used by the fungi might be considered an investment resulting in increased growth. However, on more favorable sites where supplies of nutrients and water are abundant and mycorrhizal development is less pronounced, the carbon otherwise allocated to mycorrhizal fungi can contribute instead to plant growth. Part of the growth response obtained from fertilizer applications may, therefore, come from reduced allocation of carbon to these fungi.

Ectomycorrhizal fungi can also deter infection of feeder roots by certain pathogenic fungi (Marx 1973). Under conditions of high fertility but where pathogenic fungi such as *Phytophthora*, *Pythium* and *Fusarium* are present, the protection of feeder roots from pathogens could be the major benefit to the plant (Marx 1973). With many of the ectomycorrhizal soil fungi being capable of producing antibiotics, it is even possible that a fungal symbiont may afford protection to adjacent nonmycorrhizal roots (Marx 1972).

The two aspects of mycorrhizae of most relevance to high production are the carbon needs of these fungi versus the carbon available for plant growth and the effects of mycorrhizae on limiting root diseases. Under the high fertility and high moisture conditions sought for high production, increased availability of nutrients and water resulting from the presence of mycorrhizae is probably of minor significance to the trees. Since the level of mycorrhizal development is significantly less under these conditions, more carbon is available to trees for their growth. However, because of the role of mycorrhizal fungi in deterring root disease, a presence of mycorrhizae in high production poplar plantations seems desirable and artificial inoculation may be necessary.

Environmental controls of production

Climate

Climate is probably the most important fundamental factor influencing production. Many of the environmental factors associated with climate, such as water, light, temperature, humidity, etc., are also discussed in Chapter 17 by Neuman and others. For production at the stand level, the factors of climate that seem most critical are the moisture regime, the temperature regime, and the cumulative quantity of solar radiation over the growing season. Total precipitation and humidity, as they influence the quantity of water available to plants, are the most important components of the moisture regime. However, as we are seeing in eastern Oregon and Washington, USA, irrigation is being employed on poplar plantations to overcome the water deficits of an arid region. Water requirements are addressed in a later section of this chapter. Humidity is also important with respect to poplar diseases.

Accumulated solar radiation is a product of (1) length of the growing season, (2) radiation intensity, which is influenced by both sun angle and atmospheric clarity (i.e., cloudiness and haze), and (3) daylength during the growing season, which is primarily influenced by latitude. Added complexity results from interactions of solar radiation with both temperature and vapor pressure deficits. Although data are somewhat incomplete (e.g., no information on the seasonal total PAR), an examination of a 30-year record of hours of "bright" sunshine at locations in British Columbia, Canada, with its great breadth of latitude, helps illustrate the interaction between radiation and other climatic factors. Despite a 22° spread in latitude, the most important factor in limiting hours of bright sun in the growing season seems to be cloudiness of the local climate. For a comparison of two locations at 54°N latitude, the cloudy, coastal climate of Prince

Rupert has almost a 50% longer growing season than the drier continental climate at Prince George, but Prince Rupert has 60% less growing season sunlight. The effects of longer days were also evident. Comparing two locations in dry and sunny climates, Fort St. John at 56°N latitude and Penticton at 49.5°N latitude, we find that the growing season at Penticton is about 75% longer than at Fort St. John, but Penticton receives only 46% more growing season sunlight than Fort St. John. If there is a discrepancy in growing season cloudiness at these two locations, it would probably be that Fort St. John is more cloudy, thus underestimating, in this comparison, the effect of latitude on growing season sunlight.

The temperature regime is important because of direct and indirect effects on the physiology of the plant (e.g., carbon exchange, metabolism, and water loss) and indirect effects on soil processes. Perhaps the most important physiological effect of temperature is on respiration because respiration increases exponentially with an increase in temperature (Jarvis and Leverenz 1983). Two implications of this relationship are important. The first is that where day and night temperatures differ by much, the diurnal mean temperature will underrate actual respiration compared with a climate where day and night temperatures are more similar. Although maintenance respiration acclimates to increases in temperature (Farrar and Williams 1991), temperatures above the optimum for net photosynthesis decrease productivity. Productivity of most C_3 plants is very sensitive to high daytime temperatures and high vapor pressure deficits, a situation often leading to no or negative net carbon gains (Jarvis 1994). The very high productivity for hybrid poplars in the Pacific Northwest of North America as well as for its native evergreen, conifer forests may well be because of its generally cool temperatures (as is characteristic in the more humid "westside" environment). The problem of the low summer rainfall which is also characteristic of the region seems to be compensated by the benefit of high solar radiation. Nevertheless, productivity, as it relates to *Pseudotsuga menziesii*, can be restricted in those growing seasons with below normal rainfall, occurring on the average about every seven years (Robertson and Spittlehouse 1990). Access to groundwater appears to be a critical feature of highly productive "westside" plantations. With irrigation, high production is being obtained in the high temperature, high irradiance "eastside" areas of Oregon and Washington, USA — higher production, in fact, than is being obtained in the cooler and more humid western portions of these two states. Consequently, it would appear that the poplars are able to capture CO_2 at relatively high temperatures (at least when compared to evergreen conifers) as long as there is an abundant supply of water.

The ability of poplar plantations to grow satisfactorily over a range of climates results partly from the relative adaptability of poplar species, hybrids and/or individual clones to diverse environments. However, given the wide range of climatic conditions available for poplar, success ultimately depends on how well a clone is matched to the climate. Our testing of *Populus* hybrids has revealed important differences among clones in their adaptability to climates. As would be expected, those hybrids with more southern parents show preferences for warmer and more southerly locations, whereas those with more northern parents prefer cooler and more northerly locations. The most robust clones are those that show no real preferences, doing well over a fairly broad range of temperatures (and latitudes). A new laboratory technique has been developed that is claimed to be able to identify clones of trees with superior growth (Anekonda et al. 1994) and to match cultivars or clones to appropriate temperature regimes (Criddle et al. 1995). It involves measurement of the temperature sensitivity of the rate and efficiency of respiration by the plant (Hansen et al. 1995). Using this technique, relative sensitivity of cultivars to temperature can reportedly be determined and, in addition, for those that are sensitive, the optimum temperature range for growth can be estimated. The technique has not yet been fully tested for poplars, but it shows promise in initial work with several clones (R.W. Breidenbach, personal communication).

Water relations

Poplar trees are known as preferring sites with an abundant and continuous supply of moisture and, by way of anecdotal statements, poplar trees are thought to lose copious quantities of water. A comparison of their maximum stomatal and leaf conductances with other species would certainly suggest an ability to lose tremendous quantities of water. For example, the maximum stomatal conductance for hybrid poplar (*P. trichocarpa* × *P. deltoides*) approaches 600 mmol·m^{-2}·s^{-1} while for *Quercus alba* and *Pseudotsuga menziesii* it approaches 300 and 100 mmol·m^{-2}·s^{-1}, respectively. Based on these differences and similar leaf areas per tree, one would expect large differences in individual tree water loss.

In a study of 4-year-old hybrid *Populus* ($T \times D$) trees growing in a stand and evaluated over a wide range of atmospheric evaporative demand conditions, the maximum calculated rate of stand water loss was 4.92 mm·d^{-1} (Hinckley et al. 1994). Noteably, this is a value slightly less than the maximum of 5.0 mm·d^{-1} observed by Fritschen et al. (1973) for *Pseudotsuga menziesii*. Leaf area indices were approximately the same in the two species whereas individual stomatal conductances were 3–5 times higher in *Populus* than in Douglas-fir. However, similar rates of stand transpiration could arise from different partitioning of total vapor phase conductance (crown or canopy conductance) between stomatal and boundary layer components. For example, in a conifer stand, the boundary layer component of conductance will be very high but stomatal conductance rather low. In the poplar stand, the stomatal component was rather high and the boundary layer low. As a consequence, the total conductance may have been similar in the two stands. All other things being equal (LAI, soil moisture, and net radiation), the two stands could then have similar rates of transpiration. A number of authors have suggested that once canopies close, very similar rates of transpiration may occur regardless of species, LAI, etc. (Whitehead and Jarvis 1981; Kelliher et al. 1993). As demonstrated by Meinzer and Grantz (1991), a number of different structural, morphological, and physiological properties and mechanisms may be responsible for these apparent similarities.

Nutrient relations

Poplars have a relatively high nutrient content, thus high demands for nutrients. Consequently, high rates of productivity in poplar require high nutrient uptake (Table 2). Annual uptake of nitrogen in above- and below-ground biomass in productive 4-year-old hybrids was estimated to be about 350 kg·ha^{-1}. Fourth-year uptake of nitrogen by rapidly growing plantations is probably close to the maximum, since by that age the canopy is fully developed and annual production is very high. For this reason, most sites will eventually, if not initially, require addition of fertilizer to achieve and maintain high production. Nitrogen is the major nutrient required in fertilization, but phosphorus and zinc can also be beneficial on certain sites. Lime containing calcium and sometimes magnesium can also give a response on some sites. Addressing methods to diagnose soil fertility and the fertilizer requirements for specific sites is beyond the scope of this chapter (see Heilman 1992).

Analysis of the response of six $T \times D$ hybrids to nitrogen fertilization showed a close association between response of canopy leaf area and stand volume growth (Heilman and Xie 1994). The trees were treated in spring at the beginning of their second year of growth with 167 kg·N·ha^{-1}·yr^{-1}. No response was evident that year either in increased bole growth or canopy development. A second nitrogen application of the same magnitude was made in spring at the beginning of the third growing season. That year, LAI was 58% higher on the fertilized plots than on the controls, and aboveground leafless biomass showed a 40% growth increase. The third application of nitrogen at the same rate was made at the beginning of the fourth growing season. In that fourth year, LAI in fertilized plots increased 31% over third-year values, but on control plots, it increased 71%. Consequently, the fourth year increment in dry weight of biomass on fertilized plots was only 8% higher (nonsignificant) than the increment on control plots. With increased LAI (L) on control plots, coupled with lower efficiency (ε) of the fertilized canopies to produce biomass (mean ε on fertilized plots was 5.3 Mg·ha^{-1}·yr^{-1}·L^{-1}, on control plots, ε was 5.8 Mg·ha^{-1}·yr^{-1}·L^{-1}), only a small response to nitrogen occurred in the fourth year.

The above study by Heilman and Xie (1994) also demonstrated a close relationship between annual biomass production and LAI ($R^2 = 0.96$). Similar to other broadleaf trees, the relationship of production to LAI was approximately linear up to about a LAI of 6. Above that point, productivity continues to increase with increasing LAI, but at a declining rate of increase. Control plants in the third

Table 2. Biomass, nutrient content, and estimated annual uptake of N in 4-year-old clones of *Populus* (from Heilman and Stettler 1986).

	Black cottonwood		Hybrids	
	Low-yielding	High-yielding	Robusta (Euroamerican)	$T \times D$ (Interamerican)
Number of clones studied	3	3	1	2
Aboveground leafless biomass at 4 years (mg·ha^{-1})	29–58	65–72	44	110–111
Nutrient content of above (kg·ha^{-1})				
N	95–148	169–223	241	410–420
P	14–25	30–34	41	77–105
Ca	80–181	182–205	159	268–288
Leaf fall in 4th year (mg·ha^{-1})	4.4–5.1	5.0–5.6	4.9	5.9–6.6
Nitrogen content of leaf fall in 4th year (kg·ha^{-1})	62–73	69–80	82	80–84
Annual uptake of N in aboveground biomass in 4th year including leaves* (kg·ha^{-1})	95–116	134–159	168	271–276
Annual uptake of N in above- and below-ground biomass in 4th year[†] (kg·ha^{-1})	125–150	175–210	220	350–360

*Based on leaf fall N; thus, considered 'net' uptake since considerable translocation of N occurs prior to abcission.

[†]Based on the estimate of roots containing about 30% of the nutrients present in aboveground leafless biomass.

year of growth had LAIs and annual productions as low as 2 m^2·m^{-2} and 6.5 Mg·ha^{-1}·y^{-1}. In contrast, nitrogen-fertilized stands in the fourth year had LAIs and annual productions near 13.5 m^2·m^{-2} and 70 Mg·ha^{-1}·y^{-1}, respectively. Fertilization appeared to accelerate the development of stand LAI. The average fertilized plot was more productive than the average control; however, individual control plots approached the production of the most productive fertilized plot, persumably due to site variability. The study also documented lower efficiency of the canopy in producing biomass at high levels of LAI. Consequently, even though efficiency of the canopies on fertilized plots was not, on the average, significantly reduced, lower efficiencies were evident for those canopies with the highest LAIs. Clearly, further study of the relationship between canopy development, LAI values, productivity, and nutrition of clones is needed before we can develop appropriate management regimes for increased yields of short rotation plantations.

Both species and clones within species exhibit differences in nutritional relationships. Heilman (1985) showed that clones of *P. trichocarpa* differed significantly in concentration of nitrogen in their foliage. Additionally, Curlin (1967) found differences in response to nitrogen fertilizer among clones of *P. deltoides*. Similarly, Heilman and Xie (1993) showed significant differences in response to nitrogen among six $T \times D$ hybrids.

Another aspect of nutritional difference among species and clones relates to fertilizer use efficiency. For instance, the species *P. trichocarpa* appears to be much more efficient in using nitrogen and calcium than either *P. deltoides* or hybrids between the two species (Heilman and Stettler 1986). In that study, black cottonwood clones produced between 310 and 420 kg woody biomass·kg N^{-1} contained in that biomass. This compares to only 180 kg woody biomass for the Euramerican hybrid 'Robusta' and between 260 and 270 kg woody biomass·kg N^{-1} for the $T \times D$ hybrids. Furthermore, *P. trichocarpa* produced about twice as much woody biomass·kg P^{-1} as did either the $T \times D$ or Euramerican hybrids. With respect to calcium, the Euramerican hybrid was substantially below both *P. trichocarpa* and the $T \times D$ hybrids in efficiency of woody production for this nutrient.

Summary and conclusions

Fast-growing and high-producing hybrid genotypes incorporate larger LAI, longer leaf-area duration and more favorable leaf display than slower-growing clones or parental species under SRF regimes. Maximum leaf area indices range from 6 for *P. trichocarpa* and *P. deltoides* to more than 11 for hybrid clones. The quantity and the display of the leaf area present are two of the most important variables determining light interception. Superior clones incorporate low extinction coefficients and high canopy conversion efficiencies. Accumulated biomass is found to be closely correlated with total leaf area as well as with accumulated intercepted solar radiation, and this correlation is genetically controlled.

One can see a wide range of links between the topics covered in this book and the specific topic of production physiology. In a sense, production physiology is the integrated outcome of the interaction between genetics, environment, stand history, and organ-level physiology. We have made great strides in bringing these various facets together in our understanding of the controls of stand productivity. Yet, we are just in the infancy of understanding these links. For example, how leaf area in older plantations can be easily measured (either allometrically or remotely) and how different LAIs are associated with specific rates of canopy gas exchange are still unanswered questions. Our understanding of the links between climate, including the increase in carbon dioxide levels over the last 150 years from 260 to 360 ppm, biotic stresses, soils, genotype, stand density and development and productivity, is still very tenuous. Quantifying how many of the large features of climate or site affect production can be done. It is even possible to ameliorate specific limiting factors. However, achieving a comprehensive understanding by which a client can maximize yield on a specific site by a deliberate combination of genotype and management regime is still miles away. The beauty of the *Populus* system is that its biology makes it highly amenable for study and that there is a critical mass of scientists and users who are eager to learn more about it.

References

Adams, J.B., Smith, M.O., and Gillespie, A.R. 1989. Simple models for complex natural surfaces: A strategy for the hyperspectral era of remote sensing. *In* Proceedings of the IGARSS'89: 12th Canadian Symposium on Remote Sensing, Vancouver, BC, July 1989. Vol. 1. pp. 16–21.

Anekonda, T.S., Criddle, R.S., and L.D. Hansen. 1994. Bioenergy '94 Sixth NTL Bioenergy Conference, Reno/Sparks, Nevada. Oct. 2–6, 1994. pp. 665–671.

Baret, F., and Guyot, G. 1991. Potentials and limits of vegetation indices for LAI and APAR assessment. Remote Sens. Environ. **35**: 161–173.

Cannell, M.G.R. 1989. Light interception, light use efficiency and assimilate partitioning in poplar and willow stands. *In* Biomass production by fast-growing trees. *Edited by* J.S. Pereira and J.J. Landsberg. Kluwer Academic Publishers, Dordrecht, The Netherlands. pp. 1–12.

Cannell, M.G.R., Milne, R., Sheppard, L.J., and Unsworth, M.H. 1987. Radiation interception and productivity of willow. J. Appl. Ecol. **24**: 261–278.

Cannell, M.G.R., Sheppard, L.J., and Milne, R. 1988. Light use efficiency and woody biomass production of poplar and willow. Forestry, **61**: 125–136.

Ceulemans, R. 1990. Genetic variation in functional and structural productivity determinants in poplar. Thesis Publish., Amsterdam, The Netherlands. 101 p.

Ceulemans, R., and Saugier, B. 1990. Photosynthesis. Chapter 2. *In* Physiology of Trees, *Edited by* A.S. Raghavendra. John Wiley & Sons Inc., New York & London. pp. 21–50.

Ceulemans, R., Stettler, R.F., Hinckley, T.M., Isebrands, J.G., and Heilman, P.E. 1990. Crown architecture of *Populus* clones as determined by branch orientation and branch characteristics. Tree Physiol. **7**: 157–167.

Ceulemans, R., Scarascia-Mugnozza, G.E., Wiard, B.M., Braatne, J.H., Hinckley, T.M., Stettler, R.F., Isebrands, J.G., and Heilman, P.E. 1992a. Production physiology and morphology of *Populus* species and their hybrids grown under short rotation. Can. J. For. Res. **22**: 1937–1948.

Ceulemans, R., Impens, I., Mau, F., Van Hecke, P., and Chen, S.G. 1992b. Dry mass production and solar radiation conversion efficiency of poplar clones. *In* Biomass for energy, industry and environment. *Edited by* G. Grassi, A. Collina, and H. Zibetta. Elsevier Science Publishing, New York, NY. pp. 157–163.

Chen, S.G., Shao, B.Y., and Impens, I. 1993. A computerised numerical experimental study of average solar radiation penetration in plant stands. J. Quant. Spectrosc. Radiat. Transfer, **49**: 651–658.

Chen, S.G., Ceulemans, R., and Impens, I. 1994. A fractal-based *Populus* canopy structure model for the calculation of light interception. For. Ecol. Manage. **69**: 97–110.

Chen, Z.,. Tsang, L, and Zhang, G. 1995. Microwave scattering by vegetation based on wave approach and stochastic Lindenmayer system. Microwave and Optical Techn. Letters, **8**: 30–33.

Criddle, R.S., Anekonda, T.S., Breidenbach, R.W., and Hansen, L.D. 1995. Site fitness and growth rate selection of *Eucalyptus* for biomass production. Thermoclimica Acta, **251**: 335–349.

Curlin, J.W. 1967. Clonal differences in yield response of *Populus deltoides* to nitrogen fertilization. Soil Sci. Soc. Am. Proc. **31**: 276–280.

Demetriades-Shah, T.H., Fuchs, M., Kanemasu, E.T., and. Flitcroft, I.D. 1994. Further discussions on the relationship between cumulated intercepted solar radiation and crop growth. Agric. For. Meteorol. **68**: 193–207.

Denmead, O.T., and Bradley, E.F. 1985. Flux – gradient relationships in a forest canopy. *In* The forest – atmosphere interaction. *Edited by* B.A. Hutchinson and B.B. Hicks. Reidel Publish. Co., Dordrecht & Boston. pp. 421–442.

Dickmann, D.I., and. Pregitzer, K.S. 1992. The structure and dynamics of woody plant root systems. *In* Ecophysiology of short rotation forest crops. *Edited by* C.P. Mitchell, J.B. Robertson, T. Hinckley, and L. Sennerby-Forse. Elsevier Applied Science, London and New York. pp. 95–123.

Dickmann, D.I., Pregitzer, K.S., and Nguyen, P.V. 1988. Net assimilation and photosynthate allocation of *Populus* clones grown under short rotation intensive culture: physiological and genetic responses regulating yield. Report to the U.S. Department of Energy, Michigan State University, East Lansing, MI. 86 p.

Dunlap, J.M., and Stettler, R.F. 1996. Genetic variation of *Populus trichocarpa* and its hybrids. IX. Phenology and Melampsora rust incidence of native black cottonwood clones from four river valleys in Washington. For. Ecol. Manage. In press.

Dunlap, J.M., Heilman, P.E., and Stettler, R.F. 1992. Genetic variation and productivity of *Populus trichocarpa* and its hybrids. V. The influence of ramet position on 3-year growth variables. Can. J. For. Res. **22**: 849–857.

Farrar, J.F., and Williams, M.L. 1991. The effects of increased atmospheric carbon dioxide and temperature on carbon partitioning, source-sink relations and respiration. Plant Cell Environ. **14**: 819–830.

Fassnacht, K.S., Gower, S.T., Norman, J.M., and McMurtrie, R.E. 1994. A comparison of optical and direct methods for estimating foliage surface area index in forests. Agric. For. Meteorol. **71**: 183–207.

Friend, A.L., Scarascia-Mugnozza, G., Isebrands, J.G., and Heilman, P.E. 1991. Quantification of two-year-old hybrid poplar root systems: morphology, biomass, and ^{14}C distribution. Tree Physiol. **8**: 109–119.

Fritschen, L.J., Cox, L., and Kinerson, R. 1973. A 28-meter Douglas-fir in a weighing lysimeter. For. Sci. **19**: 256–261.

Gamon, J.A., Field, C.B., Roberts, D.A., Ustin, S.L., Valentini, R. 1993. Functional patterns in an annual grassland during an AVIRIS overflight. Remote Sens. Environ. **44**: 239–253.

Gamon, J.A., Field, C.B., Joel, G., Goulden, M.L., Griffin, K.L., Hartley, A.E., Penuelas, J., and Valentini, R. 1995. Relationships between NDVI, canopy structure, and photosynthetic activity in three California vegetation types. Ecol. Appl. **5**: 28–41.

Gong, P., Pu, R., and Miller, J.R. 1995. Coniferous forest leaf area index estimation along the Oregon transect using compact airborne spectrographic imagery data. Photogram. Eng. Remote Sens. **61**: 1107–1117.

Gower, S.T, and Norman, J.M. 1991. Rapid estimation of leaf area index in conifer and broadleaf plantations. Ecology, **72**: 1896–1990.

Hansen, L.D., Hopkins, N.S., Taylor, D.K., Anekonda, T.S., Rank, D.R., Breidenbach, R.W., and Criddle, R.S. 1995. Plant calorimetry. Part 2. Modeling the difference between apples and oranges. Thermoclimica Acta, **250**: 215–232.

Harley, J.L. 1969. The biology of mycorrhiza. Leonard Hill, London, England. 334 p.

Harrington, C.A., and DeBell, D. 1996. Above- and below-ground characteristics associated with wind toppling in a young *Populus* plantation. Trees. In press.

Heilman, P.E. 1985. Sampling and genetic variation of foliar nitrogen in black cottonwood and its hybrids in short rotation. Can. J. For. Res. **15**: 1137–1141.

Heilman, P.E. 1992. Sustaining production: Nutrient dynamics and soils. *In* Ecophysiology of short rotation forest crops. *Edited by* C.P. Mitchell, J.R. Ford-Robertson, T. Hinckley, and L. Sennerby-Forse. Elsevier Applied Science, London and New York. pp. 216–226.

Heilman, P.E., and Stettler, R.F. 1986. Nutritional concerns in selection of black cottonwood and hybrids clones for short rotation. Can. J. For. Res. **16**: 860–863.

Heilman, P.E. and Xie, F. 1993. Influence of nitrogen on growth and productivity of short-rotation *Populus trichocarpa* × *Populus deltoides* hybrids. Can. J. For. Res. **23**: 1863–1869.

Heilman, P.E., and Xie, F. 1994. Effects of nitrogen fertilization on leaf area, light interception, and productivity of short-rotation *Populus trichocarpa* × *Populus deltoides* hybrids. Can. J. For. Res. **24**: 166–173.

Heilman, P.E., Ekuan, G., and Fogle, D.B. 1994*a*. First-order root development from cuttings of *Populus trichocarpa* × *P. deltoides* hybrids. Tree Physiol. 14: 911–920.

Heilman, P.E., Ekuan, G., and Fogle, D. 1994*b*. Above- and below-ground biomass and fine roots of 4-year-old hybrids of *Populus trichocarpa* and *Populus deltoides* and parental species in short-rotation culture. Can. J. For. Res. **24**: 1186–1192.

Hinckley, T.M., Ceulemans, R., Dunlap, J.M., Figliola, A., Heilman, P.E., Isebrands, J.G., Scarascia-Mugnozza, G., Schulte, P.J., Smit, B., Stettler, R.F., Van Volkenburgh, E., and Wiard, B.M. 1989. Physiological, morphological and anatomical compoments of hybrid vigor in *Populus*. *In* Structural and functional responses to environmental stresses. *Edited by* K.H. Kreeb, H. Richter, and T.M. Hinckley. SPB Academic Publishing, The Hague. pp. 199–217.

Hinckley, T.M., Braatne, J., Ceulemans, R., Clum, P., Dunlap, J., Neuman, D., Smit, B., Scarascia-Mugnozza, G., and Van Volkenburgh, E. 1992. Growth dynamics and canopy structure. *In* Ecophysiology of short rotation forest crops. *Edited by* P. Mitchell, L. Sennerby-Forsse, and T.M. Hinckley. Elsevier Applied Science, London and New York. pp. 1–34.

Hinckley, T.M., Brooks, J.R., Cermak, J., Ceulemans, R., Kucera, J., Meinzer, F.C., and Roberts, D.A. 1994. Water flux in a hybrid poplar stand. Tree Physiol. **14**: 1005–1018.

Impens, I., Van den Bogaert, J., Mau, F., Van Hecke, P., and Ceulemans, R. 1990. Biomass production in first and second year intensively cultured poplar clones as related to leaf photosynthetic performance and radiation intercepting foliage canopy. *In* Biomass for energy and industry. *Edited by* G. Grassi, G. Gosse, and G. Dos Santos. Elsevier Science Publishing, London & New York. pp. 1.439–1.445.

Isebrands, J.G., and Michael, D. 1986. Effects of leaf morphology and orientation on light interception and photosynthesis in *Populus*. *In* Crown and canopy structure in relation to productivity. *Edited by* T. Fujimori and D. Whitehead. Forestry Research Institute, Tsukuba, Japan. pp. 359–381.

Isebrands, J.G., Nelson, N.D., Dickmann, D.I., and Michael, D. 1983. Yield physiology of short rotation intensively cultured poplars. U.S. Dep. Agric. For. Serv., Gen. Tech. Rep. NC-91. pp. 77–93.

Jarvis, P.G. 1994. Capture of carbon dioxide by a coniferous forest. *In* Resource capture by crops. *Edited by* J.L. Monteith, R.K. Scott, and M.H. Unsworth. Nottingham University Press, Nottingham, England. pp. 351–374.

Jarvis, P.G., and Leverenz, J.W. 1983. Productivity of temperate, deciduous and evergreen forests. Physiol. Plant Ecol. **4**: 234–280.

Kelliher, F.M., Leuning, R., and E.-D. Schulze. 1993. Evaporation and canopy characteristics of coniferous forests and grasslands. Oecologia, **95**: 153–163.

Landsberg, J.J., and Wright, L.L. 1989. Comparisons among Populus clones under intensive culture conditions, using an energy conversion model. For. Ecol. Manage. **27**: 129–147.

Larcher, W. 1980. Physiological plant ecology, 2nd edition. Springer-Verlag, Berlin, Germany. 303 p.

Ledent, J.F. 1978. Mechanisms determining leaf movement and leaf angle in wheat (*Triticum aestivum* L.). Ann. Bot. (London), **42**: 345–351.

Lemeur, R.J.P., and Impens, I. 1981. Stand architecture, biomass production, energy flow and production efficiency in a man-made poplar ecosystem. *In* Proceedings of the XVII IUFRO World Congress, Division 1. September 6–17, 1981, Kyoto, Japan. pp. 205–225.

Linder, S. 1985. Potential and actual production in Australian forest stands. *In* Research for forest management. *Edited by* J.J. Landsberg and W. Parsons. Commonwealth Scientific Industrial Research Organization, Melbourne, Australia. pp. 11–35.

Liou, K.L. 1980. An introduction to atmospheric radiation. Academic Press, New York, NY.

Marx, D.H. 1972. Ectomycorrhizae as biological deterrents to pathogenic root infections. Annu. Rev. Phytopathol. **10**: 429–454.

Marx, D.H. 1973. Mycorrhizae and feeder root diseases. *In* Ectomycorrhizae. Academic Press, Inc. New York, NY. pp. 351–382.

Meinzer, F.C., and Grantz, D.A. 1991. Coordination of stomatal, hydraulic, and canopy boundary layer properties: Do stomata balance conductances by measuring transpiration? Physiol. Plant. **83**: 324–329.

Michael, D.A., Isebrands, J.G., Dickmann, D.I., and Nelson, N.D. 1988. Growth and development during the establishment year of two *Populus* clones with contrasting morphology and phenology. Tree Physiol. **4**: 139–152.

Milne, R., Sattin, M., Deans, J.D., Jarvis, P.G., and Cannell, M.G.R. 1992. The biomass production of three poplar clones in relation to intercepted solar radiation. For. Ecol. Manage. **55**: 1–14.

Monteith, J.L. 1965. Light distribution and photosynthesis in field crops. Ann. Bot. **29**: 17–37.

Monteith, J.L. 1977. Climate and efficiency of crop production in Britain. Philos. Trans. R. Soc. London, B **281**: 277–294.

Monteith, J.L. 1981. Does light limit crop production? *In* Limiting plant productivity. *Edited by* C.B. Johnson. Butterworth & Co., London, England. pp. 23–38.

Monteith, J.L. 1994 Validity of the correlation between intercepted radiation and biomass. Agric. For. Meteorol. **68**: 231–220.1977

Monteny, B.A. 1989. Primary productivity of a Hevea forest in Ivory Coast. Ann. Sci. For. **46** (Suppl.): 502–506.

Myneni, R.B., and Impens, I. 1985. A procedural approach for studying the radiation regime of infinite and truncated foliage spaces. Part I. Theoretical considerations. Agric. For. Meteorol. **33**: 323–337.

Myneni, R.B., Asrar, G., Wall, G.W., Kanemasu, E.T., and Impens, I. 1986. Canopy architecture, irradiance distribution on leaf surfaces and consequent photosynthetic efficiencies in heterogeneous plant canopies. Part II. Results and discussion. Agric. For. Meteorol. **37**: 205–218.

Norman, J.M. 1980. Interfacing leaf and canopy light interception models. *In* Predicting Photosynthesis for Ecosystem Models, Volume II. *Edited by* J.D. Hesketh and J.W. Jones. CRC Press, Boca Raton, FL. pp. 49–67.

Pauley, S.S., and Perry, T.O. 1954. Ecotypic variation of the photoperiodic response in *Populus*. J. Arnold Arbor. Harv. Univ. **35**: 167–188.

Ridge, C.R., Hinckley, T.M., Stettler, R.F., and Van Volkenburgh, E. 1986. Leaf growth characteristics of fast-growing poplar hybrids *Populus trichocarpa* × *P. deltoides*. Tree Physiol. **1**: 209–216.

Roberts, D.A., Green, R.O., Sabol, D.E., and Adams, J.B. 1993. Temporal changes in endmember abundances, liquid water and water vapor over vegetation at Jasper Ridge. *In* Summaries of the 4th Annual JPL Airborne Geoscience Workshop, October 25–29, 1993. Vol. 1. AVIRIS Workshop, Washington, DC. pp. 153–156.

Robertson, E.O., and Spittlehouse, D.L. 1990. Estimating Douglas-fir wood production from soil and climate data. Can. J. For. Res. **20**: 357–364.

Ruimy, A., Saugier, B., and Dedieu, G. 1994. Methodology for the estimation of terrestrial net primary production from remotely sensed data. J. Geophys. Res. **99**: 5263–5283.

Running, S.W., Loveland, T.R., Pierce, L.L., Nemani, R.R., Hunt, E.R. Jr. 1995. A remote sensing based vegetation classification logic for global land cover analysis. Remote Sens. Environ. **51**: 39–48.

Saugier, B. 1986. Productivité des ecosystèmes naturels. Biomasse Actualités, **9**: 42–49.

Scarascia-Mugnozza, G. 1991. Physiological and morphological determinants of yield in intensively cultured poplars (*Populus* spp.). Ph.D. dissertation, University of Washington, Seattle, WA. 189 p.

Scarascia-Mugnozza, G.E., Isebrands, J.G., Hinckley, T.M., and Stettler, R.F. 1989. Dynamics of light interception, leaf area and biomass production in *Populus* clones in the establishment year. Ann. Sci. For. **46** (Suppl.): 515–518.

Sellers, P.J., Berry, J.A., Collatz, G.J., Field, C.B., and Hall, F.G. 1992. Canopy reflectance, photosynthesis, and transpiration. III. A re-analysis using improved leaf models and a new canopy integration scheme. Remote Sens. Environ. **42**: 187–216.

Sinclair, T.R., and Horie, T. 1989. Leaf nitrogen, photosynthesis, and crop radiation use efficiency: A review. Crop Sci. **29**: 90–98.

Smit, B. 1988. Selection of flood-resistant and susceptible seedlings of *Populus trichocarpa* Torr. & Gray. Can. J. For. Res. **18**: 271–275.

Söderström, B. 1991. The fungal partner in the mycorrhizal symbiosis. *In* Proceedings of the Marcus Wallenberg Foundation Symposia. Ecophysiology of ectomycorrhizae of forest trees. S–791 80 Falun, Sweden, **7**: 5–29.

Spanner, M.A., Pierce, L.L., Peterson, D.L., and Running, S.W. 1990. Remote sensing of temperate coniferous forest leaf area index: the influence of canopy closure, understory vegetation and background reflectance. Int. J. Remote Sens. **11**: 95–111.

Stenberg, P., DeLucia, E.H., Schoettle, A.W., and Smolander, H. 1995. Photosynthetic light capture and processing from cell to canopy. *In* Resource physiology of conifers. *Edited by* W.K. Smith and T.M. Hinckley. Academic Press, San Diego, CA. pp. 3–38.

Stettler, R.F., Fenn, R.C., Heilman, P.E., and Stanton, B.J. 1988. *Populus trichocarpa × Populus deltoides* hybrids for short rotation culture: variation patterns and 4-year field performance. Can. J. For. Res. **18**: 745–753.

Thornley, J.H.M. 1976. Mathematical models in plant physiology. Academic Press, New York, NY.

Thornley, J.H.M., and Johnson, I.R. 1990. Plant and crop modelling. Oxford University Press, Oxford, UK.

Ustin, S.L., Smith, M.O., and Adams, J.B. 1993. Remote sensing of ecological processes: A strategy for developing and testing ecological models using spectral mixture analysis. *In* Scaling physiological processes: Leaf to globe. *Edited by* J.R. Ehleringer and C.B. Field. Academic Press Inc., San Diego, CA.

Waring, R.H., Law, B.E., Goulden, M.L., Bassow, S.L., McCreight, R.W., Wofsy, S.C., and. Bazzaz, F.A. 1995*a*. Scaling gross ecosystem production at Harvard Forest with remote sensing: a comparison of estimates from a constrained quantum-use efficiency model and eddy correlation. Plant Cell Environ. **18**: 1201–1213.

Waring, R.H., Way, J., Hunt, E.R. Jr., Morrissey, L., Ranson, K.J.,. Weishampel, J.F., Oren, R., and Franklin, S.E. 1995*b*. Imaging radar for ecosystem studies. BioScience, **45**: 715–723.

Weber, J.C., Stettler, R.F., and Heilman, P.E. 1985. Genetic variation and productivity of *Populus trichocarpa* and its hybrids. I. Morphology and phenology of 50 native clones. Can. J. For. Res. **15**: 376–383.

Whitehead, D., and Jarvis, P.G. 1981. Coniferous plantation. *In* Plant growth and water deficit. VI. *Edited by* T.T. Kozlowski. Academic Press, London, England. pp. 49–152.

Zavitkovski, J. 1982. Characterisation of light climate under conditions of intensively cultured hybrid poplar plantations. Agric. Meteorol. **25**: 245–255.

CHAPTER 19
Linking physiology, molecular genetics, and the *Populus* ideotype

Donald I. Dickmann and Daniel E. Keathley

*As we acquire more knowledge, things do not become
more comprehensible, but more mysterious.*
Albert Schweitzer

Introduction

The improvement of certain tree taxa by manipulation of their genomes has been
an integral part of the quest to increase the productivity of land for forest
products. In certain genera such as *Pinus*, *Eucalyptus*, and *Populus*, planting of
the high-yielding seed sources or cultivars that have been released by tree improvement programs has become routine. However, the success of this approach
within the genus *Populus* is almost unprecedented. The combination of a large
and varied taxon, few biological barriers to intra- and interspecific hybridization,
a large knowledge base of tree biology, ease of vegetative reproduction (except
in section *Populus* of the genus), intrinsically fast growth rate, a straight-forward
silvicultural system, which may include coppicing, and a wood of great utility
has put *Populus* cultivars in the forefront of intensive tree culture (see papers
in Mitchell et al. 1992). Perhaps in no other region of the world has the success
of this approach to silviculture been more evident than in the conifer-dominated
U.S. Pacific Northwest, where a thriving forest industry has arisen centered
around plantations of the fast-growing poplar hybrids created by the joint University of Washington–Washington State University tree improvement program
(Stettler et al. 1988, Heilman et al. 1995). This industry did not exist 15 years
ago; in fact, the very idea was considered by many experts to be laughable.

D.I. Dickmann and D.E. Keathley. Department of Forestry, Michigan State University,
East Lansing, MI 48824-1222, USA.
Correct citation: Dickmann, D.I., and Keathley, D.E. 1996. Linking physiology,
molecular genetics, and the *Populus* ideotype. *In* Biology of *Populus* and its implica-
tions for management and conservation. Part II, Chapter 19. *Edited by* R.F. Stettler,
H.D. Bradshaw, Jr., P.E. Heilman, and T.M. Hinckley. NRC Research Press, National
Research Council of Canada, Ottawa, ON. pp. 491–514.

The genetic gains in poplar productivity shown in the Pacific Northwest, as well as in many other parts of the U.S. and the world, have resulted largely from an empirical approach to tree breeding (Stettler et al. 1992). Parents from natural populations are chosen randomly or after intensive selection of open-pollinated progeny or clonal ramets for growth traits, crossed artificially, and the typically variable F_1 hybrid progeny screened for exceptional height, diameter, and volume growth. Frequently, certain F_1 genotypes show heterotic growth responses (Cain and Ormrod 1984; Heilman and Stettler 1985, Hinckley et al. 1989), which can be easily captured in *Populus* by cloning. Rarely have more than two cycles of recurrent selection and hybridization been applied to any poplar breeding population. Whereas this approach has led to impressive gains in productivity, some workers have argued that further gains in productivity are contingent upon movement of poplar improvement programs away from strict empiricism to an approach based more on theoretical knowledge of the genetic and physiological mechanisms underlying tree growth (Dickmann 1991; Dickmann et al. 1994; Kramer 1986).

An approach that can be more mechanistically oriented and that has proven successful in agriculture (e.g., Brothers and Kelly 1993; Donald and Hamblin 1983; entire issue of Field Crops Res. **26**(2), 1991) is based on the formulation of ideotypes. Dickmann et al. (1994) suggested that this approach may be important in the future of tree improvement. The concept of the crop ideotype was first proposed by Donald (1968) as "a plant model which is expected to yield a greater quantity or quality of ... useful product when developed as a cultivar." Dickmann's (1985) definition of an ideotype linked it more closely to forest tree improvement: an ideotype is a model tree that will produce an economic yield that approaches maximum in a particular environment (or on a certain site), using a prescribed cultural system, and assuming a well-defined end use for the harvested products. Few tree geneticists have formulated or published ideotypes, and to our knowledge the ideotype concept currently is not central to the operational breeding of poplars. Dickmann et al. (1994) fully discussed the pros and cons of ideotype breeding and gave examples of the application of this approach to the improvement of forest, horticultural, and multi-purpose tree crops.

Our objectives in this chapter are three-fold: (1) briefly discuss the ideotype concept; (2) present a short-rotation poplar ideotype and a protocol for its use; and (3) candidly assess the role that physiological and molecular genetics research can play in practical *Populus* improvement using ideotypes. Our intent is to stimulate thinking and discussion about the various avenues available to increase poplar productivity through genome manipulation. We consider this a worthy goal, given the projected increases in world population (and concomitant demand for wood products), and the need to reduce exploitation pressure on dwindling natural forests (Gladstone and Ledig 1990).

The ideotype concept

It has been argued that the most fruitful application of ideotypes to tree improvement can occur only if they distill state-of-the-art knowledge of the crop plant (Dickmann et al. 1994). A vague or incomplete conception of the target plant in the mind of the breeder can limit breeding goals or, in the worst case, lead to release of unsuitable genotypes. Fortunately, the genus *Populus* probably has been studied more than any other tree taxon, so a wealth of information exists upon which to base ideotypes. The trend has been to make ideotypes as complete as possible, with the intent of initiating a cycle of formulation-application-research-reformulation, and so on (Dickmann 1991, Fig. 1), leading to advances in genetic improvement as well as a deeper understanding of the biological basis of yield. While a noble and appealing aspiration, this system has not, in fact, worked very well.

The rationale for ideotype breeding was outlined by Rasmusson (1987) and Dickmann et al. (1994) for agronomic and woody crops, respectively. While certainly not the standard for most plant breeding, there are examples of successful yield enhancement by selecting for certain yield-related morphological characters in an ideotype, e.g., an upright, indeterminate architecture in *Phaseolus vulgaris* (Brothers and Kelly 1993), the spurred branching habit in *Malus domestica* (Brown 1975), or a narrow crown in *Pinus sylvestris* (Pöykkö 1993). Crop yield is the product, directly or indirectly, of the cumulative effects of single traits. The ideotype breeder seeks to select numerous genotypes that exhibit the traits defined by the model plant, thus laying a foundation for further genetic improvement. Sometimes the complex of traits interacting to produce an ideotype can be recovered as a unit during several cycles of recurrent selection (Brothers and Kelly 1993).

A potentially important and useful application of ideotypes is in the selection of elite parents or progeny from a diverse natural or hybrid population. A breeding program may generate millions of seedlings or clonal ramets each year, each with its own phenotypic characteristics. An efficient and expedient means of screening these phenotypes for further testing is essential, provided that the criteria used are highly correlated with economic yield. For example, in forest-tree improvement height has been a common selection criterion in nursery screening. However, correlations between seedling height (or other characters) and final crop yield often are poor (Zobel and Talbert 1991), leading many to view this as one of the central problems of forest-tree improvement. Such single quantitative traits could be viewed as simplistic ideotypes. A multitrait ideotype provides a compendium of yield-related characters, certain of which may be nontraditional or physiologically-based selection criteria, and represents a more comprehensive model against which individuals may be evaluated.

Way et al. (1983) caution that any single-minded breeding approach to yield improvement is unlikely to be successful. Nonetheless, it is a dictum of plant breeding that, since maintaining a specific selection intensity requires increasing the size of the breeding population as the number of traits increases, the efficiency of improvement of any one trait declines as additional traits are added. Therefore, it would be nearly impossible to assemble all the yield-enhancing characters specified in an ideotype into single genotypes through recurrent, multipletrait selection. Selection indices, which produce acceptable efficiencies when a few genetically correlated traits are manipulated (Baker 1986), are a way around this limitation, but their use in tree breeding has been problematic (Zobel and Talbert 1991). Moreover, selection indices cannot efficiently employ more than two or three traits from the complete ideotype.

Another approach, however, was proposed by Dickmann et al. (1994). An ideotype might be viewed as a *single quantitative trait*, and selections made based on the degree of fit to the ideotype as a whole. In fact, an ideotype is like any other quantitative trait (e.g., stem volume, which is scored as a function of two other independently measured quantitative traits, height and diameter), albeit more complicated: an expression of many gene loci, all working in concert to produce a continuum of phenotypes. Though heritabilities will be low if this approach is used, this limitation may be offset by screening large populations and, when possible, testing clones instead of seedling progeny. A major pragmatic problem in viewing an ideotype as a quantitative trait is the assigning of weights to each character of the ideotype, though certain statistical procedures can be used (e.g., Pöykkö 1993).

Regardless of how ideotypes are employed, the practical limitations of breeding set a limit on the number of characters that can be included in an ideotype, even though from the standpoint of understanding tree growth as many yield-related characters as possible can be included. The breeder, however, can select from a comprehensive ideotype a limited subset of genetically-correlated characters that offer the most promise of producing the desired genetic gain, or whose economic values are greatest, creating a practical *working* ideotype (Dickmann et al. 1994). The working ideotype then can be employed alone as a single quantitative trait, as one of the two variables in an independent culling procedure, or in a selection index. This approach would be particularly effective if all genotypes are screened for certain absolute requirements, such as rooting ability of cuttings and disease resistance, prior to scoring their ideotype ranking.

Pöykkö (1993) employed these ideas by formulating a working ideotype for *Pinus sylvestris* consisting of a set of morphological crown variables — branch diameter, branch angle, number of branches per whorl, length of branches, and internode length. These variables were combined as a single quantitative selection trait through a discriminate analysis, in which a canonical variable (K) was computed. This analysis combined normed crown variables of the working pine ideotype into a score value that could be used to rate phenotypes in a progeny test. The K value was then used as one trait in a selection index, along with stem mass. This work by Pöykkö (1993) is significant because it is one of the first real applications of selection based on an ideotype in forest tree breeding.

There is not necessarily one definitive ideotype for a particular crop, nor should an existing ideotype be viewed as static. Ideotype formulation has to be a dynamic process to take into account changes in biological knowledge, the environment, silvicultural practices, and commodity demand. Several models may be proposed for a given crop or for a particular crop–silviculture combination. For example, stem straightness may be an essential trait in an ideotype for poplars grown for wood, but not so when they are grown as a windbreak. As knowledge increases and crop-tree improvement programs become more sophisticated, existing ideotypes will be modified and new ones proposed. It is likely, though, that all ideotypes for a particular crop, regardless of species, or for a particular species, regardless of the crop, will have some common characteristics, e.g., high rate of growth or disease resistance (Dickmann et al. 1994; Donald and Hamblin 1983).

Besides their use in practical tree improvement, ideotypes are valuable because they stimulate thinking about plants in a holistic way. So, from a basic biological point of view, an ideotype can serve as the structural framework for a research program. In an iterative process, the ideotype both summarizes the current conception of yield-related traits of a crop plant, and points out where gaps in knowledge exist. Further research refines the ideotype by filling in these gaps, and the process repeats itself (Fig. 1).

Ideotypes for Populus

A complete crop ideotype for poplars is presented in Table 1. This ideotype outlines the many yield-related physiological and morphological traits currently thought to contribute to high biomass productivity, but it does not contain any molecular markers associated with yield. Two points are emphasized about this poplar ideotype. First, it is not the final word on the subject; ideotypes need to be continually modified as new information becomes available. Second, this ideotype does not represent a specific breeding goal; the practical limitations of tree breeding set a limit on the number of characters included in an applied ideotype, and in fact some of the traits in this ideotype may show negative genetic correlations. But these limitations do not negate the importance of the ideotype approach, as our following discussion shows.

A limited subset of characters that offer the most promise for producing the desired genetic gain, or whose economic values are greatest can be employed in a working ideotype, which can be viewed as an achievable breeding goal. In fact, more than one working ideotype could be created if different breeding goals existed, or different working ideotypes could be used in different phases of a

Fig. 1. Using a model taxon, such as *Populus*, a feed-forward cycle can develop where breeders and genetic engineers produce pedigreed material which, in turn, is studied by geneticists and physiologists. The resultant increase in biological knowledge can be used to revise ideotypes, which then can aid breeders and engineers in developing new pedigreed genotypes, and the cycle continues. Most genetic improvement programs currently do not operate in this way. Modified from Dickmann (1991).

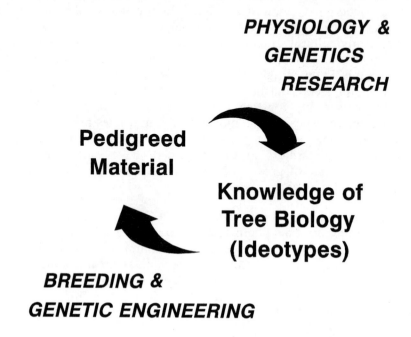

testing program. During the selection phase working ideotypes can be treated as a single quantitative trait (Dickmann et al. 1994; Pöykkö 1993).

An outline of a poplar testing program using the working ideotype approach is presented in Table 2. The breeding goals are basic to any approach to short-rotation culture of woody plants. This scenario represents one cycle of breeding, from the initial controlled crossing of a limited number of rigorously selected elite parents to eventual identification of a few genotypes for selective release or large-scale testing. Note that a major emphasis has been put on selection for resistance to *Septoria* canker and *Melampsora* rust, rooting potential of hardwood cuttings, and growth potential.

The working ideotypes proposed in this scenario are composed largely of morphological characters that are easily measurable, though, as discussed below, not all of them have always been highly correlated with volume growth. Whereas the complete poplar ideotype presented in Table 1 contains several important physiological variables and root system characteristics, these characters cannot

Table 1. A complete crop ideotype for poplar trees grown for energy or wood fiber in a high-density, unirrigated, intensive silvicultural system.

Growth and physiology:

Rapid rate of height and diameter growth

Bud flushing late enough to avoid spring frost injury

Indeterminate shoot growth with bud set just prior to first autumn frosts

Leaves with a high photosynthetic rate per unit leaf area or weight, area leaf weight, and ratio of net photosynthesis to dark respiration

High water-use efficiency (CO_2 fixed per unit of stomatal conductance)

Stomata close at moderate levels of water stress

Leaves, cambium, and fine roots osmotically adjust to gradual dehydration

Ecological characteristics:

Weak competitor (generalized "crop" ideotype)

High nutrient-use efficiency (stemwood biomass per unit of stemwood nutrient)

Effective remobilization of nitrogen and other mobile nutrients into stems and roots prior to leaf abscission

Resistant to snow and ice breakage

Tolerant of winter minium temperatures

Tolerant of common post- and pre-emergent herbicides

Resistant to major pathogens and invertebrate pests

Unpalatable to mammals

Morphology:

Relatively few, small diameter, upturned branches forming a long, narrow crown

Sylleptic branching

Rapid natural pruning of dead branches

High foliage density on branches

Relatively large, vertically oriented (erectophile) leaves in upper crown

Long seasonal leaf retention

High ratio of long (indeterminate) to short (determinate) shoots in upper crown

Light fruiting; male

Shoot/root ratio of ca. 1.5

Stem and wood properties:

Excurrent growth habit with straight, low-taper stem

Cambium active until late in the growing season

Thin bark, but resistant to winter sunscald

High density wood (>0.4 $g \cdot cm^{-3}$)

Wood low in gelatinous fibers; vessel content <20%

Table 1. (*concluded*).

Roots:

Strongly rooting hardwood cuttings

Taproot for anchorage and exploitation of deep soil nutrients and water

Many highly branched lateral and fine roots throughout the soil profile

Fine roots turn over slowly

Readily colonized by mycorrhizal fungi

Strong sink for photosynthates late in the growing season

---

be assessed in the early stages of a breeding cycle when large numbers of genotypes must be screened. The same could be said of any molecular markers that might be associated with traits in the complete ideotype. Only after the initial phases of the winnowing process presented in Table 2 are complete, and a few genotypes are identified that will be carried forward, can these more logistically problematic physiological and molecular characters be assessed. However, during these latter phases it is important that these more detailed selections be carried out, and clones fully characterized for the yield-related traits or markers comprising working ideotypes, before commercial cultivars are released for wide-scale planting.

Achieving the ideotype: can physiology and molecular genetics play a part?

Paul Kramer, the dean of American tree physiologists, summed up the view of most of his disciplinary colleagues in a 1986 paper: "... foresters must understand how trees grow, and this requires some understanding of tree physiology," and "The only way geneticists can increase growth is by providing genotypes with a more efficient combination of physiological processes for a particular environment ..." Similar sentiments were espoused by Dickmann (1991) and Riemenschneider et al. (1988). Many other tree physiologists and geneticists subscribe to Kramer's dictums, and this has lead to an increased involvement of physiologists in genetic improvement programs. The highly successful University of Washington–Washington State University poplar program is one of the most prominent examples of this inter disciplinary joining of forces (see e.g., Hinckley et al. 1989; Ceulemans et al. 1992).

While we endorse the current activity in physiological genetics research, and its link through ideotypes to practical tree improvement, certain questions nonetheless need to be asked. Is physiological research essential to continued genetic improvement of poplars and other tree genera? More specifically, are physiologists helping to create a new tree improvement paradigm or are they simply

Table 2. Genetic improvement of poplars for energy or wood fiber production: A selection protocol using working ideotypes based on current biological knowledge.

Breeding Goals: high stemwood biomass production per ha (rotations of 5–10 years); high harvest index; long-term resistance to diseases; stability across environments

Phase 1 Selection (Years 1 and 2): Nursery growth of seedling progeny

Trait A

Septoria canker or leaf spotting; no infection allowed

Trait B (Combined into a working ideotype)

Days between terminal bud break and bud set

Days of leaf retention beyond terminal bud set

Leaf size

Height

Since Trait A is all or nothing — any clone with *Septoria* infection is eliminated — selection is based primarily on the working ideotype.

Phase 2 Selection (Years 3 and 4): Stool beds established from unrooted cuttings of seedling progeny selected in Phase 1

Trait A

Septoria canker or spotting (natural infection or leaf spot bioassay (Ostry et al. 1988)) or *Melampsora* rust (any *Septoria* infection or moderate to heavy *Melampsora* infection eliminates a clone)

Trait B

Rooting of cuttings (> 90% for clone to be retained)

Trait C (Combined into a working ideotype)

Days between terminal bud break and bud set

Days of leaf retention beyond terminal bud set

Leaf size

Height

Traits B and C selected using **Independent Culling Levels** to assure that all plants meet propagation requirements (YEAR 4). Screening for molecular markers associated with traits A, B, and C initiated.

Phase 3 Selection (Years 5–10): Row-plot field plantings on multiple sites of unrooted cuttings from stool bed selections made in Phase 2

Trait A

Septoria or *Melampsora* (any *Septoria* infection or moderate to heavy *Melampsora* infection eliminates a clone)

Trait B

Survival (> 90% for clone to be retained)

499

Table 2. (*concluded*).

Trait C (Combined into a working ideotype)

Diameter^2 × Height (D^2H)

Straight, excurrent stem

Sylleptic branching

Small branch diameter

Traits B and C selected using *Independent Culling Levels* to assure that all plants meet propagation requirements (YEAR 10). Screening for molecular markers associated with trait C initiated. Evaluation of the efficacy of previously identified molecular markers for marker-aided selection.

Phase 4 Selection (Years 10 to 15 or 20): Region-wide block plantings on multiple sites of unrooted cuttings of selections from Phase 3

Trait A (Working ideotype)

Defined by a combination of molecular markers, physiological traits, and phenotypic characters

---

following along at the heels of geneticists and breeders? To put it another way, does physiological knowledge lead to new and improved genotypes or do new genotypes simply lead to the accumulation of more physiological knowledge and its incorporation into useless ideotypes? Does the feed-forward cycle in Fig. 1 really operate in practice or is the flow almost exclusively from left to right? Because the answers to these questions are not plainly obvious, or at the least, are arguable, we will explore them.

Frances Bacon concluded that knowledge is power, and as academicians we hold this famous aphorism to be almost inviolate. The future well-being of society is based in large part on the gathering and dissemination of knowledge, and this constitutes our mission as educators. Furthermore, as natural scientists our goal is to not only to increase empirical knowledge, but, more importantly, to progress towards mechanistic or theoretical understanding, to determine not only what happens in natural systems but *why* it happens, with the ultimate goal of *predicting* what happens. Ultimately, as forest scientists we want to understand the genetic–biochemical mechanisms underlying tree form and function. So it is unthinkable to argue that more knowledge of the physiological functioning of trees and its incorporation into ideotypes is unnecessary to further intellectual progress. On this level, tree physiologists stand on solid ground.

However, when the worth of physiological knowledge is judged strictly on a pragmatic level, the ground physiologists occupy becomes less stable. Will further understanding of the manner in which the poplar genome is expressed at the cellular level via enzyme-mediated physiological processes enable molecular geneticists to engineer further increases in productivity? Will moving closer to the genome by using physiological–morphological selection criteria in working ideotypes improve the efficiency of selection and enable breeders to approach the theoretical limits of yield? Or will a primarily empirical approach, which has served poplar improvement well to this point in time, serve just as well in the future? The words of Albert Schweitzer that preface this chapter caution us that there may even be a worst-case scenario; new information that does not lead to an advance in theoretical knowledge, with its attendant practical spin-offs, may not be helpful.

Identifying rate-limiting steps

Recent work on the molecular genetics of growth and development in *Populus* showed that relatively few quantitative trait loci (QTL) accounted for a large share of the genetic variance in tree volume growth (Bradshaw and Stettler 1995). The authors interpret their findings as indicating that the quantitative polygenic model of volume growth that has traditionally been accepted may not hold true in poplars, but rather that a few single genes may govern certain rate-limiting biochemical pathways leading to volume growth. That hypothesis certainly is provocative, both to poplar physiologists and breeders, but the inevitable question must be asked: what are these rate-limiting steps and what specific genes control them? In point of fact, we still don't know what they are nor do we know if selecting on these rate-limiting steps will yield gains that cannot otherwise be attained by direct selection on volume. Nonetheless, if one were to prioritize the work that tree improvement physiologists and geneticists have ahead of them, elucidating the biochemical and physiological steps limiting volume growth would appear right at the top of the priority list.

At this juncture an important point must be made, however. It is unlikely that a specific physiological process in its totality limits yield, and therefore such processes really have no utility as selection traits. Thus, their inclusion in ideotypes requires a caveat. For example, there has been much discussion in the literature over the years as to whether or not rates of photosynthesis are correlated with crop productivity (e.g., Austin 1992; Isebrands et al. 1988; Nasyrov 1978; Nelson 1988), and one can develop arguments on either side of the issue. We feel for several reasons that further examination of this issue is pointless. First, photosynthetic rate, like volume growth, is itself a quantitiative trait, controlled by many enzyme-catalyzed steps, each regulated by a gene or genes. It is hard for us to imagine that selecting for one complex quantitative trait (photosynthetic rate) to indirectly improve another complex quantitative yield trait can offer any gain in efficiency.

Second, the inherent rate maxima of physiological processes seldom are limiting to growth and yield because environmental limitations to these processes in production plantations are so pervasive. It is likely that many of the differences in photosynthesis reported among poplar clones, for example, are not expressions of variance in the genome-controlled biochemical process of photosynthesis per se but rather reflections of genotypic variation in the influence of environmental factors, morphological variables, and other physiological processes (e.g., stomatal regulation of transpiration) on photosynthesis. Third, there often is a reciprocal interpretation when considering the meaning of a physiological–growth correlation. In other words, a correlation does not necessarily imply a limitation of the dependent (growth) variable by the independent (physiological) variable, since both are chosen arbitrarily. It may be the other way around. For example, it is as legitimate to argue that some genotypes have greater rates of photosynthesis because they grow faster as it is to argue that some genotypes grow faster because they have inherently greater rates of photosynthesis. There is considerable evidence that photosynthetic rates of many crop plants, including poplars, adjust to the relative demands of rapidly growing or metabolizing carbohydrate sinks (Wardlaw 1990; Tschaplinski and Blake 1989*a*, *b*).

We do not, however, want to imply by the preceding discussion that a certain biochemical step or steps in the overall photosynthetic process, or any other physiological process, might not be rate limiting, with possible genotypic-based variation in the magnitude of this limitation. For example, Mott et al. (1986) concluded that regeneration of ribulose 1,5-bisphosphate during the carbon reduction cycle in the chloroplast was the rate-limiting step in CO_2 fixation at subsaturating light intensities. Might physiologists be better occupied looking at genetic variation in this rate-limiting step than powering up their infrared CO_2 analyzers to do yet another investigation of photosynthetic carbon uptake of potted plants? If such biochemical variation occurs and a specific genetic linkage can then be discovered, a real increase in photosynthetic potential could be realized, although the moderating influence of environmental stresses or other limiting factors might render such gains negligible.

The modification of a hybrid poplar clone for glyphosate tolerance by *Agrobacterium*-mediated transformation with genetic constructs that included the mutant aroA gene (Donahue et al. 1994) illustrates the practical power of basic genetic-physiological knowledge of a plant system. While this transformation might have eventually been conceived and accomplished strictly via empirical trial and error, the transformation was greatly facilitated by fundamental understanding of poplar physiology as well as the mode-of-action of glyphosate tolerance and its genetic basis (Riemenschneider and Haissig 1991). Of course, the fortuitous discovery of the mutant *aroA* gene and the availability of a compliant bacterial gene vector are important elements in this precedent-making work. Further engineering of traits related to control of insects, diseases, and weeds, as well as production of important specialty and pharmaceutical chemicals may well be on the horizon (Gasser and Fraley 1989). Unfortunately, the specter of horizontal gene transfer when transgenic plants are released into the natural environment (Amábile-Cuevas and Chicurel 1993), as well as regulatory, proprietary, and social constraints (Gasser and Fraley 1989), raise serious questions about the practical usefulness of such genetically-engineered tree improvement in the near future.

In the end, the time, expense, and sophistication of physiological measurements makes their use as selection criteria on a large scale very problematic in an improvement program. The height of one thousand progeny or propagules in a nursery bed can easily be measured in a day or two, but even the simplest physiological measurement could take weeks or months, provided no major equipment failures occurred. This problem is exacerbated by the continually changing environment, which biologically and statistically confounds most physiological measurements to the point of uselessness. Assessing plants in a sophisticated growth chamber can eliminate most of this environmental confounding, but then the price tag of the measurements increases even more and test population size is constrained. Furthermore, the question of genotype ×　environment interaction looms large; tree crops are not raised in growth chambers and relative performance in a controlled environment cannot be extrapolated to field performance. And finally the error associated with the measurements themselves is compounded by sampling error; seldom are physiological processes measured on a whole-tree level, so some sampling scheme is adopted which rarely produces an acceptable level of error variance. Physiologically-based rather than random sampling schemes, however, can minimize this error (Wolf et al. 1995).

Using morphological traits

It is important at this point to distinguish between the improbable gains due to direct selection for specific physiological traits per se versus possible gains due to selection for morphological traits which optimize these physiological processes. There is more practical promise using morphological traits, as evidenced by the gains made in altering the architecture of certain agronomic crops (e.g., Kelly and Adams 1987). For example, it is clear that a predominate influence on light interception and net CO_2 influx into plant canopies is the structural arrangement of these canopies (Cannell 1989). So leaf morphology, arrangement, and display, branch architecture, and shoot phenology become implicated as important determinants of yield, and much study of this subject has occurred using various *Populus* genotypes (Ceulemans et al. 1987; Dickmann et al. 1990; Hinckley et al. 1989; Isebrands and Nelson 1982; Nelson et al. 1981; Ridge et al. 1986; Wu 1994*a*). There has even been discussion of optimum architecture in terms of poplar ideotypes (Chen et al. 1994; Wu 1993). However, no single, all encompassing architectural solution has been proposed or adopted, and in reality none should be expected. The morphological diversity that exists within the woody-plant kingdom, even at the species level, indicates that optimal designs for a particular environment are multifarious. In addition, morphological traits of poplar hybrids do not always show the heterotic response often shown by volume growth, reducing the magnitude of variability available for identification and selection of superior individuals.

Correlations between morphological and phenological traits and growth in poplar genotypes have usually been low, with individual leaf area showing the most value as a selection trait (Ceulemans et al. 1987; Ceulemans et al. 1992). Indirect selection for height growth in forest trees based on indices incorporating tree morphology have not been encouraging (e.g., Morris et al. 1992; Sulzer et al. 1993). Riemenschneider et al. (1992, 1994) found that predicted responses to index selection for second-year height using leaf morphological traits as secondary selection criteria were not significantly greater than predicted responses to univariate selection using tree height alone in *P. balsamifera* and *P. trichocarpa* populations. On the other hand, Wu (1994*b*) indicated that indirect clonal selection on leaf size and area, branching capacity, and branch angle was expected to generate more genetic gain in volume than direct selection on volume alone in *P. deltoides* × *P. simonii* hybrids. Thus, there is some reason to encourage the pursuit of similar lines of investigation in the future. But first physiologists need to refine the list of key morphological selection traits included in the ideotype; they must be easy to measure, strongly linked to volume growth, and, ideally, show a heterotic response. We suggest that testing selection indices based on these morphological criteria on highly segregating F_2 or back-cross clonal progeny may be a good way to reveal the potential value of this approach. Such tests need to evaluate early use of these morphological selection indices against volume growth performance at or near rotation age, an element missing in the work by Riemenschneider et al. (1992, 1994).

Disease resistance

In most regions of the world where poplars are grown, leaf and stem diseases are the major limitations to plantation yield (Ostry et al. 1989; Ostry and McNabb 1990). For example, in the region we know best — the North Central U.S. and Southeast Canada — stem cankers caused by *Septoria musiva* and other pathogens are extremely virulent on most exotic and hybrid poplars, although the native *P. deltoides* is generally resistant. Susceptible clones rarely survive more than a few years. Any genotype being considered for plantation production must first be shown to be resistant to these cankers, as well as the rust caused by *Melampsora* spp., throughout the longest rotation envisioned. Thus, disease resistance must be a primary ideotype trait in any poplar breeding program in this region (cf. Table 2). Furthermore, the ominous outbreaks of *Melampsora* leaf rust in production plantations in the Pacific Northwest indicates that a pathogen-free condition in plantations in a particular region which may persist for many years must be considered ephemeral. Plant poplars and diseases will come.

The means to disease resistance is problematic using conventional breeding methods. Although most poplar species and hybrids show considerable variation in susceptibility to pathogens (Ostry and McNabb 1986; Pinon 1992) inheritance of resistance is complex (Hsiang et al. 1993; Prakash and Heather 1986). On the positive side, disease resistance can be reliably determined at an early age (Rajora et al. 1994), in contrast to volume growth performance, and rapid screening procedures have been developed, e.g., for *Septoria* (Ostry et al. 1988). Furthermore, Rajora et al. (1994) and Riemenschnieder et al. (1992, 1994) concluded that, although such use must be considered on a case-by-case basis, restricted selection indices could be used to identify fast growing genotypes that were also pathogen resistant. Their conclusion supports the inclusion of disease resistance in a working ideotype.

The problematic nature of conventional breeding for disease resistance supports the employment of certain biotechnological strategies. For example, somatic variation in resistance to *Septoria musiva* has been reported in explants from aseptic poplar cultures (Ostry and Skilling 1988), indicating an alternate route to the creation of disease-free genotypes. Techniques of molecular genetics also could be used to engineer transgenic poplars resistant to major pathogens. Such techniques have proven successful in plant agriculture, especially with viral diseases (Gasser and Fraley 1989), and they have begun to be employed in *Populus* (Klopfenstein et al. 1993). But again, fundamental understanding of plant–pathogen interactions at the genetic–biochemical level is a necessary prerequisite if biotechnological techniques are to succeed. Some work in this area has already been done in other plant–pathogen systems (e.g., Ryan 1990; VanEtten et al. 1989). We regard the acquiring of fundamental understanding of poplar–pathogen interactions to be one of the highest research priorities for basic physiologists, pathologists, and geneticists.

Understanding phenotypic plasticity

A phenomenon greatly confounding the analysis of genetic control and expression of ideotypic traits is phenotypic plasticity; i.e., the modification of genotypic trait expression in the phenotype by the environment. The practical manifestation of phenotypic plasticity is known in quantitative genetics as genotype × environment (g × e) interaction (Zobel and Talbert 1991). Environmental variation provokes a spectrum of responses in organisms, from complete stability across environments (nonplasticity or canalization) to continuous phenotypic change (high plasticity) (Stearns 1989). Phenotypic plasticity is now regarded as a separate trait which evolution (or tree breeders) may act upon (Eriksson 1991).

Of particular concern to tree breeders is the shift in ranking of families or clones when they are planted on different sites or in different regions. In such cases genotypes may interact for both growth and quality characteristics. A striking example of phenotypic plasticity is the relative performance of certain of the University of Washington *P. trichocarpa* × *P. deltoides* hybrids when planted in the Pacific Northwest region vs. the North Central region of the U.S. In the Northwest *Septoria* cankers are virtually unknown on these fast-growing hybrids, whereas in the North Central region stems begin to canker during the first year of growth and by age five most trees are dead. In this case there may well be g × e interaction occurring on both the host and pathogen level. The estimation of QTL effects on environmentally sensitive traits also are complicated by g × e interactions (Strauss et al. 1992). Furthermore, Namkoong (1985) showed that some traits which separately do not show any g × e interaction can do so when combined in a selection index.

Breeders have used two empirically-based approaches to reducing g × e interaction: stratifying environments and developing different breeding programs for each stratum, or selecting stable genotypes that perform well and show little interaction over a wide variety of environments (Zobel and Talbert 1991). Each approach is expensive, requiring genetic tests to be replicated over numerous sites and regions, and both have advantages and disadvantages. The simplest approach is to chose genotypes that are stable across environments, although yield penalties may result. The *Populus* cultivars 'Tristis' and 'Eugenei' illustrate a case in point. 'Tristis' was shown to be to be highly nonplastic (stable) in certain growth and physiological traits, whereas 'Eugenei' was very plastic (unstable) (Dickmann et al. 1992; Nguyen et al. 1991; Pregitzer et al. 1990). Yet 'Eugenei' grew faster than 'Tristis' in every environment in which they were grown together. The key point in the context of the current discussion is that little mechanistic or theoretical knowledge of the basis for phenotypic plasticity exists. Zobel and Talbert (1991) state that it is generally conceded that most g × e interactions are more closely related to edaphic than climatic factors, but little evidence exists to support such a statement.

Physiological work with poplars can give clues to the basis for phenotypic plasticity and its effect on yield, but no hypotheses that we are aware of can stand up to rigorous examination. For example, certain *P. trichocarpa* clones from the humid, western Washington U.S. environment showed a poor ability to control stomatal water loss (Schulte et al. 1987) or osmotically adjust (Tschaplinski and Tuskan 1994) at low leaf water potentials, whereas *P. deltoides* clones from Illinois showed tight stomatal control and osmotic adjustment. Thus, we could hypothesize that the *P. trichocarpa* clones would show high plasticity in growth and survival and the *P. deltoides* clones would be relatively nonplastic for those traits when grown in environments with contrasting water availability. However, this does not seem to be the case. Obviously, other gene-mediated physiological compensations relating to plasticity are occurring.

In our opinion the way to approach this problem is by taking a holistic approach to environmental adaptations; only by understanding the responses by the entire plant to a varying, and often limiting, environment can we hope to efficiently locate their genetic basis and, eventually, produce high-yielding genotypes with the desired levels of phenotypic plasticity. The recent increase in studies of root responses of poplars (Friend et al. 1991; Heilman et al. 1994; Nguyen et al.

1990; Pregitzer et al. 1990) and root–shoot interactions (Isebrands and Nelson 1983; Tschaplinski and Blake 1989c, d), as well as physiological studies on the basis for hybrid vigor (Ceulemans et al. 1992; Hinckley et al. 1989), for example, are important steps towards a better understanding of the physiology of the complete tree. Given that stability in growth and survival requires physiological adjustments to changing environmental and silvicultural conditions, in the foreseeable future the problem of phenotypic plasticity will continue to be important and will be dealt with using an empirical, quantitative genetics approach.

The role of molecular genetics

Conventional breeding methods yield very limited success in improving complex sets of traits, such as are found in ideotypes. Rates of improvement in multiple-trait selection programs are limited both by the need for exponentially larger population sizes to maintain a specified selection intensity as the number of traits increases and by the difficulty imposed by negative genetic correlations between desirable traits. The problem of negative correlations between desirable traits becomes more critical with each subsequent cycle of selection, and ultimately results in negative correlations between traits that were initially positively correlated, due to the fixation of pleiotropic genes (Falconer, 1989). It is possible that the inclusion of additional traits defining a *Populus* ideotype would enable greater improvement through conventional breeding, however, as outlined above, the barriers to this are substantial. Currently there is high interest in the possibility of using marker-aided selection to resolve this problem.

The use of clonal propagation in conventional breeding systems can convey the appearance of circumventing these barriers. For species that are easily cloned and where extensive areas of natural stands or plantations exists, as is true for most species in the genus *Populus*, large numbers of individual trees can be screened to establish a population that contains only individuals that meet all characteristics of the ideotype. These genotypes can then be cloned to produce planting stock for the establishment of plantations that uniformly meet all selection criteria. This yields a sudden burst of improvement that can be immediately translated into practical gains in genera, like *Populus*, that are easily propagated clonally. However, the problems inherent in breeding for multiple traits still constrain selection efficiency if conventional breeding methods are employed in efforts to attain further progress towards the ideotype. Certain additional ideotype traits, however, may be engineered into the *Populus* genome in the future using molecular genetic techniques (Gasser and Fraley 1989).

Such a clonally produced population could be used for continued breeding or seed production for the ideotype if the traits comprising the ideotype are inherited in an additive fashion and have high heritability, as was reported for tree height, phenology, leaf morphology, and resistance to damaging agents for black cottonwood (Riemenschneider et al. 1994), given that phenotypic manifestation of all desired characteristics was a prerequisite for a genotype's inclusion in the breeding population. However, it is likely that two factors would limit the usefulness of this breeding strategy. First, in spite of the example above, the levels of environmental and g × e variation in economically important traits in trees are typically large, with resultantly low or moderate heritability values for these traits. Estimation of individual tree breeding values, though a time consuming process, is the only conventional breeding method that can be used to circumvent this problem. Second, high levels of dominance variation in the desired traits, something that is also common in studies of quantitative traits, will cause erratic results from efforts to increase performance of the ideotype by artificial selection. In essence, intense selection combined with clonal propagation can allow the identification and propagation of a group of elite genotypes that all manifest the desired ideotype, but continued, systematic improvement of all traits in the ideotype is still blocked by the usual barriers to the simultaneous improvement of multiple traits using quantitative methods. The process only appears to circumvent the problems in the first generation due to the ability to achieve an extremely high selection differential when large populations are grown on relatively uniform sites.

Although alluring to contemplate, marker-aided selection and other molecular genetics techniques offer little promise for overcoming these problems. The use of molecular technology in breeding programs, though not specifically ideotype breeding in *Populus*, has been thoroughly reviewed (Bradshaw and Foster 1992, Riemenschneider et al. 1988; Strauss et al. 1992). Several points from these authors cast doubt on the efficacy of these techniques for circumventing the traditional barriers to ideotype breeding. These include the lack of inbred lines producing levels of diversity that preclude identification of QTL with general applicability as markers, the high heterozygosity of individuals used in interspecific crosses results in different alleles being associated with QTL in the progeny, g × e interaction limiting the usefulness of marker-aided selection unless QTL expression is mapped in multiple environments, and the high cost of marker-aided selection, particularly in selection schemes involving multiple traits where large numbers of individuals must be screened. Justification of this cost would be especially difficult, since the results would be largely applicable within pedigrees, but not population-wide (Bradshaw and Stettler, 1995). Overall these limitations of QTL and marker-aided selection offer little potential for practical breeding programs where cost effectiveness is critical and where the need to understand the genetic basis for an elite genotype's performance on a basic level is not nearly so important as simply identifying the top performers and then clonally replicating them.

One possible use for QTL and marker-aided selection in practical breeding programs would be to screen individual plants for inclusion in the breeding population. This could either be at the inception of the program, to assure a common genetic base, or used as a means of standardizing the genetic background as new genetic material is added to the breeding population in subsequent generations. Although this can be effectively done on a performance basis through clonal evaluation in the desired environment, the use of QTL would enable greater certainty that the population was based on a common, elite, genetic background. Clonal evaluation can only certify that all individuals are genetically elite. It lacks the power to determine that their exceptional performance is based on a common genetic background. Use of QTL in conjunction with a clonal trial would lend greater confidence that performance of all genotypes had a common genetic basis.

Is this additional confirmation of the basis for inclusion of parents in the breeding population of sufficient value to offset the additional cost of the intensive QTL identification and testing for each population and individual? Given that the most critical information, actual growth data, can be attained with most *Populus* crosses through clonal testing, a positive answer to this question is doubtful. Strauss et al. (1992), however, suggested that marker-aided selection can be used to identify QTL that result in negative correlations in desired traits and to eliminate them from the breeding population.

Another potential use of markers and QTL would be in standardizing the plant material utilized in basic tree physiology and genetics research. Currently, standardization can only be attained using clonal material. While this successfully assures genetic homology in all plants, it does not yield a method for studying the effects of specific combinations of genes in varying genetic backgrounds across differing environments. In contrast to the apparently limited usefulness of QTL and marker-aided selection in conventional tree improvement programs, the use of molecular markers to type and certify the genetic background of research material holds high promise.

The use of molecular markers to assure uniformity of research material at loci of interest would eliminate many difficulties in studying relationships between genotype and the physiological expression of important traits (Bradshaw and Foster, 1992). Tree physiology could then be studied without responses being confounded by uncontrollable genetic variation and interactions. The establishment of clonal banks varying at specific QTL could provide forest geneticists and tree physiologists with important new tools comparable to the monosomic and nullisomic lines that have been so powerfully used in elucidating the linkage between genotype and physiological response in other plant species (Sears 1953). The advantage of the QTL approach is that actual aneuploidy, and the concomitant drop in viability that is always associated with it, can be avoided. Understanding these relationships, even if practical breeding is most efficiently conducted in a conventional fashion, is essential if the ideotype goal of producing trees that approach the maximum economic yield in a particular environment (Dickmann 1985) is to be attained.

Summary and conclusions

Impressive gains in wood productivity of poplar plantations have been realized by using traditional tree improvement methods, especially the breeding of F_1 hybrids followed by selection and clonal propagation of progeny showing heterotic growth. But moving beyond this plateau has proved difficult. We feel that the ideotype concept represents not only a framework for studying the linkage between genetics and physiology, but also for moving practical improvement of *Populus* to a new level. The value of the ideotype approach lies in four areas (Belford and Sedgley 1991): (1) it encourages breeders to adopt innovative ways of choosing parents and making crosses to ensure that there is adequate variability in breeding material for yield-related traits; (2) it gives breeders a useful set of goals to make choices about the structure of their breeding program, the type of germplasm they introduce into it, and the testing strategies used to evaluate new material; (3) it generates hypotheses for further research, particularly in identifying potential ways to enhance yield; and (4) at a more physiological or molecular level, it encourages careful thought about which plant characters could contribute to improved crop yield. These potentials of the ideotype approach thus far have not been realized in tree improvement.

Certain difficult problems interfere with the implementation of the ideotype approach. Foremost among them is that few morphological or physiological traits, or their associated molecular markers, which go beyond the traditional height, diameter, pest resistance, and survival, can be used to more efficiently select within a breeding population or improved genotypes for production plantations. This statement applies even for a taxon as well studied as *Populus*. Thus, we have to agree with Riemenschneider et al. (1992) that the hypothesis that genetic selection for physiological yield components is more effective than univariate selection for an integrating trait such as stem volume is largely untested. Actually, basic physiological research has followed development of new genotypes, not led to it. We suggest that physiologists working with tree improvement programs for *Populus* and other taxa need to reevaluate their modus operandi. They should focus on identifying and understanding the genetic control of simply-measured, environmentally stable, mono- and oligogenic traits that are the rate-limiting steps in physiological processes and, therefore, highly correlated with yield. Unraveling the basis for phenotypic plasticity, which has confounded implementation of the ideotype approach, also should be given high research priority.

Advances in understanding the molecular genetics of poplars, especially the development of a linkage map for QTL mapping (Bradshaw and Stettler 1995), have been impressive. Nonetheless, QTL and marker-aided selection, in their current forms, are of questionable importance for improving breeding efficiency in *Populus*, although these techniques offer the potential for resolving the barriers to understanding the linkage between genotype and physiological expression of important traits. Thus, the continual pursuit of research into the workings of the genome by molecular biologists and physiologists is essential to the final evolution of the ideotype from a concept to a functional breeding strategy.

References

Amábile-Cuevas, C.F., and Chicurel, M.E. 1993. Horizontal gene transfer. Amer. Sci. **81**: 332–341.

Baker, R.J. 1986. Selection indices in plant breeding. CRC Press, Boca Raton, FL.

Belford, R.K., and Sedgley, R.H. 1991. Conclusions: Ideotypes and physiology: Tailoring plants for increased production. Field Crops Res. **26**: 221–226.

Bradshaw, H.D., Jr., and Stettler, R.F. 1995. Molecular genetics of growth and development in *Populus*. IV. Mapping QTLs with large effects on growth, form, and phenology traits in a forest tree. Genetics, **139**: 963–973.

Bradshaw, H.D., Jr., and Foster, G.S. 1992. Marker-aided selection and propagation systems in trees: advantages of cloning for studying quantitative inheritance. Can. J. For. Res. **22**: 1044–1049.

Brothers, M.E., and Kelly, J.D. 1993. Interrelationship of plant architecture and yield components in the pinto bean ideotype. Crop. Sci. **33**: 1234–1238.

Brown, A.G. 1975. Apples. *In* Advances in fruit breeding. *Edited by* J. Janick and J.N. Moore. Purdue University Press, West Lafayette, IN. pp. 3–37.

Cain, N.P., and Ormrod, D.P. 1984. Hybrid vigor as indicated by early growth of *Populus deltoides*, *P. nigra*, and *P. ×euramericana*. Can. J. Bot. **62**: 1–8.

Cannell, M.G.R. 1989. Physiological basis of wood production: a review. Scand. J. For. Res. **4**: 459–490.

Ceulemans, R., Impens, I., and Steenackers, V. 1987. Variations in photosynthetic, anatomical, and enzymatic leaf traits and correlations with growth in recently selected *Populus* hybrids. Can. J. For. Res. **17**: 273–283.

Ceulemans, R., Scarascia-Mugnozza, G., Wiard, B.M., Braatne, J.H., Hinckley, T.M., and Stettler, R.F. 1992. Production physiology and morphology of *Populus* species and their hybrids grown under short rotation. I. Clonal comparisons of 4-year growth and phenology. Can. J. For. Res. **22**: 1937–1948.

Chen, S.G., Ceulemans, R., and Impens, I. 1994. Is there a light regime determined tree ideotype? J. Theor. Biol. **169**: 153–161.

Dickmann, D.I. 1985. The ideotype concept applied to forest trees. *In* Trees as crop plants. *Edited by* M.G.R. Cannell and J.E. Jackson. Institute of Terrestrial Ecology, Huntington, England. pp. 89–101.

Dickmann, D.I. 1991. Role of physiology in forest tree improvement. Silva Fenn. **25**: 248–256.

Dickmann, D.I., Michael, D.A., Isebrands, J.G., and Westin, S. 1990. Effects of light interception and apparent photosynthesis in two contrasting *Populus* cultivars during their second growing season. Tree Physiol. **7**: 7–20.

Dickmann, D.I., Liu, Z., Nguyen, P.V., and Pregitzer, K.S. 1992. Photosynthesis, water relations, and growth of two hybrid *Populus* genotypes during a severe drought. Can J. For. Res. **22**: 1094–1106.

Dickmann, D.I., Gold, M.A., and Flore, J.A. 1994. The ideotype concept and the genetic improvement of tree crops. Plant Breed. Rev. **12**: 163–193.

Donahue, R.A., Davis, T.D., Michler, C.H., Reimenschneider, D.E., Carter, D.R., Marquardt, P.E., Sankhla, N., Sankhla, D., Hassig, B.E., and Isebrands, J.G. 1994. Growth, photosynthesis, and herbicide tolerance of genetically modified hybrid poplar. Can J. For. Res. **24**: 2377–2383.

Donald, C.M. 1968. The breeding of crop ideotypes. Euphytica, **17**: 385–403.

Donald, C.M., and Hamblin, J. 1983. The convergent evolution of annual seed crops in agriculture. Adv. Agron. **36**: 97–143.

Eriksson, G. 1991. Challenges for forest geneticists. Silva Fenn. **25**: 257–269.

Falconer, D.S. 1989. Introduction to quantitative genetics, 3rd ed. Longman Scientific and Technical, New York, NY.

Friend, A.L., Scarascia-Mugnozza, G., Isebrands, J.G., and Heilman, P.E. 1991. Quantification of two-year-old hybrid poplar root systems: morphology, biomass, and ^{14}C distribution. Tree Physiol. **8**: 109–119.

Gladstone, W.T., and Ledig, F.T. 1990. Reducing pressure on natural forests through high-yield forestry. For. Ecol. Manage. **35**: 69–78.

Gasser, C.S., and Fraley, R.T. 1989. Genetically engineering plants for crop improvement. Science (Washington, DC), **244**: 1293–1299.

Heilman, P.E., and Stettler, R.F. 1985. Genetic variation and productivity of *Populus trichocarpa* T. & G. and its hybrids. II. Biomass productivity in a 4-year plantation. Can. J. For. Res. **15**: 384–388.

Heilman, P.E., Ekuan, G., and Fogle, D. 1994. Above- and below-ground biomass and fine roots of 4-year-old hybrids of *Populus trichocarpa* × *Populus deltoides* and parental species in short-rotation culture. Can. J. For. Res. **24**: 1186–1192.

Heilman, P.E., Stettler, R.F., Hanley, D.P., and Carkner, R.W. 1995. High yield hybrid poplar plantations in the Pacific Northwest. Pac. NW Reg. Exten. Bull. PNW356 (revised).

Hinckley, T.M., Ceulemans, R., Dunlap, J.M., Figliola, A., Heilman, P.E., Isebrands, J.G., Scarascia-Mugnozza, G., Schulte, P.J., Smit, B., Stettler, R.F., Van Volkenburgh, E., and Wiard, B.M. 1989. Physiological, morphological and anatomical components of hybrid vigor in *Populus*. *In* Structural and functional responses to environmental stresses. *Edited by* K.H. Kreb, H. Richter, and T.M. Hinckley. SPB Academic Publ., The Hague, The Netherlands. pp. 199–217.

Hsiang, T., Chastagner, G.A., Dunlap, J.M., and Stettler, R.F. 1993. Genetic variation and productivity of *Populus trichocarpa* and its hybrids. IV. Field susceptibility of seedlings to *Melampsora occidentalis* leaf rust. Can. J. For. Res. **23**: 436–441.

Isebrands, J.G., and Nelson, N.D. 1982. Crown architecture of short-rotation, intensively cultured *Populus*. II. Branch morphology and distribution of leaves within the crown of *Populus* 'Tristis' as related to biomass production. Can J. For. Res. **12**: 853–864.

Isebrands, J.G., and Nelson, N.D. 1983. Distribution of [^{14}C]-labeled photosynthates within intensively cultured *Populus* clones during the establishment year. Physiol. Plant. **59**: 9–18.

Isebrands, J.G., Ceulemans, R., and Wiard, B. 1988. Genetic variation in photosynthetic traits among *Populus* clones in relation to yield. Plant Physiol. Biochem. **26**: 427–437.

Kelly, J.D., and Adams, M.W. 1987. Phenotypic recurrent selection in ideotype breeding of pinto beans. Euphytica, **36**: 69–80.

Klopfenstein, N.B., McNabb, H.S., Hart, E.R., Hall, R.B., Hanna, R.D., Heuchelin, S.A., Allen, K.K., Nian-Qing, S., and Thornburg, R.W. 1993. Transformation of *Populus* hybrids to study and improve pest resistance. Silvae Genet. **42**: 86–90.

Kramer, P.J. 1986. The role of physiology in forestry. Tree Physiol. **2**: 1–16.

Mitchell, C.P., Ford-Robertson, J.B., Hinckley, T., and Sennerby-Forsse, L. (*Editors*). 1992. Ecophysiology of short rotation tree crops. Elsevier Applied Science, New York, NY.

Morris, D.M., Parker, W.H., and Seabrook, R. 1992. Some considerations when selecting young jack pine families using growth and form traits. Can. J. For. Res. **22**: 429–435.

Mott, K.A., Jensen, R.G., and Berry, J.A. 1986. Limitation of photosynthesis by RuBP regeneration rate. *In* Biological control of photosynthesis. *Edited by* R. Marcelle, H. Clijsters, and M. Van Poucke. Martinus Nijhoff Publishers, Dordrecht, The Netherlands. pp. 33–43.

Ostry, M.E., and McNabb, H.S. 1986. Populus species and hybrid clones resistant to *Melampsora*, *Marssonina*, and *Septoria*. USDA For. Serv. Res. Pap. NC-272.

Ostry, M.E., and McNabb, H.S. 1990. Minimizing disease injury to hybrid poplars. J. Environ. Hortic. **8**: 96–98.

Ostry, M.E., and Skilling, D.D. 1988. Somatic variation in resistance of *Populus* to *Septoria musiva*. Plant Dis. **72**: 724–727.

Ostry, M.E., McRoberts, R.E., Ward, K.T., and Resendez, R. 1988. Screening hybrid poplars in vitro for resistance to leaf spot caused by *Septoria musiva*. Plant Dis. **72**: 497–499.

Ostry, M.E., Wilson, L.F., McNabb, H.S., and Moore, L.M. 1989. A guide to insect, disease, and animal pests of poplars. U.S. Dep. Agric. For. Serv. Agric. Handb. 677.

Namkoong, G. 1985. The influence of composite traits on genotype by environment relations. Theor. Appl. Genet. **70**: 315–317.

Nasyrov, Y.S. 1978. Genetic control of photosynthesis and improving crop productivity. Annu. Rev. Plant Physiol. **29**: 215–237.

Nelson, C.J. 1988. Genetic associations between photosynthetic characteristics and yield: Review of the evidence. Plant Physiol. Biochem. **26**: 543–554.

Nelson, N.D., Burk, T., and Isebrands, J.G. 1981. Crown architecture of short-rotation, intensively cultured *Populus*. I. Effects of clone and spacing on first-order branch characteristics. Can J. For. Res. **11**: 73–81.

Nguyen, P.V., Dickmann, D.I., Pregitzer, K.S., and Hendrick, R. 1990. Late-season changes in allocation of starch and sugar to shoots, coarse roots, and fine roots in two hybrid poplar clones. Tree Physiol. **7**: 95–105.

Pinon, J. 1992. Variability in the genus *Populus* in sensitivity to *Melampsora* rusts. Silvae Genet. **41**: 25–34.

Pöykkö, T. 1993. Selection criteria in Scots pine breeding with special reference to ideotype. Rep. Found. Forest Tree Breed. No. 6, Helsinki, Finland.

Prakash, C.S., and Heather, W.A. 1986. Inheritance of resistance to races of *Melampsora medusae* in *Populus deltoides*. Silvae Genet. **35**: 74–77.

Pregitzer, K.S., Dickmann, D.I., Hendrick, R., and Nguyen, P.V. 1990. Whole-tree carbon and nitrogen partitioning in young hybrid poplars. Tree Physiol. **7**: 79–93.

Rajora, O.P., Zsuffa, L., and Yeh, F.C. 1994. Variation, inheritance and correlations of growth characters and *Melampsora* leaf rust resistance in full-sib families of *Populus*. Silvae Genet. **43**: 219–226.

Rasmusson, D.C. 1987. An evaluation of ideotype breeding. Crop Sci. **27**: 1140–1146.

Ridge, C.R., Hinckley, T.M., Stettler, R.F., and Van Volkenburgh, E. 1986. Leaf growth characteristics of fast-growing poplar hybrids *Populus trichocarpa* × *P. deltoides*. Tree Physiol. **1**: 209–216.

Riemenschneider, D.E., and Haissig, B.E. 1991. Producing herbicide tolerant *Populus* using genetic transformation mediated by *Agrobacterium tumefaciens* C58: A summary of recent research. *In* Woody plant biotechnology. *Edited by* M.R. Ahuja. Plenum Press, New York, NY. pp. 247–263.

Riemenschneider, D.E., Haissig, B.E., and Bingham, E.T. 1988. Integrating biotechnology into woody plant breeding programs. *In* Genetic manipulation of woody plants. *Edited by* J.W. Hanover and D.E. Keathley. Plenum Press, New York, NY. pp. 433–449.

Riemenschneider, D.E., McMahon, B.G., and Ostry, M.E. 1992. Use of selection indices to increase tree height and control damaging agents in 2-year-old balsam poplar. Can J. For. Res. **22**: 561–567.

Riemenschneider, D.E., McMahon, B.G., and Ostry, M.E. 1994. Population-dependent selection strategies needed for 2-year-old black cottonwood clones. Can J. For. Res. **24**: 1704–1710.

Ryan, C.A. 1990. Protease inhibitors in plants: genes for improving defenses against insects and pathogens. Annu. Rev. Phytopath. **28**: 425–449.

Schulte, P.J., Hinckley, T.M., and Stettler, R.F. 1987. Stomatal responses of *Populus* to leaf water potential. Can. J. Bot. **65**: 255–260.

Sears, E.R. 1953. Nullisomic analysis in wheat. Amer. Nat. **87**: 245–252.

Stearns, S.C. 1989. The evolutionary significance of phenotypic plasticity. BioScience, **39**: 436–445.

Strauss, S.H., Lande, R., and Namkoong, G. 1992. Limitations of molecular-marker-aided selection in forest tree breeding. Can. J. For. Res. **22**: 1050–1061.

Stettler, R.F., Fenn, R.C., Heilman, P.E., and Stanton, B.J. 1988. *Populus trichocarpa* × *Populus deltoides* hybrids for short-rotation culture: variation patterns and 4-year field performance. Can. J. For. Res. **18**: 745–753.

Stettler, R.F., Bradshaw, H.D., Jr., and Zsuffa, L. 1992. The role of genetic improvement in short rotation forestry. *In* Ecophysiology of short rotation tree crops. *Edited by* C.P. Mitchell, J.B. Ford-Robertson, T. Hinckley, and L. Sennerby-Forsse. Elsevier Applied Science, New York, NY. pp. 285–308.

Sulzer, A.M., Greenwood, M.S., and Livingston, W.H. 1993. Early selection of black spruce using physiological and morphological criteria. Can. J. For. Res. **23**: 657–664.

Tschaplinski, T.J., and Blake, T.J. 1989*a*. Photosynthetic reinvigoration of leaves following shoot decapitation and accelerated growth of coppice shoots. Physiol. Plant. **75**: 157–165.

Tschaplinski, T.J., and Blake, T.J. 1989*b*. The role of sink demand in carbon partitioning and photosynthetic reinvigoration following shoot decapitation. Physiol. Plant. **75**: 166–173.

Tschaplinski, T.J., and Blake, T.J. 1989*c*. Water relations, photosynthetic capacity, and root/shoot partitioning of photosynthate as determinants of productivity in hybrid poplar. Can. J. Bot. **67**: 1689–1697.

Tschaplinski, T.J., and Blake, T.J. 1989*d*. Correlation between early root production, carbohydrate metabolism, and subsequent biomass production in hybrid poplar. Can. J. Bot. **67**: 2168–2174.

Tschaplinski, T.J., and Tuskan, G.A. 1994. Water-stress tolerance of black and eastern cottonwood clones and four hybrid progeny. II. Metabolites and inorganic ions that constitute osmotic adjustment. Can. J. For. Res. **24**: 681–687.

VanEtten, H.D., Matthews, D.E., and Matthews, P.S. 1989. Phytoalexin detoxification: Importance for pathogenicity and practical implications. Annu. Rev. Phytopath. **27**: 143–164.

Wardlaw, I.F. 1990. The control of carbon partitioning in plants. New Phytol. **116**: 341–381.

Way, R.D., Sanford, J.C., and Lakso, A.N. 1983. Fruitfulness and productivity. *In* Methods in fruit breeding. *Edited by* J.N. Moore and J. Janick. Purdue University Press, West Lafayette, IN. pp. 353–367.

Wolf, A.T., Burk, T.E., and Isebrands, J.E. 1995. Evaluation of sampling schemes for estimating instantaneous whole-tree photosynthesis in *Populus* clones: a modeling approach. Tree Physiol. **15**: 237–244.

Wu, R-L. 1993. Simulated optimal structure of a photosynthetic system: implication for the breeding of forest crop ideotype. Can. J. For. Res. **23**: 1631–1638.

Wu, R-L. 1994*a*. Quantitative genetics of yield breeding for *Populus* short-rotation culture. II. Genetic determination and expected selection response to tree geometry. Can. J. For. Res. **24**: 155–165.

Wu, R-L. 1994*b*. Quantitative genetics of yield breeding for *Populus* short-rotation culture. III. Efficiency of indirect selection on tree geometry. Theor. Appl. Genet. **88**: 803–811.

Zobel, B., and Talbert, J. 1991. Applied forest tree improvement. Waveland Press, Prospect Heights, IL.

CHAPTER 20
Trends in poplar culture: some global and regional perspectives

L. Zsuffa, E. Giordano, L.D. Pryor, and R.F. Stettler

Introduction

The purpose of this chapter is to give the reader a sense of the diversity in which poplar culture is practiced in several regions of the world. It begins with a brief historical introduction, then provides an overview of major planting types and their diverse purposes, and in the second part illustrates different approaches to harness poplar's range of favorable attributes in a small sample of geographical contexts. The authors' intent was not to conduct a *tour de force*, but rather to give some perspectives on past and current trends in poplar culture, their socio-economic settings, and the opportunities and challenges they may present for the future.

General overview

Poplars (*Populus* L.) grow naturally on a variety of sites from boreal to sub-tropical, and from mountainous to riparian. At times they form large stands, as still nowadays in boreal forests and along major river valleys, and at other times small stands, lines, and groups of trees.

Poplars have been useful to humans and cultivated since historical times. Poplars grow fast, are easy to propagate, and can be grown on many types of sites in the forest, as well as in the open landscape. They serve as an excellent source

L. Zsuffa. Faculty of Forestry, University of Toronto, Toronto, ON M5S 3B3, Canada.
E. Giordano. DISAFRI, Università di Tuscia, Via. S. Camillo de Lellis 3, I-01100 Viterbo, Italy.
L.D. Pryor. Forestry and Landscape Consultant, 69 Endeavour Street, Red Hill, Canberra ACT 2603, Australia.
R.F. Stettler. College of Forest Resources, University of Washington, Box 352100, Seattle, WA 98195–2100, USA.
Correct citation: Zsuffa, L., Giordano, E., Pryor, L.D., and Stettler, R.F. 1996. Trends in poplar culture: some global and regional perspectives. *In* Biology of *Populus* and its implications for management and conservation. Part II, Chapter 19. *Edited by* R.F. Stettler, H.D. Bradshaw, Jr., P.E. Heilman, and T.M. Hinckley. NRC Research Press, National Research Council of Canada, Ottawa, ON. pp. 515–539.

of a wide range of wood products, especially in temperate zones. Poplars also play a significant role in environmental protection and improvement, especially in protecting land from wind and flood erosion, in remedying contaminated groundwater, in the safe disposal of sewage sludge, and in improving the carbon balance of the atmosphere. These fast-growing trees can augment forest diversity and enrich the open landscape, providing it with aesthetic appeal through a variety of forms.

In the International Poplar Commission's book on poplars and willows (FAO 1979), the spread of cultivation of these species is described as follows: "When the forests of the plains of the old world were cleared and converted to agriculture, many parts of the land were left aside because no worthwhile crops could be expected from them. However, it was there that poplars and willows were successful in establishing themselves. People came to look at these areas of natural growth to satisfy their needs for timber and fuelwood. Because of the value of the trees and the proven ease of propagating them by cuttings, people started planting them near their homes and around their fields to furnish wood and provide greenery and shade."

In Central Asia, the Near-East, and the Mediterranean region, poplars have been closely associated with agriculture since antiquity. For many centuries poplar provided timber, withes, and fuel. Additionally, poplar leaves and twigs were used as forage and bedding for animals. Even today, from Kashmir to the Atlantic in the Old World and throughout the New World, poplars mark the neighborhoods of human habitations. Early poplar culture made use of locally available trees: almost everywhere, these were various forms of *Populus nigra* (Zsuffa 1974) and *P. alba*. However, the introduction of *P. deltoides* from North America into Europe during the 17th century resulted in natural hybrids (*P. deltoides* × *P. nigra* = *P.* ×*euramericana* syn. *P. canadensis*), which in time revolutionized concepts of poplar cultivation.

During the first years of this century, poplar cultivation had spread in Europe with the support of industry. Voices began to be raised on all sides deploring the general state of its disorder, the ignorance of tree planters, and the ravages caused by insects and diseases. Scientists began to take interest in poplar; in France, Dode (1905); in England, Henry (1914); in Italy, Jacometti (1933); in Germany, von Wettstein (1933); in the United States, Schreiner and Stout (1934); in Canada, Heimburger (1936); and in The Netherlands, Houtzagers (1937). By 1947, there was enough momentum in poplar culture, and enough need perceived for its study, that an International Poplar Commission was founded under the aegis of the Food and Agriculture Organization (FAO) of the United Nations. The efforts of this Commission have led to much progress in the cultivation and utilization of poplars and willows and to important agreements on nomenclature, registration of clones, varietal control, and the exchange of germ plasm. Since then, additional organizations have joined to foster and coordinate research on poplar, such as the International Union of Forest Research Organizations (IUFRO), and the International Energy Agency (IEA), through several of their working groups.

Common types of plantings

A variety of planting types is practiced in different countries depending on needs, tradition, tree varieties, sites, and economics. We may group them according to their major purpose and distinguish between production plantations, protection plantings, and poplars in landscape use, while keeping in mind that these functions often overlap.

Common deployment patterns within plantations include monoclonal stands, mosaics of monoclonal blocks, mosaics of clonal rows, and intimate single-tree mixtures of seedlings of various genotypes (Zsuffa et al. 1993).

Production plantations

A widely-practiced method for growing poplar is in monoclonal stands at medium spacing (3–4 m) in 20–30 year rotations. Prerooted stock and unrooted poles or cuttings are used for planting, and some form of site preparation and tending is practiced. Such forms of plantations are found in Northern and Central Europe, and in parts of North America, including Canada, with aspen varieties as well as Euramerican, Interamerican, and balsam poplar hybrid clones. Examples of these plantings are in traditional French and German poplar culture (Zsuffa et al. 1993), and in poplar plantations in Eastern Canada (Zsuffa and Barkley 1984). Another form of poplar cultivation in Southern and Central Europe is in widely spaced (5–8 m) monoclonal plantations with intensive cultural treatment. Large-sized, prerooted, or unrooted-pole stock is planted on well-prepared sites. The trees are harvested for high quality products at 10–15 years of age. This management type is common in Italy (Piccarolo 1959).

By contrast, balsam poplars are often planted within the forest, especially in mixture with other species. These poplars can complement the natural regeneration and plantings of pure species, including conifers. Examples of these plantings are in Germany (Weisgerber 1983). Aspens, too, are well-suited for growing on forest soils. The management is usually extensive, often making use of natural regeneration of existing stands, such as occurs in much of northeastern and western North America. Under artificial regeneration, aspen planting stock is usually seedlings because cuttings root poorly. Planting is in pure stands, and the rotation is 40–50 years. Examples of these plantings are in Germany and Northern Europe (Hessmer 1951).

A traditional form of poplar planting in Asia, the Mid-East, and North Africa is with fastigiate (columnar) poplar clones, usually of *P. nigra* and *P. alba* origin, grown in very dense spacings (1 m) in 10–20 year rotations. The trees are an important source of timber for rural construction and firewood, and leafy branches taken from the trees are used for animal feed (FAO 1979).

More recent forms of monoclonal plantations have developed in North America for the purpose of short-rotation (7–10 years) production of fiber (Armson 1983; Zsuffa and Barkley 1984; Boysen and Strobl 1990; Heilman et al. 1995). Typically, unrooted cuttings of easy-to-propagate clones are used for medium-density spacing (2.5–4 m). The planting sites are carefully surveyed, prepared, and the plantations are tended by cultivation and chemical weed control. The plantings are in mosaics of monoclonal blocks (often less than 5 ha in size) and the clones are matched to sites. After harvest, the site is either regenerated from coppice or replanted to new stock. Biomass plantations with poplar and willow push the rapid cycling even further. These are dense monoclonal plantings (0.3–1.5 m spacing) of cuttings planted on carefully prepared sites, well-cultivated and maintained, harvested frequently (every 1–5 years), and subsequently regenerated by coppice growth. The whole aboveground portion of the biomass, sometimes including the foliage, is harvested and used as a source of fiber, energy, and food. Good examples are the poplar and willow energy plantations in Sweden (Christersson et al. 1993).

Finally, an important context for poplar plantings is agroforestry, i.e., the joint culture of trees with agricultural crops. Plantings vary from lines along field boundaries and irrigation channels to block plantations with wide spacing (4 m and more), in combination with sugarcane, maize, wheat, soybean, pulses, mustard, and ginger, etc. Intercropping typically begins with light-demanding plants (e.g., maize, wheat) and proceeds after 3–4 years to more shade-tolerant crops (e.g., medicinal plants). Trees are harvested after 10–12 years. Examples abound in Asia, especially in northern India (e.g., Chaturvedi 1982) and China (see Fig. 1).

Protection plantings

Plantings of poplars for environmental protection and improvement, as their primary role, have existed in different parts of the world for many centuries. They have served for stream bank protection and rehabilitation, and for windbreaks and shelterbelts. More recently, plantings have been installed for safe disposal of wastewater and sewage sludge and for remediation of contaminated groundwater. Now fast growing poplars are used for renewable energy and for improving the carbon balance of the atmosphere.

Poplars have several advantages over other tree species for windbreaks: the ease of establishment and fast growth, and the possibility, through cultivar selection, to manipulate at will the width and length of the tree-crown. In the Mid-East and North Africa, fastigiate poplars are often grown in single or multiple rows in dense (1–2 m) spacing. In Russia, Ukraine, and Canada (on the prairies), poplars are an important component of the windbreaks.

Fig. 1. Widely-spaced *P. ×tomentosa* (3-yr-old) interplanted with a wheat crop, growing in an irrigated plantation in the Shandong Province of the People's Republic of China. (Photo by L. Pryor.)

In Saskatchewan, Canada, the Indian Head Tree Nursery (established by the Canadian Government as part of the farm rehabilitation program) has supplied planting stock and provided technical assistance to farmers for shelterbelt plantings since 1903. Soil erosion by wind was the main problem. Poplars had a prominent role in these plantings, especially hybrids with frost-hardy local selections and exotic balsam poplars (Cram 1960, 1968). More recently poplars have been planted in shelterbelts as the outside row, with the inner row(s) being composed of longer-lived species. Poplar breeding work for windbreaks in the prairie provinces of Canada resulted in local hybrids between natural and introduced trees of Aigeiros and Tacamahaca poplars. Regional test plantings of these poplars were conducted since 1964. Only two poplar clones have consistently demonstrated the ability to form effective and reliable shelterbelts on the prairies: 'Walker' (selected from open pollinated seedlings of *P. deltoides* var. *occidentalis*) and 'Northwest' (a *P. balsamifera* × *P. deltoides* var. *occidentalis* hybrid). A new clone, 'Assiniboine' has been selected from an open pollinated family of 'Walker' poplar and was introduced in 1987 (Schroeder and Lindquist 1989). Ronald (1980) reported on special crosses and selections made for shelterbelts and landscaping on the prairies, such as the 'Tower Poplar' (*P. alba* × *P. tremula* 'erecta').

Planting poplars along stream banks helps protect the bank from erosion, reduce sediment and pollution from agricultural chemicals, and helps to improve water quality (Boysen and Strobl 1990). Poplar's fast growth, extensive root system, and ease of establishment can provide immediate erosion control and bank stability. The same features are taken advantage of in the use of poplar to check soil erosion resulting from sheep and cattle grazing on hilly land, as in New Zealand (FAO 1979).

Perhaps the newest role for which poplar seems to hold much promise is in the disposal of municipal wastewater and sewage sludge. Special plantations of fast-growing hybrids have been established for this purpose and have shown high uptake rates for nitrogen and other minerals (Henry 1991). This has led to active studies on the suitability of poplar in the broader context of phytoremediation, i.e., the use of plants as clean-up organisms for the absorption and conversion of toxic pollutants from air, water, and soil (e.g., Adler 1996). Eventually, genetically engineered clones may be developed that display greater efficiency for specific target pollutants.

Poplars in landscape use

Several poplar species and varieties seem to be favored as plantings of single trees or small groups for landscaping (FAO 1979). Some examples of the most common cultivars are: *P. nigra* cv. *italica* (Lombardy poplar) in temperate zones throughout the world, *P. nigra* var. *thevestina* in Asia and the Mediterranean region, *P. ×euramericana* cvs. Robusta, Serotina, Regenerata, and Marilandica throughout Western and Central Europe, *P. ×euramericana* cv. Eugenei (Carolina poplar) in North America, *P. alba* cv. Bolleana in temperate and subtropical zones throughout the world. Many times these trees also provide the benefit of timber for rural structures, firewood, pollards, and forage (Fig. 2).

Euramerican hybrid clones are frequently used in row plantings along ditches, roads, and farm boundaries (Fig. 3). The spacing is usually wide (4–10 m) and rooted stock is planted in single or double rows. The trees are cut at 20–40 years of age and replaced by new planting. This form of row plantings is common throughout Europe (Zufa 1961).

Some regional perspectives

Europe

Importance of poplar culture

Poplar cultivation has a long-standing tradition in Europe, especially in those countries with limited forest resources. In the twentieth century, even countries endowed with extensive forest lands have given more consideration to poplars because of the widespread need for raw material for industry, particularly the

Fig. 2. Lumbermill in Kirsehir, Anatolia (Turkey), utilizing primarily the widely-planted fastigiate cultivars of *P. nigra*, visible in the background. (Photo by R. Stettler.)

Fig. 3. Row planting in the fertile agricultural Po Valley, one of the prime poplar growing regions of Italy.

pulp and paper industry. In fact, new Euramerican hybrid clones of the same basic cross type (*P. deltoides* × *P. nigra*) that had occurred spontaneously following the introduction of cottonwood from North America to Europe in the 17th century have received renewed attention because of their high biomass production under short-rotation culture.

The historical expansion of the Euramerican hybrids partly explains why the native European species, *P. nigra*, *P. alba*, and *P. tremula*, fell into neglect. Beginning in the 1920s in most European countries new research institutes were founded to address the taxonomy of poplars, to generate information on varieties and clones of interest, and to improve cultivation as well as utilization techniques. Thus, important contributions to the development of poplar culture in Europe were made by experiment stations in Italy, the Netherlands, Belgium, Germany, France, and Spain. With the spread of poplar plantations, pests and diseases appeared causing major damage to the plantings. These included spring defoliation caused by *Pollaccia elegans*, rust infection by *Melampsora* spp., leaf brown spot by *Marssonina brunnea*, bark necrosis by *Cryptodiaporthe populea*, bacterial canker by *Xanthomonas campestris* pv. *populi*, and poplar mosaic virus; together these pests began to critically affect the further expansion of poplar culture (Anselmi 1991).

After World War II, the urge for greater international cooperation led in 1947 to the foundation of the International Poplar Commission within the FAO. Research activities of the different countries were then directed at the identification and description of species and cultivars in use and at studies on their ecological and biological characteristics (reproductive biology, phenology, growth potential, vegetative propagation ability, etc.). A large amount of genetic material was exchanged among countries, with the aim of identifying clones with broad adaptability to different environments. To this end, the Commission fostered the establishment of a network of poplar arboreta in Germany, France, Italy, Turkey, Belgium, and the Netherlands, to study the genetic variation among poplar genotypes of different species and geographic origins. The collection of natural germ plasm was concentrated mainly on *P. deltoides*, *P. tremuloides*, and *P. trichocarpa*, but later was also extended to the native European species *P. nigra*, *P. alba*, and *P. tremula*. Significant results were obtained in selecting clones resistant to various pathogens and in mitigating stress conditions by means of appropriate cultural techniques, derived from a better understanding of ecophysiology. For example, an aspect of great importance for modern poplar culture is the determination of appropriate planting densities to control competition among trees for nutrients and light. The great diversity in climatic and soil conditions make it necessary to study in detail the environmental adaptability of candidate clones under stressed and unstressed plantation regimes.

In more recent years, the greater availability of land for tree plantations has given a new direction to poplar culture in the context of environmental and landscape conservation. As a consequence, new interest has been directed to the preservation and improvement of the three native species. For this purpose, specific collections and comparative field trials have been established (Avanzo et al. 1985; Bisoffi 1989; Lefèvre et al. 1992).

The energy issue that is becoming increasingly important at the end of this century will renew interest in the cultivation of poplars, although the available information on appropriate materials and cultural techniques is still quite limited. Energy and the environment will be new objectives that poplar cultivation and breeding should not overlook in the future.

Poplars and recent European legislation

European agricultural policy underwent major changes in the last decade (Anonymous 1987); the reason for this was the overproduction of some traditional agricultural crops like cereals that caused problems on the international market. In addition, the transformation of dairy farming from extensive to intensive management and the rapid improvement of its production levels meant higher allocations to the preservation of excess dairy products. This meant that European agricultural policy had to formulate new regulations to favor the reduction of land area devoted to agriculture, which was hardly sustainable under international market conditions. The abandonment of agricultural crops, however, brought up the problem of soil protection and the conservation of soil fertility. Furthermore, several countries of the European Union (EU) suffer from a high production deficit in forest timber, that could partly be solved by the use of fast-growing species and agroforestry plantations. Accordingly, several financial incentives have been introduced by the EU that favor the reforestation of idle agricultural land. In some cases, the financial support from the European Union can be augmented by national subsidies. For example, in Italy, there is specific legislation that guarantees financial support for new plantations.

On average, the contributions for new poplar plantations vary between 2500 and 3000 ECU ha^{-1} (1 ECU = European Count Unit = 1.3 US$) depending on the clone utilized, with higher contributions for clones that are more resistant to *Marssonina brunnea*. This incentive also aims at the improvement of new poplar plantations but, at the same time, finds an obstacle in current environmental policy. In many European countries, riverbanks and lake shores are subject to specific protection laws. In Italy, the Environmental Act sets severe restrictions on the utilization of trees located within a 300 m wide belt along streams. This is an important constraint because most of the land generally devoted to poplar culture falls within these limits. Another conflict stems from the policy towards restoring the natural vegetation of streamsides, particularly for land included in natural parks. Hence, cultural practices are subject to specific authorization procedures and logging operations are limited. In Italy, another disincentive for the expansion of poplar culture is the significant price increase (about three-fold) in leasing public land along the riverbanks. If these obstacles are not overcome, it is difficult to expect a further increase in poplar culture along streamsides. Another factor is the restrictive attitude shown by river catchment authorities, which strictly enforce the regulation that prohibits growing large trees along riverbanks, since they would impede water runoff during floods (a disadvantage?). All in all, there seems to be a need for better coordination at the international level of the specific laws and regulations pertaining to poplar cultivation.

Prospects for European poplar culture

In almost all European countries, a reduction in the acreage cultivated with poplars has been observed during the last two years, due to widespread harvesting that has not been compensated by new plantations. In Italy, for example, this reduction amounted to 7%. Yet the timber produced has good technological properties because of the use of new clones, the application of improved cultural techniques, and effective treatments against pests. Average yields at 10–15 year rotations have been about 200 $m^3 \cdot ha^{-1}$ of stemwood, which is even more significant considering that in Italy poplar makes up 45% of the national timber production (Lapietra et al. 1994).

Analysis of trading of poplar timber in Europe indicates new trends, with a reduced presence of France on the international markets and an increasing involvement of central and eastern European countries, particularly Hungary. The progressive reduction of poplar wood production expected in the near future is unlikely to be balanced by an increase in the total standing volume of poplar plantations which currently is the lowest recorded in the last five years. For the Italian Po Valley, a forecast based on 1989–1994 production data projected a decrease in standing volume of poplar (age > 6 yr) to a low value in 1995 of 4.0×10^6 m^3, to be followed by an increase to 5.5×10^6 m^3 by the year 2000. The 1995 low in harvested timber caused major problems for the wood industry.

An expansion of poplar plantations can be anticipated following the diffusion of new industrial techniques; in fact, while the market for plywood and packing material is likely to remain stationary, an upturn in the use of poplar wood for structural purposes can be expected. Testing is underway on the suitability of poplar for the construction of laminated beams and laminated boards. The mechanical properties of laminated boards are promising for their inclusion among standard products as they offer the same structural strength of other tree species with greater wood density. The possibility to use poplar wood for the production of laminated beams will also open a market for lower-quality poplar timber. As a consequence, poplar plantation management may become less intensive, reducing costs and environmental impact, and tree breeding can concentrate on the improvement of such traits as resistance to biotic and abiotic stresses. Another new sector for the utilization of poplar wood is energy production; intensive studies of this option are underway, especially in Northern Europe. An Italian project foresees poplar wood to fuel a 12-megawatt power plant, with an hourly consumption of 7 tons of wood.

New Zealand

As with other countries of the southern hemisphere, poplar does not occur naturally in New Zealand. But it has been widely planted since about 1850, at first as some four or five clones followed by increases in clone diversity at intervals up to the present time. This diversity has been achieved largely by imports from poplar culture centers in the northern hemisphere, particularly in Europe, supported by an active breeding program in New Zealand based on the exotic imports. This work was pioneered in the 1960s by Van Kraayenoord at Palmerston North and further developed with the setting up of the National Plant Materials Center nearby in Aokautere in 1973. Even though experience was limited initially to a handful of clones, it soon became apparent that several poplars were well-suited to local conditions of climate and soil and they were used widely for horticultural protection and then later for soil erosion control. This latter application had some novel features such as planting poplar poles deeply on unstable soils in relatively remote areas where they were delivered by helicopter. The development has culminated in a system of agroforestry with grass and trees as the main components. Enthusiastic farmers have been especially involved in this development.

Poplar for wood production, though fully feasible, has never developed as an operating industry in the face of established and very extensive *Pinus radiata* forest plantations as well as substantial eucalypt plantations for hardwood production. The special characteristics of poplar wood, favored in other parts of the world, seem never to have been of critical importance to the New Zealand wood industry. Attempts to market poplar wood for veneer through some of the larger New Zealand wood product companies led to the statement that "they could use poplar if sufficiently large quantities were available at a price not dearer than radiata pine, and if continuity of supply were assured" (Van Kraayenoord 1993). In 1993 one mill in Canterbury producing medium-density fiber board reported using 50% poplar to reduce the matt thickness (Wardrop 1993). Plantations established by the match industry to produce match splints foundered on business grounds unrelated to forestry considerations and poplar itself.

The gene pool collection

In the 1960s, faced with numerous clones and much misidentification, a careful cataloguing of all the material then existing in New Zealand was undertaken and a collection of correctly-named clones established in Hawkes Bay (Van Kraayenoord 1993). A further gene pool (Populetum) was assembled near Napier between 1975–1980.

The value of the Hawkes Bay collection was recognized when in 1973 two rust species entered the country. The effect on susceptible clones was devastating, but there were clones resistant to those rust fungi. Identification and propagation of rust-resistant clones was accomplished in a year or two and of clones resistant to other diseases in later years.

Damaging agents

A peculiar problem of New Zealand poplar culture is the presence of a marsupial, the possum *Trichosurus vulpecula*, introduced some time ago from Australia. In a manner similar to the rabbit in Australia, it has multiplied and emerged as a voracious eater of native plants as well as exotic poplar. Among the clones available some have been unpalatable to possums, thus providing a way to minimize the damage in poplar plantations and to provide a base for development of new clones carrying the resistant characters. Domestic honey bees, too, can be damaging to poplar. Some clones have sticky buds that attract bees which harvest the surface material and appear to take cuticle with it. Restricted growth and, in extreme cases, a witches-broom-like condition causes dead tops in affected clones. In these and other cases, as for example in the search for wind-resistant material, the gene pool collection has provided a valuable reservoir of diversity, from which to select suitable clones.

Future industrial wood production

Although there is, up to the present, no significant industry based on poplar wood utilization, studies indicate that there is a good case for more sustainable land management by including poplar, essentially in an agroforestry program. This could cover substantial land areas. Eyles (1993) suggests that poplar use in this way will provide the next breakthrough in agricultural development; about a third of the North Island would be involved in such a step. The economic imponderables of policy, finance, and industrial commitment make forecasting the future of poplar in New Zealand difficult and uncertain. Wilkinson (1993) considers that the reestablishment of New Zealand membership in the International Poplar Commission could favorably influence such a development.

Australia

Poplar culture in Australia has followed closely the pattern of other southern-hemisphere countries. With no native poplars, a small number of clones was introduced mainly from Europe in the mid-1800s and additions were made to these from time to time in the subsequent century. By 1960 a strong economic demand arose for poplar wood, triggered by the match companies' need for raw material for processing into match splints. From the clones already in Australia, in addition to others well-known in Europe at that time and newly imported to Australia, plantations were established in Victoria and New South Wales. These plantations were based either on rainfall or, in Victoria, on irrigation (Fig. 4). Fortunately, one clone of *P. deltoides* proved adaptable to either situation and

Fig. 4. An irrigated plantation of the euramerican clone I–214 (8-yr-old) in Yarrawonga, Victoria, Australia. (Photo by L. Pryor.)

planting progressed. Economic factors in the match industry led to the closing down of most plantation projects by 1990 but there is some indication of renewed business interest more recently.

Combating poplar rust

In 1973 there was a major alarm caused by the sudden spread of poplar rust (*Melampsora medusae*) following accidental introduction to the country. *P. deltoides* is especially susceptible to this rust. The problem was compounded by the introduction also of *M. larici-populina* to which clones of *P. nigra* or its hybrids are often susceptible. The second rust species became abundant in 1974 which meant that virtually all the clones then in cultivation were at risk. The consequent alarm in the poplar wood growing industry led to strenuous efforts to ameliorate the trouble. By then numerous clones from a wide genetic base had been established in trial plots. Combined selection for resistance and for easy clonal propagation allowed the rapid development of highly resistant material which quickly came into plantation use. Regardless, additional pathogens are likely to find their way into the country and constant monitoring will be necessary with that expectation.

Adapting poplar to lower latitudes

While parts of southeast Australia are in suitable latitudes and have appropriate soils and climate for a number of clones important in poplar culture in Europe, additional land is readily available in the more northern parts of New South Wales between latitudes 30 and 27°. Such latitudes were found to be too low to suit many of the then tested and available clones. Without long photoperiods in summer, growth rhythms become disordered; in extreme cases growth ceases and buds are set even though temperature and water needs are met for continued shoot growth.

Several ways of overcoming this deficiency were developed so that in a short time clones were produced that restored growth performance to the level necessary for commercial plantations. Some poplar species have a natural range to areas below 30° and provenances from such sites are day-length neutral. The most notable of these is *P. deltoides* which occurs in southeast Texas, USA, to about latitude 27°. Clones derived from such provenances continue to show freedom from day-length control. In addition the feature is heritable. Another approach was to use for breeding the Chilean semi-evergreen poplar, *P. nigra* 'Chile,' which itself is day-length neutral. It is a male clone, probably a somatic mutant, and seems to transmit photoperiod insensitivity in dominant fashion. Numerous clones were developed from crosses between this parent and *P. deltoides*, many ortets of which were unaffected by day length. These ortets retained desirable silvicultural features of *P. deltoides* and supplied much material for selection of clones for commercial use (Fig. 5).

There are probably opportunities for similar developments with species of the Section *Populus,* within which some semi-evergreen clones are known to exist. For example, one Mexican species in this Section, *P. guzmanantlensis*, extends at least to 17°N latitude. If found to have silvicultural value, this species may help in extending the range of poplar cultivation into the tropics (Vasquez and Ceuvas 1989).

Clones that are not affected by day length appear to require planting sites at higher elevation in tropical latitudes, but there are no physiological studies that help to explain this need. If innovations that allow extension of poplar to lower latitudes are realized, much greater areas in Australia could provide sites suitable for poplar culture. The hitherto untested *P. mexicana* of the new Section, *Abaso*, established by Eckenwalder (1977), might also be involved.

Populus euphratica

For the future it is worth considering possible developments based on *P. euphratica*. The species is widespread in the Mediterranean, the Near East, central Asia and western China. It endures very high temperatures, considerable salinity on some sites, and periodic waterlogging. While the species ususally occurs in shrubby thickets on large rivers as a riparian species, recent studies

Fig. 5. Six-year old *P. deltoides* × *P. nigra* "Chile," Clone 65/31, growing in Myrtleford, Victoria, Australia. (Photo by L. Pryor.)

have identified stands of trees of substantial size and good form in some localities in central Asia (Wang et al. 1995).

Interspecific hybrids between *P. euphratica* and *P. deltoides* (Pryor and Willing 1982) and between *P. euphratica* and *P. simonii* have been produced with the aid of stigma manipulation. Other hybrids succeeded with *P. nigra* var. *thevestina*, using irradiated 'mentor pollen' in mixture (Wang et al. 1995). A recent preliminary study (Marcar et al. 1995) indicates that an F_1 hybrid of *P. euphratica* × *P. deltoides* displays tolerance to salinity and waterlogging to an extent comparable with the *P. euphratica* parent. Thus, breeding could be successful in producing desirable combinations.

There may also be prospects for the development of *P. ilicifolia*, a species found near the equator in Kenya on the Tana River below about 1300 m elevation and referred to often as *P. euphratica* (Jestaedt 1986). The species propagates readily from stem cuttings but nothing has been reported to date on its crossability (Oballa 1996).

Trends and prospects

Economic fluctuations in the last 10 years have led to an almost complete close down of the match industry. Although safety matches are still a significant, if diminished, commodity in the country, they are now imported. The production of poplar veneer is still undertaken in a very limited way for use as facing in some products otherwise based on *Pinus radiata*, and if the resource existed there are indications that the industry might revive. Current manufacture is based on scattered small plantings on private property. The most likely method for renewal would be in agroforestry systems mainly on rather small individual land holdings. The necessary clones are available and plantation techniques are known and tested on a commercial basis. So, revival of the industry would be easy if the necessary financial commitment were made. In the event that such development takes place, support by a modest research program would be important to facilitate continuing improvements in materials and methods, as well as to meet unforeseen setbacks from potential pathogens and pests.

India

India has two or three species of indigenous poplar of which the best known and most widespread is *P. ciliata*, found naturally at elevations from 1300 to 2800 m in the western Himalaya of northern India (Chauhan and Khurana 1992). Some use has been made of the wood of the species from natural stands, and small areas of plantation also have been established. But sites available for planting have in general been at lower elevations than where it grows naturally and the results disappointing.

In recent decades planting has been stimulated by the West Indian Match Company, WIMCO, in their endeavors to secure a local resource of raw material for match splints. Land was available in the Terai zone at low elevation, along the Himalayan foothills for an agroforestry system. The latitude of this zone is mostly below 30°N, so day-length conditions are not suited to most of the fast-growing clones commonly raised in Europe and which had been disappointing in preliminary trials. Some years ago WIMCO gained access to clones developed for low latitudes in Australia, which became known as "Australian Hybrids." About 9000 ha of plantations were established in the Terai with this material by 1988, within a planned goal of about 20 000 ha (Chaturvedi 1982; Jones and Lal 1989). Results were good but not without some specific problems related to patchy hardpan in the soil. It appears this Indian initiative has been successful (Sagwal 1986).

At the same time one must recognize that in India much of the land suitable for poplar culture, and the irrigation water needed for it, are in high demand for food production. Agroforestry systems that combine the two offer an attractive solution for certain regions (Chaturvedi 1982). Genetic improvement also may expand the opportunities. For example, breeding with *P. euphratica* may take advantage of its salinity tolerance (Marcar et al. 1995) and its capacity to endure very hot and dry conditions, given adequate water supply. Since it also hybridizes with *P. deltoides* (Pryor and Willing 1982) and is reported to breed successfully with other species of *Aigeiros* and *Tacamahaca* (Wang et al. 1995) it may be improved in tree form and wood quality. The combination may allow poplar culture in India to be extended into areas of greater salinity and lesser value for food crops.

People's Republic of China

The place of China in the poplar world is in many ways unique. There is a great diversity of poplar species and subspecies, most of which are endemic and relatively little-known outside the country (see Eckenwalder, Chapter 1, for a discussion of pertinent taxonomy). The habitat diversity is comparable with North America east of the Rocky Mountains and the geographic dimensions are similar. Most of the indigenous species have local use but silvicultural activity is centered on only a few, together with some imported material, mainly of the Aigeiros section. Of this Section *P. nigra* may occur naturally in western China; historical records are equivocal as to whether it is indigenous there or not.

There have been massive tree-planting phases in the last 50 years and the indigenous *P. simonii*, a balsam poplar, has featured prominently in these through the use of mainly three locally-selected clones. Relatively recent introductions of *P. deltoides* have performed well and interest has been sustained in the use of various clones. Several well-known *P. ×euramericana* cultivars, such as I–214 and cv. Lux have also been employed. In areas south of the Yangtse River, which generally is about 30° latitude, day-length effects become significant and successful clones require the same capacity as those performing well in similar latitudes in the southern hemisphere. Some recent trials with material developed in Australia for such sites have shown promise but not without complicating factors at times, such as heartwood discoloration. Current experiments will provide much valuable information as to future options in the southern part of the country.

Populus ×tomentosa Silviculture

One of the most striking poplar programs in China in the last 15 years is based on the indigenous *P. ×tomentosa*, mainly at latitudes 30–40° in Shandong and adjoining provinces. An intensive development program has been undertaken nationally, directed by Professor Zhu Zhiti of the Beijing Forestry University. *P. ×tomentosa*, generally treated as a species, is actually a hybrid between *P. alba* and probably *P. adenopoda*, belonging to Section *Populus* (Eckenwalder, Chapter 1). It has long been cultivated in the provinces of Shandong, Hebei, Shanxi, and Shaanxi and there is some doubt that genuine natural stands still exist. Variation present in natural stands, however, is likely to be largely preserved since traditional propagation has been based on root suckers from local sources.

The development of *P. ×tomentosa* has been challenging. It does not grow in the nursery readily from stem cuttings, so a method has been developed of grafting it onto a small piece of stock of *P. simonii* or one of its hybrids. If the graft junction is buried below the soil surface the scion may form its own roots after a time. This is a labor-intensive operation but one that can be absorbed into the current labor market at a competitive cost. There has also been some promising backcross breeding with *P. alba* cv. Bolleana. Bolleana roots readily from stem cuttings in the nursery and the capacity is transmitted to hybrids with *P. ×tomentosa*. Plantations of the best clones have good stem and branching form as well as rapid growth. The wood has been shown suitable for a range of uses characteristic of poplar, from sawn wood to veneer and pulp.

The Shandong plantations of the North China Plain are in an agroforestry layout with irrigated wheat and some other crops between widely spaced lines or pairs of lines of trees often about 50 m apart (Fig. 1). The soil is a deep loess usually exceeding 3 m, at which depth a water table is often reached. Heavy and increasing demands for irrigation water have already caused anxiety, and there is evidence of water table lowering.

Pacific Northwest of North America

The Pacific Northwest region of North America is famous for its extensive forests of evergreens. These forests are unique in their degree of conifer dominance and contrast with the deciduous hardwood or mixed hardwood–conifer stands commonly found in the North Temperate Zone. Dry, warm summers, typical for this region, seem to be a key factor for the scarcity of deciduous hardwoods, confining them largely to riparian zones and moist hillsides where their high transpirational demands can be met (Waring and Franklin 1979). Management of these forests has traditionally been practiced so as to maximize the commercially valuable conifers and has treated hardwoods as "weed" species. In sharp contrast to this tradition, the past 15 years have seen a rapid emergence of poplar culture as a new form of tree production, primarily on agricultural land, and now occupying approximately 15 000 ha of plantations in the region between southern Oregon and British Columbia (Heilman et al. 1995). Carried by seven major forest-product companies in Canada and the United States, this development seems to have reached full stride and is anticipated to expand to about twice the current acreage over the next five years. How did this come about and what were the driving forces?

There were at least four partly independent developments that in the late 1970s combined to launch poplar culture in the region. One was the industrial concern about highly fluctuating pulp prices and the dwindling supply of fiber. The Northwest pulp and paper industry had long been dependent on sawmill residues as its primary fiber source. As a consequence, the lumber market had a decisive influence on the availability of chips and their price (Cellier 1987). With a systematically declining sawmill industry one could foresee an eventual shortage of raw material and a need for substitutes. A source of hardwood fiber that would broaden the range of products and give access to the high-quality paper market seemed especially desirable. During the same time, research at the Washington

State University (WSU) Research Centers in Mt. Vernon and Puyallup had explored the concept of short rotation culture with the fastest-growing native hardwood, black cottonwood (*P. trichocarpa*). Special attention was given to its cultural needs on typical farmland, such as type of planting stock, chemical and mechanical weed control, spacing and harvest cycle (Heilman et al. 1972; Heilman and Peabody 1981). It became evident that cottonwood culture for pulp production held promise — but even greater promise if the genetic stock could be improved. This happened to be the focus of another, separate, research project at the University of Washington (UW) in Seattle, where studies on the reproductive biology of black cottonwood, begun in the mid-1960s, had generated a number of fast growing *P. trichocarpa* × *P. deltoides* (T × D) hybrids. Tests with these hybrids at Puyallup showed their clear superiority to the native cottonwood in growth and form. The fourth and final thrust came from the US Department of Energy (DOE). As a result of the energy crisis, the DOE in 1978 launched a research initiative on the potential role of woody crops as a future source of liquid fuel. In response, the UW and WSU researchers decided to combine their studies and, with funding from DOE, to test the production potential of their hybrids under short-rotation intensive culture. By 1981 the results looked promising enough that two companies (Crown Zellerbach Corp., and Mt. Jefferson Farms) started field trials with these hybrids on their own land. Scaling up began soon thereafter and by 1991 additional industries had joined the new venture.

What features characterize poplar culture in the Pacific Northwest?

Favorable growing conditions

Climate and soils of the region provide an excellent environment for the growth of poplars. West of the Cascade Range, winters are mild and wet, summers warm and dry but with cool nights. The frost-free season in the lower valleys averages about 200 d and allows assimilation to extend into mid-November. Commercial yields of T × D hybrids grown in 7–8 year rotations in the lower Columbia River Valley range from 17 to 21 $Mg \cdot ha^{-1} \cdot yr^{-1}$ of oven-dry woody biomass (Schuette 1995). East of the Cascades, the climate in the fertile Columbia River Basin is sunnier and drier. With only about 50 mm of precipitation during the growing season, irrigation is mandatory but results in growth rates that exceed those west of the Cascades by 30–50%. To date, T × D hybrids have given the most consistent production, but additional selections under testing include *P. trichocarpa* × *P. maximowiczii*, *P. trichocarpa* × *P. nigra*, and *P. deltoides* × *P. nigra* hybrids.

Industrial scale

The majority of poplar plantations are owned by forest-products companies and operated on an industrial scale, e.g., 4400 ha in the case of James River Corp., 5500 ha by Boise Cascade Corp., with plans to expand to 8000 ha, and a similar dedicated acreage by Potlatch Corporation (see Fig. 6). Industrial ownership has permitted greater up-front investments in land and equipment and to approach poplar culture in an innovative spirit. With the success of these operations local farmers have also become interested in growing hybrid poplar on their own land. Several companies support this through land-owner assistance programs, i.e., providing planting stock, advice on culture, and a market for the crop.

Advanced technology

An existing agricultural infrastructure combined with the large scale of operations have helped to develop a sophisticated technology for poplar culture. This is evident in site selection, site preparation, weed control, assessment of cultural treatments and soil amendments, tracking of growth, harvest, and subsequent conversion. Culture east of the Cascades involves enormous installations to provide growing trees with water, fertilizer, and pesticides in the right amount and at the right time via trickle irrigation (see Fig. 7). Thus, one of the companies in its annual expansion of plantations by 1200 ha required 4000 km of irrigation lines to be added to its existing network. They had to be imported from Israel. Another company has its harvesting operations so carefully tuned to the changing needs of its nearby pulp mill that the time required to process standing trees to finished paper products is less than 3 d.

Short rotation

To date, most poplar culture in the Northwest has been geared to pulp production. For this purpose, plantations are grown at a spacing of 1300–2000 trees·ha^{-1} and harvested in 7–8 year rotations. Faster growth east of the Cascades may permit cycles of 5–6 years. The short hardwood fiber is typically mixed with longer conifer fiber in the manufacture of tissue paper, towels, high-grade newsprint and high-quality bond paper. A growing interest in higher-value structural wood products, such as oriented strandboard and laminated veneer, may tend to extend the cycles to 10 years for part of the operations in the future. Short rotations allow early assessment of plant materials and treatments, favor experimentation, provide rapid rewards for success, and make for steep learning curves. They have great appeal to practitioners, researchers, and corporate managers alike.

Closely integrated research and practice

Because of the early involvement of UW/WSU researchers in the development of Northwest poplar culture, practice and research have been closely coordinated between companies and the universities from the beginning. Apart from a systematic breeding, testing, and selection program, intensive studies have focused on the genetics and physiology of productivity, nutrition, and disease resistance. All companies have been engaged in operational research, some have initiated breeding of their own. One company has estimated increases in yield of 31% over the past 10 years thanks to hybrid improvement and refinement of cultural

Fig. 6. A hybrid poplar plantation of the Potlatch Corporation in the semiarid region of eastern Oregon near Boardman, USA. The plantation forms part of a contiguous block of 1800 ha and is irrigated via drip lines. The trees, *P. trichocarpa* × *P. deltoides* hybrids in their second year, are grown for pulp and will be harvested in another 3–4 years. (Photo by Potlatch Corp.)

Fig. 7. Pump and filter station of the Ice Harbor Fiber Farms (Boise Cascade Corporation) near Pasco, eastern Washington, USA. Water from the Snake River is pumped, filtered, and delivered by booster pumps via 5500 km of trickle tubes to 1800 ha of hybrid poplars. Fertilizer and insecticides are added to the water when needed. (Photo by R. Stettler.)

techniques. The sustained support from federal and state agencies, especially the Biofuels Feedstock Development Program of DOE, has played a major role in this development. Two new industry/university ventures, the Poplar Molecular Genetics Cooperative, located at UW, and the Tree Genetic Engineering Research Cooperative, located at Oregon State University, have introduced the latest biological techniques to the genetic manipulation of poplar and are financed by a membership of companies and agencies from around the world. There is an active exchange of germ plasm (pollen, clones, pedigrees) with research laboratories and production programs on an international scale.

In summary, poplar now constitutes a new crop in the Pacific Northwest with undoubtedly a bright future. Beyond adding to the regional spectrum of land-use systems, utilization of marginal farmland, and sparing of native forest, it is also recognized for several environmental benefits: its greater soil and nutrient conservation (vs. agricultural crops), its low bleaching requirement in pulping (vs. conifers), its suitability for wastewater and sewage sludge utilization, and its potential for bioremediation. Hybrid poplar culture has also stimulated research in the native species, *P. trichocarpa*, its ecology, genetic diversity, and its role in riparian ecosystems. Indeed, the ultimate challenge to scientists and practitioners in the region will be to develop a poplar domestication program that will take full advantage of the native genetic resource while at the same time making every effort to conserve it. The rapid rise of poplar culture in the Pacific Northwest has also stimulated a new look at the potential of poplar in other regions of the United States, with emphasis on energy production in the Lake States, and on pulp and wood products in the South.

Global outlook

Globally, it seems safe to say that the classical role of poplar culture in providing a source of timber to areas deficient in forest will be maintained and will become even more significant in the future. The demands for timber will increase, and new legislation in many countries will further restrict tree cutting in existing forests. Poplar trees grown in various forms of culture on land outside of the forest will become a very significant new source of timber and other wood products. Genetically improved varieties of poplar, intensively grown as currently practiced in the Northwestern United States, are but one of the examples of what is coming in the next decades.

More recently, energy shortages have focused attention on short rotation forestry with fast-growing tree species for production of fuel, a form of renewable energy, worldwide. In the Northern Hemisphere the climatic and site conditions favour the growth of poplars and willows in such energy plantations. We foresee these plantations to play an increased role in the coming decades. Intensive research and demonstration programs already under way in Europe and North

America are a forerunner of this new aspect of poplar culture. Related to this role of poplar culture is its potential in improving the carbon balance of the atmosphere.

The environmental role of poplar culture will also increase. This applies primarily to its role in cleaning the groundwater, lakes, and rivers from the pollution of excess fertilizers. In this it has to complement the natural and near-natural populations of *Populus* still in existence in riparian settings. In fact, it will be important to show that plantation culture and the conservation and restoration of natural stands are mutually supportive (e.g., as repositories of germ plasm) and highly compatible, if properly conducted. Finally, we see great promise in the future use of poplar in the rapidly expanding field of phytoremediation, especially in the disposal of municipal wastewater, but also in more sophisticated systems aimed at specific pollutants.

Acknowledgements

We gratefully acknowledge the translation of one of the contributions to this chapter by Giuseppe Scarascia-Mugnozza, and the critical review and valuable suggestions provided by Don Dickmann.

References

Adler, T. 1996. Botanical cleanup crews: Using plants to tackle polluted water and soil. Sci. News, **150**: 42–43.

Anonymous. 1987. Forest policy in the European Community — Commission of the European Communities Com. (87) 621, Bruxelles.

Anselmi N. 1991. Pathological problems in poplar plantations in Italy. *In* Proceedings of the International Energy Agency — Task V Meeting on Exchange of Genetic Material, Pest/Disease Management, and Joint Trials of *Alnus, Populus,* and *Salix,* 22–27 August, 19xx, Ames, Iowa, USA. *Edited by* B. Richard, R.D. Hall, R.N. Hanna, and O. Nyong. Iowa State University Printing Service, Ames, IA. pp. 10–18.

Armson, K. (*Editor*). 1983. New forests in eastern Ontario. Sci. Technol. Ser. Vol. 1. Queen's Printer for Ontario, Toronto, ON.

Avanzo, E., Bisoffi, S., Gras, M.A., and Mughini, G. 1985. Breeding strategies for "Aigeros" poplars adopted in Italy. Genet. Agrar. **39**: 308–311.

Bisoffi, S. 1989. Recent developments of poplar breeding in Italy. *In* Proceedings of a Meeting of IUFRO Working party 52.02.10 Hann. Munden, Germany. pp. 18–45.

Boysen, B., and Strobl, S. (*Editors*). 1990. A grower's guide to hybrid poplar. Ontario Ministry of Natural Resources, Toronto, ON. Forest Resource Development Agreement Publ. 148 p.

Cellier, G.A. 1987. The feasibility of hybrid poplar in the Pacific Northwest. M.Sc. dissertation, University of Washington, Seattle, WA.

Chaturvedi, A.N. 1982. Poplar farming in U.P. (India). U. P. For. Bull. No. 45.

Chauhan, P.S., and Khurana, D.K. 1992. Growth performance of different provenances of Himalayan poplar. *In* Proceedings of the 19th Session of the International Poplar Commission, 22–25 September, 1992, Zaragoza, Spain. pp. 687–693.

Christersson, L., Sennerby-Forsse, L., and Zsuffa, L. 1993. The role and significance of woody biomass plantations in Swedish agriculture. For. Chron. **69**: 687–693.

Cram, W.H. 1960. Performance of seventeen poplar clones in south-central Saskatchewan. For. Chron. **36**: 204–208.

Cram, W.H. 1968. Poplars for grassland plantings. *In* Growth and utilization of poplars in Canada. *Edited by* J.S. Maini and J.H. Cayford. Can. Dep. For. Rural Dev., For. Br. Publ. **1205**: 113–115.

Dode, L.A. 1905. Extraits d'une monographie interdite du genre *Populus*. Mem. Soc. His. Nat. Autun. **18**. 73 p.

Eckenwalder, J.E. 1977. North American cottonwoods (*Populus*, Salicaceae) of the sections Abaso and Aigeiros. J. Arnold Arbor. Harv. Univ. **68**(3): 193–208.

Eyles, G. 1993. Poplars, the next breakthrough in agricultural development. *In* A potential for growth. *Edited by* B. Bulloch. Manaaki Whenua — Landcare Research New Zealand Ltd., Private Bag 11052, Palmerston North, NZ. pp. 17–25.

FAO. 1979. Poplars and willows. FAO Forestry Series No. 10. Food and Agricultural Organization of the United Nations, 1979, Rome, Italy. 328 p.

Heilman, P.E., and Peabody, D.V., Jr. 1981. Effect of harvest cycle and spacing on productivity of black cottonwood in intensive culture. Can. J. For. Res. **11**: 118–123.

Heilman, P.E., Peabody, D.V., Jr., DeBell, D.S., and Strand, R.F. 1972. A test of close-spaced, short-rotation culture of black cottonwood. Can. J. For. Res. **2**: 456–459.

Heilman, P.E., Stettler, R.F., Hanley, D.P., and Carkner, R.W. 1995. High yield hybrid poplar plantations in the Pacific Northwest. PNW Extension Bulletin No. 356, Revised Edition. WSU Coop. Extension, Pullman, WA. 42 p.

Heimburger, C. 1936. Report on poplar hybridization. For. Chron. **12**: 285–190.

Henry, A. 1914. Note on *Populus generosa*. Gardeners Chron. **1914**: 257–258.

Henry, C.L. 1991. Nitrogen dynamics of pulp and paper sludge to forest soils. Water Sci. Technol. **24**(3/4): 417–425.

Hessmer, H. 1951. Das Pappelbuch. Verlag des Deutschen Pappelvereins, Bonn, Germany.

Houtzagers, G. 1937. Het Geslacht *Populus* in verband met zijn Beteekenis voor de Houtteelt. Veeman & Zonen, Wageningen, The Netherlands. 291 p.

Jacometti, G. 1933. Il Miglioramento di Pioppo. Torino S.A.F. Publ. Rome, Italy.

Jestaedt, M. 1986. Die Tana-River-Pappel in Kenya. Holzzucht **40**(1–2): 14–15.

Jones, N., and Lal, P. 1989. Commercial poplar planting in India under agroforestry system. Commonw. For. Rev. **68**(1): 19–26.

Lapietra, G., Coaloa, D., and Chiarabaglio, P.M. 1994. Rapporto Annuale sulla Pioppicoltura. Cellul. Carta **45**(3): 2–8.

Lefèvre, F., Faivre-Rampant, P., Villar, M., and Teissier du Cros, E. 1992. *Populus nigra* resource preservation in France. *In* Proceedings of the 19th Session of the International Poplar Commission, 22–25 September, 1992. *Edited by* A. Padro. Deputacion General de Aragon, Zaragoza, Spain. pp. 729.

Marcar, N., Banks, J., Crawford, D., and Pryor, L. 1995. Salt tolerance of some poplar clones: a preliminary experiment. *In* Symposium on Poplar Biology and its Implications for Management and Conservation, August 1995, Seattle, WA.

Oballa, P.O. 1996. Status of *Populus ilicifolia* in Kenya. Kenya Forestry Research Institute, Occasional Paper No. 3. 36 p.

Piccarolo, G. 1959. Pioppicoltura. ENCC, Roma. Publ., Rome, Italy.

Pryor, L.D., and Willing, R.R. 1982. Growing and breeding poplar in Australia. L.D. Pryor and R.R. Willing, Red Hill, Australia.

Ronald, W.G. 1980. Tower poplar. Can. J. Plant Sci. **60**: 1055–1056.

Sagwal, S.S. 1986. Grow poplars for profit. Indian Farming, **36**(9): 27–31.

Schreiner, E.J., and Stout, A.B. 1934. Description of ten new hybrids. Bull. Torrey Bot. Club **61**: 449–460.

Schroeder, W.R., and Lindquist, C.H. 1989. Assiniboine poplar. Can. J. Plant Sci. **68**: 351–353.

Schuette, W.R. 1995. Growth and yield of hybrid poplars at the Lower Columbia River Fiber Farm. *In* Proceedings of International Poplar Symposium, 20–25 August, 1995, Seattle, WA. pp. 114.

Van Kraayenoord, C.W.S. 1993. Poplar Growing in New Zealand — Past to present. *In* A potential for growth. *Edited by* B. Bulloch. Manaaki Whenua — Landcare Research New Zealand Ltd., Private Bag 11052, Palmerston North, NZ. pp. 9–16.

Vasquez, A., and Ceuvas, R. 1989. Una Nueva Especie Tropical de *Populus* (Salicaceae) de la Sierra de Manantlan, Jalisco, Mexico. Acta Bot. Mexicana, **8**: 39–45.

von Wettstein, W. 1933. Die Kreuzungsmethode und die Beschreibung von F1-Bastarden bei *Populus*. Pflanzenzüchtung, **1933**: 597–626.

Wang, S., Chang, B., and Li, H., 1995. Euphrates poplar forest. Chinese Academy of Forestry, Beijing. 1012 p.

Wardrop, P. 1993. Current prospects for a sustained growth and yield industry. *In* A potential for growth. *Edited by* B. Bulloch. Manaaki Whenua — Landcare Research New Zealand Ltd., Private Bag 11052, Palmerston North, NZ. pp. 26–31.

Waring, R.H., and Franklin, J.F. 1979. Evergreen coniferous forests of the Pacific Northwest. Science (Washington, DC), **204**: 1380–1386.

Weisgerber, H. 1983. Wuchsverhalten und Anbaumöglichkeiten einiger neu zum Handel zugelassener Balsampappeln und Aspen. Holzzucht, **37**: 2–11.

Wilkinson, A. 1993. Poplar commissions — Guiding forces for industry development. *In* A potential for growth. *Edited by* B. Bulloch. Manaaki Whenua — Landcare Research New Zealand Ltd., Private Bag 11052, Palmerston North, NZ. pp. 3–8.

Zsuffa, L. 1974. The genetics of *Populus nigra* L. Ann. For. (Zagreb) **VI**(2). 53 p.

Zsuffa, L., and Barkley, B. 1984. The commercial and practical aspects of short rotation forestry in temperate regions: A state of art review. *Edited by* H. Egneus and A. Ellegard. Bioenergy, **84**(I): 39–57.

Zsuffa, L., Sennerby-Forsse, L., Weisgerber, H., and Hall, R.B. 1993. Strategies for clonal forestry with poplars, aspens and willows. *In* Clonal forestry II. *Edited by* M.R. Ahuja and W.J. Libby. Springer-Verlag, Berlin, Heidelberg, Germany. pp. 91–119.

Zufa, L. 1961. The cultivation of poplars in alignments. Topola No. 20-21. Beograd (in Yugoslav, with English summary).